5

Molecular Sieves

Science and Technology

Editors: H. G. Karge · J. Weitkamp

Molecular Sieves
Editors: H. G. Karge · J. Weitkamp

Recently Published and Forthcoming Volumes

Characterization II
Editors: Karge, H. G., Weitkamp J.
Vol. 5, 2006

Characterization I
Editors: Karge, H. G., Weitkamp J.
Vol. 4, 2004

Post-Synthesis Modification I
Editors: Karge, H. G., Weitkamp J.
Vol. 3, 2003

Structures and Structure Determination
Editors: Karge, H. G., Weitkamp J.
Vol. 2, 1999

Synthesis
Editors: Karge, H. G., Weitkamp J.
Vol. 1, 1998

Characterization II

Editors: Hellmut G. Karge · Jens Weitkamp

With contributions by
R. Aiello · F. Bauer · J.-L. Bonardet · J. Fraissard
R. Fricke · A. Gédéon · G. Giordano · H. G. Karge
A. Katovic · I. Kiricsi · Z. Kónya · H. Kosslick
J. B.Nagy · G. Pál-Borbély · S. Sealy · M.-A. Springuel-Huet
F. Testa · Y. Traa · J. Weitkamp

Springer

Molecular Sieves – Science and Technology will be devoted to all kinds of microporous crystalline solids with emphasis on zeolites. Classical alumosilicate zeolites as well as microporous silica will typically be covered; titaniumsilicate, alumophosphates, gallophosphates, silicoalumophosphates, and metalloalumophosphates are also within the scope of the series. It will address such important items as hydrothermal synthesis, structures and structure determination, post-synthesis modifications such as ion exchange or dealumination, characterization by all kinds of chemical and physico-chemical methods including spectroscopic techniques, acidity and basicity, hydrophilic vs. hydrophobic surface properties, theory and modelling, sorption and diffusion, host-guest interactions, zeolites as detergent builders, as catalysts in petroleum refining and petrochemical processes, and in the manufacture of organic intermediates, separation and purification processes, zeolites in environmental protection.
As a rule, contributions are specially commissioned. The editors and publishers will, however, always be pleased to receive suggestions and supplementary information. Papers for *Molecular Sieves* are accepted in English. In references *Molecular Sieves* is abbreviated *Mol Sieves* and is cited as a journal.
Springer WWW home page: springer.com
Visit the Molecular Sieves home page at springerlink.com

Library of Congress Control Number: 2006923496

ISSN 1436-8269
ISBN-10 3-540-30457-6 Springer Berlin Heidelberg New York
ISBN-13 978-3-540-30457-9 Springer Berlin Heidelberg New York
DOI 10.1007/b58179

Springer is a part of Springer Science+Business Media

springer.com

Cover design: WMXDesign GmbH, Heidelberg
Typesetting and Production: LE-TEX Jelonek, Schmidt & Vöckler GbR, Leipzig

Printed on acid-free paper 02/3100 YL – 5 4 3 2 1 0

Editors

Dr. Hellmut G. Karge
Fritz Haber Institute
of the Max Planck Society
Faradayweg 4–6
14195 Berlin
Germany

Professor Dr.-Ing. Jens Weitkamp
Institute of Technical Chemistry
University of Stuttgart
70550 Stuttgart
Germany

Molecular Sieves
Also Available Electronically

For all customers who have a standing order to Molecular Sieves, we offer the electronic version via SpringerLink free of charge. Please contact your librarian who can receive a password or free access to the full articles by registering at:

springerlink.com

If you do not have a subscription, you can still view the tables of contents of the volumes and the abstract of each article by going to the SpringerLink Homepage, clicking on "Browse by Online Libraries", then "Chemical Sciences", and finally choose Molecular Sieves.

You will find information about the

- Editorial Board
- Aims and Scope
- Instructions for Authors
- Sample Contribution

at springer.com using the search function.

Preface

In the preface to Vol. 4, it was stressed that characterization of molecular sieves is an indispensable prerequisite for the evaluation of results in the synthesis, modification, and application of these microporous and mesoporous materials. Thus, Vol. 2 grouped together contributions to structure analysis of molecular sieves, whereas Vol. 4 was particularly devoted to characterization by spectroscopic techniques. Logically, Vol. 5 is intended to complement the preceding volume in that it covers a variety of non-spectroscopic techniques for the characterization of zeolites and related materials. Thereby, some of the contributions are specifically focused on methods of characterization such as chemical analysis, thermal analysis, pore-size characterization using molecular probes or ^{129}Xe NMR, which are, of course, illustrated by a wealth of applications. Two other chapters deal specifically with the characterization of important molecular sieve systems, viz. coke on zeolites and isomorphously substituted molecular sieves.

The first chapter, co-authored by R. Fricke and H. Kosslick, provides, in an exhaustive manner, methods of chemical analysis of (microporous) aluminosilicates, aluminophosphates and related molecular sieves. These analytical methods are exemplified by a large number of cases and, very importantly, the chemical procedures and treatments are meticulously described. To the best of our knowledge, no comparable compendium of chemical analysis of zeolites and related substances has ever been available before.

In her contribution, G. Pál-Borbély shows the great potential of thermal analysis for the characterization of molecular sieves and processes occurring with them as far as they are accompanied by changes in weight and/or heat effects. Such processes are, for example, dehydration, dehydroxylation, deammoniation, phase transition, structure collapse, decomposition of occluded complexes, and oxidation or reduction of framework constituents. In this context, applications of thermogravimetry (TG), derivative thermogravimetry (DTG), differential thermoanalysis (DTA), differential scanning calorimetry (DSC), and also more recent developments such as tapered element oscillating microbalance (TEOM) measurements are discussed.

With respect to the understanding and possible applications of structured microporous and mesoporous materials, the knowledge of their pore sizes is of paramount importance. There are several approaches to determining pore sizes, for example, via scattering techniques, electron microscopy or ^{129}Xe NMR (see also the fourth chapter). In this volume, however, Y. Traa, S. Sealy and J. Weitkamp describe and critically discuss the characterization of pore sizes using probe molecules, i.e., by techniques based on either adsorption or catalytic test reactions. The intense investigation of the relationship between a catalytic test reaction, viz. shape-selective hydrocracking of C_{10} cycloalkanes, and the effective pore width of zeolites finally led to the introduction of the spaciousness index (SI), which, since then, has found widespread acceptance. A versatile method of characterizing molecular sieves was developed with the help of ^{129}Xe NMR. One of the pioneers of this remarkable characterization technique, J. Fraissard, has contrubuted the fourth chapter of this volume. Here, he explains the fundamentals of the method and demonstrates its capability of studying the porosity of zeolites, the effect of cation exchange, the distribution of adsorbed phases, the behavior of zeolite-supported metals, phenomena of Xe diffusion, and special features of other microporous (pillared clays, heteropolyoxometalate salts, activated carbons) and mesoporous solids (M41S and SBA materials).

Catalyst deactivation as a consequence of the undesired deposition of carbonaceous materials plays an important role in many catalytic processes on micro- and mesoporous solids. Coke formation on zeolites and the effect of several features of coke deposition, such as shape selectivity, acidity of the catalyst, location, mechanism and kinetics of coke build-up, activity of and selectivation by coke are dealt with in the fifth chapter written by F. Bauer and H.G. Karge. However, the focus of this contribution is laid on the characterization of coke formed on zeolites by spectroscopic and non-spectroscopic techniques and the relationship derived therefrom between the nature of the coke and the conditions of its formation.

The last chapter is devoted to the characterization of very interesting derivatives of the common molecular sieves, i.e., the isomorphously substituted molecular sieves. The notorious main problem here is whether or not the hetero-element (such as B, Ga, Fe, Ti, V, Zn, Co) is unambiguously incorporated into the framework of the porous material. The authors, J. B.Nagy, R. Aiello, G. Giordano, A. Katovic, F. Testa, Z. Kónya, and I. Kiricsi provide a large body of experimental results of successful isomorphous substitution and a great number of cases where the position of the isomorphously introduced hetero-element (before and after additional treatment) can be identified by sophisticated physico-chemical investigations.

Most likely, the important art of characterizing micro- and mesoporous structured materials will turn out to have not been exhaustively covered by Vols. 4 and 5. Thus, it could well be that additional characterization techniques will be dealt with in a future volume of this book series.

April 2006

Hellmut G. Karge
Jens Weitkamp

Contents

Contents of Volume 4

Characterization I

Editors: Hellmut G. Karge, Jens Weitkamp
ISBN: 3-540-64335-4

Mol Sieves (2007) 5: 1–66
DOI 10.1007/3829_001

Published online: 23 March 2006

Chemical Analysis of Aluminosilicates, Aluminophosphates and Related Molecular Sieves

H. Kosslick[1] (✉) · R. Fricke[2]

[1]Leibniz-Institut für Katalyse e. V. an der Universität Rostock, Außenstelle Berlin (ehemals ACA), 12489 Berlin-Adlershof, Germany
kosslick@aca-berlin.de

[2]Institut für Angewandte Chemie, 12489 Berlin-Adlershof, Germany

Abbreviations

AAS	Atomic absorption spectroscopy
AES	Atomic emission spectroscopy
AS	Auger electron spectroscopy
a.u.	Arbitrary units
CP	Cross polarization
CHN	Simultaneous analysis of carbon, hydrogen, and nitrogen by combustion
CN	Coordination number
DTA	Differential thermoanalysis
Erio T	Colored indicator
EDTA	Ethylendiaminetetraacetic acid sodium salt-complexing agent
ED	Energy disperse
EPMA	Electron probe micro analysis
ESCA	Electron spectroscopy for chemical analysis (synonym for XPS, UPS)
EXAFS	Extended X-ray absorption fine structure
FTIR	Fourier transform infrared
HPLC	High-pressure liquid chromatography
HF-generator	High-frequency generator
HTP	High temperature peak
I	Intensity
ICP	Inductive coupled plasma
ICP-AES	Inductive coupled plasma (activated) atomic emission spectroscopy
IR	Infrared
ISS	Ion scattering spectroscopy
LTP	Low temperature peak
M	Metal nucleus (NMR)
MAS	Magic angle spinning
Me^+	Metal cation
NMR	Nuclear magnetic resonance
PVA	Polyvinyl alcohol
PTFE	Polytetrafluorethylene
Q	Radius ratio of atoms of two elements
Q^n	n-fold connectivity of the metal nucleus Q giving rise to NMR resonance ($n = 0$–4)
r_{Me}	Atomic radius
$r_{\mathrm{Me-O}}$	Metal-oxygen distance
Si(4Me)	Silicon atom (usually in tetrahedral oxygen coordination) connected to four metal atoms via oxygen bridges
TG	Thermogravimetry
TGA	Thermogravimetric analysis
TMS	Tetramethylsilane
TPD	Temperature-programmed desorption
TPDA	Temperature-programmed desorption analysis
TO_4-tetr.	Metal (T)-oxygen tetrahedra
T – O – T bridge	Metal-oxygen-metal bridge
u.c.	Unit cell
UPS	Ultraviolet photoelectron spectroscopy
UV-Vis	Ultraviolet visible
$V_{\mathrm{u.c.}}$	Unit cell volume

WD	Wavelength disperse
(1 + 1)	Relative contributions of two components to a mixture
XRD	X-ray diffraction
XPS	X-ray photoelectron spectroscopy
XRF	X-ray fluorescence spectroscopy
δ	Chemical shift

1 Introduction

During the last years, the synthesis and detailed investigation of zeolites including mesoporous materials has obtained a tremendous impetus. A large number of new structures and compositions of zeolitic frameworks has been developed. Some of them have already found applications such as catalysts, adsorbents or detergents.

Zeolites are inorganic crystalline compounds that are related to tectosilicates. Their lattice usually consists of three-dimensional, four-fold interconnected TO_4 tetrahedra. They are linked via $T-O-T$ bridges. Most natural and the first synthetic zeolites are aluminosilicates. The substitution of silicon by aluminum in framework positions leaves a negative framework charge. Therefore, the framework can be considered as a polymeric anion. The negative framework charge is compensated by cations, often sodium, which is a part of most synthesis recipes. The pores of these zeolites are filled with water. The general compositional formulae of these alkali ion-containing zeolites is:

$$Me_x{}^+[(AlO_2)_x(SiO_2)_y]^{x-} \cdot zH_2O$$

The interstitial cations are exchangeable (in principle) by, all other kinds of cations. A great diversity of zeolite composition exists. Nowadays, aluminum was substituted by other metals like Ge, Fe, Co, etc. Also, new, silicon-rich zeolite structures were obtained using organics as templates. Therefore, in addition to water, these new types of zeolites contain these template molecules. Zeolites and molecular sieves with completely new framework compositions were synthesized on the basis of aluminophosphates, gallophosphates, and nitridophosphates. New molecular sieves consisting at least in part of octahedrally coordinated framework atoms are also now known. Some zeolites may contain additional anions like chloride or sulphate.

The knowledge of the chemical composition of zeolites is the key for the understanding of their physico-chemical properties and the bases for the consideration of "structure-property-action(application)-relationships".

The methods applied for the characterization of the zeolites have received a high technological standard by the introduction of computer-controlled spectrometers or other equipment allowing better and more detailed insight into the properties of these solids. Therefore, numerous papers have been written in connection with the application of modern characterization methods and their use.

The contents of the present chapter deals with the *quantitative analysis* of the composition of zeolite frameworks. This seems to be a necessary addition to other characterization methods because in many papers the data available are limited to the composition of the synthesis gel only. In addition, induction coupled plasma (ICP), atomic absorption spectroscopy (AAS), and other similar methods are often the only methods used for the quantitative analysis of the solids.

The intention of the authors was not to focus mainly on the description of these modern methods but to gather and summarize some of the main *wet-chemical* recipes that are already well known for a long time. Unfortunately, however, the description of these wet-chemical methods are more or less dispersed over a large number of chemical handbooks. Often, the advantages and limits of these methods for the quantitative analysis of various molecular sieves (zeolites, aluminophosphates, etc.) are unknown. Therefore, it is our hope that the description of those methods, which are directed to the quantitative analysis of elements that form the framework of the most important zeolites and other micro- and mesoporous materials, may help to find the right method for the analysis of the appropriate element.

In the first part of the paper an overview of routes and methods for the determination of the chemical composition of zeolites is given, starting with the preliminary preparation of samples including their dissolution by different techniques. The classical wet chemical way is described and illustrated by analysis instructions. This conventional method is not only of historical interest, but necessary as a reference, and is useful if only a partial analysis (one or two elements) is required. An overview of the possible determination methods of elements will be given including appropriate references. Modern physical routine methods, such as atom absorption spectroscopy (AAS), inductive plasma coupled-atom emission spectroscopy (ICP-AES), and X-ray fluorescence spectroscopy (XRF) and their useful application and limitations/problems, are described in the next chapter.

In the second part of the paper attention is directed to the spatial distribution of elements. This becomes more and more important because of its influence on zeolite properties. This includes (a) determination of the silicon/metal framework ratio (mainly Si/Al) by IR, NMR, NH_3-TPD and NH_4^+-exchange capacity, and (b) the problem of differentiation between framework and extra-framework species by means of acid leaching and photoelectron spectroscopy (ESCA). The latter is closely related to the quantitative determination of catalytically active sites and will be treated in other parts of this series.

The quantitative determination of organics by thermal and CHN analysis is also considered.

2 Sample Preparation

The use of modern instrumental analytical techniques like atomic absorption and emission spectrometry and X-ray fluorescence spectrometry has considerably enhanced the accuracy and reproducibility of analytical measurements. Often, errors or standard deviations given in papers consider only those of the analytical method that has been applied. Due to the high level of instrumental techniques, however, nowadays the accuracy of the chemical analysis is increasingly determined by the accuracy of prior to measurement operations such as sampling, preparation of the sample, determination of ignition loss, and fusion of samples. In general, the sample passes through these essential steps before its composition will be determined by the different techniques.

2.1 Sampling

The problem of sampling, the choice of a representative part that reflects the mean chemical composition of the material, is mainly restricted to cases were natural zeolites obtained from deposits are under study or a zeolite is obtained with a second phase, e.g., cristobalite. Usual synthesis products obtained by lab-scaled crystallization consist of crystals of relative uniform composition and sizes of 0.1–10 μm. They are soluble and need no further mechanical decomposition before disintegration.

2.2 Drying and Stable Hydrated Form

Zeolite synthesis products are washed until the washing water is neutral in order to remove the excess alkali and templates. This should be done carefully because under supersaturated hydrothermal conditions and concentration gradients crystalline by-products can appear. Extra-zeolitic silica precipitates, if sometimes present on crystal surfaces, can be removed by careful treatment with HCl or 2 M NaOH solution. Sodium aluminosilicates or other zeolites containing only water in the cavities are obtained in the complete cation exchanged form. They will be repeatedly exchanged with NaCl solution, if necessary at ambient temperature, to obtain the stoichiometrically correct form. This holds especially for aluminum-rich sodium aluminosilicates like zeolite A, faujasites, mordenite, and others. Samples are then dried. Some authors prefer the complete removal of water by calcination at 400 °C for 3 h; others dry the samples at elevated temperatures of 40–60 °C to remove excess wash water and loosely bound zeolitic water. In view of possible dealumination at higher temperatures, which might change the sample with respect to the framework composition upon combustion of organics at high temperatures, the latter procedure is preferable. Also, template containing microporous materials are treated in this manner. The dry sample cake is broken and slightly milled in an agate mortar to obtain a powdered crystalline material. This crystalline powdered sample is transferred to a desiccator containing a saturated ammonium chloride solution (constant water vapor pressure). Then the sample is saturated with water under reduced atmospheric pressure for 1–3 days until adsorption equilibrium is achieved.

2.3 Weighing

A typical sample weight for complete wet chemical analysis (two parallel runs for each sample) is 0.5–1 g. If only one or a fraction of elements present in the zeolite should be determined or modern instrumental methods are used, down to approximate 0.05 g of the sample is necessary. Weighing of the sample should be carried out as quick as possible to minimize errors which might arise by weight changes due to the adsorption of moisture, by drift of the balance, etc. The sample is directly weighed into a platinum dish. The disadvantage is the high weight of platinum compared to the sample weight, which leads to an increased weighing error. Therefore, the sample can also be weighed on a special "weighing paper", which ensures that all the sample is removed from the paper and transferred to the dish.

2.4 Loss of Ignition

Different temperatures of ignition are recommended for different materials. Usually, silica- and alumina-containing minerals, rocks, clays, etc., are ignited at 1200 °C, if possible. These "dead burned" samples are of lowest hygroscopy making an accurate weighing possible. Microporous materials lose alkaline cations at this high temperature. Therefore, zeolite samples are ignited in platinum crucible at 800 °C, where alkaline loss is negligible. Rapid weighing of the cooled samples ensures sufficient accuracy.

A conceivable procedure might be the following:

Weigh 0.05 to 1 g *of microporous material into a platinum disk. Then heat the sample in an electric furnace from room temperature to* 800 °C *and hold the sample at this temperature for 0.5 to* 1 h*. Then quickly transfer the platinum disk with the sample into a desiccator (containing zeolite NaA or KA as drying agent) and then cool down to room temperature (0.5 to* 1 h*). Determine the weight loss. This material can be used for fusions, too.*

To accelerate the procedure, samples are sometimes preheated over the flame of burners. With respect to the possible water content of microporous materials, this should be avoided. In addition, the sample should not be transferred into hot ovens. Sudden water eruption may destroy the sample and may lead to undesired sample losses. It is worth mentioning that the use of a platinum crucible always requires thermal treatment in oxidizing atmosphere. In reducing atmosphere at high temperature, some oxides, for instance iron, can be reduced and form alloys with platinum. A reaction with carbon is also possible. The heated platinum should be always handled with platinum-tipped tongs to keep the platinum as clean as possible.

3 Fusion Techniques for Elemental Analysis

In general, three kinds of fusion techniques [1–3] are applied before the determination of elements with instrumental methods can be carried out:

- dissolution of the sample by acid treatment at room or ambient temperature
- acidic and basic fusion by heating in suitable crucibles and
- fusion under reductive conditions (normally, however, not necessary for microporous materials).

Dissolution of samples occurs with various acids such as HF, H_2SO_4, HNO_3, HCl, HJ, H_3PO_4, and Br_2. All the acid fusions can be carried out in autoclaves and also in a microwave oven.

Basic and acidic fusions are carried out in melts of Na_2CO_3, K_2CO_3, $LiBO_2$, NH_4NO_3, $Na_2B_4O_7$, $Li_2B_4O_7$, H_3BO_3, $K_2S_2O_7$, $KHSO_4$, NH_4HSO_4 and KHF_2.

Fusion under reductive conditions uses Pb or NiS as a collector and are applied for the enrichment of gold and platinum metals.

Prior to elemental analysis, microporous materials have to be dissolved or transferred into soluble compounds. Therefore, the fusion solution must contain the determined elements in the same quantities as the original sample. During this operation, contamination of the sample from fusion agents or crucible components should be negligible. Fusion agents and materials released from the vessel walls are the main sources of contamination. Another problem is the blank value of the solutions. This value and its variation determines the detection limit of the analytical method. Therefore, fusion is the main source of error. Best results are obtained with platinum crucibles or Teflon-lined autoclaves. Care has to be taken of the purity of fusion compounds and acids. If necessary, the acids should be purified by distillation under vacuum, starting with analytical-grade acids. The acidic fusion solution should not be stored for a prolonged time in glass vessels. Depending on the contact time and on the prior treatment of the vessels, 0.05 to 10 mg/L of different elements can be released [4].

3.1 Dissolution in Acids

Problems may also arise from the different volatility of compounds of elements, i.e., $SiF_4 \uparrow$, $BF_3 \uparrow$, or from hardly soluble residues or accompanying minerals. Small residues are difficult to detect by eye. Nevertheless, they may contain large amounts of trace elements (low degree of substitution) like Ti, Zr, Cr, Ni, Co, Zn. It is known that chromium spinels are difficult to dissolve in $HF/HClO_4$ mixtures. This can be overcome by the use of HF/H_2SO_4. This may cause a loss of lead or barium precipitated as sulphates which, however, can be redissolved by treatment with water or diluted nitric acid.

Boron- and zirconium-containing minerals are difficult to dissolve in open systems (evaporation in uncovered vessels). They should be treated in closed Teflon-lined autoclaves (no loss of volatile $BF_3 \uparrow$). Loss of boron during evaporation can also be avoided by the addition of polyalcohols like mannite and glycerol.

Hydrofluoric acid exhibits a very high complexing ability and, consequently, dissolves the frameworks of microporous materials very rapidly. Therefore, the instrumental analysis of fluoride-containing stock solutions requires the application of HF-resistent nebulizer systems. Otherwise, intensive corrosion of the apparatus occurs. High fluoride concentrations may also be the reason for the appearance of systematic errors in the atomic absorption spectroscopy (AAS) and atomic emission spectroscopy (AES). In some cases, fluoride interference can be avoided by the addition of sulphuric acid. Hydrogen fluoride is carefully evaporated by heating to 150 °C after sample treatment in autoclaves. Likewise, poorly soluble minerals can be successfully fused using this technique and high fluoride concentrations are prevented.

Many as-synthesized microporous materials contain organic templates in their pore system, which may also cause analytical errors. Therefore, calcined samples should be used for analysis provided that calcination is not accompanied with the formation of poorly soluble minerals. If as-synthesized samples are analyzed, the organic should be removed by nitric acid treatment at elevated temperature. The sample should be slightly warmed to prevent explosive decomposition. Thereafter, the residue is dissolved in HF/H_2SO_4. Residual carbon does not disturb the analysis.

Although more laborious than the HF treatment (Sect. 4.2), dissolution in H_3PO_4 is a very effective way to fuse zeolites, especially if the sample contains volatile boron or weakly soluble minerals. The reason is the high complexing ability of pyrophosphoric acid. On heating, orthophosphoric acid undergoes successive condensation reactions, giving pyrophosphoric and metaphosphoric acid, so-called "condensed" or "strong" phosphoric acid:

$$H_3PO_4 \rightarrow H_6P_4O_{13} \text{ and } H_4P_4O_{12} \quad (1)$$

This oligomeric phosporous acid has a very high complexing ability and also forms stable silica compounds SiP_2O_7 [5]. The potential of this fusion or the use of phosphorous acid in combination with other acids has been underestimated so far. One disadvantage of this fusion is, however, the high viscosity of the resulting solution. Evaporation of the fusion solution should be carried out in such a way that the H_3PO_4 is not lost. The addition of some phosphoric acid to other acid-fusion mixtures makes the fusion more effective.

3.1.1 Procedures

Dissolution in H_3PO_4

Typically, 50 mg *of ground sample is treated with* 3.5 mL *of conc.* H_3PO_4 *in a platinum crucible. Heat slowly on a sand bath or hot plate to* 300 °C *while shaking at regular intervals of* 15 min. *The released water should have enough time to evaporate. Solution should be completed within* 2 h. *If not (metal sulfide), the combined use of* 0.25 mL $HClO_4$ *(70% in water) and* 3.5 mL H_3PO_4 *is necessary. The obtained solution is cooled and diluted with demineralized water to 100 or* 200 mL. *To prevent precipitation of titanium* ($Ti(OH)PO_4$), 0.5 mL H_2O_2 *is added before dilution.*

Dissolution in HF/H_2SO_4

Transfer 50–100 mg *of the sample into a platinum crucible and add* 1–2 mL *of HF (40% in water) and* 2–4 mL *conc.* H_2SO_4. *Gently shake the fusion mixture. Cover the crucible and heat on a sand bath up to* 180–200 °C. *Continue thermal treatment for* 2 h *or more. Then evaporate the excess HF by increasing the temperature until dryness is reached. Cool down to room temperature and dissolve the residue with conc. HCl or conc.* HNO_3. *Dilute with demineralized water and make to volume.*

Dissolution in $HF/HClO_4/HNO_3$

Transfer 50 to 100 mg *of the sample into a platinum crucible. Add, little by little,* 1–2 mL *of conc. HF,* 2–3 mL *of conc.* $HClO_4$ *and* 2–3 mL *of conc.* HNO_3. *Heat slowly on a plate or sand bath, starting at slightly elevated temperature, to* 180–200 °C *in a time interval of* 2–3 h. *Evaporate by prolonged heating at increased temperature. Cool the fusion and dissolve in diluted acid.*

3.2
Acidic and Basic Fusions

Acidic and basic fusions require mixing of the sample with the fusion compounds. First, part of the fusion mixture is deposited on the bottom of the (platinum) crucible. Then the sample will be mixed with the main part of the fusion mixture and transferred to the crucible. Finally, the mixture should be covered by the last portion of the pure fusion compound. If fusion mixtures, for example KOH, release water or gases like CO_2 or SO_2, the fusion mixture is slowly heated to melt to evaporate all volatile contaminants and then cooled. The sample is put onto the cold solid melt cake and again heated to the fusion temperature. After melting, the fusion is held at the fusion temperature for a certain time. The cooled (but still warm) melt cake is dissolved by the successive addition of as small as possible portions of water and/or diluted acid (HNO_3, HCl) to ensure complete dissolution. The remaining non-soluble residues will be dissolved in 1 mL of conc. HCl or HNO_3 and 5 mL of water and again heated to 180 °C for ca. 1 h.

In general, the fusion melts can react either acidic or basic. Basic compounds release oxygen according to the following types of reactions:

$$CO_3^{2-} \rightarrow CO_2 + O^{2-} \quad (2)$$

$$4BO_2^{-} \rightarrow B_4O_7^{2-} + O^{2-} \quad (3)$$

$$O^{2-} + Al_2O_3 \rightarrow 2AlO_2^{-} \quad (4)$$

Pyrosulphates undergo an acidic reaction:

$$2KHSO_4 \rightarrow K_2S_2O_7 + H_2O\uparrow \quad (5)$$

$$O^{2-} + S_2O_7^{2-} \rightarrow 2SO_4^{2-} \quad (6)$$

$$Al_2O_3 \rightarrow 2Al^{3+} + 3O^{2-} \quad (7)$$

Substances that behave acidic are fused with basic melts: silica and aluminosilicates. Oxides of Al, Cr, Ti, Fe, Ge, Mn, Mo, Sn, V, W, Zr and others behave amphoteric. They are fused in acidic and basic melts. But acidic tetraborate melts are also used for the fusion of silicates for X-ray fluorescence analysis (XRF). The material readily decomposes if the silicates contain high amounts of alkaline oxides. Pure dense silica (quartz) dissolves slowly, and its dissolution is completed only after the addition of another metal oxide. Mixtures of meta- and tetraborate can be applied universally. Acidic melts of potassium pyrosulphate are used to fuse hardly soluble oxide minerals of Be, Cr, Fr, Ti, W, Zr and of rare earth elements. For the latter two, the corresponding sodium salts are preferred. Ammonium pyrosulphate is used to fuse sulfides (in quartz or borosilicate crucible). In the following paragraphs some examples of fusion are given in brief (compare also the chapter wet chemical analysis).

3.2.1
Procedures

3.2.1.1
Fusion in KOH

Transfer 0.5 to 1 g *of KOH into a nickel or zirconium crucible (not platinum!) and heat slowly to melt on a plate or sand bath to evaporate residual water and carbon dioxide. Cool the melt. Put* 50–100 mg *of the sample onto the KOH cake and heat to* > 350 °C *until a clear melt is formed. Cool down and dissolve the melt in water and subsequently treat with diluted HCl. Make the combined solutions to volume.*

3.2.1.2
Fusion in $LiBO_2/Li_2B_4O_7$

Transfer a mixture of 100 mg *of the sample and* 500–1000 mg *of the fusion mixture (meta-to-tetraborate* = 4 : 1*) into a platinum crucible. Heat the crucible in an electric furnace to* 1000 °C. *Swirl carefully at certain time intervals. After 0.5 to* 1 h, *the dissolution is completed. Dissolve the cake in diluted HCl or* HNO_3. *If fusion is incomplete, decrease the meta-to-tetraborate ratio and increase the amount of fusion mixture.*

3.2.1.3
Fusion in $K_2S_2O_7$

Transfer a fusion mixture containing 50–100 mg *of the sample and* 1–3 g *of pyrosulphate into a platinum crucible. Heat to* 450 °C *and hold the temperature for* 30 min. *Cool down and dissolve the cake with diluted HCl, diluted sulphuric acid or tartaric acid* (10% *in water).*

4
Determination of the Bulk Chemical Composition

4.1
Wet-Chemical Analysis

Although the wet-chemical analysis of microporous materials is laborious compared to modern instrumental analytical techniques, there exist some reasons to consider this classical chemical analysis in more detail. The advantage of this analytical method is that the sample is decomposed and all constituents are chemically separated from each other so that interferences are minimized. The element content is directly determined, for example, by the weight of a defined compound.

Table 1 List of fundamental classical chemical analytic literature

Books

Fresenius W, Jander G (eds) (1953) Handbuch der Analytischen Chemie, Bd. I–VII, and new editions. Springer, Berlin Heidelberg New York

Halman M (ed) (1972) Analytical Chemistry of Phosphorous Compounds. Wiley-Interscience, New York. In series: Chemical Analysis, vol. 37

Radke CJ (ed) (1950) Analytical Chemistry of the Manhattan Project. McGraw-Hill Book Company, Inc. In: National Nuclear Energy Series, vol. VIII

van Nieuwenberg CJ, Böttger W, Feigl F, Komarovsky AS, Srafford N (1938) Tables of Reagents for Inorganic Analysis. Akademische Verlagsgesellschaft m.b.H., Leipzig

Kolthoff IM, Sandell EB (1943) Textbook of Quantitative Inorganic Chemistry. The Macmillan Company, New York

Maisson et Cie (1974) Chemie Analytique Quantitative (I+II). Paris

Müller G-O (1968) Praktikum der Quantitativen Chemischen Analyse. S. Hirzel Verlag, Leipzig

Bennett H, Reed RA (1971) Chemical Methods of Silicate Analysis. Academic Press, London

Journals

Fresenius Zeitschrift für Analyische Chemie
The Analyst
Talanta

First, this technique is needed if new compounds are to be analyzed unambiguously. Second, it is of interest if modern (and expensive) methods are not available, and, third, if the analysis of only one component is required. Procedures given in the following paragraphs may be verified or cut according to one's own needs. A list of fundamental literature of classical chemical analysis is given in Table 1.

4.1.1 Metal Silicates

4.1.1.1 Course of Wet-chemical Analysis of Metal Silicates

Fuse the sample with a fusion mixture of Na_2CO_3/H_3BO_4 *or potassium pyrosulphate. Dissolve the cold melt in hydrochloric acid. Eventually, add drops of hydrosulphuric acid to prevent hydrolysis of titanium salts. Now evaporate the solution until a stiff silicon dioxide gel forms. At low silicon content (< 20%), the solution hardly gels. Coagulate the silica by addition of polyethylene oxide. Filter off the precipitate and wash the solid first with HCl followed by demineralized hot water. Transfer the filtrate into a platinum crucible and ignite before and after HF treatment. The difference in weight gives the gravimetric silica content. Fuse the residue of the HF treatment again and put its solution to the* *filtrate from the silica determination. Make the combined solutions to volume and use for the following determinations:*

Aliquots are used for the spectrophotometric determination of iron with phenanthroline chloride after reduction with hydroxylamine, spectrophotometric determination of titanium using hydrogen peroxide and H_3PO_4 to prevent interference with iron.

Determination of soluble rest-silicon by spectrophotometric measurement of either the yellow color using ammonium molybdate or the blue color after reduction of the silicomolybdate if interference with phosphate occurs.

Other aliquots are used for the spectrophotometric determination of phosphate as molybdatophosphate or as vanadatophosphate, and the determination of manganese by oxidation to permanganate by potassium periodate.

A further aliquot is used for the volumetric determination of alumina after increasing the acidity by the addition of HCl/H_2SO_4. Iron and titanium are precipitated with cupferron. The precipitate and excess reagent is removed by extraction with chloroform and determined separately. The solution is neutralized with ammonia, an excess of EDTA is added, and the solution is buffered. The solution is boiled for complete complexation of the aluminum, and non-reacted EDTA is back-titrated with $ZnCl_2$ solution.

In a similar way, bivalent transition metal elements can be determined. For aluminous materials, lime and magnesia the determination is also made from this stock solution after complexing the oxides of the ammonia group with triethanol amine.

4.1.1.2 Determination of Silica

Some microporous materials like aluminum-rich silicates or some silicon-substituted aluminophosphates contain HCl-soluble silicon, which allows the determination of the silica content by precipitation with 20% HCl in water as silicon dioxide (gel method).

Procedure

*The sample is weighed into a platinum disk and three times evaporated with hydrochloric acid until it is completely dry (*30 min *after dryness) on a sand bath. Before each addition of acid, the sample should be thermally treated until caustic. Allow to cool and drench the residue with* 5 wt % *HCl, filtrate, and wash the filter cake with hot water to dissolve the salts. Place the filter with the residue* (SiO_2) *into a weighed Pt crucible, heat and dry the residue with caution and char the filter paper at* 1000 °C. *The remaining* SiO_2 *is weighed after cooling down in a desiccator to room temperature (ca.* 1 h*).*

Control

The evaporation with HF leaves no residue. High-silica zeolites of the ZSM-5 type as well as aluminophosphates with relative high silicon content contain silicon that is not or poorly soluble in HCl. These compounds have to be fused. The silica content is determined by the coagulation method as follows:

Transfer ca. 0.5 g *of weighed sample into a platinum dish. Carefully mix the material with the fusion mixture (five times the weight of the sample). Then cover the dish with a platinum lid. Heat on a mushroom burner with a low flame under fan until gas evolution is finished. Then continue heating under a full flame for* 10–15 min *until a flowing and red-heat melt appears. Immerse the platinum dish containing the melt in cold distilled water. Put the dish into another platinum crucible and add some distilled water and* 40 mL *of hydrochloric acid* (1 + 1) *and cover with a watch glass dish. After dissolution of the cake, the crucible is withdrawn and carefully rinsed with a minimum amount of required water. The fusion is complete if no scrape is felt during stirring (by hand) with a glass stick. Boil off the solution until it is dry on a water bath and continue heating for an additional* 30 min *until caustic, to enclose all the silica. Take up the residue with* 30 mL *hydrochloric acid* (1 + 1) *and warm it (covered) on the hot water bath for an additional* 10 min. *Precipitate the silicic acid from this solution by adding* 5 mL *of Polyox coagulation agent and continue stirring with a glass stick for* 5 min. *Age* 10 min *and filtrate the solution through a filter paper of medium pore size (yellow etiquette) and wash the solid residue with hot water to remove all the chloride. If large amounts of other elements are present (alumina, iron oxide, lime, or magnesia), a mucous filter should be used. Fold the filter and transfer it to the platinum crucible used for the fusion. Cautiously char the filter paper over the low non-colored flame. Then ignite the residue at* 1200 °C *for* 30 min. *Cool down the crucible in a desiccator and weigh the crucible content. Now wet the residue with* 1 mL *of distilled water and add 10 drops of sulphuric acid* (1 + 1) *and ca.* 10–20 mL *of hydrofluoric acid* (40 wt % *in water). Evaporate again until dryness as described above, ignite at* 1200 °C, *cool down in the desiccator and again weigh the crucible. The difference to the former determination represents the amount of fine-silica.*

Then fuse the residue with 0.5 g *of potassium pyrosulphate or another fusion mixture. Dissolve the cold melt cake with some water or (diluted) hydrochloric acid and add the solution to the former filtrate.*

Fusion mixture: caustic sodium carbonate-sodium potassium carbonate (1 + 1), sodium carbonate-borax (336 + 221, calcined at 400 °C, and carefully ground).

Coagulation reagent: Polyethylene oxide: Add 250 mg of polyethylene oxide to 100 mL of distilled water under magnetic stirring until it is dissolved (molecular weight ca. 600 000 – BDH Chemicals-Ltd. Poole England, Union Carbide "Polyox" resins).

4.1.1.3 Determination of Soluble Silica

Due to the general solubility of silica as monomeric silicic acid in water, part of the silicon (ca. 1%) always remains in the solution. It can be determined spectrophotometrically either as yellow silicomolybdate, or after reduction according to the molybdenum blue method. The molybdenum blue complex has a ca. ten times higher extinction coefficient. Hence, the latter method is more sensitive and allows the determination of traces of silicon. Additionally, the molybdenum blue method has to be used if phosphate interferes with the silicon determination by the formation of the likewise yellow molybdatophosporic acid.

Molybdic acid reacts with monomeric silicic acid and forms the yellow silicomolybdic acid according to:

$$7Si(OH)_4 + 12H_6Mo_7O_{24}{\cdot}4H_2O + 126H_2O \rightarrow 7H_8Si(Mo_2O_7)_6{\cdot}28H_2O \quad (8)$$

The intensity of the yellow acid is measured and corresponds to the amount of silicon that has to be added to the gravimetrically determined amount of silicon to give the total silicon content of the sample.

Procedures

Silicomolybdate Method
Add up to 5 mL *of stock solution to* 20 mL *ammonium molybdate solution and make to* 50 mL *volume by addition of distilled water. Allow the components to react for* 10 min *and measure the absorption at* 410 nm. *The adjusted solution should contain* $\leq$ 0.05 mmol *of silicon* [6].

Reagents
Dissolve 10 g *of* $(NH_4)_6Mo_7O_{24} * 4H_2O$ *in* 90 mL *of water, add* 4.7 mL *conc. ammonia solution (*28 wt % *in water), and make to volume (*100 mL*). Add this solution to* 500 mL *of water that was acidified by the addition of* 200 mL *of* 0.75 *molar sulfuric acid in water. This acidified molybdate solution is used for silicon determination.*

Molybdenum Blue Method
Dilute an aliquot of the stock solution containing ca. 0.15 to 0.8 μmol *of silicon to* 20 mL. *Add* 3 mL *of molybdate solution. Allow it to react for* 10 min *and add* 15 mL *of the reducing agent and make to* 50 mL *volume with distilled water. After* 3 h, *the intensity of the absorption at* 810 nm *is measured.*

Ceric ions, fluoride, germanium, and vanadium give serious interference, whereas phosphorous is masked by oxalic acid [6, 7].

Reagents
The molybdate solution is prepared by dissolving 10 g *of* $(NH_4)_6Mo_7O_{24}$ $4 \cdot H_2O$ *and* 30 mL *of conc. hydrochloric acid* (36 wt% *in water) in water. The resulting mixture is made to* 500 mL *volume.*

The reducing agent is prepared in the following way:

Prepare a solution containing 2 g *per* 100 mL *metol (5-amino-2-hydroxytoluene sulfate) and* 1.2 g *per* 100 mL *sodium sulfite in water. Add to this solution* 60 mL *of oxalic acid* (10 wt% *in water),* 120 mL *sulfuric acid* (1 + 3) *and* 20 mL *of water and store this solution.*

The filtrate of the silica precipitation or the residue of the hydrofluoric treatment is used for determination of Me(III) elements as well as of mono- and bivalent cations.

4.1.1.4 Determination of Aluminum

The gravimetric determination of aluminum as Al-oxinate is of high accuracy and has been proven to be a reliable method in silicate analysis. Besides, also a complexometric determination is possible.

Procedures

Gravimetric
Neutralize the acidic stock solution with ammonia against methylene-red and acidify with conc. hydrochloric acid. Now add 4 mL *of hydrochloric acid* (1 + 1). *Mix this solution with* 2 mL *of hydroxyaminhydrochlorid and* 2 mL *of o-phenanthroline and warm up to* 40–50 °C. *At this point, the volume of the solution should be ca.* 250 mL. *Add the required amount of 8-oxychinoline* (< 30 mg Al_2O_3 – 10 mL; 30–65 mg Al_2O_3 – 20 mL). *Slowly add* 40 mL *of ammonium acetate under stirring. Filter off the resulting precipitate. Wash and dry.*

1 g of Al oxinate contains 58.73 mg of Al.

Reagents
Dissolve 50 g *of* $NH_2OH \cdot HCl$ *in warm water and make to volume* (100 mL). *If necessary, filtrate this solution.*

Add 50 g *of 8-oxychinoline to* 120 mL *of pure acetic acid (Eisessig), dissolve it under warming and fill up with water to a volume of* 700 mL. *Filtrate the cold solution and make to* 1 L *volume.*

Indicator
Solution of 0.1 g o-phenanthroline in 10 mL of pure acetic acid (1 + 1).

Combined Titration of Fe, Al and Ti

Adjust the acidified stock solution with the addition of urotropine to pH = 5–6 against pH-paper. Add the 0.05 molar standard complexon(III) solution in excess and boil for 5 min. *Cool the solution under running cold water and add* 0.5 mL *of benzidine solution and* 1 mL *of ferro-ferricyanide solution as indicator. Titrate back with a 0.05 molar standard solution of zinc chloride. The consumption of complexon solution is equivalent to the amount of Fe, Ti, and Al, or to their sum. The end of titration is reached if no further deepening of the blue color is achieved. Add sodium fluoride (*3 wt% *in water) and boil for* 5 min *in order to destroy the aluminum complex and to release the corresponding amount of complexon into solution. Determine the released complexon again by back- titration with 0.05 molar standard solution of zinc chloride.*

Additional Solution

Mixture (1 + 1) of 0.1% solution of ferro(II)cyanide in water and 0.1% solution of ferri(III)cyanide in water.

Determination of Alkaline Earth and Transition Metals

Alkaline earth and most of the bivalent transition metals can be directly titrated with a standard complexon(III) solution or alkalimetrically determined by back-titration of protons of non-reacted complexon(III) that was added in excess. For details, see Schwarzenbach [8].

Determination of Magnesium

Mg^{2+} ions can be directly complexometrically determined in the alkaline stock solution by titration against Erio T as indicator. In the same way, Zn, Cd, Sc, La, Ga, In and rare earth metal ions can be determined.

Procedure

Neutralize 5 mL *of stock solution with some drops of ammonia (*13.5 M *in water) and make to* 25 mL *volume by addition of water. Acidify with* 0.5 mL *of HCl (*2 M *in water), add* 0.5 mL *of conc. ammonia, and make to* 50 mL *volume (pH =* 10.3–10.5*). Add ca.* 50 mg *of the indicator (mixture of* 1 wt% *of Erio T in NaCl finely milled in a mortar) and titrate with EDTA solution (*0.01 M*) until the color changes from red to blue.*

Interfering ions like Fe^{2+}, Fe^{3+}, Ni^{2+}, Cu^{2+}, Co^{2+}, Ti^{4+} *and ions of platinum metals can be masked with KCN or triethanolamine.*

Determination of Calcium

Procedure

Add 5 mL *of triethanolamine (*1 : 1 *diluted with water) to the acidified solution to mask aluminum. Add the indicator (Fluorexon, a fluoresceine complex,* 1 wt% *in KCl) and titrate with standard complexon-solution (EDTA) to the appearance of a green color.*

4.1.2
Analysis of Alumino- and Gallophosphates

4.1.2.1
Course of Wet-Chemical Analysis

The determination of silicon is carried out best separately from the determination of the other constituents.

Procedure
For the determination of Al, Ga, P and cations, dissolve a part of the sample in acid and remove the silicon by repeated evaporation with HF followed by evaporation with conc. H_2SO_4 *until dryness in order to completely remove fluoride and chloride. Dissolve the residue in diluted* HNO_3 *and make to volume. Or, transfer the combined filtrates of the silica determination (compare Sect. 4.1.1, metal silicates) into a flask and aliquot with water to* 200 mL *and use an aliquot of* 25 mL *of this solution for Al determination. If the alumino-or gallophosphate is poorly soluble, fuse the sample with KOH.*

Determine aluminum gravimetrically as oxinate. Also complexometric determination with EDTA is possible. However, in concentrated solutions, interference with phosphate may appear, leading to an underestimation of the aluminum content of up to 5%. Determine phosphorous colorimetrically as ammonium vanadatophosphate. If interference occurs, separate the transition elements.

4.1.2.2
Determination of Silica

*Fuse the sample in acid, i.e., HCl (*10% *in water),* HNO_3, H_2SO_4 *or take another fusion if necessary and determine silica according to the gel method gravimetrically or use the coagulation method. Silica can also be determined as weight loss upon HF treatment, provided the residue is converted completely into oxides by evaporation with* H_2SO_4 *and* HNO_3.

4.1.2.3
Determination of Aluminum

In the presence of phosphorous, aluminum should be determined gravimetrically as oxinate or oxide. Also complexometric titration is possible after masking interfering elements or after separation of the aluminum.

Procedure
Acidify the stock solution to pH = 1–2 *with hydrochloric acid (*1 + 1*). Add the 0.1 molar standard complexon solution in excess (with respect to the Al content) and boil for* 20 min *in order to complex the aluminum. Add the indicator*

(xylenol orange) to the cold solution. Now back-titrate with a standard solution of lead nitrate or acetate the non-consumed complexon.

1 mL of titratable non-consumed complexon solution (0.1 molar) corresponds to 2.7 mg Al.

Indicator: xylenol orange in potassium nitrate (1 + 100).

4.1.2.4 Separation of Aluminum

The separation of aluminum is considered in detail in volume III of the *Handbuch der Analytischen Chemie* (see Table 1) and is summarized in brief here.

Under acidic conditions, the precipitation of aluminum as oxinate allows a quantitative separation from the alkaline earth metals as well as In. Phosphate does not disturb.

Fe, Ti, Zr, V, Sn, Bi, Nb, Ta, Ga, U can be separated from Al by precipitation with cupferron in 1–2 N acidic solution. Eventually, some tartaric acid (1.5) has to be added.

Tartrate ions mask Al, Fe, and Cr, allowing precipitation of Ni as diacetyldioxime. Cu, Mg, Zn, and Cd can be precipitated and separated from Al in alkaline tartrate-containing stock solution as oxinate. 10–20 mL of 2 N NaOH have to be added to 100 mL of stock solution.

4.1.2.5 Determination of Gallium

Make the acidic solution to volume and add 8-oxychinoline solution (5 wt% *in ethanol) in excess. Heat on a water bath to* 70–80 °C *and neutralize by dropwise addition of dilute ammonia solution. Age at* 70 °C *for* 1 h *and cool down. Filter through a porcelain funnel after* 1–2 h *ageing at room temperature. Weigh the precipitate after drying at* 120 °C [9].

4.1.2.6 Separation of Gallium

Zn, Cd, Co, Ni, Mn, Be, Tl, alkaline earth and also alkaline metals can be separated from gallium by precipitation of gallium with tannin. Al, Cr, In, and rare earth metals can be separated by precipitation with cupferron.

Separation of Gallium from Iron

Procedure

Take an aliquot of the stock solution and add ammonia until a steady turbidity appears. Add the reagent for precipitation. Wait a few hours and filtrate. Wash the precipitate with cold acetic acid (1 + 1) *and subsequently with water. Char the precipitate in a furnace by heating to red heat. Weigh the formed* Fe_2O_3 *or dissolve in acid and determine otherwise. Gallium can be determined in the filtrate as usual.*

Reagent

Freshly prepared and filtrated solution of 1 g a-nitroso-b-naphthol in 50 mL of acetic acid (1 + 1).

Precipitation of Gallium with Tannin

Acidify the stock solution weakly with acetic acid and add ammonium nitrate to a final concentration of ca. 2%. Heat to boiling and add dropwise under stirring the tannin (1 : 9 in water) until complete precipitation is achieved. In general, ten times the tannin with respect to the gallium content is sufficient, but at least 0.5 g. *Filter after precipitation and wash with hot water to which some ammonium nitrate and acetic acid was added. The filtrate is clear and of weak yellow color. Dry the precipitate and then char. Weigh the formed* Ga_2O_3. *The precipitation should be carried out twice after dissolution of the gallium oxide, because the gallium tannin precipitate adsorbs accompanying metals.*

Precipitation of Gallium with Cupferron

Neutralize an aliquot containing 10–300 mg *of gallium with ammonia and acidify with sulphuric acid until the acid concentration is nearly* 2 N *in a total volume of* 100–500 mL. *Add the cupferron under vigorous stirring at room temperature in excess (per* 0.1 g *Ga add* 1 g *of cupferron, stoichiometric reaction requires* 0.6 g *of cupferron) and warm the solution on a water bath. A cloudy precipitate forms which readily agglomerates. Press with a rubber wiper to a crystalline paste. Filter through a filter paper on a platinum cone. Add* 1–2 mL *of the reagent to the turbid filtrate to complete precipitation and filtrate again (no tarnishing of the filtrate after* 1 h*). Or, better, cool down the solution under running water for* 10 min *(or in a refrigerator) and filter. Wash with* 2 N H_2SO_4 *until all of the chloride is removed, dry, and char. Determine the gallium by weighing the formed* Ga_2O_3.

Weighing of Ga_2O_3

Char the precipitate in a porcelain crucible in oxidizing atmosphere. Heat slowly to reduce loss by volatilization. Ga_2O_3 *is not volatile until* 2000 °C. *However,* Ga_2O *begins to evaporate at* 660 °C *quantitatively. Red-heat causes corrosion of crucible walls but no weight change. Chloride causes weight loss due to* $GaCl_3$ *evaporation. The gallium oxide, thermally treated at* 1000–1200 °C, *is not hygroscopic. Avoid the use of Pt crucibles, because under reducing conditions, Pt-Ga-alloys form.*

4.1.2.7
Determination of Phosphorous

Phosphate is determined in the filtrate of the silica determination or after evaporation of silicon with HF and removal of excess fluoride by evaporation with conc. H_2SO_4 and conc. HNO_3. Determination takes place either spectrophotometrically or gravimetrically as $NH_4MgPO_4\downarrow$ after masking the aluminum with citric acid or after a procedure proposed by Dahlgren.

Procedures

Spectrophotometrical
Transfer an aliquot of the phosphate solution containing 0.25–6 mg P_2O_5 *into a* 100-mL *graduated flask and add* 30 mL *of the reagent solution, make to volume, and then shake. Measure the absorbance after* 5 min *at* 440 nm.

Reagent:

1 : 1 : 1 mixture of HNO_3 (1 + 2), ammonium vanadate (2.5 g in 500 mL boiling water, acidified with 20 mL conc. HNO_3 after cooling and made to 1 L volume), ammonium molybdate (5 wt % in water). It is necessary to keep the given sequence of mixing!

Silicon does not disturb if the content is below 1 mg per 100 mL of stock solution. More than 200 mg of sulphate and 20 mg of fluoride interfere and have to be removed prior to measurement by evaporation with conc. HNO_3. Fusion of samples can result in the formation of pyrophosphate which is not detectable. Therefore, dissolve products of fusion refinement carefully in acid, i.e., HNO_3.

After Dahlgren
Take the filtrate of the silica determination or dissolve the sample in 20 mL *conc. nitric acid. Eventually, heat* 15 min *to boiling and make to* 100 mL *volume. The* 100 mL *of the aliquot should contain* 15–20 mL *conc.* HNO_3. *Add* 100 mL *of a chinolate-acetone mixture (72 + 28) to the cold solution. Heat the resulting solution in a water bath and maintain it at boiling for ca.* 30 s. *Cool the beaker in cold water for* 2–3 h. *Then filtrate the solution through a G4 sintered-glass mat. Dry the mat at* 275 °C *before using. Carefully rinse the beaker with small portions of water with the aid of a rubber wiper. Dry the precipitate at* 275 °C *and weigh the cold product after ca.* 2 h. *Clean the mat with hydrochloric acid (1 + 1) or conc. ammonia solution.*

1 g of precipitate contains 13.998 mg P.

Additional Solution
Prepare the chinolate solution in the following way: Dissolve 70 g *of ammonium molybdate in* 150 mL *water and add this solution to a solution of* 60 g *citric acid in* 150 mL *water (solution A). Now dilute* 35 mL *of nitric acid (conc.) in* 100 mL *of water and add* 5 mL *of chinoline (solution B). Combine solu-*

tions A and B under stirring. Store the chinolate solution overnight. Filter the solution through a G4 sintered-glass mat and keep it in a polyethylene flask.

4.1.2.8
Determination of Transition Metals (Zn, Cu, Ni, Co, and V)

Besides the complexometric determination by titration with EDTA solution against Erio T indicator described above, copper and nickel can be accurately determined gravimetrically. Zinc is preferentially determined by complexometric titration like magnesium at a pH of 10, whereby 1 mL of 0.01 M EDTA solution corresponds to 0.6538 mg Zn. (Buffer solution pH = 10 : 70 g NH_4Cl, 570 mL conc. NH_4OH in 1000 mL water [8]).

Procedures

Electrogravimetric Determination of Copper
Dilute an aliquot of the stock solution containing up to 32 mg *of copper with water to* 70–80 mL *and acidify by addition of* 3 mL *of* 18 M H_2SO_4. *Electrolysis is carried out at a voltage of* 2–2.5 V *and an electric current density of* 0.2–0.6 A dm^{-2} *at* 40 °C. *Precipitate the copper on a Pt-net cathode. Electrolysis is complete after* 30–60 min. *Weigh the dry Pt cathode. Prior to electrolysis, clean the Pt-net in acid, wash in water and ethanol and dry under slight warming to constant weight.*

Gravimetric Determination of Nickel
Transfer 10 to 20 mL *of the stock solution containing ca.* 6–60 mg *of Ni into a* 400-mL *glass vessel. Acidify by addition of* 2 mL *of diluted HCl (*2 M *in water) and make to* 200 mL *volume with deionized water. Now add* 20–30 mL *of ammonia (*0.2 M *in water) to achieve a pH of 3–4. Heat to boiling on an electric heating plate and add slowly, with the help of a burette,* 25 mL *of the reagent solution. Adjust the pH at 8–9 by dropwise addition of diluted ammonia. Stir the covered solution for* 10 min *with a magnetic stirrer and allow to precipitate. Decant after* 30 min *and filtrate through a glass-mat. Wash the precipitate with* 10 mL *of hot water and subsequently with* 20 mL *of cold water. Dry at* 110–120 °C *for* 2 h. *Continue drying until the weight is constant in steps of ca.* 45 min. *One milligram of precipitate corresponds to* 0.2032 g *Ni.*

Determination of Ni besides Zn^{2+}, Mn^{2+}, *and* Fe^{2+} *is possible in weak acetic acid solution (dropwise addition of* 2 M *acetic acid in water) buffered with* Na/NH_4Ac *(*1 + 1, 2 M *in water). Addition of ca.* 20 mL *tartaric acid (*15 g *in* 100 mL *solution in water) to the mineral acidic solution masks* Fe^{3+}, Al^{3+}, *and* Cr^{3+} *effectively.*

Reagent: ca. 0.1 M Na-dimethylglyoximate solution prepared by solution of ca. 16 g of the chemical in warm water, eventually filtrated, and diluted to 1 L volume.

Complexometric Determination of Cobalt
Neutralize 5 mL *of the stock solution and make to* 25 mL *volume with water. Add* 0.5 mL *of* 2 M *HCl and* 100 mg *of the indicator (*1 wt% *of murexide in NaCl, finely milled). Drop ammonia (*2 M *in water) into the red-pink solution until it just becomes yellow. Dilute with water to* 50 mL *volume (pH =* 7.5–8.5*) and titrate with* 0.01 M *EDTA solution from yellow to pink.*

1 mL 0.01 M *EDTA corresponds to* 0.5893 mg *Co.*

Potentiometric determination of vanadium
Dissolve ca. 20 mg *of the sample in* 70 mL *of sulfuric acid (*1 + 1*) at* 140 °C *under flowing dry nitrogen for* 20 min. *Dilute this solution to* 250 mL *with water. The hot solution is titrated under flowing nitrogen (Titrierautomat Schott TPC 2000) with n/100* $Ce(SO_4)_2$ *solution until the color changes from blue to yellow. Lead and barium precipitate as sulfates. However, these residues can be dissolved by treatment with water or diluted nitric acid.*

Indicator: ferroine, e.g., tri-1,10-phenanthroline-iron(II) ions. The change of the oxidation state of iron in the complex caused by the excess of Ce^{4+} *changes the optical absorption* [10].

4.1.3 Nitrido-zeolites

In these microporous frameworks the SiO_2 tetrahedra are replaced by PN_2^--units. The negative framework charge is balanced by different cations located in the interstitial void volume. Also, anions (Cl^-) are included [11]. For instance, a sodalite structure of the formula $Zn_7[P_{12}N_{24}]Cl_2$ was synthesized. Generally, there also exist attempts to replace the atoms of the anionic framework by other elements like As and Sb. Therefore, the importance of analysis of today's less common elements in microporous materials will increase in the near future.

P – N-zeolites are chemically highly resistent and thermally stable. To avoid the formation of volatile phosphanes, fusion is carried out under oxidative conditions with diluted hydrosulphuric acid at 190 °C for 2 days in Teflon-lined autoclaves. Also conc. nitric acid can be used to fuse poorly soluble phosphides, arsenides, antimonides, and sulfides in autoclaves [12].

The oxidative fusion of P – N-zeolites yields ammonium hydrogen phosphate. The cation is determined directly in the fusion solution by complexometric titration. Phosphate is determined by the molybdate method. The nitrogen, now present as ammonium, can be determined photometrically as indophenol (Berthelot reaction) and the chloride by argentometry.

The discovery of P – N-zeolites is an example of the importance of classical wet-chemical analysis (standards) for the determination of new compounds, before routine methods as ICP-AES could be applied.

4.1.3.1
Determination of Nitrogen by the Berthelot Reaction

In the Berthelot reaction, ammonium ions react with a phenol under suitable conditions, giving an indophenol dye. The absorption of the dye appears in the range of 630–720 nm. The method is sensitive and specific to ammonium ions. Ammonium ions can be directly determined in the stock solution of the fusion or after pretreatment, such as the Kjeldahl digestion. Solids can be directly measured after decomposition with acid. The residue is filtered off before handling. Sodium salicylate and phenol are the phenolic compounds mainly used. Sodium dichloro-S-triazine-2,4,6-trione (NaDDT) or sodium hypochlorite are used as hypochlorite sources. Sodium nitroprusside $[Fe(CN)_5NO]^{3-}$ is added as a catalyst. The reaction proceeds at 25–100 °C and is sensitive to the pH value. The catalyst stabilizes probably the monochloramine formed after combining components and takes part in the reaction. The probable reaction sequence according to Krom [13] is shown in Fig. 1.

Interfering metals are complexed with EDTA. Organic amines depress the indophenol reaction. In general, interferences are diminished by sample dilution. Ammonium exchanged zeolites can be directly determined after acid

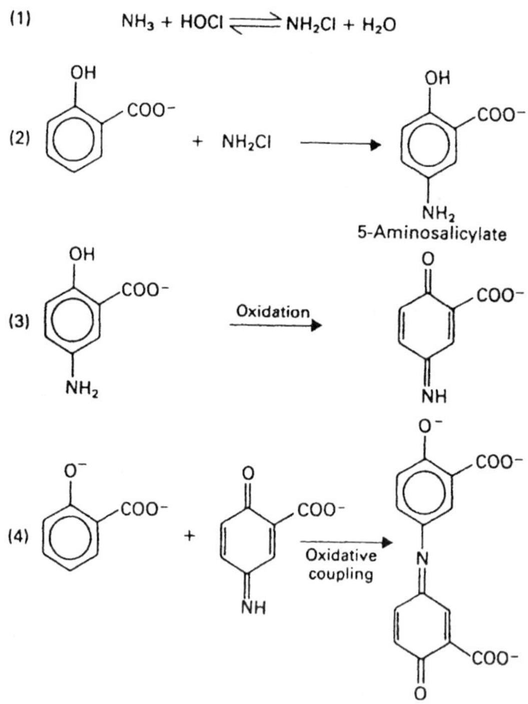

Fig. 1 Reaction scheme of the Berthelot reaction [13]

Table 2 Reagents, final concentrations, and sensitivities (extinction *e*) of some methods based on the Berthelot reaction [13]

Phenolic compound	Final conc./ $g L^{-1}$	Hypo-chlorite	Final conc./ $g L^{-1}$	Catalyst/ $g L^{-1}$	Optimum pH	e' 10^4/ $L mol^{-1} cm^{-1}$
Phenol	12	NaOCl	1.0	40	11.3–11.7	1.23
Phenol	5.6	NaOCl	0.167	27	11.4–11.8	2.04
Phenol	8.3	NaOCl	0.22	83	11.5–13.0	0.87
Phenol	0.9	NaOCl	0.004	11	10.4	2.01
Na salicylate	40	NaOCl	0.745	1400	12.0	2.04
Na salicylate	10.2	NaDDT	0.3	72	12.8–13.1	1.87
Na salicylate	17	NaDDT	5.0	120	–	1.23
o-Cl-phenol	0.45	NaOCl	0.045	2	10.5–11.5	2.01

fusion as it was done already with clays [14]. NO_3^- and NO_2^- can also be determined if a reduction step is involved prior to analysis [15].

A comprehensive survey of the Berthelot reaction is available [16] that also summarizes some reaction conditions as given in Table 2.

4.2 Instrumental Methods of Chemical Analysis

Today, atom absorption (AAS) and atom emission spectrometry (AES) as well as X-ray fluorescence spectrometry (XRF) are most commonly used for elemental analysis of substances in addition to a great variety of different methods available. All three methods are also suitable for the analysis of microporous materials. They allow the determination of many different elements known to constitute the frameworks of microporous and mesoporous materials. Their sensitivity is high enough to ensure an accurate quantitative determination of these elements. Due to their practical importance for "zeolite chemists", these three methods will be introduced in brief. Prior to analysis, samples have to be prepared, handled, and fused as described above. For atomic spectrometry, samples are mainly dissolved in concentrated acids and then diluted. This diluted sample solution is used for AAS and AES. In the case of X-ray fluorescence determination, samples are fused in melts of lithium tetraborate.

Fusion in Hydrofluoric Acid

Procedures

Transfer ca. 100 mg *of the sample into a platinum crucible. Moisten the sample with conc. sulfuric acid and add* 3–5 mL *of hydrofluoric acid* (40 wt % *in water) drop by drop. Attention, the solution may foam. Heat the solution until dryness to remove the silicon as volatile* $SiF_4 \uparrow$. *Dissolve the residue in*

Wetting 500 mg of sample with 3 mL of water and addition of 4 mL conc. $HClO_4$ and 15 mL conc. HF

⇩

Evaporation in PTFE or Pt crucible to dryness, 3 times repetition

⇩

Evaporation with 23 mL conc. HCl

⇩

Dissolution of the residue in 20 mL 6 N HCl

⇩

Heating to clear solution
⇨ If not clear: Fusion of the residue with 0.5 g NaOH / Ag crucible/10 min. and dissolution of the melt cake in 20 mL 6N HCl

⇩

Making to 100 mL volume

Fig. 2 Course of fusion of poorly soluble rare earth-containing samples for ICP-AES determination [24]

hydrochloric acid (1 + 1) *and warm a little; if necessary, filtrate the solution. Transfer this solution into a volumetric flask and make to volume* (50 mL).

This solution is used for AAS or ICP-AES determinations of elements, except for silicon.

Or, *weigh ca.* 10 mg *into a dry Teflon beaker. Add* 5 mL *of conc. hydrofluoric acid. Dilute with water, and then transfer the solution into a Teflon volumetric flask. Rinse the Teflon beaker with a small portion of distilled water and combine it with the solution. Make to volume.*

This solution can be directly used for ICP-AES determination of elements including silicon, provided the spectrometer is equipped with an HF-resistent Teflon device. Otherwise, to protect from spectrometer damage, the samples have to be evaporated until all fluoride is removed (Fig. 2).

4.2.1 Atomic Absorption Spectrometry (AAS)

Nowadays, AAS has obtained a wide range of application in analytical laboratories [17–19]. Much methodic work has been done to develop this method:

spectrometers are cheap in comparison to the other competing methods. Several techniques of AAS, which differ in the type of atomization, have been developed so far. In the *hydride-technique*, elements are converted to their hydride compounds. It uses the ability of the elements of the IVth, Vth, and VIth main group of the periodic system of elements to react with "nascent" hydrogen. This reaction yields gaseous hydrides, which can be analyzed. Hence, this method is restricted and used to separate and enrich As, Sb, Bi, Se, Te and (Sn). The stock solution is combined with a solution of $NaBH_4$. Likewise, Zn and (hydrochloric) acid are used for the determination of As, Sb, and Se.

To avoid the disadvantage of the flame burner system, the electrical heating of samples in a *graphite-tube furnace*, closed and open was developed. It allows the direct determination of solid samples and is suitable for investigations combined with other methods. Atomization of a solid sample is realized in a lightening arc. Graphite, which covers the sample, acts as the electrode. The *cold-vapor* technique is applied for the determination of easily volatile compounds (for example mercury) and shows low sensitivity.

From the different AAS techniques available, *flame AAS* is the preferentially used method (Fig. 3). It covers the previously mentioned advantages of AAS and allows the determination of the most interesting elements. The function of the flame in AAS spectrometry is the atomization of sample solutions. The sample solution is pneumatically directly sprayed into the flame. Several flames containing different feeds are used. The flames differ in their temperature. Higher temperatures are especially necessary to atomize less volatile elements. The AAS spectra of atomized elements are recorded as a function of the wavenumber. The physical background is as follows:

Light absorption occurs by the excitation of the outer-shell electrons, which then occupy states of higher energy levels. These excitations are quantified. According to the laws of quantum mechanics, only "discrete" energy levels can be occupied and hence, only discrete amounts of energy can be absorbed by the atom. Consequently, the atoms of each element show a characteristic absorption feature with respect to possible excitation energies and probabilities. As a result, a characteristic atomic absorption spectrum for each element is registered, which consists of many spectral lines of different intensities belonging to the possible energetic states and transition probabil-

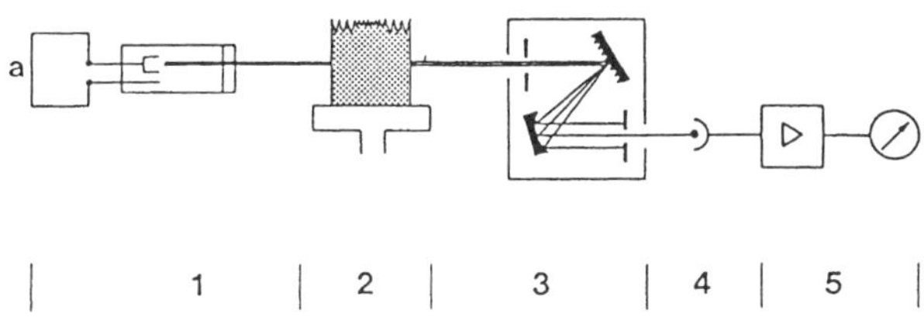

Fig. 3 Schematic representation of an atom absorption spectrometer [17] 1 – radiation source, 2 – flame, 3 – monochromator, 4 – detector, 5 – electronics

ities of electrons. An example for the possible electron transitions from the ground state to higher energy levels is described in the term scheme (Fig. 4). The excitation energies of electron transitions in the atoms cover the UV-Vis range of the electro-magnetic spectrum ranging from far UV to the near IR (200–1000 nm or 50 000–10 000 cm^{-1}, where 8066.5 cm^{-1} correspond to an energy of 1 eV).

The light absorption (*A*) depends on the specific extinction (absorption) coefficient of the electron transition (characteristic wavenumber), the concentration of the element (*c*), and the thickness of the sample (*d*). According

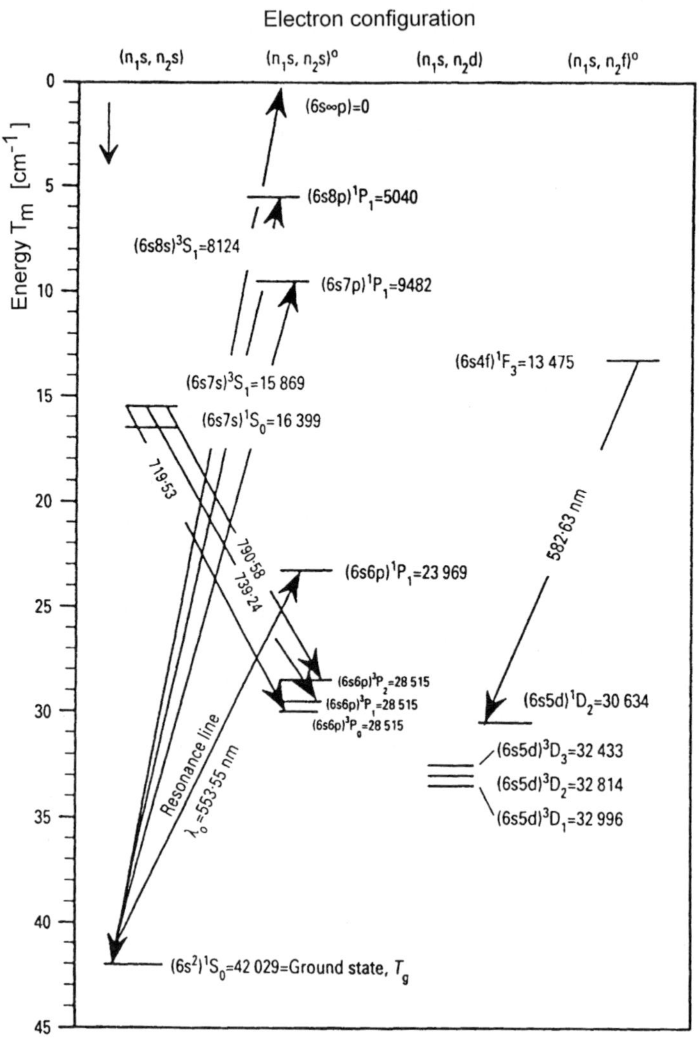

Fig. 4 The energy level diagram for barium [19]

Table 3 Detection limits of elements ($mg\,L^{-1}$) in flame AAS in air/C_2H_2 or N_2O/C_2H_2

< 0.001	0.001–0.01	0.01–0.1	0.1–1	1–5	> 10
Na, Mg	Li, K, Rb, Cs	Ba, Al, Ga, In	B, Si, Ge, Sn		P
	Be, Ca, Sr	Ni, Mo, Sb, Bi	As, Se, Te		
	Cd, Co, Mn	Sc, Ti, V, Pb	Y, Zr, Nb, Hg	Sn, Ta, W	
	Zn, Cr, Fe	Eu, Er, Ho,Yb	Lu, Tb	La, Nd, Gd, Pr	U
	Cu, Ag, Rh	Pd, Ru	Au, Os, Ir, Pt,		

The lower limit of normal detection range is one magnitude larger than the detection limit. Concentrations may differ by 2 orders of magnitude in AAS, but by 5–7 orders of magnitude in ICP-AES. As a rule, practically achieved detection limits of AES are similar to AAS.

to the law of Lambert-Beer, the light absorption at a certain frequency is proportional to the amount of the considered element in the sample. The following relation exists between these parameters:

$$A \equiv \log I_o / I_D = k \cdot c \cdot d\,, \tag{9}$$

with I_o – intensity of incoming light (constant wavenumber) and I_D – intensity of the transmitted light.
k is the proportionality factor and contains the intrinsic extinction coefficient and experimentally founded corrections.

Most of the elements can be determined qualitatively as well as quantitatively by AAS. The existence of atom-specific absorption lines in the spectra allows the simultaneous determination of different elements present in the sample. This is almost ensured in zeolite chemistry, because the sensitivity of AAS is high compared to the element content in zeolites and the amount of available sample. This high sensitivity allows the quantitative determination of elements even of small samples of 50 mg and less. AAS detection limits for some elements are summarized in Table 3.

To register the whole spectral range of electron transitions, different light sources are needed. Hence, it is necessary to change light sources during the measurement or to repeat the measurement if the spectrometer does not allow the combined use of different light sources during one run. Distortion of detection can occur due the superposition of lines.

4.2.2 Atomic Emission Spectrometry (ICP-AES)

In the case of AES [20–24], the intensity of the light emitted by the atoms is registered. Light emission is caused by the transition of excited electrons from their different states to the ground state and other low-lying energetic states. Again, emission lines exist that are characteristic of the elements under study.

The emission intensity is proportional to the concentration of the element in the sample and is used for the quantitative determination. However, the overall emission intensity depends on the transition probability of electrons from the ground to the excited state. In AES, the electrons are thermally excited. In contrast to AAS, the flame or plasma has two functions. First, it is used for atomization and, secondly, it supplies the energy of excitation. At flame temperature, most of the atoms stay in the ground state, and only a low fraction is excited. Therefore, in general, AES is expected to be of lower sensitivity than AAS.

In AES, all elements are excited simultaneously without the need of a light source. Thus, emission of different elements can easily be registered at the same time. AES is the method of choice if multi-element analysis has to be performed. Concentration differences of 5 to 7 orders of magnitude can be managed. AAS is more suitable for determination of low concentrations, alkali metals, and single-element analysis and allows the determination of concentration intervals of 2 to 3 orders of magnitude (linearity of intensity versus concentration). The present opinion is that optimum results are obtained by the combined use of AAS and AES.

The lower sensitivity of AES can be partially compensated by the use of an inductive coupled plasma instead of a flame (ICP-AES). It allows a drastic increase of the excitation temperature to 6000–8000 K as shown in Figs. 5 and 6. At this temperature the gas is converted into the state of a plasma. A plasma is a lightening-charged gas. This contains atomized elements and also ions and electrons, but the internal charge is completely balanced. Thus, the plasma behaves as electroneutral.

In ICP-AES, the liquid fusion solution is sprayed into a stream of argon. This stream passes the plasma burner. This burner consists of a quartz-glass tube that is surrounded by a high-frequency induction coil. The flowing gas is then inductively heated and converted to a plasma state in the tube with the aid of an HF-generator.

Due to the high temperature in the plasma, elements are more efficiently excited than in a flame, which leads to a higher sensitivity. Also, hardly excitable elements undergo remarkable excitation and can now be determined. Another advantage is that the plasma contains only the noble gas as a carrier, the fused sample and water, but no oxygen. Therefore, the danger of the formation of oxides, which hardly undergo dissociation, is low in the plasma as compared to a flame. Elements with high affinity to oxygen (like alkaline earth and rare earth metals, boron, silicon and others) can be very effectively atomized. As ionized atoms absorb and emit radiation at other wavenumbers than excited atoms, the ionization of atoms in the plasma is a real source of disturbance. This negative influence is in part overcome by the addition of ionization buffers like potassium chloride. The high electron density in the plasma shifts the dissociation equilibrium nearly completely towards the atomic state. Therefore, despite the high plasma temperature, the

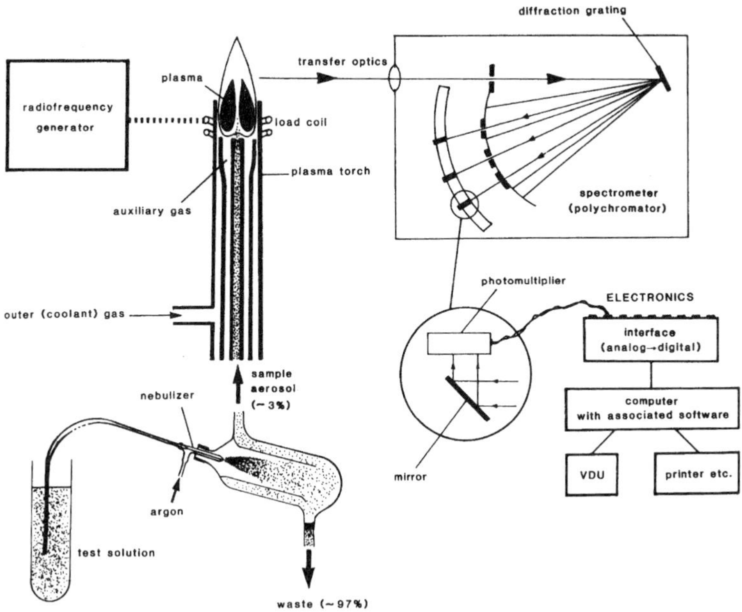

Fig. 5 Schematic diagram of a conventional simultaneous ICP system [21]

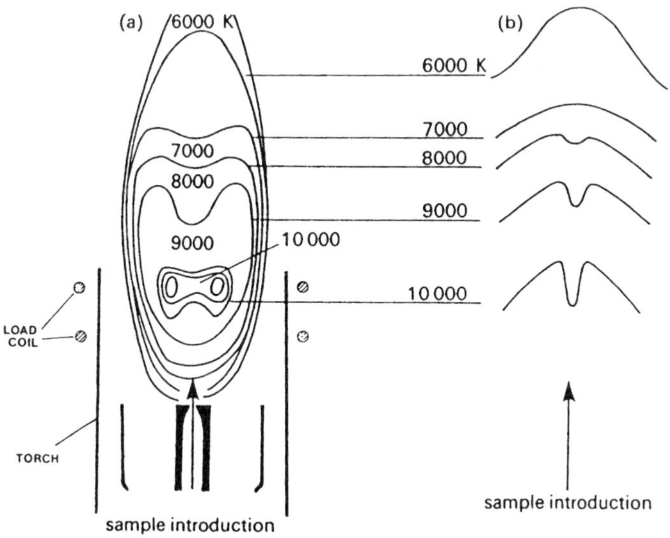

Fig. 6 **a** Approximate temperatures, in Kelvin, in the ICP flame. **b** Cross section view of temperature distribution within the ICP flame at varying heights. Sample is injected through the flattened base of the ICP. The temperature of the injector "tunnel" is significantly cooler than the surrounding plasma in the lower regions of the ICP. The temperature profile broadens out and only high in the tailflame of the ICP does the temperature profile resemble a conventional source, decaying progressively from a central point [21]

Table 4 Concentrations (wt %) of elements in stock solutions of zeolite fusions determined by ICP-AES in two different laboratories

Sample	Si		Al		Na		Fe	
NaA	16.47	15.06	15.50	14.48	13.00	12.55	0	0
ERI	30.06	–	7.37	7.90	0.06	0.06	0.02	0.03
MOR	35.42	–	5.51	5.80	0.07	0.06	0.05	0.05
HS-30	40.95	–	2.07	2.23	0.07	0.06	0.01	0.01
DAY	42.74	–	0.76	0.36	0.11	0.06	0	0

Table 5 Test of the reproducibility of ICP-AES results obtained by six different groups participating in a ring analysis of Albtalgranite (DFG-KA 1), oxide content in wt% [23, 24]

Oxide	I	II	III	IV	V	VI	$\overline{x}^*$	s/%**
Al_2O_3	15.0	14.8	–	13.7	13.8	13.1	14.1	5.7
Fe_2O_3	3.09	3.09	–	2.95	3.02	2.87	3.00	3
MgO	1.62	1.61	1.51	1.58	1.56	1.56	1.57	2.6
Na_2O	3.39	3.33	3.35	3.36	3.33	3.28	3.34	1.2
K_2O	4.56	4.64	4.83	4.60	4.66	4.56	4.64	2.2
TiO_2	0.52	0.51	0.52	0.51	0.5	0.5	0.51	2.0

* mean value of determination
** standard deviation (accuracy)

major fraction of elements is not ionized. Only alkaline atoms undergo substantial dissociation, which explains the lower sensitivity of ICP-AES towards these elements.

Flame temperature allows only excitations requiring energy equivalent to visible and near-UV light, but not higher energy excitations which require the energy of far-UV light quanta.

Analytical results illustrate the accuracy and correctness of AES analysis. Examples are given in Tables 4 and 5.

4.2.3
X-ray Fluorescence Spectroscopy (XRF)

Nowadays, XRF is a well-established method for the qualitative and quantitative analysis of microporous and mesoporous materials [25]. The X-ray region covers the part of the electromagnetic spectrum between 10 and 100 Å or 100 to 0.1 keV, respectively, and is neighbored by the UV-region on the low-energy or long-wavelength side. When X-rays pass matter, their intensity is attenuated due to the interaction with electron shells of atoms. Three main processes contribute to this attenuation: photoelectric absorption, inelastic

(Compton) and elastic (Rayleigh) scattering, and pair production ($e - e^+$, only by high-energy radiation).

X-ray fluorescence is caused by photoelectric absorption [26–28]. Due to the interaction of X-rays, electrons are expelled from the inner electron shells of atoms and leave vacancies. A thus-excited ionized atom is intrinsically unstable. Immediately, a decay to a more stable electronic state occurs. The excited atom relaxes by transition of electrons from outer (high-energy) shells to the vacancy in the lower-energy shell. This electron transition from a higher to lower energetic state is accompanied by the emission of radiation (Fig. 7). The wavelength of an emitted X-ray quantum is equivalent to the energy difference between the two states of the electron. Due to the laws of quantum mechanics, only certain states of energy are allowed for the electrons. As a result, each element exhibits its own characteristic fluorescence spectrum according to the allowed energetic states of electrons.

The fluorescence spectrum consists of several lines corresponding to the different possible electron transitions (Fig. 8). The wavenumbers of emitted X-rays are characteristic of the element, which does not exclude interference of lines of different elements due to the manifolds of possible electron transitions. Under comparable experimental conditions, the fluorescence intensity is proportional to the concentration of the element in the sample.

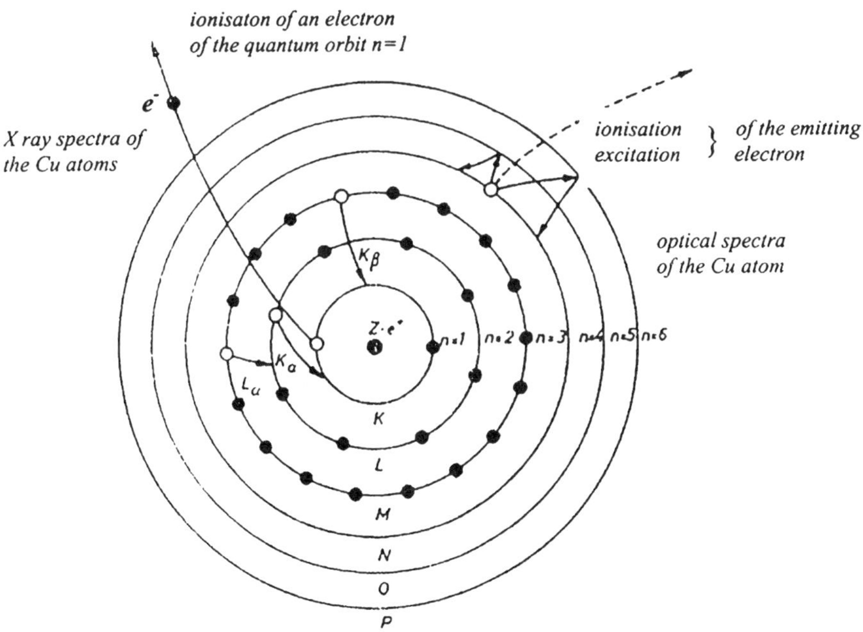

Fig. 7 Generation of the optical spectra and the X-ray spectra for the case of the Cu atoms (the transitions marked do not appear simultaneously, but in an order in succession) [29]

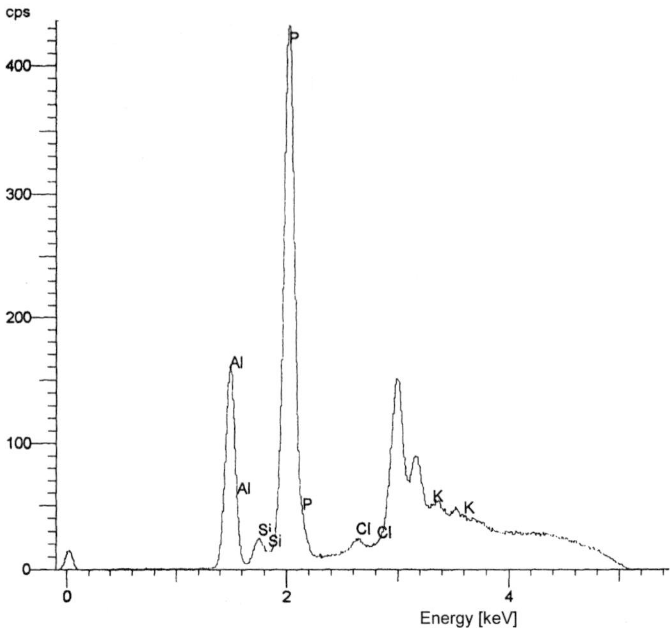

Fig. 8 XRF spectrum of SAPO-11 (PVA wafer) [Kosslick, unpublished]. Fixed conditions: very light elements, acquisition lifetime = 40.0 s, realtime = 74.5 s, tube current = 900 μA

The probability that a certain transition will occur is governed by quantum chemical selection rules. Therefore, X-ray lines of an element differ in intensity. Within a series of lines (transitions of electrons from different excited states to one energetic level as the K, L, M, etc. shell) (Fig. 9), the fluorescence yield increases with the atomic number Z. Hence, atoms with $Z < 11$ are hardly detected by XRF. Within a series of lines of one element, the line intensity decreases in the order K > L > M series, i.e., the energetically next higher electrons with respect to the vacancy show the largest transition probability to the electron vacancy. Therefore, these lines are preferentially used for quantitative determination if no interferences require the choice of other lines. Generally, the excitation radiation must possess an energy greater than the threshold value ranging from 1 to 20–30 keV.

X-ray fluorescence yields are lower than unity due to simultaneously proceeding competing processes mentioned above. Additionally, the Auger-electron emission affects fluorescence yields. It is caused by the absorption of radiated fluorescence by outer-shell electrons of the same atom, which are than expelled. This process becomes increasingly important for lighter elements and prevents their XRF detection.

The X-ray fluorescence can either be detected in a wavelength dispersive (WD) or energy dispersive (ED) mode. In WD detection, the collected fluorescence radiation is diffracted and the radiation intensity is analyzed as

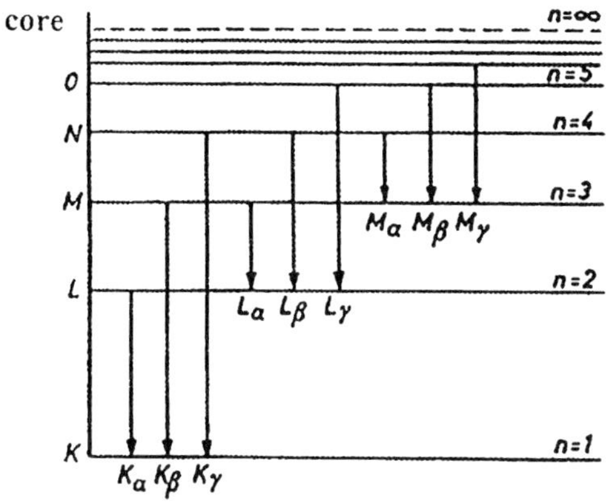

Fig. 9 Electron transitions of the K, L, M line series

a function of the wavelength. It operates like a usual spectrometer. ED is of high sensitivity, precision, and allows the simultaneous detection of a variety of elements. Routine spectrometers use this technique.

In ED detection, the whole fluorescence intensity (independent of the wavelength of X-ray photons) is registered by a semi-conductor detector. The differentiation of the different wavelengths is made electronically on the basis of the different energies. ED allows on-line identification and determination of a variety of elements in a few seconds. However, lower spectral resolution and stronger background effects decrease the accuracy. ED is used in electron probe micro analysis (EPMA), because it is a simple small volume equipment and is preferred when semi-quantitative analysis is required.

Interferences may disturb the determination. Scattering of exciting X-rays may give rise to enhanced background signals. Also, matrix effects can occur due to different penetration depths of X-rays in different materials. Their magnitudes differ with the elemental composition and the structure/state of material.

Nowadays, the intensity of lines of X-ray fluorescence spectra can be mathematically calculated based on physical models of the whole fluorescence process with high accuracy. In principle, this fundamental approach allows the direct prediction of sample composition without any standard or reference. However, each material, matrix and composition requires an independent calculation. Therefore, empirical matrix corrections and calibrations of signal intensities by reference samples of known composition are used in routine analysis of different samples. This is connected with a loss in accuracy but enhances the reliability.

Calibration of the spectrometer takes place with reference samples of known composition that contain the same elements as the stock material in desired concentrations. Several reference samples are prepared by mixing the appropriate amounts of standard materials. Due to high concentration and presence of various elements, the X-ray fluorescence intensity is no longer linearly proportional to the element content. A correction is therefore necessary by using a series of reference samples of different composition.

After calibration of the spectrometer, knowing the loss of ignition and exact sample weight, a direct quantitative determination of elements is possible. The detection limit of most elements is ca. 1–10 ppm [30]. The accuracy described by statistical data evaluation is 0.1–1%. The correctness, based on comparison with reference samples, is better than 0.5%. Determinations close to the detection limit of an element are connected with lower accuracies of 4–10% (standard deviation).

4.2.3.1 Sample Preparation

Direct determination. The composition of metals and alloys can be directly determined from samples without the use of an additional support. Also, wafers of microcrystalline samples can be determined directly without loss of substance. However, it is our experience that matrix effects become more important due to different structures and compositions of materials as well as textural effects. This leads to a substantial loss of accuracy.

Determination in polyvinyl alcohol (PVA) matrix. PVA wafers can be easily and rapidly prepared. Since the microcrystalline sample is not fused, matrix effects render more difficult the use of standards and the direct comparison of results obtained for different zeolites. PVA wafers can be successfully applied for the determination of the constituents of zeolites modified by ion exchange, where the fluorescence intensities of lines of zeolite framework constituents can be used as internal standard. Another requirement is the homogeneity of the sample, i.e., the element distribution.

In this way, we have determined the cation composition of alkaline- and alkaline earth-exchanged A-type zeolites. Wet-chemically analyzed A-type zeolites were used as standards. The relative standard deviation even for the less sensitive sodium was better than 3%. The fluorescence intensity was directly proportional to the concentration of the element in the wafer.

Glassy lithium tetraborate matrix. Fusion in lithium tetraborate allows complete dissolution of solid samples. Elements are homogeneously distributed in the wafer and embedded in a unique matrix due to the high complexing ability of the surrounding borate. Therefore, high accuracy determinations of elements are possible using this fusion technique.

Procedures

Preparation of PVA Wafers

Grind 0.25–0.5 g *of the sample in an agate mortar and mix with* 2–5 g *of PVA. Grind again for* 10 min *until the mixture is homogeneous. Press a pellet by in-*

creasing the pressure to 160–180 bar, *maintain for a minute and release the pressure slowly.*

Preparation of Glassy Lithium Tetraborate Wafers
Mix 0.5 g *of the sample with* 6.0 g *of* $Li_2B_4O_7$ *and grind in an agate mortar to homogeneity for ca.* 10 min. *Use a platinum crucible. For preparation of wafers (easy removal of wafers), the crucible should have a diameter of ca.* 3–5 cm *on the ground and the walls should have an opening angle of ca. 30 deg with respect to the ground. Add a drop of ammonium bromide solution (*10% *in water) into the empty platinum crucible and add the mixture to be analyzed. Shake a little to obtain a smooth surface and an equal sample thickness. Transfer this mixture into an electric furnace and heat to* 1050 °C *(red-hot). Remove after every* 3 min *using an extra-long platinum-tipped tongs (use heat-resistant gloves) and slew with care (keep slanting in each direction slowly and back fast) to remove air bubbles. Thereafter, remove from the furnace and cool the crucible by standing on a plate. Remove the wafer by shaking while slightly tapping.*

5
Determination of Anions

Some anions such as fluoride, chloride, bromide, sulphate can be detected by XRF. Halogenides can also be determined gravimetrically by precipitation of their silver salts or by argentometry.

Procedure
Add a 0.1 normal silver nitrate standard solution in excess to the stock solution. Filtrate or centrifugate the solution and back-titrate the non-precipitated silver with 0.1 N *solution of NaCl. The colloidal silver halogenide coagulates at the end-point of filtration (Gay-Lussac). Or, back-titrate with ammonium rhodanide and use some drops of ammonium iron(III)sulphate acidified with nitric acid as indicator:*

$$AgNO_3 + NH_4CNS \rightarrow NH_4NO_3 + AgCNS \downarrow \quad (10)$$

After the silver is fully consumed, the red-brown color of formed Fe(CNS)3 *indicates the end of the titration. Alternatively, use dyes like eosine or fluoresceine as indicators. The precipitating colloidal silver halogenide interacts with the dye and influences its color.*

Sulphate can be determined gravimetrically as $BaSO_4 \downarrow$. Nitrate can be determined by the Berthelot reaction after reduction to ammonia (vide supra, 4.15). Complex oxidic anions like phosphate, tungstate, molybdate, etc., can be detected spectrophotometrically (compare Sect. 4.1.2).

Another elegant method for the determination of anions and also cations uses ion chromatography [31]. In general, ion chromatography represents

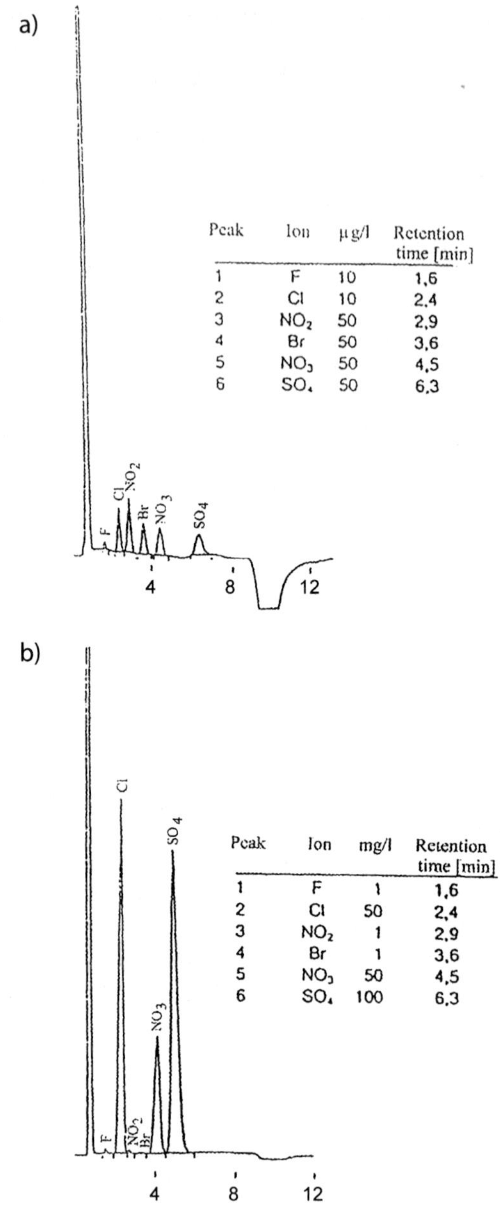

Peak	Ion	µg/l	Retention time [min]
1	F	10	1,6
2	Cl	10	2,4
3	NO_2	50	2,9
4	Br	50	3,6
5	NO_3	50	4,5
6	SO_4	50	6,3

Peak	Ion	mg/l	Retention time [min]
1	F	1	1,6
2	Cl	50	2,4
3	NO_2	1	2,9
4	Br	1	3,6
5	NO_3	50	4,5
6	SO_4	100	6,3

Fig. 10 The sensitivity of the conductivity detection is shown by the ion chromatograms for the lower (**a**) and the upper (**b**) limiting range [32]

a liquid-chromatographic separation often carried out under enhanced pressure (HPLC-conditions) (Fig. 10). The column contains a silica-support that is coated with an organic ion exchanger. Water and other solvents can be used as carriers and aromatic carboxylic acids like benzoic acid, salicylic acid, citric acid, boric acid/mannite, phthalic acid, etc., as eluents. The detection is carried out either conductometrically or by higher-sensitivity UV- and amperometric (AgCl-electrode) detectors. Ionic chromatography is routinely used for the determination of F, Cl, Br, J, NO_2, NO_3, phosphate, sulphate, thiosulphate, rhodanide, chromate, borate, BF_4^- and others in water. Also, alkaline and alkaline earth ions as well as ammonium and Mn, etc., can be determined with high accuracy. The detection limits using conductometric detection lie in the range from 10 to 100 μg/L. Linear signal intensity vs. concentration dependencies are found in the concentration range between 0.5 and 100 mg/L. The application of this method for zeolitic ion and anion analysis requires the separation of the considered ions from framework components, templates, and excess elements because they enhance the background and cause interferences. Notwithstanding, the method requires the use of accurate standards and blank measurements.

6 Determination of Organic Inclusions

The synthesis of new types of microporous materials using organics as templates but also the field of new applications of zeolites, where organic molecules of specific properties are embedded in the microporous structures, requires the quantitative determination of organics in these types of materials. The loss of ignition, sufficient for the determination of zeolitic water of organic-free zeolites like LTA, FAU, MOR and other aluminum-rich structure types, does not allow to separate exactly the water and organic content. Separation and quantitative determination is possible by extraction after dissolving the zeolite, by CHN analysis and thermal analysis. In principle, all the three methods can give reliable and equivalent results.

Extraction

Procedure
Dissolve 0.1–1 g *of the microcrystalline material in* 20–50 mL *of* 0.1 M *hydrochloric acid or* 10–20 mL *diluted or concentrated hydrofluoric acid at room temperature under magnetic stirring. Neutralize with sodium hydroxide or ammonia solution. Extract the resulting solution three times with* 100 mL *of toluene or chloroform or another suitable solvent. Evaporate the solvent and weigh the organic residue.*

Results are in good agreement with CHN analysis provided the organic is soluble in the solvent and does not undergo side reactions, such as polymerization, or forms poorly soluble carbocations.

CHN-Analysis

Typically, 0.1–1 mg of sample of known loss of ignition are transferred into a stannum container that is brought into a combustion tube. The sample is heated to 1030 °C in a helium flow and subjected to a flush-combustion by injecting 5 mL of pure oxygen into the helium carrier gas. The heat of combustion increases the actual temperature of the sample to ca. 1700 °C, ensuring total sample decomposition and oxidation of included organics as well as water desorption and dehydroxylation of silanol groups. Combustion gases pass an oxidation tube (e.g., doped with $Co_3O_4/Cr_2O_3/Ag$) and, thereafter, a Cu reduction tube. The formed H_2O vapor, CO_2, and N_2 is separated in a GC column filled with Poropack QS. A heat conductivity detector is used for registration.

Thermal analysis

In the TG-DTA experiment, the sample is subjected to temperature-programmed heating in air (or argon) in order to thermally desorb water and decompose organic compounds [33–36]. TG can also be coupled with gas chromatography, mass spectrometry, and IR spectroscopy [37]. Water desorption and decomposition of different organics proceed at different temperatures giving rise to characteristic weight losses (steps in the TG curve), which can be easily quantified. Loosely bound water of template-containing zeolites desorbs on heating up to ca. 150 °C.

Water desorption is a strong endothermic process, whereas oxidative template decomposition/desorption is exothermic above 300 °C. Therefore, the

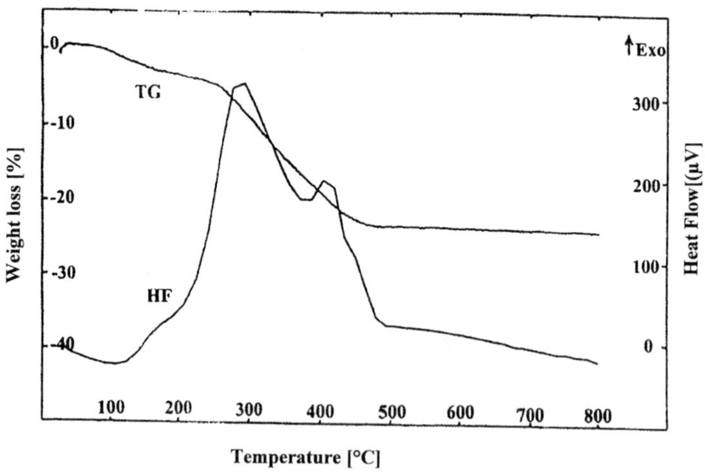

Fig. 11 TG-DTA curves of N-methylindolizine on zeolite NaX [Kosslick, unpublished]

simultaneous registration of the DTA signal gives additional hints for the assignment of weight losses observed in the TG curve to different contained compounds (Fig. 11).

The sample is heated to 800 °C to ensure complete decomposition and desorption of the organic. Modern TG-DTA devices need ca. 10 mg of sample for analysis at a heating rate of 10 K min^{-1}. Otherwise, a sample weight of ca. 100 mg and a heating rate of 5 K min^{-1} is chosen. Normally, the measurement is carried out in air. Problems may arise with extra-large crystals of a size > 20 μm and unidimensional pore systems, e.g., SAPO-5, ZSM-12. Here, the oxidative decomposition of the organic may lead to the formation of stable coke species, which can be hardly removed from the pores. We have observed an underestimation of up to 20 wt % of organic in the system n-propylamine/AlPO-5 using air as a carrier gas. The use of argon diminishes coke formation drastically. The presence of structural defects leading to internal pore blocking may also affect thermoanalytical results. In this case, the sample has to be decomposed.

7
Differentiation Between Framework and Non-Framework Species

Within a sample, elements may be distributed in different ways. Differences in element distribution may occur in the outer and inner parts of a crystal caused by changes of the gel composition during the course of crystallization. Crystals of different size and composition may be formed.

Most interesting is the location of elements at framework and non-framework sites. Non-framework species may be ionic or neutral, highly dispersed in the channels and cavities of the crystal lattice, or located outside the crystals as extra-particles or located at the crystal surface. Extra- or non-framework species may have an important impact on the catalytic properties of samples. They may carry different types of acid sites. The substitution of 4-valent framework atoms like Si by 3-valent atoms like Al, Ga, Fe leaves one excess negative charge in the framework per substituted atom. This charge can be compensated by one proton that acts as a Brønsted acid site. The same happens by substitution of Al or Ga in frameworks of AlPOs and GaPOs by 2-valent cations. The catalytic properties of microporous materials are determined to a large extent by the Si/Me(III) framework ratio of silicates or the Me(III)/Me(II) framework ratios of alumino- and gallophosphates.

The total content of elements is not sufficient for characterization because extra-framework species are also often present. Extra-framework species may act as Lewis acid sites, redox-centers, or be involved in hydrogen-transfer reactions. For example, non-framework gallium species play an important role in the aromatization of hydrocarbons. In contrast, in the case of Ti-zeolites, it is framework titanium that is assumed to be the active oxidation site.

Other examples are the tuning of acidity of Y-type zeolites by dealumination or the creation of "super-acid" sites by mild steaming. Therefore, it is of considerable interest to determine the metal location in microporous materials in a simple way to support the catalyst manufacturing process.

The relation between framework and non-framework species is obtained by comparison with the total metal content of the sample.

7.1 Framework Composition

7.1.1 Cation Exchange

A suitable method for the determination of the content of substituting atoms that cause negative charges in the framework is the cation exchange. The number of exchangeable cations is directly related to the content of substituting framework atoms. However, elements that leave no framework charge after incorporation into framework sites are not detectable. This happens if interchanged framework atoms have the same charge, i.e., substitution of Si by Ti(IV), or by incorporation of Me(III) atoms into Q^3 groups in silicates. In the latter case, the charge is balanced by the release of a hydroxyl group, which may schematically described as follows:

$$3[(SiO)_3SiOH] + M^{III}Cl_3 \rightarrow 3[(SiO)_3SiO]M^{III} + 3HCl \quad (11)$$

Cation exchange can be carried out continuously by steady flow exchange, or discontinuously in a batch-exchange, whereby the exchange solution is several times replaced by new solution.

Due to the different site selectivity and possible ion sieving effects of microporous materials, the ions used should fulfill some criteria: In general, the cation should be able to penetrate the crystal. In addition, inner sites (sodalite cage) should be accessible. The cation should fit well into the oxygen rings and should have a high affinity to these sites in competition with other cations (at least if it is offered in excess). The cation should be stable against hydrolysis. In practice, ammonium ion exchange and cation exchange with sodium and potassium are used. The achievement of complete cation exchange has to be checked independently by chemical analysis.

Ammonium Exchange

Procedure (see Fig. 12)
Arrange a set of devices according to Fig. 12. Put 0.5–2 g *of the zeolite on a frit within the glass mat and cover it with a PTFE sieve. Position the tube of the dropper funnel so that it touches the ground of the sieve. Place a Pt electrode*

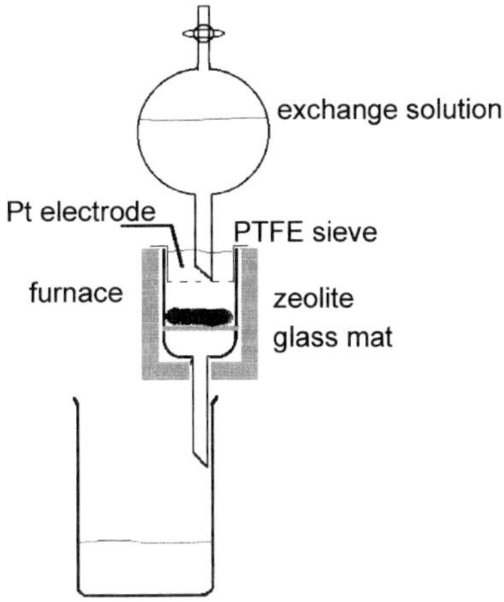

Fig. 12 Apparatus for continuous NH_4 ion exchange (according to H.-G. Jerschkewitz, private communication)

into the exchange solution and connect it through a relay with the heating of the furnace (or a water thermostat).

Fill the mat wholly with the exchange solution by opening of the tap. Close the tap again. Heat the exchange solution up to 70–80 °C.

This arrangement allows the overnight use of the apparatus. While the exchange solution flows through the mat, new solution is supplied "automatically" from the funnel whenever the solution level drops under the opening of the tube. After consumption of the exchange solution, the furnace or other heating devices are automatically switched off via the relay. The funnel tube should have an opening of 7–8 mm and its end should be sloped. The exchange is carried out with a 0.2 M solution of ammonium acetate. Approximately 250 mL exchange solution pass the sample over 4–6 h.

After exchange, wash the sample and dry at 120 °C. *Determine the ammonium content by Kjeldahl distillation with* 2 M *solution of sodium hydroxide. The overdistilled ammonia is determined by titration. Simultaneously, determine the silicon (silicates) or Al, Ga content (alumino-or gallophosphates). The ammonium content can also be determined by TPD or CHN analysis.*

This method was applied for the determination of the Si/Al ratio of medium- to large-pore zeolites such as ZSM-5, mordenite, ZSM-12, and Y-type zeolites [38, 39]. For medium-pore zeolites, this method operates quantitatively. For the large-pore and open faujasites and ZSM-20, a systematic underestimation of 30–40% of sites was found, which gave too high silicon-to-aluminum ratios [40]. An ammonium ion exchange capacity of only 1.7 mmol g^{-1} was found for zeolite ZSM-20, however, ^{29}Si NMR data gave an Si/Al ratio of 4.3 (3.14 mmol g^{-1} Al). Obviously, the exchange was not complete. Complete ammonium exchange in the small sodalite cages is only achieved by repeated refluxing with 0.5 M of ammonium nitrate solution. At least three exchanges are necessary for zeolite ZSM-20 to remove also hardly accessible inner cations. The last exchange is carried out with diluted solution to remove occluded excess ammonium from the cavities. Thereafter, the sample is washed (stirred) three times with warm water and filtered off or centrifuged.

Alkaline Cation Exchange

Procedure
Stir 100 mg *of sample with* 10 mL *of a* 0.1 M *solution of sodium or potassium chloride at* 60–80 °C *or heat under reflux in a micro-apparatus for* 1 h. *Offer 3–5 times new exchange solution. Filter off or centrifuge the residue and wash repeatedly with small portions of deionized water to remove excess salts. Dry the sample and analyze the cation content after fusion. The concentration of the exchange solution should not be higher to avoid salt inclusions.*

7.1.2
Temperature-Programmed Desorption (TPD) of Test Molecules

Ammonia

TPD of ammonia (NH_3-TPD) and FTIR-TPD have been successfully applied to the evaluation of nature and concentration of acid sites (Lewis or Brønsted) of metal-substituted micro- and mesoporous molecular sieves and zeolites [41–44] (see also Volume 6, Chapter 4). Ammonia reacts stoichiometrically with the protons of Brønsted sites:

$$Z-H + NH_3 \rightarrow Z^{-}NH_4^{+} \qquad (12)$$

The use of this method requires fully ion exchanged zeolites to give correct results on framework Me-content.

NH_3-TPD

Procedure
0.2–0.5 g *of the H-form of the sample are placed in a reactor tube. Adsorbed water is removed by heating the sample in a stream of pure helium up to the final temperature of the TPD run* (400–500 °C) *with a heating rate of* 10 K min^{-1} *and a gas flow of ca.* 2.5 L h^{-1}. *After cooling to* 100 °C, *the sample is saturated with ammonia in a helium stream containing ca.* 3 vol% NH_3, *which requires* 20–60 min. *The sample is then flushed with pure helium at* 120 °C *for* 2–3 h *to remove physisorbed and weakly bound ammonia. Under these conditions, complete desorption of* NH_3 *from silanol groups is achieved (blank test with Aerosil). The TPD run starts with a heating rate of typically* 10 K min^{-1}. *Am-*

monia desorption is followed by a thermal conductivity detector. The desorbed ammonia is trapped in 0.01–0.1 N H_2SO_4 *and determined by titration.*

Representative TPD profiles of microporous materials are shown in Fig. 13. Using TPD of test molecules like ammonia requires some care because desorption of weakly bound species may overlap with the decomposition of the stoichiometric complex, i.e., the decomposition of ammonium ions occurs. Another difficulty may arise by adsorption of ammonia on strong Lewis sites that cannot be differentiated by the NH_3-TPD profile alone. This leads to an overestimation of the Me(III) framework content. On the other hand, ammonia might be desorbed from weak Brønsted sites at low temperature in open pore zeolites like faujasites.

Quantitatively correct results are obtained for medium-pore zeolites like ZSM-5, ZSM-12, and large-pore mordenite and SAPO-5 [45]. The TPD profile of these materials consists of a low and a high temperature peak (LTP and HTP, respectively). The LTP area corresponds to ammonia weakly bound to Lewis sites, the HTP area to ammonia desorbed from Me-framework sites giving rise to Brønsted acidity. LTP and HTP ammonia can be separated using a temperature ramp in the desorption run near the minimum of the desorption profile between LTP and HTP. A simple graphical evaluation leads to similar results. Comparison with the chemical composition of Al-ZSM-5

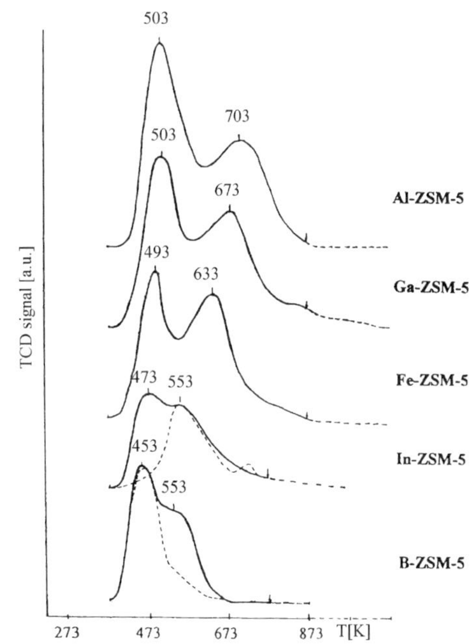

Fig. 13 TPD of ammonia (TPDA) plots of the isomorphously substituted ZSM-5 samples (*full lines*) and decomposition plots of the ammonium form of In- and B-ZSM-5 (*dashed lines*) recorded at continuous heating (heating rate 10 K/min) [45]

containing only framework aluminum confirms the potential of the method. Absence of Lewis-type aluminum-containing sites was checked by FTIR of ammonia. The presence of strong Lewis sites has to be independently checked by IR spectra of adsorbed NH_3 or pyridine. In the TPD run, strong Lewis sites release ammonia above 500 °C which results in an overlap with the HTP arising from strong Brønsted acidity.

A linear correlation exists between NH_3 desorption in the NH_3-TPD run and the FTIR-TPD profile of ammonium ions of fully loaded dealuminated faujasite type zeolites and the intergrowth ZSM-20 [46] (Fig. 14).

However, a systematic underestimation by ca. 30–40% of the Al framework content by quantitative evaluation of desorbed ammonia of ca. 30–45% is observed resulting in apparently increased Si/Al ratios [46, 47]. The ammonia adsorption capacity determined by TPD was 1.7 and 2.6 mmol g^{-1} for H-ZSM-20 and HY, respectively (Fig. 15). The corresponding ^{29}Si NMR data were 3.14 and 4.17 mmol g^{-1} framework Al. The reason for these differences

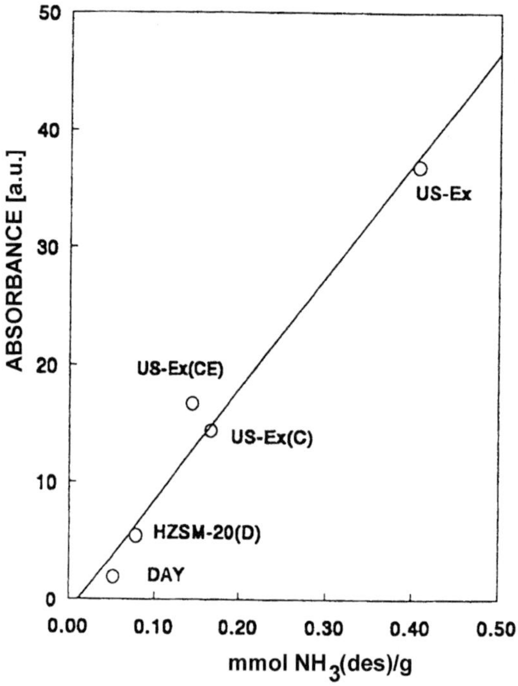

Fig. 14 Correlation between the amount of NH_3 desorbed during TPD and the normalized integrated absorbance of the NH_3 bending modes in the region between 1300 and 1700 cm^{-1} [46]. Samples: **a** US-Ex: NaY twofold exchanged, steamed and extracted with HCl, **b** US-Ex(C): sample **a** additionally calcined at 800 °C in air, **c** US-Ex(CE): sample **b** EFAL species were additionally extracted by HCl, **d** DAY (Degussa), **e** ZSM-20: synthesis conditions see [46]

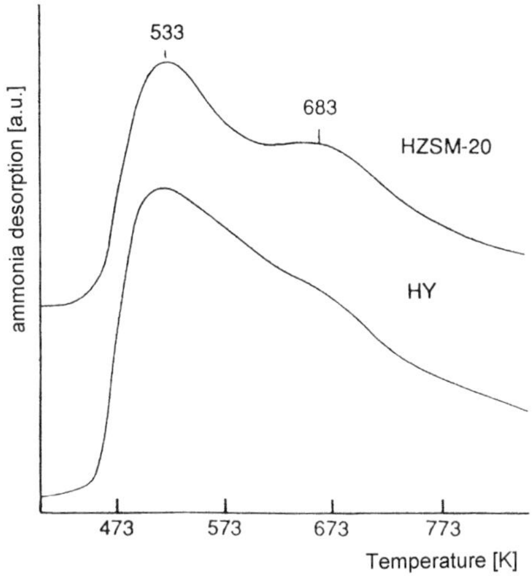

Fig. 15 TPDA profiles of the HZSM-20 and HY samples (heating rate 10 K/min) [47]

is the weak bonding of ammonium ions located in the large cavities [48]. Complete coverage is achieved only after contacting faujasites with pure ammonia at atmospheric pressure [49]. In contrast to medium-pore zeolites, the whole desorption profile corresponds to Brønsted ammonia. In a similar way, several authors have recently used this technique for the characterization of, e.g., acidic OH groups in zeolites of different types [50], cubic mesoporous aluminosilicates [51], or lanthanum-containing mesoporous aluminosilicates [52].

In summary, ammonia TPD is a useful method for the determination of framework metals that give rise to Brønsted acid sites. However, features like incomplete coverage of sites, overlap of different desorption peaks corresponding to different sites, and contribution of strong Lewis sites have to be taken into account.

n-Propylamine

Gorte et al. [53, 54] have shown that well-defined adsorption complexes, corresponding to one molecule per framework Me(III), can be observed for a range of simple organic molecules in substituted zeolites. For instance, in the case of 2-propanol and propylamine, molecules in excess of one/Me(III) can be removed unreacted by evacuation or heating to low temperature. However, molecules associated with framework Me(III) in 1 : 1 stoichiometry desorb in the TPD run as propene and water or ammonia at ca. 132 °C. This loss is easily detected in the TPD apparatus or on a thermobalance coupled with a mass spectrometer. Different medium-pore Al-containing zeolites and also silica-rich B, Al, Ga, and Fe-ZSM-5 were tested (Fig. 16).

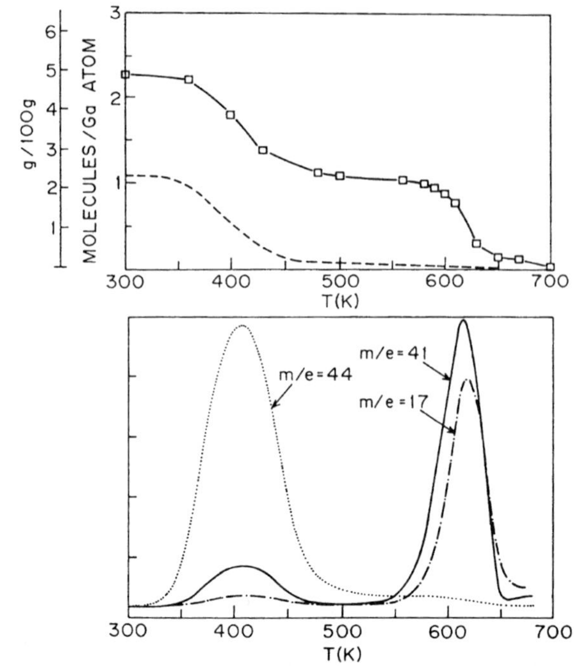

Fig. 16 TGA results (above) for 2-propanamine on H-[Ga]-ZSM-5 (*full line*) and ZSM-5 ($SiO_2/Al_2O_3 = 880$) (*dashed line*) along with TPD results for the same samples (*below*). On both samples, weakly adsorbed 2-propanamine (m/e = 44) desorbed unreacted at low temperatures. On H-[Ga]-ZSM-5, a second desorption feature, corresponding to a coverage of one molecule/Ga, desorbed as propene (m/e = 41) and ammonia (m/e = 17) [54]

The H-form of zeolites was prepared by exhaustive ion exchange with ammonium nitrate solution. The dried ammonium-exchanged zeolite was heated in a vacuum to 425 °C. Then, 17 mg of the zeolite in the H-form were spread over the flat sample pan of a microbalance to avoid bed effects in ad- and desorption. The heating rate for the TPD-TGA experiments was maintained at 10 K min^{-1}. The TPD-TGA experiments were carried out simultaneously in a high-vacuum chamber equipped with a microbalance and a mass spectrometer. The sample was activated at 400 °C in vacuum, cooled down and exposed to 0.4 to 2 kPa of the adsorbate of interest at 22 °C for 5 min. After exposure, the samples were evacuated and TPD-TGA measurements were performed. The saturation uptakes were 11–12 g per 100 g for 2-propanol and 6–11.5 g per 100 g for 2-propanamine. Unreacted 2-propanamine desorbs in a separate desorption step below 130 °C. 2-propanamine in excess was very difficult to remove by evacuation at room temperature and coverages higher than one/Ga were always present at the beginning of the desorption run.

Adsorption of ammonia, 2-propanol, and 2-propanamine provide a convenient way to identify framework Me(III) atoms in high-silicon zeolites. The adsorption of these molecules appears to be insensitive to gallium outside the zeolite framework.

One difficulty in using TPD to determine site concentrations is that desorption of weakly bound species may overlap with the decomposition of the stoichiometric complex, as was found for ammonia and 2-propanol. The presence of the 1 : 1 complex was easiest to quantify with 2-propanamine in that this complex desorbs at a considerable higher temperature (330–380 °C) than is observed for physically adsorbed species (130 °C). Additionally, Brønsted sites catalyze the decomposition of 2-propanol and 2-propanamine.

7.1.3
Lattice Expansion

Due to the different sizes of atoms and ions (Table 6), the isomorphous substitution of framework atoms by atoms of other elements leads to changes of the mean T – O distances in the framework which affect the wavenumbers of lattice vibrations and the unit cell parameters (see also Volume 2, Chapter 1 and Volume 4, Chapter 1 of the present series).

If the stress introduced by incorporated metal atoms is (in part) compensated by changes of T – O – T angles and tetrahedra tilting, the mean change in the T – O distances becomes smaller or even zero [57]. Resulting effects on the wavenumber shift of vibration bands and lattice parameters become smaller, if any change is observed. With ZSM-5 and MeAPOs, changes in unit

Table 6 Electronegativities, ionic radii, and radii ratios of substituting atoms after Allred et al. [a], Pauling [b], Shannon et al. [c], and Behrens et al. [d]

Element	Electronegativity Allred	Pauling	Radius [c]/Å r^{z+}_{Me}	Distance/Å $r^{*}_{Me\text{-}O}$	Radius ratio $Q = r_{Me}/r^{2-}_{O}$
Si^{4+}	1.74	1.90	0.26	1.62	0.2
B^{3+}	2.01	2.04	0.11	1.47	0.08
Al^{3+}	1.47	1.61	0.39	1.75	0.29
Ga^{3+}	1.82	1.81	0.47	1.83	0.35
Fe^{3+}	1.64	1.83	0.47 [d]	1.83 [d]	0.35
In^{3+}	1.49	1.78	0.62	1.98	0.46

* for $r^{2-}_{O} = 1.36$ Å

[a] Allred AL, Rochow EG (1958) J Inorg Nucl Chem 5:264; E.J. Little, Jr EJ., Jones MM (1960) J Chem Educ 37:231;

[b] Pauling L (1967) The Nature of the Chemical Bond, 3rd edn. Cornell Univ. Press, Ithaka, New York;

[c] Shannon RD, Prewitt CT Acta Crystallogr B25:1046 and B26:925; Shannon RD Acta Crystallogr A32:751;

[d] Behrens P, Kosslick H, Tuan VA, Fröba M, Neissendorfer F (1995) Microporous Mater 3:433

cell parameters and wavenumbers of lattice vibrations are not unambiguously related to the framework composition.

In conclusion, relationships found in some zeolites can hardly be applied to other zeolite structures without detailed verifying that their validity is maintained. Considering the variety of zeolite structures known today, there is no general answer to the question of how to monitor quantitatively (and also qualitatively) the framework atom substitution.

Relationships between composition, unit cell parameters and wavenumbers of lattice vibrations were preferentially established for sodium-exchanged hydrated zeolites. It is worth mentioning that different cations and states of hydration may cause considerable deviations from these relationships. Due to electrostatic cation-framework interaction and hydrogen-bonding of water with the framework oxygen atoms as well as of hydration of cations, the framework is distorted in different ways [58].

IR Lattice Vibration Spectra

The IR lattice vibration spectra of crystalline microporous materials in the spectral range of 1200–400 cm^{-1} consist of four main absorptions. The absorption bands in the range of 1200–900 cm^{-1} are assigned to anti-symmetric T – O – T stretching vibrations (T = P, Si, Ge, Al, Ga, Mg etc.), the absorption near 800 cm^{-1} to symmetric T – O – T vibrations. A band at 600–500 cm^{-1} is assumed to correspond to a structural sensitive framework vibration (double-ring bands). The absorption band of TO_4 bending modes appears at ca. 450 cm^{-1}. These bands shift with changes in the mean T – O distances after substitution [59]. It could be shown that there exists a relationship between

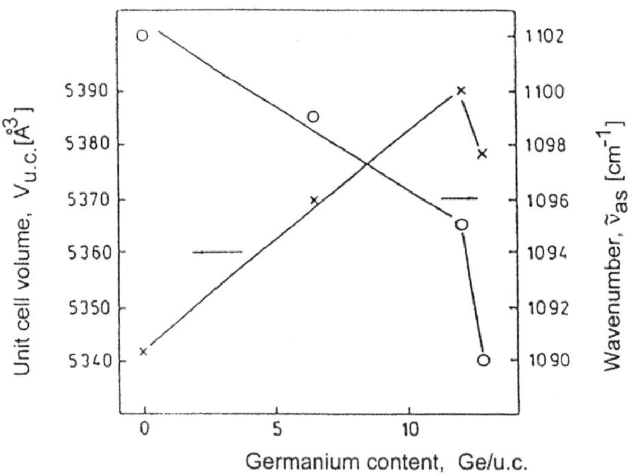

Fig. 17 Dependence of the unit cell volume ($V_{u.c.}$) and the wavenumber of the asymmetric T – O – T vibration band on the germanium content of calcined Ge-ZSM-5 [56]

the wavenumber of a specific absorption band and the number of substituted atoms for some zeolites [60]. This relationship allows the accurate determination of Si/Al framework ratios (Fig. 17). The best results were obtained for the shift of the symmetric T – O – T vibration, because this band is relatively narrow allowing a precise determination of the wavenumber. Of course, low degrees of substitution cause only small band shifts. Therefore, the estimation of high Si/Me ratios from lattice IR spectra is highly doubtful, especially, if broad absorption bands complicate the accurate determination of the location of the band maximum.

Quantitative evaluation is possible in the Si/Al range from 1 to 10–20. However, most experience in the use of IR data for the determination of the framework composition is based on results obtained for faujasite type zeolites where substitution changes cause changes in the mean T – O distance. The wavenumbers ν of the anti-symmetric (ν_{as}) and symmetric (ν_s) lattice vibrations depend on the number of aluminum framework atoms N_{Al} the in the following way [61]:

$$N_{Al} = 0.766(1086 - \nu_{as}) \text{ and} \tag{13}$$

$$N_{Al} = 1007(838 - \nu_s) \,. \tag{14}$$

In contrast, quantitative substitution of Si by Ge in the framework of zeolite ZSM-5 causes only comparatively low changes of IR data (Fig. 17). Instead of a wavenumber shift, the appearance of decoupled Si – O – Ge vibrations was observed.

Unit Cell Parameters (XRD)

The change of lattice parameters allows the accurate determination of the framework composition of faujasites [60, 62, 63]. Recently, a similar relationship was found for Ti-substituted silicalites [64]:

$$V_{u.c.} = 2110.4x + 5339 \tag{15}$$

where $V_{u.c.}$ is the unit cell volume and $[x\text{TiO}_2\cdot(1 - x)\text{SiO}_2]$ the framework composition. It was based on the assumption that all titanium is incorporated into the framework until the reflections of anatase appear in the XRD pattern. Maximum lattice expansion is found for a Ti content of 2.5% ($V_{u.c.}$ ca. 3396 Å^3). However, a similar maximum expansion of 3400 Å^3 was found for a titanium content of only about 1% [65].

Although relationships such those as indicated above were reported for high-silicon zeolites like substituted ZSM-5, a comparison of well-characterized substituted Al-, Ga- and Ge-ZSM-5 [56, 66] showed that these relationships cannot unambiguously be used for these types of zeolites. In some cases, a decrease in the unit cell volume was observed for Ga-ZSM-5 with increasing framework gallium content.

Using X-ray powder diffraction and EXAFS, the influence of the cation loading and the ion exchange with Mg on the symmetry of the crystal lattice and on the unit cell as well as on crystal structure parameters of a FER zeolite has been studied [67]. It could be shown that after cation exchange the space group changed from $P2_1/n$ to Imnm in the as-synthesized form. A linear increase of the unit cell parameters with growing extent of Mg exchange could be proved by the authors.

7.1.4
Solid-State MAS NMR

At present, solid-state NMR measurements are possible of nuclei of the main framework components of microporous materials, i.e., ^{29}Si, ^{31}P, ^{27}Al, ^{71}Ga, ^{11}B, ^{17}O together with a series of other elements, which are in part present as cations [68]. Some of these nuclei, however, exhibit only a low relative abundance (e.g., ^{17}O). Others carry a quadrupole moment giving rise to large line-broadening effects, which makes their evaluation difficult. Therefore, the number of routinely studied nuclei is relatively small and most studies deal with Si, P, B, Al, and, more recently, with Ga (compare, however, also Volume 4, Chapter 2 of the present series).

The application of magic angle spinning (MAS) technique allows to diminish line broadening caused mainly by chemical shift anisotropies of the ^{29}Si nucleus. But also dipolar interactions between the Si-nucleus and other framework atoms may contribute to line broadening. Highly resolved ^{29}Si NMR spectra with narrow Si-signals are obtained under the essential application of magic angle spinning. The ^{29}Si chemical shift values depend on the effective charge of the considered Si-nucleus or on the electronegativity of its ligands [68–71]. Therefore, the connectivity of the SiO_4 tetrahedra in the framework and the number and kind of T-atoms bound via oxygen-bridges to the silicon atoms affect the chemical shift of silicon. As a result, separate ^{29}Si-signals appear in the ^{29}Si-NMR spectrum belonging to differently bound framework silicon atoms (e.g., Si(nAl), $n = 0–4$) (Fig. 18). Similar shifts were observed after Ga-substitution [74, 75]. In practice, a distribution of chemical shift values of Si-nuclei of the same nearest neighbors are observed for different types of zeolites because changes of the Si – O – T angles and Si – O distances also modify the location of the Si-NMR signals. The electronegativity of the bridging oxygen atoms increases with increasing Si – O – T angle and decreasing Si – O distance.

From the intensity ratio of signals belonging to different types of silicon, the Si/Al framework ratio of metal silicate zeolites can be determined:

$$\mathrm{Si/Al} = \sum I_n / \sum 0.25 n I_n \,, \tag{16}$$

with

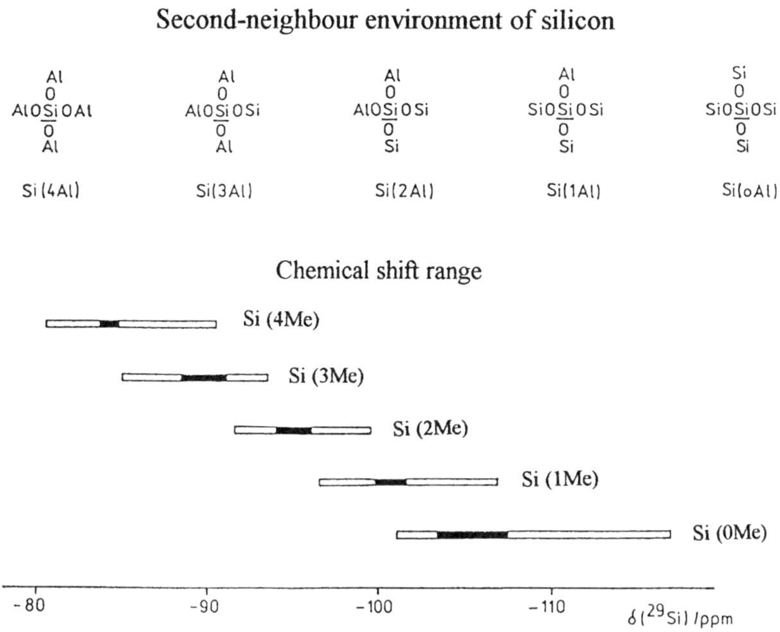

Fig. 18 Chemical shift range of ^{29}Si-NMR in dependence on the second-neighbor environment of silicon [68]

I_n – intensities of the Si (nAl) peaks and summation from $n = 0$ to $n = 4$.

This relation is independent of the zeolite structure type. For the determination of the Si/Al ratio, spectra have to be decomposed to determine the intensities of the different signals. Since, however, the spectra may be simulated by different models (the summary of different lines and line intensities), the determination of Me framework content of silicate zeolites becomes increasingly more difficult for Si/Me ratios larger than 10, and the results may contain a systematic error.

Signal broadening due to framework distortions or the existence of different crystallographic sites of identical environment and interference with Q^3 signals of $(SiO)_3SiOH$ groups, which appear at ca. – 103 ppm referenced to TMS, may render the evaluation of the framework composition of high-silicon zeolites difficult. For instance, the signal at – 106 ppm in the Si-NMR spectrum of Ga-ZSM-5 did not change although the gallium framework content was altered considerably [76]. The reason was the overlapping of the Si(1Ga) signal with the signal of the internal silanol groups located at – 103 to – 105 ppm, which increases with decreasing gallium content (Fig. 19).

The 100% natural abundance of the ^{27}Al nucleus makes Al-NMR very sensitive to small amounts of framework and also non-framework aluminum [68, 71]. Even small concentrations are detectable. However, the nu-

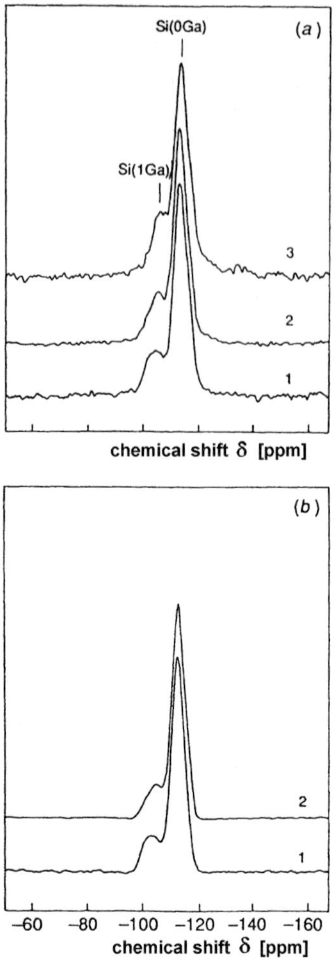

Fig. 19 **a** ^{29}Si MAS NMR spectra of as-synthesized samples of Ga-ZSM-5 referenced to TMS; 1, (Si : Ga = 100), 2, (Si : Ga = 50), 3, (Si : Ga = 25). (**b**) ^{1}H-^{29}SiCP/MAS NMR spectra of as-synthesized samples of Ga-ZSM-5; 1, Si : Ga = 100, 2, (Si : Ga = 50) [76]

clear quadrupole moment of Al (spin = 5/2 system) leads to quadrupolar interactions with the electric field gradient arising from the non-symmetric charge distribution around the detected atom. They give rise to strong line broadening and line shifts with respect to the corresponding isotropic chemical shift value. These effects make the quantitative evaluation of ^{27}Al- as well as of ^{71}Ga-NMR spectra difficult. But different chemical shifts of framework (tetrahedral) and non-framework (octahedral) Al allow a simple differentiation between these two types of species.

New insights in the local structure of protons or T atoms can be achieved by the application of new two-dimensional NMR techniques like HETCOR (two-dimensional heteronuclear correlation NMR), which are attracting increasing attention [77].

^{27}Al- and ^{71}Ga-NMR spectra of zeolites are relative simple, because the Löwenstein rule allows only the existence of $Al(OSi)_4$ and $Ga(OSi)_4$ units in the framework. The signal of framework Al appears at ca. 50 ppm and that of non-framework Al at about 0 ppm (with respect to $Al(NO_3)_3$ in water). An additional signal at 30 ppm corresponds to a third species, probably to penta-coordinated Al. The intensity ratios of the different signals are related to the content of the Al species.

The same happens with Ga, where the ^{71}Ga-NMR lines of framework and non-framework Ga appear at 160 and 0 ppm (with respect to $Ga(NO_3)_3$ solution in water). Available ^{71}Ga MAS NMR data are rare in comparison with those of aluminum, which makes the interpretation of spectra difficult. Recently, a correlation between ^{27}Al and ^{71}Ga NMR chemical shifts, δ, of structurally analogous aluminum and gallium compounds having only oxygen in the first coordination sphere was proposed [77]:

$$\delta_{71\text{Ga}} = 2.83(\delta_{27\text{Al}}) - 4.50 \,. \tag{17}$$

This relationship allows the prediction of ^{71}Ga chemical shifts from ^{27}Al chemical shifts of related Al compounds and, based on this comparison,

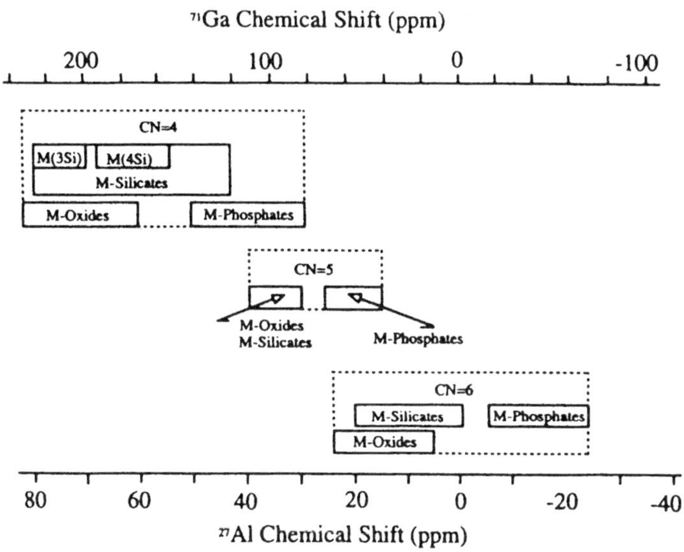

Fig. 20 Approximate ^{27}Al and ^{71}Ga NMR chemical shift ranges for aluminum and gallium compounds having only oxygen in the first metal coordination spheres. In the case of M-silicates (CN = 4), the term M(4Si) refers to tectosilicates and M(3Si) to phyllosilicates. In the case of M-oxides, for CN = 6 this term also encompasses the metal hydroxides and oxyhydroxides [77]

a better understanding and interpretation of observed gallium NMR spectra (Fig. 20).

Linear relations between the Ga-(Al-) framework content and Ga-(Al-) MAS NMR signal intensities were found for Ga,Al-ZSM-5 zeolites [78]. However, the state of material (as-synthesized or calcined) as well as the presence of counter ions influence the signal intensities considerably. Thus, only within one series of similarly pretreated samples containing the same counter ions are linear relations really observed (Fig. 21). Therefore, predictions of framework and non-framework species from Al- and Ga-NMR data are only semiquantitative.

This is well illustrated by the evaluation of ^{29}Si-, ^{27}Al-, and ^{71}Ga-NMR spectra of Ga,Al-ZSM-20 [79] obtained for the as-synthesized, calcined, and hydrogen form. Because of the low Si/Me range in ZSM-20, *viz.* 3.4–5, an accurate determination of this ratio is possible by ^{29}Si MAS NMR (Fig. 22).

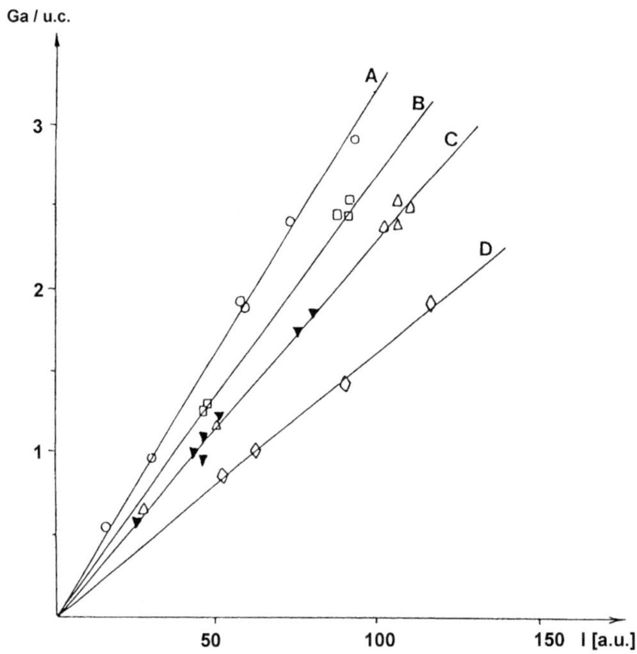

Fig. 21 Correlation between the Ga concentrations (in Ga per unit cell of zeolite) in various as-synthesized (Ga)-MFI zeolites, as determined by chemical analysis and the corresponding normalized ^{71}Ga-NMR line intensities (in arbitrary units). Correlations **A**, **B** and **C** are, respectively, derived for samples synthesized in the presence of F ions (O), Na^+ ions (□) and TPA^+ ions (*open triangles* △) for gallosilicates and *filled triangles* (▼) for gallo-aluminosilicates). Samples as described in Tables 15-1, 15-2 and 15-4 of [78]. Correlation **D** is derived for calcined, NH_4-exchanged and water-equilibrated samples (◇), as described in Table 15-6 of [78]

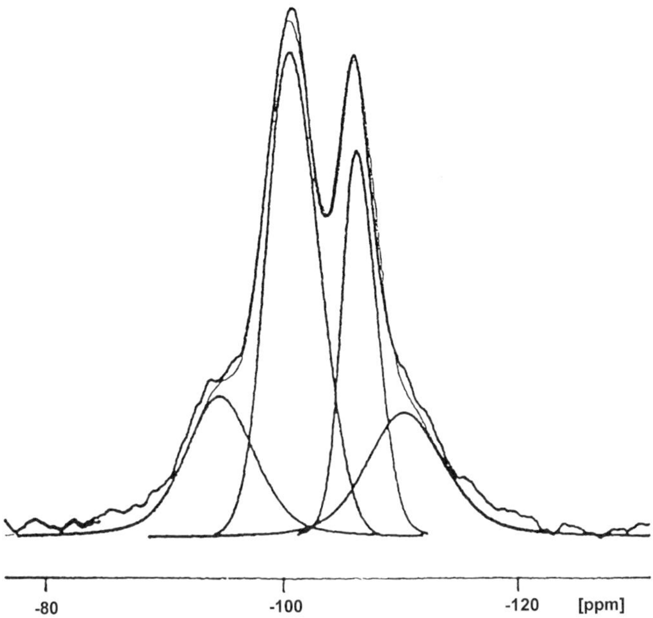

Fig. 22 Deconvoluted ^{29}Si MAS NMR spectrum of H[Ga,Al]-ZSM-20 [79]

Above Si/Me=10 a quantitative evaluation of this ratio from intensity ratios of signals becomes increasingly uncertain.

In contrast, ^{27}Al- and ^{71}Ga-NMR data deviate largely from the composition derived from ^{29}Si-NMR data as well as from the analytically determined chemical composition (Figs. 23 and 24, Table 7). These deviations cannot be diminished by saturation of samples with water vapor, which makes the surrounding of the Al- and Ga-nuclei more symmetric. Interestingly, great deviations are also found for signals belonging to the *non-framework* atoms.

In conclusion, the evaluation of quantitative data from line intensities of Al- and Ga-NMR spectra should be considered with caution and is of high risk. A further drawback is the existence of tetrahedrally coordinated non-framework gallium as was recently revealed by EXAFS studies for a series of Ga-containing ZSM-5 samples with different amounts of framework and non-framework gallium, which could not be differentiated in the ^{71}Ga-NMR spectra [80].

Substitution of Ge for Si in AST-type zeolites has been studied by ^{19}F and ^{29}Si MAS NMR combined with XRD studies [81]. The results indicate that fluoride ions are located in D4R sites with different composition of GeO_4 and SiO_4 tetrahedra (4Ge, 4Si) and (2Ge, 6Si). An ordered distribution of Ge in the framework is proposed from these results.

In a similar way, Vidal-Moya et al. [82] studied the distribution of fluorine and germanium in the new zeolite structure ITQ-13.

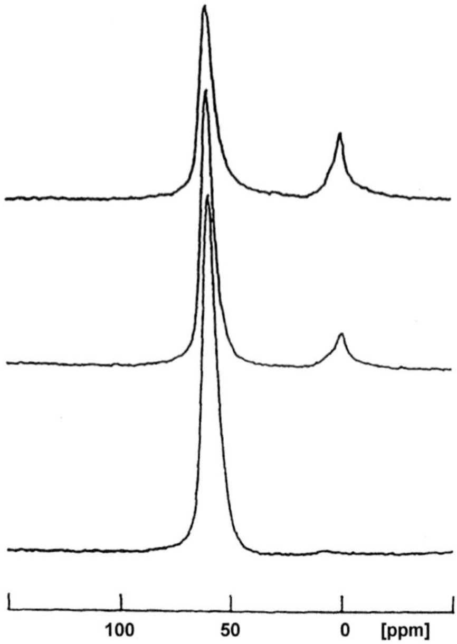

Fig. 23 ^{27}Al MAS NMR spectra of zeolite [Ga,Al]-ZSM-20 recorded in the absolute intensity mode (from the bottom to the top: as-synth., calcined, and H-form) [79]

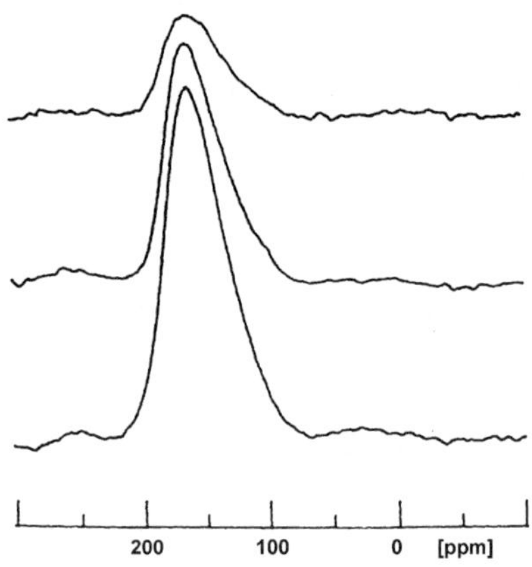

Fig. 24 ^{71}Ga MAS NMR spectra of zeolite [Ga,Al]-ZSM-20 recorded in the absolute intensity mode (from the bottom to the top: as-synth., calcined, and H-form) [79]

Using ^{27}Al MAS NMR, the reversibility of the dealumination-realumination process has been studied by Oumi et al. in the case of zeolite beta [83]. The authors could show that dissolved Al species were easily reinserted into the framework of the zeolite at a pH value lower than 7.

Table 7 Summary of NMR results of [Ga,Al]-ZSM-20 [79]

[Ga,Al]-ZSM-20	from ^{29}Si-NMR spectra Si/(Al+Ga)	number of (Al+Ga) per sodalit unit	Ch. A.**	relative signal intensities* ^{27}Al NMR FAL	FAL+EFAL	^{71}Ga NMR tetrahedral
as-synth.	3.6	5.25	3.5 Al 1.75 Ga	100	100	100
calcined	4.2	4.6		74	83	67
H-form	5.2	3.9	3.0 Al*** 0.9 Ga	59	74	29

FAL: framework aluminum
EFAL: extra-framework aluminum
* The NMR measurements were done in the absolute intensity mode (Al mode) in order to get the relative intensities by comparison of their integrated NMR signals
** Determined by chemical analysis
*** calculated under the assumption that 0.5 Al are released from the framework as observed on [Al]-ZSM-20

In a more extensive investigation, the dealumination process in zeolites beta, mordenite, ZSM-5 and ferrierite has been studied by Muller et al. using ^{1}H, ^{27}Al and ^{29}Si NMR [84]. According to these results, the extent of dealumination is directly related to the concentration of Brønsted acid sites. However, there are additional factors like the structure type, the Si/Al ratio, the crystal size and the concentration of defect sites that influence the dealumination process of the individual zeolite samples.

^{27}Al-, ^{71}Ga- and ^{11}P MAS NMR spectra of microporous alumino- and gallophosphates have been reported [85, 86]. Line broadening of Al-signals could be reduced by recording double-rotation spectra. ^{11}P NMR signals are narrow and well separated and might be interpreted quantitatively. However, the low extent of substitution of Al by other elements such as Co, Ni, Mg and the uncertainty of the real structural state connected with a lack of substituted reference MAPOs strongly limits the number of available quantitative studies. NMR data obtained from magnesium-substituted AlPO-20 (sodalite structure type) seem to prove, however, the principle potential of NMR studies for the quantitative determination of the framework composition of substituted AlPOs and GaPOs [87].

7.2 Non-framework Species

7.2.1 Leaching

Exchanged ions, extra-framework species or deposited atoms of elements are more labile than framework constituents. A simple extraction with diluted hydrochloric or nitric acid is used to remove extra-framework species from the sample. The filtrate of extraction can be directly analyzed by AAS, ICP-AES or other wet-chemical or volumetric methods of determination. In principle, other acids or complexon can also be used for extraction.

Strong or complexing acids and agents may also dissolve framework constituents. Dissolution of non-framework silica can be carried out by stirring at elevated temperatures in 1–2 M solution of sodium hydroxide. In any case, it has to be checked separately whether the leaching is accompanied by dissolution of the sample, extraction of framework constituents (dealumination), or loss of crystallinity. Leaching conditions have to be optimized and differ from zeolite to zeolite due to the different stability of structures.

Procedure

Treat ca. 100 mg *of sample with* 2 mL *of a* 0.2 molar *solution of HCl in water under magnetic stirring and keep for* 1 h *at* 60 °C. *Filtrate the solution and wash the residue with small portions of deionized water. Dilute and make to volume* (50 mL*).*

Using this method, Monque et al. [88] have determined the non-framework gallium content of ion-exchanged and Ga-impregnated MFI-type zeolites (Table 8). Framework gallium and sample crystallinity was not affected by this treatment. Also extra-framework aluminum can be removed from zeolites by acid leaching [63]. A similar procedure is the following one.

Table 8 Chemical composition results obtained by the total digestion and extractive sample treatment [88]

Sample	Total digestion		HCl extraction	
	Ga(%)	Si(%)	Ga(%)	Si(%)
4%Ga/SiO_2	4.00	–	4.02	–
0.5%Ga/[Al]ZSM-5	0.49	–	0.48	–
[Ga]ZSM-5(38)	2.3	35.3	< 0.18	< 0.2
2%Ga/[Ga]ZSM-5(38)	4.35	35.2	2.05	< 0.2
[Ga]ZSM-5(38) HClextr	2.28	35.2	–	–

Procedure
Stir ca. 100 mg *of sample with* 50 mL *of* 0.05 M *solution of HCl or* HNO_3 *for* 1 h *at* 50 °C. *Centrifuge and make to volume.*

This treatment allows removal of non-framework titanium without affecting framework titanium of Ti-silicalite [65].

7.2.2
X-ray Photoelectron Spectroscopy (XPS)

XPS is one of the spectroscopic methods that are limited to the analysis and characterization of the surface region of a solid (compare also Volume 4, Chapter 6 of the present series). In principle, XPS works as follows:

A solid sample is irradiated by an X-ray source, which causes low-lying electrons to leave the surface. The energy spectrum of these electrons can be measured and used for the determination of the binding energies of electrons. The binding energies are characteristic of the elements and are influenced by the surrounding of atoms from which the electrons are released. The latter allows, in some cases, the differentiation between framework and non-framework atoms. As the depth of penetration of the X-rays comprises only few atom layers, XPS can give information about the qualitative and quantitative composition of these layers only. A depth profile, however, can be obtained by sputtering the surface with an appropriate gas.

Relevant characteristic data are [89]

- detected elements: $Z > 3$
- analyzed surface (beam area): μm^2 to mm^2
- analyzed thickness (depth): 0.5 to 3 nm.

The application of XPS requires evacuation of the gas phase down to about 10^{-8} Pa. It is evident that this high-vacuum treatment may simultaneously have an influence on some properties of the sample under investigation, i.e., on valency or rearrangement of the surface layer of the sample due to possible partial desorption. This has to be taken into account, especially when binding energies should be measured, but is less important when only the chemical composition of the surface region is investigated.

One of the most serious problems that have to be taken into account for zeolites is the *charging* of the insulating sample due to the escape of electrons from the surface effected by X-rays (XPS) or ultraviolet radiation (UPS). This leads to a loss of any sample-independent reference level. However, this problem can be overcome (or at least minimized) by using the C(1s) binding energy of carbon impurities [90] always present in the samples or by various pretreatments (gold layer, ion irradiation, specimen bias, pulsed primary beam etc.). The right choice will depend on the aim of the measurement.

Because the detection of electrons from the sample surface is not limited to a special geometric shape of the surface, XPS, UPS and Auger spectroscopy

(AES) can also be applied for the investigation of zeolite powders [90]. Usually, however, these methods are applied for the determination of the binding energies of elements forming the zeolite lattice. The reason for their good suitability for *qualitative* analysis is the high sensitivity of these methods.

Concerning the *quantitative determination* of the chemical composition of the surface layers with or without sputtering, XPS can only be regarded as a method which is able to give a rough estimation of the composition [91]. Of course, it is possible to calibrate the spectrometer by signal processing (background corrections, signal deconvolution etc.) as is also possible with other spectroscopic methods. Nevertheless, two points have to be taken into consideration which limit the use of these methods,

a) the absolute concentration of the elements to be analyzed quantitatively should not be lower than 3–5%,
b) the relative error of the determination of the concentration of elements is estimated to be not better than 10%.

XPS spectra were used to differentiate between framework and non-framework titanium in Ti silicalite [65]. The presence of non-framework titanium is evidenced by the signal at 458 eV in the Ti 2p spectra (Fig. 25). After leaching, this signal disappears, revealing removal of non-framework titanium. Because the bulk chemical composition determined by ICP-AES does

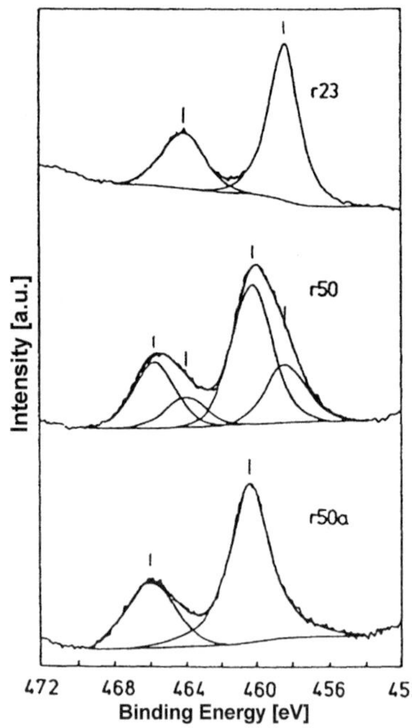

Fig. 25 Ti 2p photoelectron spectra of calcined (r23, r50) and acid-leached (r50a) Ti-silicates [65]

not change after leaching, evidence is given for a location of non-framework Ti species near the crystal surface. On the other hand, framework Ti (signal at 460.3 eV) is not affected by leaching. After leaching, the Si/Ti surface ratio has increased, and correlates now with the unit cell volume (Table 9).

Gross et al. [92] have used XPS measurements for the investigation of the surface composition of dealuminated Y zeolites. A series of NH_4NaY zeolites of varying ammonium content have been used as starting material for thermochemical treatment and extraction with EDTA which resulted in different degrees of dealumination. The Al/Si and Na/Si ratios were obtained from Al2p/Si2p and $NaKL_{2,3}$/ Si2p peak area ratios. The error was estimated to be lower than 20%. Non-framework aluminum is enriched at the external crystal surface leading to a Si/Al ratio< 1. After thermal treatment at 815 °C the Si/Al an framework ratio is increased from 0.41 to 1.36.

Mohamed et al. [93] have shown by a combined XPS and catalytic investigation of barium-loaded MFI zeolites that the Ba/Si ratio determined by XPS as a function of the total barium loading shows a maximum at about 4. At greater loadings, the surface Ba/Si ratio is lower than the bulk Ba/Si ratio. They were able to correlate the olefin selectivity in the conversion of methanol with the surface Ba/Si ratio.

In a combined XPS and ISS (ion-scattering spectroscopy) investigation of different zeolite structures published by Grünert et al. [94], the authors derive detailed conclusions on the differences between the surface and bulk composition of the H and Na forms of the zeolites and on the dealumination processes within the crystal lattice.

The surface composition of various types of zeolites (NaA, HY, NH_4-erionite, Na-mordenite) was also studied by Kaushik et al. [95]. The composition of the bulk and surface (expressed as Si/Al ratio) of most of the zeolites was nearly equal to what did not meet the results of ^{29}Si MAS NMR.

Applying XPS spectroscopy, the dispersion and formation of bimetallic Pt – Pd particles on PtPd/H-beta zeolites has been studied by Fiermans et al. [96]. By careful analysis of the intensity changes of the Pd 3d and Pt 4d photolines, the authors come to the conclusion that segregation of Pd particles occurred towards the surface of zeolite beta.

Table 9 Si/Ti surface atomic ratios before and after acid leaching and unit cell volume of silicalites with different amounts of lattice titanium [65]

Sample	Unit cell volume/Å^3	Si/Ti surface atomic ratio leached	unleached
silicalite	5322	–	–
r100a	5337	167	91
r50a	5363	143	91
r63a	5400	100	56

The dispersion of Ga on H-ZSM-5 and H-MOR has been investigated by Garcia-Sanchez et al. by applying XPS in combination with ICP, NMR and FTIR spectroscopy [97]. On the zeolite samples modified by CVD of trimethylgallium, the dispersion depends on the kind of treatment and on pore blocking effects.

Also in the case of K-L zeolites, XPS has been successfully applied to study the dispersion of Pt species. In their investigation, Zheng et al. could demonstrate that Fe has a stabilizing effect on the dispersion of Pt in Pt – Fe/KL zeolites [98].

References

1. Heinrichs H (1989) Laborpraxis Dezember 1140
2. Heinrichs H (1990) Laborpraxis Januar/Februar 20
3. Dolezal J, Powondra P, Sulcek Z (1968) Decomposition techniques in inorganic analysis. Elsevier, New York
4. Tschöpel P, Kotz L, Schulz W, Veber M, Tölg G (1980) Fresenius Z Anal Chem 302:1
5. Hannaker Oh, Hou Qing-Lie (1984) Talanta 31:1153
6. Guignard J, Hazebrouk GJ (1966) Chem Phys 63:1351
7. Mullen JP, Riley JP (1955) Anal Chim Acta 12:162
8. Schwarzenbach G (1956) Die komplexometrische Titration. Ferdinand Enke Verlag, Stuttgart
9. Jerschkewitz H-G (1961) Analytische Chemie, Teil I: Theoretische Grundlagen und qualitative Halbmikroanalyse. Teubner, Leipzig
10. Petzold W (1955) Die Cerimetrie E. Merck AG (ed) Verlag Chemie, Weinheim
11. Schnick W, Lücke J (1992) Angew Chem 104:208
12. Buresch O, von Schnering HG (1984) Fresenius Z Anal Chem 319:418
13. Krom MD (1980) Analyst 105:305
14. Fraser AR, Russell JD (1969) Clay Miner 8:229
15. Kudeyorov V (1970) Khim Sel'sk Khoz 8:708
16. Searle PhL (1984) Analyst 109:549
17. Welz B (1983) Atomabsorptionsspektrometrie. Verlag Chemie, Weinheim
18. Haswell SJ (ed) (1991) Atomic absorption spectrometry. Elsevier, Amsterdam
19. Encyclopedia of analytical science (1995) Academic Press, Harcourt Brace & Comp Publ, London, vol 1 p 195 (Fig. p 202)
20. Slickers K (1992) Die automatische Atom-Emmissions-Spektralanalyse. Buchvertrieb K.A. Slickers, Gießen
21. Thompson M, Walsh JN (1989) Handbook of inductively coupled plasma spectrometry, 2nd edn. Blackie, Glasgow
22. Montaser A, Golightly DW (1992) Inductively coupled plasmas in analytical spectroscopy, 2nd edn. VCH, New York
23. Welz B (ed) (1986) Fortschritte in der atomspektrometrischen Spurenanalytik. Verlag Chemie, Weinheim
24. Erzinger J, Heinschild HJ, Stroh A (1986) In: Welz B (ed) Fortschritte in der atomspektrometrischen Spurenanalytik, vol 1. Verlag Chemie, Weinheim, p 211; Erzinger J, Heinschild HJ, ibid., vol 2, p 333
25. Godelitsas A, Charistos D, Tsipis A, Tsipis C, Fillipidis A, Triantafyllidis C, Manos G, Siapkas D (2001) Chem-A Eur J 7:3705

26. Bertin EP (1978) Introduction to x-ray spectrometric analysis. Plenum Press, New York
27. Jenkins R (1988) X-ray fluorescence spectrometry. Wiley, New York
28. Van Grieken RE, Markowicz AA (eds) (1992) Handbook of x-ray spectrometry. Marcel Dekker, New York
29. Scholz O (1972) Atomphysik kurz und bündig. Vogel-Verlag, Würzburg, p 93
30. Knippenberg WF (1974/75) Philipps Techn Rdsch 34:298
31. Weiss J (1985) Handbuch der Ionenchromatographie. Wiley-VCH, Weinheim
32. Bogenschütz G, Loyall U, Schäfer J (1990) LaborPraxis, March, p 120
33. Franklin KR, Lowe BM (1988) Thermochim Acta 127:319
34. Minchev Chr, Valtchev V, Mintova S (1995) Thermochim Acta 264:59
35. Bedard RL, Bowes CL, Coombs N, Holmes AJ, Jiang T, Kirkby SJ, Macdonald PM, Malek AM, Ozin GA, Petrov S, Plavac N, Ramik RA, Steele MR, Young D (1993) J Am Chem Soc 115:2300
36. Kosslick H, Tuan VA, Fricke R, Jedamzik J, Lanh HD (1991) J Therm Anal 37:2631
37. McClennen WH, Buchanan RM, Arnold NS, Dworzanski JP, Meuzelaar LC (1993) Anal Chem 65:2819
38. Kürschner U, Jerschkewitz H-G, Schreier E, Völter J (1990) Appl Catal 57:167
39. Lischke G, Eckelt R, Jerschkewitz H-G, Parlitz B, Schreier E, Storek W, Zibrowius B (1991) J Catal 132:229
40. Kosslick H, Richter M, Tuan VA, Fricke R, Storek W (1992) J Chem Soc Faraday Trans 88:2421
41. Bhatia S, Beltrami DD (1990) Catal Today 3:309
42. Lohse U, Parlitz B, Patzelova V (1989) J Phys Chem 93:3677
43. Karge HG, Dondur V, Weitkamp J (1991) J Phys Chem 95:283
44. Szynkowska MI, Lesniewska E, Paryjczak T (2003) Pol J Chem 77:657
45. Berndt H, Kosslick H, Martin A, Lücke B (1994) Microporous Mater 2:197
46. Miessner H, Kosslick H, Lohse U, Parlitz B, Tuan VA (1993) J Phys Chem 97:9741
47. Kosslick H, Berndt H, Hoang DL, Martin A, Miessner H, Tuan VA, Jänchen J (1994) J Chem Soc Faraday Trans 90:2837
48. Baranski A, Dal T, Galuszka J (1979) Bull Acad Polon Sci, Ser Sci Chim 27:363
49. Lok BM, Marcus BK, Angell CL (1986) Zeolites 6:185
50. Hunger B, Heuchel M, Clark LA, Snurr RQ (2002) J Phys Chem B 106:3882
51. Li G, Kan QB, Wu TH et al. (2003) Stud Surf Sci Catal 146:149
52. Wallau M, Melo RAA, Urquieta-Gonzales EA (2003) Stud Surf Sci Catal 146:745
53. Kofke TJG, Kokotailo GT, Gorte RJ, Farneth WE (1989) J Catal 115:265
54. Kofke TJG, Kokotailo GT, Gorte RJ (1989) J Catal 116:252
55. Kofke TJG, Gorte RJ, Kokotailo GT (1989) Appl Catal 54:177
56. Kosslick H, Tuan VA, Fricke R, Peuker Ch, Pilz W, Storek W (1993) J Phys Chem 97:5678
57. Depmeier W (1984) Acta Crystallogr B40:185
58. Flanigen EM, Khatami H, Szymanski HA (1971) Adv Chem Ser 10:201
59. Flanigen EM (1976) In: Rabo JA (ed) Zeolite Chemistry and Catalysis. ACS Monograph 171, Washington, DC, p 80
60. Fichtner-Schmittler H, Lohse U, Miessner H, Manek HE (1990) Z Phys Chem, Leipzig, 271:69
61. Sohn JR, De Canio SJ, Lunsford JH, O'Donnell DJ (1986) Zeolites 6:225
62. Breck DW, Flanigen EM (1968) Molecular sieves. Soc Chem Ind London, p 47
63. Breck DW (1974) Zeolite molecular sieves. Wiley, New York, p 104
64. Millini R, Massara EP, Perego G, Bellussi G (1992) J Catal 137:497

65. Grohmann I, Pilz W, Walther G, Kosslick H, Tuan VA (1994) Surf Interface Anal 22:403
66. Kosslick H, Tuan VA, Fricke R, Peuker Ch (1992) Ber Bunsenges Phys Chem 96:1761
67. Morell H, Angermund K, Lewis AR, Brouwer DH, Fyfe CA, Gies H (2002) Chem Mater 14:2192
68. Engelhardt G, Michel D (1987) High-resolution solid-state NMR of silicates and zeolites. Wiley, Chichester
69. Lippmaa E, Mägi M, Samoson A, Tarmak M, Engelhardt G (1981) J Am Chem Soc 103:4992
70. Engelhardt G, Radeglia R (1984) Chem Phys Lett 108:271
71. Klinowski J (1991) Chem Rev 91:1459
72. Griffiths JM, Griffin RG (1993) Analysis Chimica Acta 283:1081
73. Simon A, Gougeon RD, Paillaud J-L, Valtchev V, Kessler H (2001) Phys Chem Chem Phys 3:867
74. Newsam JM, Jacobson AJ, Vaughan DEW (1986) J Phys Chem 90:6858
75. Thomas JM, Klinowski J, Fyfe CA, Anderson MW (1983) ACS Symp Ser, Washington, DC, 218:159
76. Kosslick H, Tuan VA, Parlitz B, Fricke R, Peuker Ch, Storek W (1993) J Chem Soc Faraday Trans 89:1131
77. Bradley SM, Howe RF, Kydd RA (1993) Magn Reson Chem 31:883
78. Gabelica Z, Mayanez C, Monque R, Galiasso R, Giannetto G (1992) Synthesis of microporous materials. In: Occelli ML, Robson HE (eds) Molecular sieves, vol 1. Van Nostrand Reinhold, New York, p 190
79. Kosslick H, Tuan VA, Walther G, Miessner H, Fricke R, Storek W (1996) In: Occelli ML, Kessler H (eds) Synthesis of microporous materials: zeolites, clays, and nanostructures, Chap. 17. Marcel Dekker, New York
80. Behrens P, Kosslick H, Tuan VA, Fröba M, Neissendorfer F (1995) Micropor Mater 3:433
81. Wang YX, Song JQ, Gies H (2003) Solid State Sci 5:1421
82. Vidal-Moya JA, Blasco T, Rey F, Corma A, Puche M (2003) Chem Mater 15:3961
83. Oumi Y, Mizuno R, Azuma K et al. (2001) Micropor Mesopor Mat 49:103
84. Muller M, Harvey G, Prins R (2000) Micropor Mesopor Mat 34:135
85. Grobet PE, Martens JA, Balakrishnan I, Mertens M, Jacobs PA (1989) Appl Catal 56:L21
86. Müller D, Jahn E, Ladwig G, Haubenreisser U (1984) Chem Phys Lett 109:332
87. Prasad S, Haw JF (1996) Chem Mater 8:861
88. Monque R, Parisi A, Gonzalez S, Gianetto G (1992) Zeolites 12:806
89. Darque-Ceretti E, Aucouturier M (1995) Analysis 23:49
90. Moretti G (1994) Zeolites 14:469
91. Hercules DM (1996) Fresenius J Anal Chem 355:209
92. Gross Th, Lohse U, Engelhardt G, Richter KH, Patzelová V (1984) Zeolites 4:25
93. Mohamed MH, Abdillah MM, Abbas NM, Siddiqui AB (1995) Appl Surf Sci 90:409
94. Grünert W, Muhler M, Schröder KP, Sauer J, Schlögl R (1994) J Phys Chem 98:10920
95. Kaushik VK, Bhat SGT, Corbin DR (1993) Zeolites 13:671
96. Fiermans L, De Gryse R, De Doncker G et al. (2000) J Catal 193:108
97. Garcia-Sanchez M, Magusin PCMM, Hensen EJM et al. (2003) J Catal 219:352
98. Zheng J, Schmauke T, Roduner E et al. (2001) J Mol Catal A-Chem 171:181

Mol Sieves (2007) 5: 67–101
DOI 10.1007/3829_002

Published online: 20 December 2005

Thermal Analysis of Zeolites

Gabriella Pál-Borbély

Institute of Chemistry, Chemical Research Center, Hungarian Academy of Sciences,
Pusztaszeri út 59–67, 1025 Budapest, Hungary
borbely@chemres.hu

1 Introduction

In general, thermal analysis can provide important information on the temperature-dependent properties (heat capacity, thermal stability) of materials and on thermally induced processes (phase transition, decomposition, etc.). Thermal analysis is advantageous in that it quickly gives a general view of the thermal behavior of a material under various conditions and requires a small amount of sample.

The fundamentals of the four most frequently applied methods—thermogravimetry, differential thermal analysis, differential scanning calorimetry, and dilatometry—are described and illustrated with examples of applications in the field of preparative chemistry in [1]. As to zeolite investigations by microcalorimetry, a comprehensive review with 69 references has been published recently [2].

In the case of thermogravimetry (TG), the change in the weight of a sample is measured as a function of temperature (or of time in the case of isothermal measurements). Measurements in inert or reactive gases, even in vacuum, are possible over a wide temperature range with a constant heating rate or using different non-linear temperature programs. The derivative thermogravimetry (DTG) indicates the rate of the mass loss (dm/dt). Differential thermal analysis (DTA) is based on the measurement of the temperature difference, ΔT, between the sample and an inert material (reference) during a temperature program. The temperature can be measured either near or directly inside the sample and reference. At thermal equilibrium the oven, the sample, and the inert material have the same temperature. When a thermal event occurs in the sample during the heating process, the stationary state will be disturbed. The temperature of the sample can be higher (exothermic reaction) or lower (endothermic reaction) than the temperature of the inert material. ΔT is plotted against temperature or time. Differential scanning calorimetry (DSC) monitors the change in enthalpy (dH/dt). In the technique of dilatometry changes in the length of the sample are examined during the heating process.

Thermoanalytical methods were often used simultaneously and combined with complementary methods. In this field the main attention was paid to the decomposition reactions of solids accompanied by gas evolution. Many experimental techniques combining the thermogravimetry with evolved gas analysis [gas chromatography (GC), mass spectrometry (MS)] were developed. These methods led to conclusions drawn about the reactions due to the observed changes in weight. The widespread simultaneous application of TG and DTA enabled a specific change in mass to be attributed to a thermal event. Many papers on characterization of zeolites and clays were published, in which emphasis was laid on the advantages of combining simultaneous thermoanalytical methods (TG-DTA) with other complementary techniques (e.g. MS, GC, XRD, IR). Fundamentals and fields of application of temperature-resolved X-ray and neutron diffraction methods were presented by Epple [3].

Gimzewski [4] proposed the application of high-pressure thermogravimetry for the measurement of gas sorption on solids and for the study of solids (for instance catalysts) under industrially realistic conditions with respect to temperature and pressure.

Langier-Kuzniarowa [5] has recently reviewed the application of thermal methods for the study of clay minerals and rocks. From this comprehensive work it turned out that a large number of papers dealt with sorption properties of minerals and with organo-clay complexes. Pillared and intercalated clays were frequently subjected to thermal examinations. TG and DTA were used to investigate kinetics and thermodynamic parameters of dehydration and dehydroxylation of clay minerals.

Thermal analysis has also been shown to be a valuable technique for the characterization of zeolites. Combined TG and DTA provided information on the dehydration behavior and stability of zeolites. The decomposition of organic molecules (ions) incorporated in the zeolitic channel system during the synthesis has been extensively studied since nitrogen-containing organic compounds are being used in the synthesis of high-silica zeolites. The temperature range in which the organics are removed from the cages and channels of zeolites is of great significance for the application of these materials in industrial processes (catalysis, adsorption).

In a review of predominantly the author's own papers, Gabelica et al. [6] presented very illustrative examples of applications of thermoanalytical techniques for the investigation of synthesis and properties of pentasil zeolites. Weight losses (TG) and heat effects (DTA) due to the oxidative decomposition of tetrapropylammonium ions filling the channel structure of the zeolite were related to the amount of ZSM-5 formed during the crystallization. Thermal analysis was shown to be a very accurate technique for discriminating between loosely bound TPA ions and those interacting strongly with a zeolite-type framework. Thus, very small X-ray amorphous particles of ZSM-5 zeolite could be detected in the early stages of the crystallization process. Thermal analysis could be used to monitor slight changes in the free pore volume of some modified zeolites by measuring the adsorption and desorption of small hydrocarbon molecules. Furthermore, the technique made it possible to monitor the formation and removal of coke deposits.

The present review covers literature of the last 15–20 years dealing with the application of thermoanalytical methods in the field of zeolite chemistry. The aim is to show that thermal analysis is an efficient tool for physical-chemical characterization of zeolites.

2
Characterization of Zeolites by Thermal Analysis

Thermogravimetry was applied to measure the amounts of water adsorbed on sodium, calcium, and potassium forms of faujasites with different Si/Al ratios [7]. The number of water molecules per u.c. was found to increase with decreasing Si/Al ratio.

Differential scanning calorimetry was used to study the thermal effects during adsorption and desorption of water on zeolite NaA in the form of powder or thin layers [8]. The amounts of water adsorbed after preceding activation and at different temperatures were thermogravimetrically measured, and the appropriate adsorption heats were determined by DSC. On the basis of the relation between both characteristics, the mass of the zeolite in thin layers deposited on metals was obtained from the DSC results.

The dehydration of a new microporous cesium silicotitanate (SNL-B) molecular sieve and subsequent readsorption of water on it were measured by TG [9]. An observed difference between the amount of desorbed and readsorbed water was attributed to the irreversible dehydration of a second phase present as an impurity.

Defect sites created during the dealumination of Na-Y with $SiCl_4$ vapor and subsequent acid washing were characterized by IR, ^{29}Si (CP) MAS NMR spectroscopy, and TG [10]. A rapid weight loss up to 373 K revealed by the TG curve of the dealuminated Y zeolite (see Fig. 1) was undoubtedly due to the release of physisorbed water, while the slower weight loss between 573 and 973 K could be attributed to the loss of water formed as a result of the healing of defects through condensation of silanol groups. Steaming resulted in a very low sorption capacity of dealuminated Y for water and in the healing of defects. Accordingly, the weight loss of the sample due to dehydration and dehydroxylation of silanol groups was found to be very small. The zeolite became hydrophobic.

An attempt was made to quantify the hydrophobicity (H) of zeolites by thermogravimetric analysis [11]. The value of H was related to the amount of the adsorbed water determined by TG and defined as the ratio of the weight loss at 423 K to that at 673 K. Weitkamp et al. [12] pointed to the inadequacy of the H parameter, and they proposed for the characterization of the hydrophobic properties of modified zeolites the use of the Hydrophobicity Index (HI) which was based on competitive adsorption of toluene and water from the gas phase. HI was defined as $X_{\text{toluene}}/X_{H_2O}$, where X is the loading i.e., the mass of adsorbed compound per mass of dry adsorbent.

High-siliceous zeolites of the ZSM-5 type show hydrophobic/organophilic properties. TG/DTG/DTA was used for quick characterization of the affinity of zeolites to organics [13]. A new concept concerning the affinity, A_T, of the perfect micropore surface in silicalite-1 to an adsorbed organic compound was developed. This quantity was defined as $A_T = T_d - T_b$, where T_d is the temperature of the weight loss peak in the DTG curve and T_b the boiling point of the organic compound. A_T values were used to compare

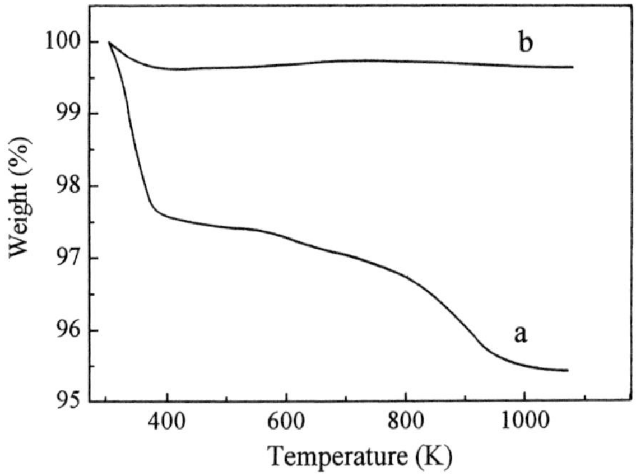

Fig. 1 TG curves of zeolite Y **a** dealuminated with $SiCl_4$ vapor at 673 K followed by acid washing, **b** after subsequent steaming at 993 K (reproduced from [10])

the affinity order of 27 organic compounds. In later papers, the interactions between the framework of high-siliceous hydrophobic FAU [14] and ferrierite type zeolite [15] and 25 adsorbed organic compounds were studied in a similar way by TG/DTG/DTA. Hydrocarbons, alcohols, organic acids, alkylamines, and esters were applied as adsorbates. The A_T values of organics adsorbed on siliceous FAU and silicalite-1 were found to be similar in their graduation. These zeolites, built up exclusively from SiO_4 tetrahedra, had a stronger affinity to saturated hydrocarbons than for unsaturated ones and exhibited hydrophobic behavior towards alkylalcohols and organic acids. Special sorbate/framework and sorbate/sorbate interactions were found for amines. It was pointed out that the A_T values might be useful to estimate the feasibility and effectiveness of separation processes in practice.

Usually, an exothermic peak in the high-temperature interval of the DTA curve was found to hint at a loss of crystallinity [16–19]. The temperature at which the exothermic peak associated with lattice collapse appeared in the DTA curve was used to characterize the thermal stability of faujasites as a function of the Si/Al ratio, cation type (La^{3+}, Ca^{2+}, Na^{+}) and degree of ion exchange [20].

The thermal stability of zeolite Y dealuminated hydrothermally (USY) and by Al-extraction with $(NH_4)_2SiF_6$ (SY) or EDTA (EY) was also studied by DTA [21]. The onset temperature of the exothermic DTA peak was considered as the temperature of crystal collapse. Figure 2 shows that the thermal stability of SY samples increased almost linearly with the decrease of their Al/(Al + Si) ratios. The framework of NH_4-Y (Si/Al = 2.7) collapsed at about 1090 K while SY (Si/Al = 6.8) was stable up to 1387 K. Dealumination by H_4EDTA solution resulted in a smaller increase in the thermal stability of the zeolite. From the point of view of framework stabilization, the substitution of silicon for aluminum by $(NH_4)_2SiF_6$ treatment proved to be more effective than the extraction of aluminum with H_4EDTA.

The exothermic lattice breakdown of zeolite Na-A, which leads to typical doublet peaks in the DTA curve at around 1200 K, was employed to check the success of crystallization of zeolite Na-A from synthesis mixtures at various pH values [22]. The TG and DTA curves of a fully crystallized zeolite Na-A are presented in Fig. 3.

Also, phase transition and destruction of the structure of natural heulandites and clinoptilolites were investigated by differential thermal analysis [23]. A relationship was found between the chemical composition (cation form) and the thermal behavior of clinoptilolites. Differential scanning calorimetric measurements of ethanolamine- and ethylenediamine-silica sodalite pointed to temperature-induced phase transitions in the temperature range of 100–300 K [24].

Dilatometry and derivative dilatometry combined with DTA are good methods for measuring and continuous recording of lattice deformations,

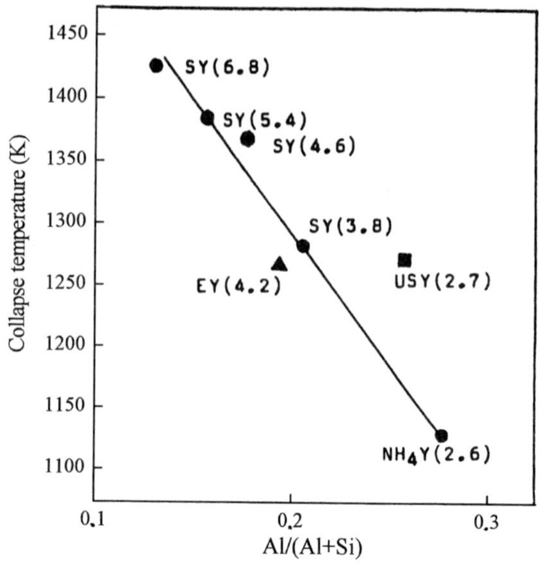

Fig. 2 Crystal collapse temperatures determined by DTA as a function of Al/(Al + Si) ratio of parent Y zeolite (NH_4-Y) and varieties dealuminated hydrothermally (USY) and by treatment with $(NH_4)_2[SiF_6]$ (SY) and H_4EDTA (EY). The figures in parentheses indicate the Si/Al ratio (reproduced from [21])

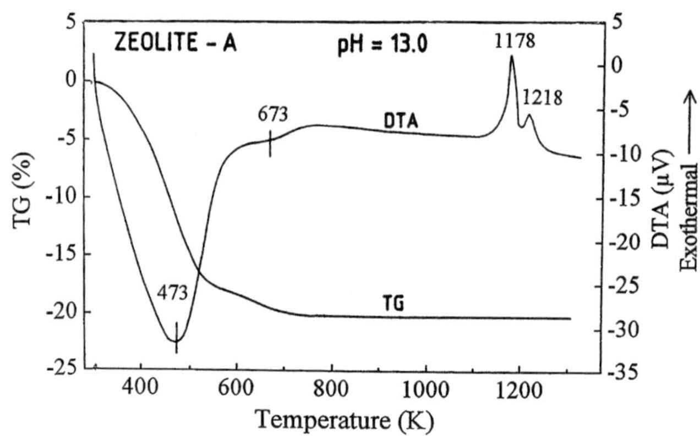

Fig. 3 TG and DTA curves of zeolite Na-A synthesized at pH = 13 (reproduced from [22])

metaphase formations, and phase transitions during the thermal dehydration of zeolites, provided these transitions are accompanied by a pronounced expansion or contraction. Ullrich et al. [25] demonstrated that the dehydration of minerals of the natrolite group is accompanied by an expansive process, while that of stilbite and heulandite proceeds under contraction, depending on the water bonding in the zeolite lattice.

Catalyst deactivation is most frequently due to the formation of carbonaceous residues in acid-catalyzed hydrocarbon reactions. Techniques and apparatus suitable for the investigation of coke formation and their application illustrated with examples were reviewed by Karge [26]. It was shown that thermogravimetric measurements and TGA combined with GS and MS could provide useful data concerning the amount of coke deposits and their H/C ratios. In the procedures presented in [26], the products of the coke combustion (H_2O, CO_2) were trapped and analyzed by GS and the weight loss was measured simultaneously. The TG technique was not only used to determine the H/C ratio of coke but also to trace the coke formation passing the gaseous reactant over the catalyst in situ in the microbalance. In this case, the amount of carbonaceous deposits was monitored via the weight gain of the catalyst. As an example, Fig. 4 shows the weight gain vs. time on stream due to in situ coke formation during ethylene conversion on a ZSM-5-based catalyst at reaction temperatures increasing from 425 to 675 K. Also, the H/C ratios after 6 h on stream are indicated.

Significant influence of reactant dilution and chemical nature of the carrier gases on composition and concentration of coke residues deposited on zeolite Beta during the catalytic disproportionation of cumene was found to be reflected by the thermograms and ^{13}C NMR spectra of the thus-deactivated catalysts [27]. The nature of coke (so-called soft or hard coke) and the removal of these deposits from the catalyst were found to be strongly influenced by the reaction conditions (e.g. temperature, space velocity) during the aromatization of propane on MFI type H-gallosilicate [28]. The results of TA supported that, in contrast to the high-temperature residues, the coke formed at low temperature was volatile. A combined technique consisting of TG and

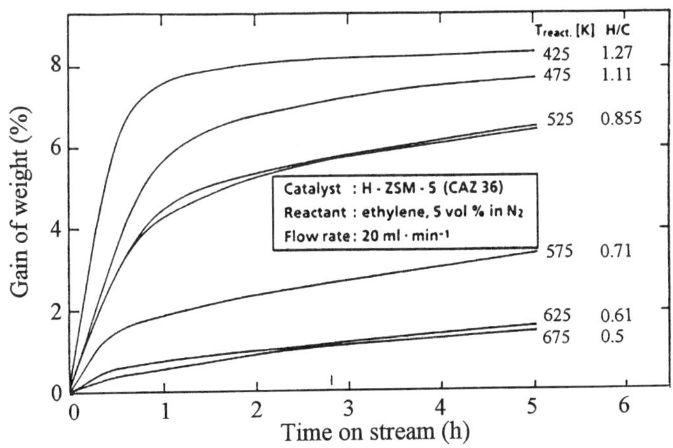

Fig. 4 Weight gain due to coke formation during ethylene conversion over H-ZSM-5 at different reaction temperatures (reproduced from [26])

continuous detection of evolved products by IR was applied to study the removal of carbonaceous deposits previously formed during the conversion of methanol to hydrocarbons on zeolite H-ZSM-5 [29]. In different gas streams (e.g. helium, hydrogen, and propane) the effect of thermal treatments on the removal of coke and the changes during these treatments in the nature of coke were investigated.

The formation of carbonaceous deposits during the alkylation of isobutane with 1-butene over La,M-Y (M = alkali or alkaline earth element) was studied by TA in [30]. Two types of carbonaceous material could be distinguished: (i) deposits at the pore mouth removable by desorption at temperatures below 573 K and (ii) olefinic oligomers located in the internal pore space resulting in a decrease of the BET surface and in a deactivation of the catalyst.

In another study [31] coked Mo/H-ZSM-5 zeolite previously used as a catalyst for the dehydro-aromatization of methane was subjected to different temperature-programmed thermal treatments in reducing and oxidizing atmospheres. Then TG was applied to determine the amount of coke after the respective treatments and to gain in this way information on the burning-off process.

Until the mid-nineties conventional dynamic balances played a decisive role as a central measuring instrument in thermoanalytical methods. However, the need for direct mass determination in weightless conditions (e.g. in space vehicles) led to the development of a new type of so-called TEOM (tapered element oscillating microbalance) systems. The mass of the sample located at the tip of an oscillating quartz element is measured by the vibrational frequency of the system. This new instrumental technique became commercialized in recent years and instantly utilized in thermoanalyzers. In conventional microbalance reactors a large part of the feed bypasses the catalyst sample which makes it difficult to confirm gradientless differential operation. In contrast, in the TEOM reactor all the fed reactant (or adsorbate) passes the catalyst (adsorbent) bed even at high flow rates of the carrier gas similarly to the flow conditions in a catalytic fixed-bed reactor (or adsorption column). Thus, external mass and heat transfer limitations play a minor role in model equations for the determination of kinetic data from mass change patterns operated in the non-linear range. Chen et al. [32] highlighted the advantages of this new measuring principle in a thermoanalytical study of the deactivation process caused by coking during the oligomerization of ethene over H-ZSM-5. Zhu et al. applied the TEOM technique in a series of investigations concerning adsorption/desorption characteristics of light alkanes [33–36] and unsaturated C4 hydrocarbons [37] on silicalite-1 and DD3R zeolite, respectively, determined by both steady-state and transient uptake and desorption measurements.

TG and DTG studies revealed the formation of weak bonds between extra-framework aluminum species in dealuminated H-ZSM-5 and acetylacetone applied for visualization of "NMR invisible" aluminum [38].

It is known that cations play a fundamental role in zeolite crystallization. Aiello et al. [39] used various thermoanalytical techniques to study the dehydration of aluminosilicate gel precursors containing Na^+ or K^+ ions. In the case of K^+-containing gels one-step low-temperature dehydration was observed, while in the presence of Na^+ ions a second peak appeared in the DTG and DSC curves at a temperature close to the dehydration temperature of the final zeolitic product. TG/DTG and DSC results confirmed the role of the sodium ion as a "structure-forming" cation and revealed that the presence of sodium ions in the synthesis mixtures induced, contrary to potassium, the formation of structural subunits or even more complex structures.

3 Chemical Modification of Zeolites Traced by Thermal Analysis

In only a few papers thermoanalytical methods were used to study the chemical modification of zeolites by gaseous reactants. The stoichiometry of the reaction between mordenite and phosgene was established by thermogravimetry [40].

Thermogravimetry combined with titration of the evolved HCl was applied to measure the temperature range of the solid-state ion exchange between H-zeolites and metal-chlorides and to determine the ion-exchange degree [41]. The simultaneous application of the two techniques made it possible to distinguish between changes in weight caused by desorption of physisorbed water and by solid-state reactions generating volatile product.

Besides the results of other techniques, TG/DTG data obtained during temperature-programmed reduction (TPR) of mixtures consisting of H-zeolite and Ga_2O_3 or In_2O_3 provided additional evidence for the replacement of zeolitic protons by Ga^+ or In^+ ions in the so-called reductive solid-state ion exchange process [42–45]. The thermal analysis of a mixture of Ga_2O_3 and H-ZSM-5 carried out after pretreatment in vacuum in a hydrogen flow revealed a relatively fast process accompanied by a weight loss at about 850 K [42]. In contrast, no change in the weight of the mixture was observed in an inert gas atmosphere. In H_2, the reduction of Ga_2O_3 resulted in water, the evolution of which was monitored by TG. Results of XRD, XPS, and IR spectroscopy suggested that in the reductive thermal treatment gallium was transferred into the zeolite. Later, In_2O_3 instead of Ga_2O_3 was mixed with the H-form of zeolites. These mixtures and simultaneously pure In_2O_3 were subjected to temperature-programmed reduction in a H_2 atmosphere. The temperatures at which weight losses and DTG peaks were observed due to the evolution of water as a reaction product differed significantly for In_2O_3 in the pure form and for In_2O_3/H-zeolite (e.g., H-Y, H-ZSM-5) mixtures [43]. The same phenomenon was reported in [44]. DTG curves are demonstrated in Fig. 5. The high-temperature DTG peak at about 730 K was attributed to the

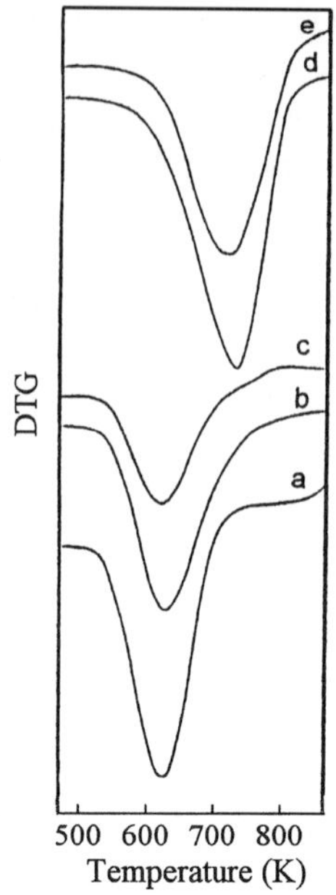

Fig. 5 DTG curves registered during TPR of $NH_4Na\text{-}Y/In_2O_3$ mixtures (NH_4/In corresponds to (a) 1, (b) 2 and (c) 3); (d) denotes In_2O_3 and (e) a NaY/In_2O_3 mixture ($Na/In = 1$) (reproduced from [44])

reduction of In_2O_3 to metallic indium. In contrast, the reduction was found to proceed in mixtures with H-zeolites at lower temperature (between 530 and 700 K) via In_2O according to the overall transformation

$$In_2O_3 + 2H^+Z^- + 2H_2 \longrightarrow 2In^+Z^- + 3H_2O\,. \tag{1}$$

The replacement of protons by In^+ cations was evidenced by IR spectroscopy [44, 45]. It was shown that In^+ cations once incorporated by RSSIE (involving hydrogen or the template as reducing agents) can be oxidized to InO^+ and again reversibly reduced to In^+, both species occupying cation sites during the redox process.

4 Thermal Removal and Decomposition of Templates Occluded in Zeolites

Thermal analysis was widely applied for characterization of zeolites. Undoubtedly, however, the main objects of thermal studies were zeolites containing organic compounds. Organic molecules may be chemisorbed on the Brönsted and Lewis sites of zeolites, physisorbed in their pore system, or entrapped in their channels and cavities during the synthesis.

Ohgushi et al. [46] investigated the ion exchange of Na^+ for tetramethylammonium and benzyltrimethylammonium ions in ZSM-5. Thermal analysis provided information on the amount and the state (counter cations or ion pairs) of organic ammonium compounds.

Many organic agents promote the synthesis of high-siliceous zeolites. These organic molecules may play a structure-directing role during the nucleation and crystal growth or they can be considered as void fillers which, accommodated in the micropore systems, stabilize the zeolite framework. Thermal analysis is an indispensable tool to characterize the gel precursors and the products obtained during the hydrothermal synthesis.

Thermal analysis of tetramethylammonium (TMA) and tetraethylammonium (TEA) aluminosilicate gels and of the zeolite precursors obtained from them by crystallization was carried out to obtain information on the environment of the organic components [47]. From TMA systems ZSM-5 or ZSM-39 were formed depending on the TMA content. TG and DTG curves of the products demonstrated that the TMA occluded in pure ZSM-5 decomposed in one step at about 813 K, while the decomposition of organic compounds entrapped in ZSM-39 zeolite showed three steps between 743 and 883 K. The different decomposition features made it possible to calculate from TG and DTG curves the relative amount of ZSM-5 and ZSM-39 in products containing both zeolites. From the TEA system the only product was ZSM-5. Thermogravimetry combined with differential thermal analysis supported the presence of TEA^+ ions in two different environments in the ZSM-5 framework. The low-temperature step ($\sim$ 663 K) in the TG curve could be attributed to the release of TEA^+ ions compensating the charge of defect sites (SiO^-), whereas the high-temperature step ($\sim$ 763 K) was associated with the decomposition of TEA^+ ions neutralizing the negative framework charges ($Si-O-Al^-$).

Zhao et al. [48] prepared large single crystals of ZSM-5 with *n*-tripropylamine (TRIPA) and TEA in the presence of fluoride ions. TG analysis showed that the numbers of TRIPA and TEA per u.c. are nearly identical (about 4/u.c.) and independent of the aluminum content.

Howden [49] used a thermobalance coupled to a titrator for examination of the thermal behavior of ZSM-5 synthesized with a small amount of tetrapropylammonium cations (0.4 TPA/u.c.) and with different α,ω-diaminoalkanes with alkane chains constituted of 3–8 carbon atoms. It was observed that diaminopropane and diaminobutane could be removed at relatively low temperatures (at around 520 K). The pentane-1,5-diamine probably cracked before leaving the zeolite, i.e. evolution of the nitrogen-bearing (alkaline) fragment of the organic compound took place at a lower temperature than the removal of the bulk of the template molecule. When hexane-1,6-diamine (HEXDA) was used as a template, the zeolite also released the alkaline nitrogen fraction before the hydrocarbon fraction of the organic molecule, but this process took place at significantly higher temperatures (above 673 K). The results indicated that diaminoalkanes with short hydrocarbon chains are not strongly held in the ZSM-5 structure. HEXDA could be regarded as a templating agent. The longer octane-1,8-diamine favored the formation of the straight channels of ZSM-11. The channels were found to be not completely filled. Carbon-to-nitrogen molar ratios determined from TA data indicated that the organic template molecules had undergone partial deamination during the synthesis.

Applying thermoanalytical techniques, Franklin et al. [50] found that silicalite-1 synthesized in the presence of HEXDA contained about 8 HEXDA/u.c. entrapped in the channel crossings of the silicalite-1 structure. This amount seemed to be sufficient to fill almost all of the void space. A comparative TG study of the thermal behavior of HEXDA- and TPA-silicalite-1 showed that the thermal decomposition of HEXDA occurred in three steps over a wide temperature range. In contrast, the weight loss due to removal of tetrapropylamine from TPA-silicalite took place in a much narrower temperature interval.

A much smaller amount of hexane-1,6-diol was incorporated in zeolite ZSM-5. Hexane-1,6-diol is believed to act as a hydrophobic void filler [51].

There are debates in the literature about the structure-directing activity of alcohols and nitrogen- or oxygen-containing heterocycles though an effect surpassing that of simple pore filling was attributed to *p*-dioxane applied in the synthesis of omega zeolite [52, 53].

The amount and decomposition of ethyleneglycol, ethanolamine, and ethylenediamine molecules enclathrated in silica sodalite were investigated by TG coupled to MS [24]. All these organic compounds, notwithstanding their different chemical properties, acted as structure-directing agents for silica sodalite synthesis. This study provided additional support to the thesis of Gies et al. [54] that chemical properties of structure-directing molecules play only a minor role in the crystallization of silica frameworks compared with their size and shape.

Though a wide variety of organic molecules have been successfully applied for synthesis of ZSM-5, the tetrapropylammonium ion proved to be the most powerful structure-directing species. Thermal methods combined with other techniques provided unambiguous evidence for the incorporation of the different organic molecules into the structure of ZSM-5 during the synthesis and gave useful information on the thermal reactivity of the guest and host compounds. Nevertheless, most of the relevant thermoanalytical studies dealt with the investigation of TPA-ZSM-5. TPA fits the ZSM-5 channel structure especially well. The "as-synthesized" sample contains in nearly every channel intersection one TPA species. Applying thermal analysis, Crea et al. [55] found that in silicalite-l prepared with different amounts of template the number of enclosed TPA species per u.c approached in most of the samples, independently of the initial TPABr concentration in the gel, the theoretical value of 4 TPA/u.c expected for full occupation of all channel crossings. The authors observed three characteristic endothermal peaks during programmed heating of such a sample in an inert gas flow. On the basis of the data obtained by thermal analysis and ^{13}C-NMR spectroscopy, they assigned in another paper [56] the three DTA peaks to different TPA species. The DTA peak at about 653 K was attributed to the release of "strained" TPA ions located in the outer shell of the crystals. The peak at about 698 K was ascribed to the elimination of "inner-strained" TPA ions occluded in the zeolitic channels. In both cases TPA ions may exist as counter-ions of SiO^- defect groups. Finally, the peak at $\sim$ 748 K was attributed to more relaxed TPA ions compensating the charge of SiO^- defect groups or, in aluminum-containing samples, i.e. in ZSM-5 zeolites, the framework charges $(Si - O - Al^-)$ associated with tetrahedrally coordinated aluminum atoms. It was assumed that the TPA^+ ions retained in the solid after heating at about 673 K were in a "relaxed" status contrary to the initial strained form.

Thermal analytical techniques were also applied for the characterization of MFI-type zeolites and borosilicates prepared from TPA-OH and TPA-fluoride containing gels with different Si/Al and Si/B ratios [57]. Besides the quantitative overall estimation of TPA^+, useful information could be obtained from the position of the endothermic peaks recorded in the inert gas atmosphere by DTA and DSC with respect to the chemical environment of different types of TPA^+ ions. Depending on the Si/M^{3+} ratio of the zeolite, the template might have been present as an ion pair (TPA^+OH^- and TPA^+F^-) and/or as TPA^+ counter-ion compensating the negative charge of the framework generated by trivalent elements. Interactions between $\equiv Si - O^-$ groups and TPA^+ might have also occurred. The ion pairs more loosely bonded to the framework were suggested to decompose at lower temperature than the counter cations. Occluded tetrapropylammonium fluoride proved to be more stable and to decompose at temperatures higher by 45 K than those observed with hydroxide. In zeolites with high Si/Al ratios (> 1000) the observed thermal effects shown in Figs. 6 and 7 (endothermic peaks between 653 and 803 K) arose from the decomposition of ion paired TPA^+ and, if lattice defects were present, of SiO^-TPA^+ species. A single peak at about 753 K in the DTA curve is typical of aluminum-rich (Si/Al $\sim$ 11) samples prepared in alkaline- or fluoride-containing media. Thus, this single peak was attributed to the decomposition of TPA^+ compensating framework charges. Samples with intermediate Si/Al ratios exhibited a continuous transition of the decomposition peaks, corresponding to the concentration of the different types of TPA^+ species.

In the sequel several papers were published on the decomposition of organic templates (tetra-, tri-, di- and mono-*n*-propylammonium cations) incorporated in MFI-type zeolites prepared in the presence of either OH^- or F^- anions [58–61]. In these papers, thermal analytical methods combined with other techniques (mass spectrometry (MS), gas chromatography (GS), ^1H–^{13}C CP MAS NMR, ^1H–^{15}N CP MAS NMR and IR spectroscopy) pro-

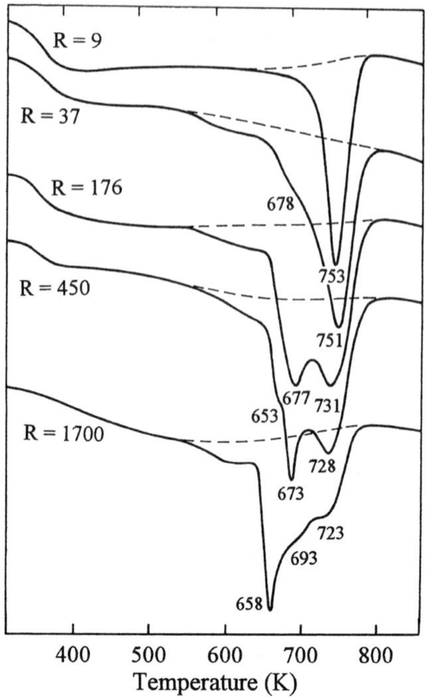

Fig. 6 DTA curves of as-synthesized MFI type zeolites prepared in alkaline media (in the presence of TPA-OH). R = Si/Al ratio; heating rate 10 K min^{-1}; argon flow (reproduced from [57])

vided more information about the mechanism of the thermal decomposition of templates by identification of the volatile products. Sometimes earlier expectations were found to be inconsistent with newer results. Here, one of these papers will be reviewed in detail.

In [61] the solids, remaining after partial thermal decomposition of Pr_4NF-MFI, Pr_3NHF-MFI, and Pr_4NOH –MFI precursors, were examined by ^1H–^{13}C CP MAS NMR and IR spectroscopy (in these cases the zeolites were heated separately in an inert gas atmosphere in the thermoanalyzer to different temperatures corresponding to characteristic points of the DTA and DSC curves). As has been shown already, the decomposition of the organic cations in a highly siliceous Pr_4NF-MFI precursor took place in two temperature ranges (see Fig. 7). After a small endothermic shoulder, both the DTA and DSC curves exhibited a strong peak with a minimum at about 693 K followed by a second broader one, displaying a shoulder at 768 K and a minimum at 793 K. It was proved by a ^{13}C NMR study of the precursor heated previously at 733 K that after the first decomposition step mainly Pr_3NH^+ remained in the solid. That is in contradiction with an earlier suggestion [57] according to which TPA ions decomposed at about 700 K into tri- and dipropylamine and propene. Since, in the case of large-size zeolite crystals, the observed weight

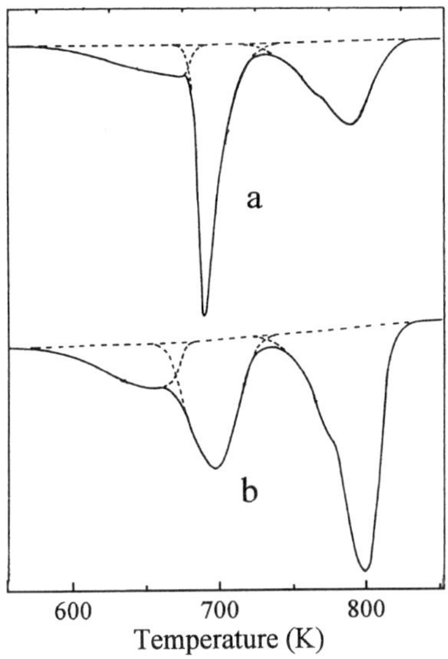

Fig. 7 DSC curves of (a) non-powdered (large crystals) and (b) finely ground siliceous (TPA)-MFI precursor prepared in non-alkaline fluoride-containing medium; argon flow, heating rate 6 K min^{-1} (reproduced from [61])

loss and DTA and DSC peak areas (Fig. 7, curve a) were greater than those derived from the stoichiometry of the reaction

$$Pr_4N^+ \longrightarrow Pr_3NH^+ + C_3H_6\,, \qquad (2)$$

it was suggested that the formation of propylene resulted in strong tensions causing fissures inside large crystals from which also part of the template escaped in the form of tripropylamine. In contrast, weight losses and heat effects due to the first degradation step were found to correspond to the stoichiometry of Eq. 2 when a finely ground sample was subjected to the same heat treatment (Fig. 7, curve b). The authors, therefore, assumed that evolution of propene did not cause significant cracks in small crystals. The second peak with a shoulder was ascribed to the overall degradation of tripropylammonium ions. The decomposition of the Pr_3NHF-MFI precursor resulted in DTA and DSC peaks similar in shape and temperature range to the second peak of ground Pr_4NF-MFI. Among the volatile products, propene was detected in significant amounts by GC and MS which pointed to an initial decomposition mechanism involving Hofmann-type elimination followed by β-elimination. The ^{13}C NMR spectra of the high-siliceous Pr_4NOH-MFI precursor proved to be different. In this material the template started to decompose at lower temperatures than Pr_4NF-MFI but part of the Pr_4N^+ remained in the intracrystalline space (in a relaxed state) until completion of the decomposition. In aluminum and gallosilicate precursors, the Pr_4N^+ ions were predominantly counter cations of the negative lattice charges. Thus, the Pr_4N^+ was bonded to the framework by stronger interactions, and its degradation started at higher temperatures and was characterized by a single peak. Also in these cases, a four-stage Hofmann-type decomposition was suggested, with the first stage as the rate-determining step.

Tavolaro et al. [62] reported on the application of thermal analysis to characterize both the initial gel and the crystalline silicalite-1 obtained in TPA^+ and fluoride-containing media. The effect of grinding of the large crystals was shown by TG and DTA. The total weight loss ($\sim$ 4 TPA/u.c.) was not affected by grinding, however, changes in the DTA curve pointed to alterations in the elimination of the template. According to the interpretation of the authors, the first weight loss step involved, at least partly, also the elimination of the second propylene molecule. In the ground sample the third elimination step was shown to occur easily leading to better-resolved DTA and DTG peaks in the temperature range 733–833 K.

Intense mechanical force (ball milling) exerted on as-synthesized ZSM-5 was found to cause amorphization of the zeolite [63]. The ball-milled sample gradually lost weight upon heating.

Yi and Ihm [64] synthesized high-silica ZSM-5 (Si/Al $\sim$ 175) under atmospheric pressure (at 363 K) in the presence of TPABr in an alkaline medium. Thermogravimetric analysis and differential thermal analysis were carried out (in air) to determine the binding state of TPA^+ ions and to monitor the crystallization in this way. As the crystallinity increased, the exothermic peak (at $\sim$ 513 K) due to "free" TPA^+ in the intermediate (amorphous) phase disappeared and the intensity of peaks attributed to TPA ions trapped in the channels of ZSM-5 increased.

Quaternary organic ammonium cations other than TPA (TMA^+, TEA^+, etc.) were used for the synthesis of offretite-erionite type zeolites [65], zeolite omega [16], EU-12 [66], Phi [67], ferrierite [68] and ZSM-20 [17]. In each case, the thermal decomposition of the template was investigated. TG and DTA data confirmed the presence of TMA cations in the gmelinite cages of MAZ and in the sodalite cages of SOD and were used to calculate the amount of these cation species and, in this way, the amount of the zeolitic components in the prepared products [69]. In a more detailed work [70], the incorporation of TMA cations and *n*-hexane derivatives into the pore system of mazzite-type zeolite was studied by TG and DTA in an oxidative atmosphere. The decomposition of TMA cations occluded in gmelinite cages was found to occur in the temperature range from 770 K to 830 K depending on the nature of the mazzite samples. The product prepared from a sodium aluminosilicate hydrogel (Si/Al = 5) containing TMA^+ as an organic template exhibited a sharp weight loss at 810 K assigned to the oxidative decomposition of TMA cations incorporated in the gmelinite cages (Fig. 8A). However, an additional exothermic peak attributed to the decomposition of 1,6-diaminohexane molecules occluded in the 12-MR channels appeared at about 610 K in the DTA curve of a pure mazzite sample crystallized in the presence of TMA, 1,6-diaminohexane, and 18-crown-6 ether (Fig. 8B). Neither a weight loss step nor a DTA peak indicative of the decomposition of 18-crown-6 ether was observed. At lower aluminum content of the synthesis gel, SOD was formed as a co-crystallized phase. The third high-temperature weight-loss step observed at about 860 K was ascribed to the decomposition of TMA cations located in sodalite cages of the impurity phase (Fig. 8C). The TG and DTA curves of the zeolite synthesized in the presence of TMA and 1,6-hexanediol reflected only a single exothermal process due to the decomposition of TMA, since the DTA signal appeared at the same temperature as in the case of mazzite prepared in the presence of only TMA. From the absence of

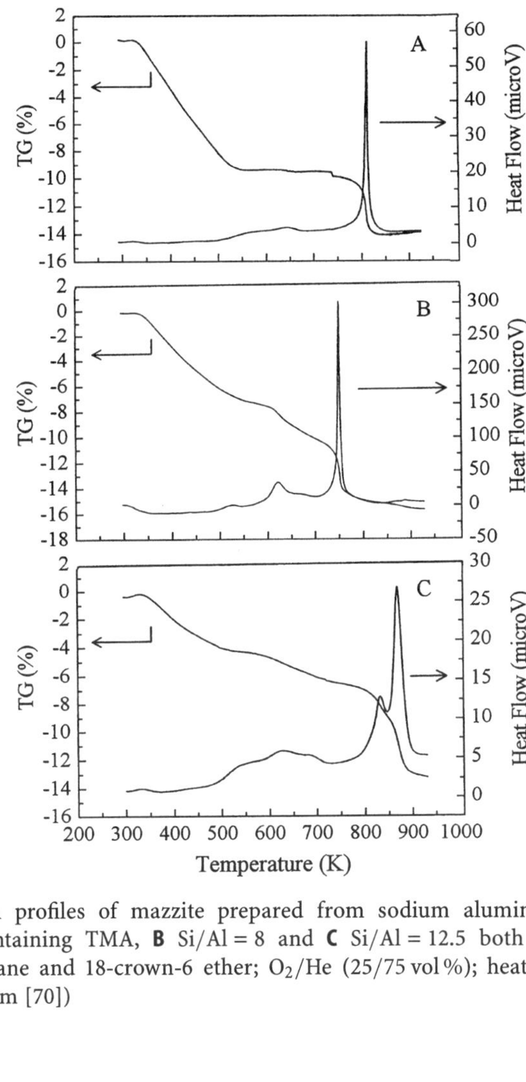

Fig. 8 TG/DTA profiles of mazzite prepared from sodium aluminosilicate hydrogel **A** Si/Al = 5 containing TMA, **B** Si/Al = 8 and **C** Si/Al = 12.5 both containing TMA, 1,6-diaminohexane and 18-crown-6 ether; O_2/He (25/75 vol%); heating rate 5 K min^{-1} (reproduced from [70])

any additional exothermic DTA peak and TG step the conclusion was drawn that 1,6-hexanediol was not occluded in the pore system. It was evidenced by ^{13}C CP-MAS NMR spectroscopy that (i) TMA^+ was present in the gmelinite cages of all mazzite samples mentioned above; (ii) 1,6-diaminohexane in non-protonated form was occluded in the 12-MR channels of the respective sample; and (iii) 1,6-hexanediol was not incorporated. TG data were used to quantify the contents of the respective compounds occluded in the pore system.

Thangaraj et al. [71] demonstrated that ^{13}C CP-MAS NMR spectroscopy jointly used with thermal analysis may give valuable information on the location and the possible configuration of dibenzyldimethylammonium ions in zeolite EU-1. In cases where the organic compound acted as a template, a geometrical fit between the configuration of the organic molecule and the topology of the zeolite was anticipated.

Five quaternary ammonium salts were tested in the presence of piperazine as templates for the synthesis of silica molecular sieves [72]. The amounts of the quaternary ammonium cations in the products were calculated from the weight losses between 573–873 K, and thermal analysis was also used to determine the equilibrium water vapor uptakes of the calcined samples.

Different thermal decomposition features were found for silicalite-2 crystals with needle-like and ovate-shaped morphology, prepared in the presence of TBA [73].

Thermogravimetric analysis in combination with N_2 adsorption measurements gave information about the state of trimethyl-cetyl-ammonium hydroxide encapsulated in the ZSM-35 structure [74].

A comparative study [75] of the structural and thermal properties of TMA-, calcined Na, H- and H-ferrierite using high-temperature X-ray diffraction and TG measurements showed a correlation between unit cell dimensions and the degree of both the dehydration of the zeolite and the decomposition of the template. The decomposition of TMA resulted in a decrease of all orthorhombic unit cell parameters. In the case of ferrierite, changes (contraction) in the unit cell parameters thermally induced above 850 K were found to be irreversible.

Schreyeck et al. [76] prepared a new highly siliceous layered precursor of FER-type zeolite in a fluoride medium in the presence of 4-amino-2,2,6,6,-tetramethylpiperidine as a template. The dehydration of the precursor and the removal of organic cations compensating $\equiv Si-O^-$ groups in the framework were monitored by TGA and DTA.

Ferrierites with different Si/Al ratios were prepared by recrystallization of aluminum-containing kanemites intercalated with piperidine as a template [77]. TG was applied to calculate the total amount of the occluded piperidine and piperidinium ions compensating the negative lattice charges in the products. It was found that the total of piperidine molecules and piperidinium ions in the samples amounted to about 4/u.c., independently of the Si/Al ratio.

Zeolite Beta is usually crystallized in alkaline media in the presence of Na^+ and TEA^+ cations as structure-directing agents. Weight losses of the products registered in air at temperatures between 493 K and 623 K have been attributed to the removal of TEA-OH, and those observed above 623 K, to the decomposition of the TEA^+ cations. Perez-Pariente et al. [78] found a linear correlation between the content of TEA^+ (weight loss above 623 K) and the degree of crystallinity of samples prepared with various techniques. The TG weight loss above 493 K, representing the sum of organics occluded in the samples, proved to be a function of the molar SiO_2/Al_2O_3 ratio of the zeolite. The pores of crystals with high SiO_2/Al_2O_3 ratios (66 and 86) were found to be totally filled with organics (90% of which was TEA-OH). In zeolites with higher aluminum content the amount of TEA^+ increased, however, the negative lattice charge could be shown to be never exclusively compensated by TEA^+ but partly also neutralized by sodium ions.

Perez-Pariente et al. [79] found that in air the decomposition of TEA in the Beta zeolite occurred in three exothermic steps associated with weight losses, while Hedge et al. [80] observed in the DTA curve of an as-synthesized Beta zeolite four distinct exothermic peaks. In [79] the authors also reported on TA results obtained under nitrogen. The decomposition of the template took place in two endothermic processes corresponding to the pyrolysis of TEAOH and TEA^+ cations.

Camblor et al. [81] reported that in the TG curve of as-synthesized [Al]-BEA zeolite four weight loss steps (I–IV) could be clearly distinguished. The endothermic process I was ascribed to the desorption of water and process II proceeding between 423–573 K to the degradation and combustion of TEA^+ cations associated with SiO^- groups in defect sites. The exothermic high-temperature processes III and IV were attributed to the decomposition and combustion of TEA^+ cations balancing the charge of framework $Al(OSi)_4^-$ entities and deposited coke, respectively.

Vaudry et al. [82] studied the thermal behavior of as-synthesized Beta zeolites with different Si/Al ratios, the DTG curves of which are shown in Fig. 9. The contents of water, TEA-OH, and charge compensating TEA^+ cations were calculated from TG results. Weight loss data and DTG curves were interpreted in the following way:

The amount of occluded tetraethylammonium hydroxide (revealed in Fig. 9 by peak II of sample (a) containing 4.6 Al/u.c.) decreased with increasing aluminum content and approached zero in the case of aluminum-rich zeolites (sample (c) in Fig. 9, containing 8.3 Al/u.c.). The complete filling of the channel space with TEA^+ corresponded to six molecules per u.c. In the case of sample (b) containing 5.8 Al/u.c., the amount of TEA^+ counterions approximated this value. When the aluminum content of the framework exceeded 6 Al/u.c., hydrated sodium ions occupied a part of the cation sites to preserve charge balance. Consequently, the water content (peak I) of the zeolite increased and, because of the space requirement of the hydrated

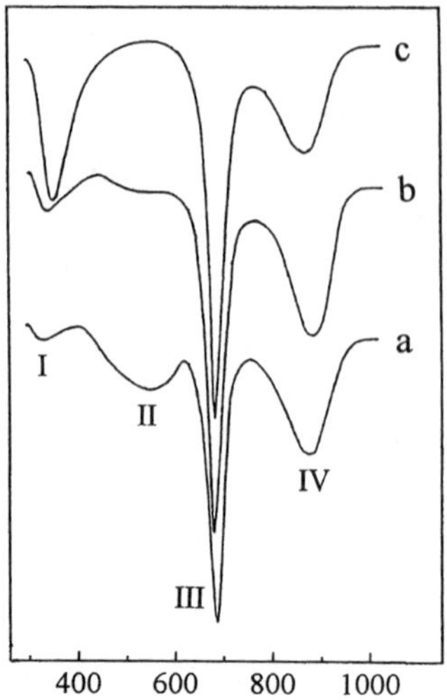

Fig. 9 DTG curves of as-synthesized (TEA)-Beta zeolites containing **a** 4.6, **b** 5.8 and **c** 8.3 Al/u.c.; I, II, III, IV: see text (reproduced from [82])

sodium ions, the amount of the TEA^+ ions decreased (peak III, IV). Above 723 K the aluminum-rich sample (c) retained a decreased amount of aminic residues (IV), suggesting a lower strength of the acid sites liberated during the calcination.

Recently, Beta zeolite containing 10.3 Al atoms per tetragonal unit cell has been synthesized [83]. TG data showed that Beta zeolites with Si/Al < 7.5 (Al/u.c. > 7.5) contained only one type of TEA species but two types were found in samples with Si/Al > 7.5.

Usually the interpretation of the various thermal events in as-synthesized Beta zeolite was similar to that proposed by Parker for TEA-MFI [84]. It was based on the hypothesis that tetraalkylammonium ions associated with acid sites of a zeolite are more stable than the other species. Bourgeat-Lami et al. [85] disagreed with such an interpretation of the TG/DTA decomposition patterns of TEA-Beta. They proposed a decomposition mechanism based on the identification of the volatile products by TG/DTA coupled to mass spectrometry and their quantification by titration and on the characterization of the zeolitic product obtained at different decomposition temperatures (^{13}C NMR and IR spectroscopy). According to these authors, TEA ions decomposed in the temperature range 473–623 K, regardless of their nature (occluded ion pairs or counter-ions), in a single step to ethylene and triethylamine. Part of the latter might have remained bound to the acidic sites of the zeolite and decomposed at higher temperatures to lighter amines by sequential Hofmann elimination reactions. A fraction of ethylene might have reacted on the acid sites yielding aliphatic and aromatic hydrocarbons. The desorption of the so-called hydrocarbon residues required a temperature higher than 773 K.

Kanazirev and Price [86] found that oxygen essentially influences the formation of residues during the thermal decomposition of TEA in zeolite Beta precursors. In a comparative study of this process monitored by thermoanalytical techniques, the experiments were carried out in both He and He/O_2 atmospheres. Up to about 650 K, thermal effects were found to be not influenced by the gas atmosphere but at higher temperatures, significant differences between thermoanalytical curves measured in inert and oxidative atmosphere were observed (see Fig. 10). In Fig. 10 the weight losses are related to the sample weight at 403 K at which desorption of weakly bound water was complete. Stable starting conditions were much better met by this reference point than by the initial weight, since the materials lost adsorbed water very rapidly when placed in a dry gas flow. The TG curves measured in inert and oxidative atmosphere began to diverge near 730 K and the organic residue retained in the oxidative atmosphere at this temperature was relatively stable up to about 800 K. At about 730 K, O_2 was obviously involved in processes which led to the formation of organic residues (the mechanism is not yet clear). The amount of the residue was found to depend on the concentration of oxygen in the atmosphere surrounding the sample and probably on the Si/Al ratio. In the final stage of the thermal treatment the organic residues burned off. Also, the thermal decomposition of 1-propylamine (1-PA) adsorbed on calcined BEA exhibited similar features. In contrast, in the absence of O_2 the formation of residue was strongly suppressed. In He flow the weight losses distinguished in the TG curves of as-synthesized (TEA)-Beta and 1-PA on Beta zeolite were assigned to

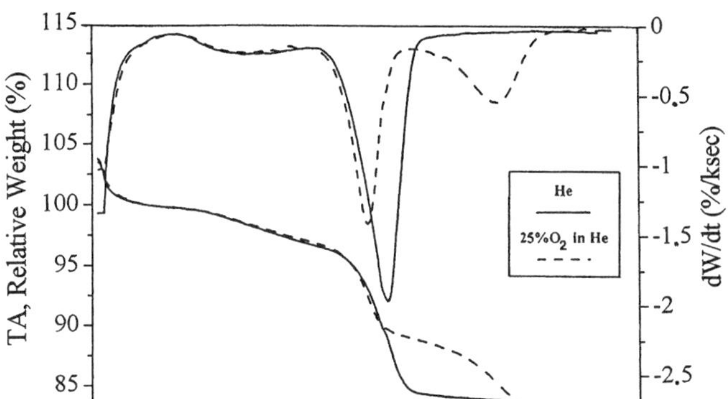

Fig. 10 Thermogravimetric analysis of zeolite Beta synthesized in the presence of TEA-OH from a gel with SiO_2/Al_2O_3 ratio = 20. Heating rate 10 K min^{-1} in a flow of pure He (—) and 25% O_2 in He (- - -) (reproduced from [86])

- desorption/decomposition of weakly bound TEA or desorption of amine which is either occluded in channels or bound as a ligand in second level solvation shells;
- decomposition of TEA cations or alkylammonium ions;
- decomposition of organic residues, a step which seems to be negligible with regard to the above-mentioned relevant observations of the authors.

It should be noted that the interpretations, given by Kanazirev and Price [86] and by Bourgeat-Lami et al. [85] to explain the decomposition feature of TEA-Beta, are contradictory in some regards.

5 Characterization of Isomorphously Substituted Zeolites

A great number of papers deal with synthesis and physico-chemical properties of isomorphously substituted zeolites. The synthesis of such materials generally requires the presence of organic compounds (templates) in the reaction mixture. Thus, it is evident that thermoanalytical techniques offer a variety of possibilities for the investigation of such materials. Kosslick et al. [87] presented excellent examples of how to get reliable information even on framework characteristics from thermoanalytical data. They carried out TG and DTA measurements of [Ga]-ZSM-5 prepared in an alkaline medium with TPA bromide and observed in the as-synthesized precursor material during heating in air four thermal events which were assigned to the following processes:

I. water desorption from zeolite pores at about 373 K;
II. decomposition (oxidation) of the embedded template between 633–723 K;
III. oxidation of coke resulting from the template above 773 K; and
IV. dehydroxylation.

Process II was found to proceed in [Ga]-ZSM-5 at somewhat higher temperatures than in the pure silica analogue of ZSM-5, and this difference in the maximum decomposition temperature proved to be correlated, at least up to 4 Ga/u.c., with the gallium content of the sample. This temperature shift illustrated by DTA data in Fig. 11 was regarded as an indication of incorporation of trivalent gallium into tetrahedral framework sites, since such an isomorphous substitution of silicon by gallium must result in the creation of negative lattice charges and, hence, in stronger ionic interactions between the template and the zeolitic framework. Obviously, the formation of coke originating from the catalytic cracking of template decomposition products, e.g., propene, on strongly acidic Brönsted sites also evidences the incorporation of gallium into framework positions. Furthermore, it was shown (as illustrated in Fig. 12) that onset and course of the crystallization process of gallium-varieties of ZSM-5 could be easily detected and monitored by the intensity of

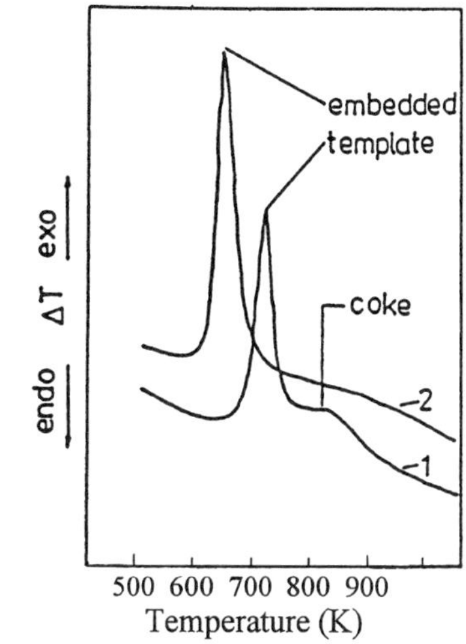

Fig. 11 DTA curves of as-synthesized (1) gallosilicate analogue of ZSM-5 (Si/Ga = 20) and (2) silicalite-1 in air. Heating rate: 5 K min^{-1} (reproduced from [87])

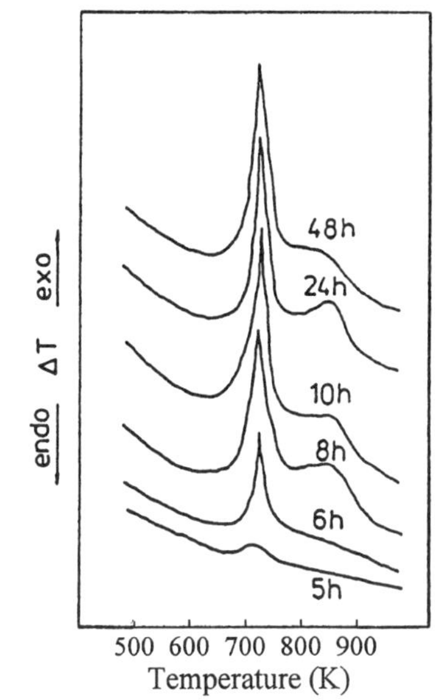

Fig. 12 Differential thermal analysis of TPA-[Ga]-ZSM-5 after different time of crystallization. Si/Ga = 25; heating rate: 5 K min^{-1}; atmosphere: air (reproduced from [87])

the DTA peak at about 743 K associated with process II. For samples isolated from alkaline crystallization mixtures at increasing crystallization times, this intensity reflected the amount of embedded template which is a function of the zeolite content (degree of crystallinity).

Gabelica et al. [6] found also for ZSM-5 samples a good correlation between the weight loss due to decomposition of intracrystalline TPA and XRD crystallinity.

The amount of the TPA^+ ions, trapped in the crystals during the synthesis of MFI-type zeolites prepared in fluorine-containing media with up to 3 Ga/u.c. in the framework, was measured by TG [88]. DTA results indicated that under N_2 in the gallium-rich sample the decomposition of TPA cations occurred in one step at about 753 K. In contrast, two-step decomposition (at 693 and 753 K) was found in the case of samples containing less Ga. Similar to the assignment of DTA peaks of Al-MFI [57], the high-temperature peak was attributed to TPA^+ $(Si-O-Ga)^-$ ion pairs and the low-temperature peak to TPA^+ ions linked to SiO^- defect sites or TPA^+ F^- (or OH^-) ion pairs. However, the authors pointed out that decomposition of TPAF might contribute to the high-temperature step, since plots of TPA involved in the high-temperature decomposition step vs. the Ga content deviated at small Ga values somewhat from the straight line towards higher amounts of TPA. A dependence of the temperature of DTA peaks on the gallium content was not found.

Studying the thermal behavior of ferrisilicate analogues of ZSM-5, Borade [89] found a linear relationship between the X-ray crystallinity on the one hand and the weight losses due to desorption of water (dehydration) and decomposition of TPA ions on the other hand. With increasing crystallinity of the sample the weight loss due to dehydration decreased, while that due to decomposition of organic cations increased.

The amount and the decomposition of hexamethyleneimine as the template in [Fe,Al]-MCM-22 were investigated by TA [90].

Thermoanalytical curves (DTA/TG) of pure Ga-substituted Nu-23 zeolite prepared in the presence of cetyltrimethylammonium bromide indicated that the template was occluded in the pores of [Ga]-Nu-23 and decomposed around 973 K [91].

Crystalline gallosilicate ($Si/Ga \approx 10$) [92] and iron-silicate ($Si/Fe = 9$) [93] analogues of mordenite have been synthesized using tetraethylammonium bromide as a template. Substitution of gallium or iron for aluminum resulted in a shift of the exothermic weight changes, associated with the loss of occluded TEA-bromide and TEA^+ lattice cations in air, to lower temperatures. In line with these observations, the exothermic DTA peak was found [94] to be slightly shifted towards lower temperatures upon substitution of iron for aluminum in the ZSM-11 framework.

In DTA curves of [Ga]-ZSM-22 [95] and [Fe]-ZSM-22 [96], both prepared in the presence of 1-ethyl-pyridinium bromide as the template, the peaks associated with the oxidative decomposition of the occluded organic material and of organic ions strongly interacting with sites created by incorporation of Ga^{3+} and Fe^{3+} ions were shifted to lower temperatures compared to their position in the thermogram of the respective Al analogue, i.e. organic species were held less strongly in the isomorphously substituted zeolite varieties. Accordingly, IR spectroscopy and TPD of NH_3 showed that Brönsted acid sites present in the ferrisilicate analogue of ZSM-22 are weaker than those present in the proper zeolite [97].

Also, in Ga- [98] and Fe-varieties [99] of zeolite Beta the interactions between TEA^+ cations and the framework were found to be significantly weaker, i.e. thermal effects due to TEA^+ decomposition were observed at lower temperatures than in the case of the Al-containing zeolite. Other kinds of TEA species (TEA hydroxide occluded in the pore system or TEA^+ associated with defect sites) could not be distinguished by thermal analysis. The total amounts of organic material, found to be between 5.5 and 6.5 TEA per u.c., were determined by thermogravimetry. The decomposition of TEA^+ ions in [Ga]-Beta in air and N_2 were compared in [100].

Borade and Clearfield [101] described a modified synthesis route (crystallization in methanolic medium) for ferrisilicate with the framework topology of zeolite Beta and reported detailed results concerning the physical-chemical properties of samples with various Si/Fe ratios (7.5–30). Valuable information with respect to the incorporation of iron into the Beta framework could be obtained from TGA experiments. The assignment of weight losses to different TEA species was based on earlier findings [78, 80, 102]. The most important observations were that the weight loss between 323–633 K related to desorption of occluded TEAOH species decreased and that due to the decomposition of TEA^+ species (in the temperature range 633–823 K) increased with increasing Fe content in the solids. The DTG and TG curves obtained in the presence of nitrogen clearly showed that the pyrolysis of TEA^+ cations interacting with the zeolite framework occurred at lower temperatures in the case of [Fe]-Beta samples compared to [Al]-Beta. The shift (40 K) in the DTG peak maximum towards lower temperatures points to the presence of negatively charged FeO_4 tetrahedra in the zeolite framework and is indicative of weaker interaction of charge-compensating TEA^+ cations with the framework of [Fe]-Beta than with that of [Al]-Beta. The acidity of [Fe]-Beta and [Al]-Beta zeolites was investigated by TGA of NH_4^+-exchanged samples. In [Fe]-Beta, NH_4^+-ions are decomposed at lower temperatures than in [Al]-Beta zeolite which points, in accordance with the electronegativity concept, to a lower acid strength of the Brönsted sites in [Fe]-Beta.

Camblor et al. [103] concluded from the thermal behavior of zeolitic varieties of the Beta-type containing various amounts of Ti that this element occupied tetrahedral framework positions.

Tuel and Ben Taarit [104] compared the thermal behavior of a titanium-silicalite with MEL (silicalite-2)-structure synthesized in the presence of tetrabutylphosphonium (TBP) ions with that of silicalite-2 prepared with TBA-hydroxide as the template. The numbers of TBA^+ and TBP^+ ions per u.c. were estimated from thermoanalytical data. The TG and DTG curves clearly revealed that the elimination of TBP requires higher temperatures than that of TBA. The thermal decomposition of organic molecules occluded in titanium silicate with

MFI structure prepared with mixtures of TBP-hydroxide and tetraethylphosphonium (TEP) hydroxide was found to occur in only one step [105]. Though, as revealed by ^{13}C and ^{31}P NMR spectroscopy, both TBP^+ and TEP^+ cations were incorporated in the zeolite structure during crystallization.

6 Characterization of Zeolites by Thermal Analysis of Adsorbed Alkylamines and Alcohols

Parillo et al. [106] examined the adsorption features of various alkylamines in H-ZSM-5 (Si/Al = 35) using thermal analysis combined with TPD and IR spectroscopy. Simultaneous TPD (MS) and TG measurements (under vacuum) showed that ethylamine, 1- and 2-propylamine adsorbed on the zeolite in excess of the aluminum content (expressed in molarities) desorbed unreacted below 500 K (MS signals characteristic of alkylamines were detected), while at higher temperatures alkenes and ammonia were formed simultaneously as decomposition products. Their amounts corresponded to the decomposition of one alkylamine molecule per bridged hydroxyl associated with framework aluminum atoms. The absence of the IR band at about 3605 cm^{-1} provided evidence for the protonation of the alkylamines. The decomposition of the protonated species was found to occur, depending on the nature of the alkyl group, between 500–575 K for *t*-butylamine, 575–650 K for 2-propylamine, 650–700 K for 1-propylamine, and 650–725 K for ethylamine.

A detailed study of adsorption and desorption processes in the 2-propylamine/zeolite and 2-propanol/zeolite systems using combined TPD-TGA methods was published by Kofke et al. [107, 108]. It was established that 2-propanol and 2-propylamine formed well-defined 1 : 1 adsorption complexes with acidic framework sites not only in H-ZSM-5 [107] but also in H-ZSM-12 and H-mordenite [108]. Proton transfer was found to be the driving force for strong binding of the adsorbate.

Kofke et al. [109] examined the adsorption of 2-propanol and 2-propylamine on H-[Ga]-ZSM-5 using simultaneous TPD(MS)-TGA in high vacuum. 1 : 1 adsorption complexes of adsorbates with the hydroxyl sites associated with tetrahedrally coordinated framework Ga ions were identified. The TPD-TGA results concerning the thermal behavior of 2-propylamine in H-[Ga]-ZSM-5 are shown in Figs. 13A and B. The weakly adsorbed 2-propylamine desorbed unreacted in a well separated low-temperature step. The weight loss between 570–650 K corresponded to a coverage of one alkylamine molecule/Ga, and TPD provided evidence for the unimolecular acid-catalyzed decomposition of the alkylamine on Brönsted sites under formation of propene and ammonia. No second thermal effect was observed for 2-propylamine on ZSM-5 with a SiO_2/Al_2O_3 ratio of 880.

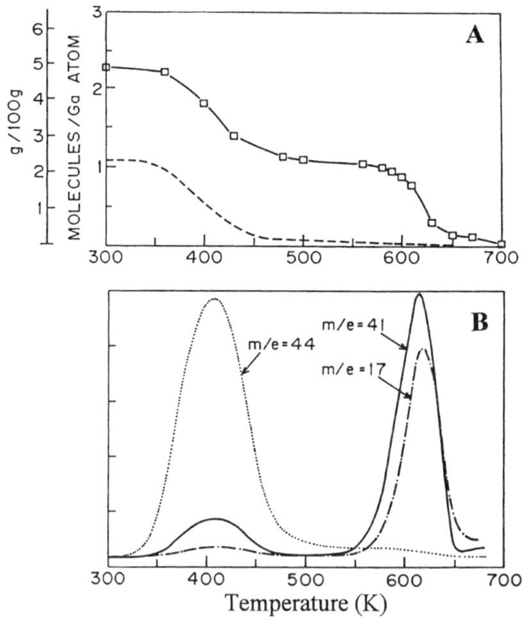

Fig. 13 TG (**A** *full line*) and TPD-MS (**B**) curves of 2-propylamine on H-[Ga]-ZSM-5. Adsorption of 2-PA at 295 K; heating rate: 10 K min^{-1} under vacuum. Signals m/e = 44, m/e = 41 and m/e = 17 correspond to the fragments $[CH_2-CH_2-NH_2]$, $[CH_3-CH=CH]$, and NH_3 indicative of 2-propylamine, propene, and ammonia, respectively. For comparison, the TG curve of 2-propylamine on high-siliceous ZSM-5 is also shown (**A**, *dashed line*) (reproduced from [109])

Kanazirev et al. [110] applied other experimental conditions for thermal analysis of adsorbed propylamines than Gorte et al. [106–109]. The experiments were carried out not in high vacuum but under atmospheric pressure in a He flow. Despite this alteration of the procedure similar TA features were observed for both 1- and 2-propylamine in H-MFI zeolite. The weight losses corresponding to the plateau of TG curves between 550–640 K were indicative of the formation of stoichiometric 1 : 1 complexes of propylamines with zeolitic protons. However, on comparison of the results with data published earlier in [107, 108] it became evident that additional to the high-temperature process not one but two thermal effects, reflected by weight loss steps and DTA peaks, occurred at low temperatures below 550 K. It was suggested that in high vacuum the very weakly bound propylamine was removed before the TA measurement was started. In accordance with previous data reported in [107] it was found that nearly two 1-propylamine (1-PA) molecules per framework Al were retained in H-ZSM-5 (Si/Al = 20) at 400 K. To explain this adsorption feature of 1-PA, Kanazirev et al. [110] assumed that a second 1-PA molecule was attached to each propylammonium ion formed in the zeolite, constituting a primary "solvation shell", the breakup of which gave rise to the second low-temperature DTA peak.

Thermal analysis (TG, DTG) was applied in a detailed study of alkylamine guests ordering in the H-form of zeolite hosts [111]. Thermoanalytical curves showing typical features of 1-PA adsorbed in H-MFI samples with different Si/Al ratios are presented in Fig. 14. The "plateau" region (575–625 K) in the TG curves is interpreted as to be associated with primarily adsorbed species (PAS), anchored directly to the active sites (protons) in the host. The two effects at lower temperatures are attributed to the elimination of

- nearest neighbor species (NNS) attached to PAS; and
- weakly bound species (WBS).

Quantification of the three adsorbed 1-PA species by TG showed that for the MFI materials with framework SiO_2/Al_2O_3 ratios ≥ 40 roughly equal numbers of 1-PA molecules were bound as PAS and NNS, i.e. each PAS interacted with another amine molecule to form a NNS. With increasing Al content in the zeolite, the amount of both PAS and NNS rose at the expense of that of WBS. When the SiO_2/Al_2O_3 ratio was < 40, steric constraints in the pore system evoked by high pore filling began to limit the number of NNS.

Biaglow et al. [112] used TGA and temperature-programmed desorption (TPD) to compare the adsorption and desorption behavior of 2-propylamine and 2-propanol on steamed and chemically dealuminated faujasites. The 2-propylamine adsorbed on Brönsted sites decomposed in the same temperature range as observed in the case of H-ZSM-5. The quantity of amine decomposed in the high-temperature region was regarded as a measure of

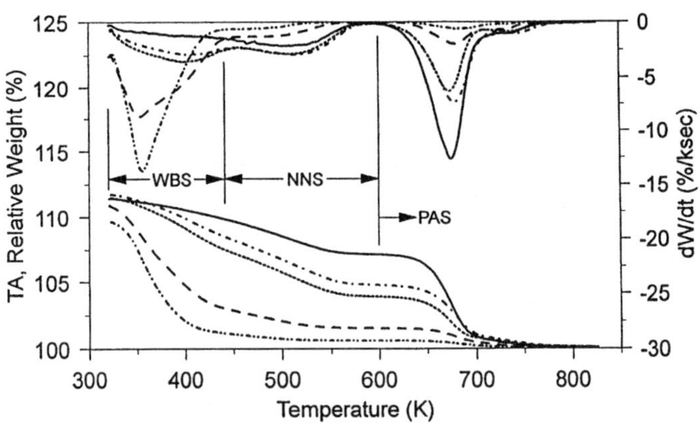

Fig. 14 TG and DTG curves of 1-propylamine (1-PA) adsorbed on H-ZSM-5 with SiO_2/Al_2O_3 ratios of 25.9 (—), 39.5 (- · -), 48.7 (·····), 127 (- - -), and 329 (-·····). The SiO_2/Al_2O_3 ratios were obtained from the weight loss due to the decomposition of 1-PA above 625 K. Heating rate: 5 K min^{-1} in He flow. PAS: primarily adsorbed species, NNS: nearest neighbor species, WBS: weakly bound species (reproduced from [111])

the Brönsted acid site concentration. In the temperature range between 450 and 600 K, unreacted 2-propylamine was only desorbed from faujasite but not from high-silica zeolites. It was supposed that this 2-propylamine interacted with the less strongly acidic hydroxyls giving rise to the low-frequency band at 3540 cm^{-1} in the IR spectrum [113].

Two different adsorption/desorption features of 1-propylamine on H-MFI (Si/Al = 20) zeolite were observed by combined DTA/TG/MS depending on the temperature (323 or 593 K) at which 1-PA was contacted with the zeolite [114]. At 593 K, conversion of 1-PA to dipropylamine (DPA) occurred which was retained in cationic form. Consequently, the major desorption feature shifted to a higher temperature compared to the run following the adsorption at 323 K. In addition, the sample weight in the plateau region (between 550 and 675 K) of the TG curve was found to correspond to 1.7 1-PA molecules/framework Al instead of one 1-PA/Al as generally obtained. The conversion of 1-PA to DPA was supported by the similarity of TA curves measured after contact of H-ZSM-5 with the mono-alkylamine at 593 K and with DPA at 323 K.

Recently, the interaction of mono-alkylamines (C_1–C_5) with ETS-10, a zeolite-like titanosilicate with Ti(IV) in an octahedral environment, was studied by TG combined with other techniques [115].

These examples demonstrate that TGA alone gives information about the temperature ranges in which different processes (reactions) take place in zeolitic samples, but additional "complementary" methods (MS, GC, IR, NMR) are in general indispensable for the assignment of weight losses to particular desorbed molecules or decomposed organic constituents. Thus, TA offers possibilities for the quantification of organic compounds weakly adsorbed in zeolites or associated with acidic framework sites and, hence, of the Brönsted acid sites themselves.

7
Thermal Analysis of Alkylamines Adsorbed on Metal Ions Located in Cationic Positions of the Zeolite Framework

TPD/TGA measurements of 2-propylamine adsorbed on copper-exchanged ZSM-5 (Si/Al = 35) and silicalite samples were performed by Parrillo et al. [116]. (The H-form of the zeolite and the silicalite-1 were treated with cupric acetate solutions.) The results were compared with those obtained on H-ZSM-5. 2-propylamine adsorbed on H-ZSM-5 showed a two-step desorption feature. Below 500 K, only unreacted 2-propylamine desorbed from the sample, between 575 and 650 K the protonated alkylamine decomposed to propene and ammonia. In the TPD/TGA curves of 2-propylamine adsorbed on a partially (Cu = 0.186 mmol g^{-1}) and a highly (Cu = 0.743 mmol g^{-1}) ion-exchanged Cu-ZSM-5 new TPD peaks and weight-loss steps were obtained, one at 500–550 K due to desorption of unreacted 2-propylamine, the other at 650–800 K. The high-temperature process was ascribed to the decomposition of 2-propylamine strongly interacting with those copper cations which compensated the negative framework charges of the zeolite associated with tetrahedrally coordinated aluminum. This process was not observed when copper was introduced into silicalite-1 in a smaller (but not negligible) amount than in the case of H-ZSM-5 (Fig. 15). The release of unreacted 2-propylamine observed in the case of all Cu-exchanged samples between 500–550 K was suggested to be due to copper species not associated with negative charges generated by framework aluminum. TPD/TGA proved to be a useful combination of thermal analytical methods for the characterization of the nature of copper in zeolites.

High-temperature treatment of a mechanical mixture of CuO and H-MFI zeolite in the absence of oxygen resulted in the incorporation of copper ions into cation positions of the zeolite. Thermal analysis (TG combined with MS) of 1-propylamine adsorbed on mixtures of CuO/H-MFI previously treated under different conditions was applied to determine the degree of consumption of zeolitic protons during the ion exchange process [117]. CuO/H-ZSM-5 previously heated in helium up to only 848 K and kept at this temperature for 0.5 h showed a desorption feature for 1-propylamine very similar to

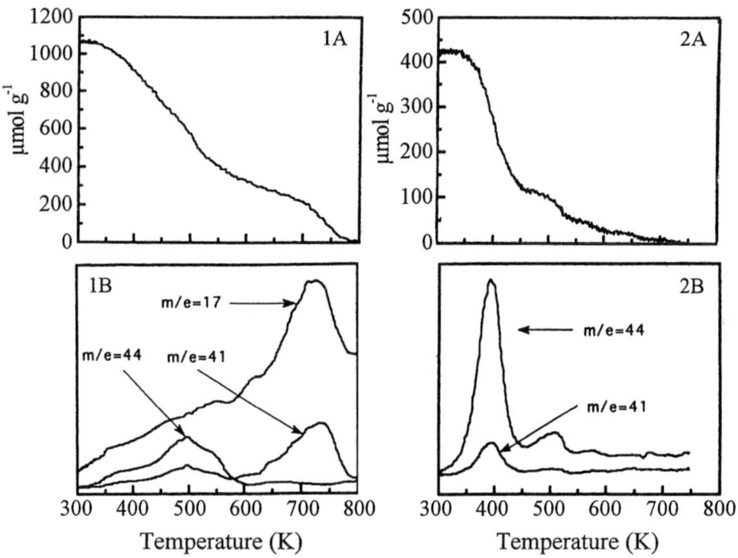

Fig. 15 TG (A) and TPD-MS (B) curves of 2-propylamine adsorbed (1) on highly copper-exchanged ZSM-5 and (2) on Cu-silicalite-1. Heating rate: 20 K min^{-1} under vacuum. Signals m/e = 44, m/e = 41, and m/e = 17 correspond to the fragments $[CH_2 - CH_2 - NH_2]$, $[CH_3 - CH = CH]$, and NH_3 indicative of 2-propylamine, propene and ammonia, respectively (reproduced from [116])

that observed in the case of H-ZSM-5 [106]. At about 670 K ammonia and propene were found to evolve simultaneously as decomposition products of 1-propylammonium cations. However, the thermal analysis of 1-propylamine adsorbed on a mixture pretreated at 973 K for 16 h in He pointed to a different behavior. Namely, the weight loss at about 670 K was much smaller than in the case of the sample pretreated at 848 K. MS data revealed that most of the ammonia eluted prior to propene. Nitriles (C3 and C6) and heavier residues were formed. The data strongly pointed to the replacement of protons by copper ions. Auto-reduction of CuO to Cu^+ was suggested to be involved in this solid-state ion-exchange process.

Kanazirev et al. [110, 118] reported on the influence of gallium cations on the desorption-decomposition features of propylamines in ZSM-5 zeolites. Mixtures of Ga_2O_3 and H-ZSM-5 prepared by the impregnation technique or by grinding of the solid components were subjected to reductive solid-state ion exchange. Thus, Ga^+ ions were introduced into cationic positions of the zeolite. TG results showed that, after adsorption of 1-propylamine at 323 K, more than one sorbate molecule per gallium cation was strongly held on reduced Ga-MFI at a temperature as high as 550 K. Furthermore, MS data clearly revealed that ammonia was released at temperatures about 160 K lower than those observed in the case of Ga-free H-ZSM-5. Thus, the release of ammonia and propene occurred in two separated steps, which indicated the occurrence of bimolecular transalkylation of 1-propylamine to dipropylamine. At higher temperatures, dehydrogenation reactions led to nitrile products. At 673 K propionitrile and propene were the main products. The decomposition of 2-propylamine in Ga-MFI was found to be completely different. Steric constraints do not allow that more than one 2-propylamine molecule is coordinatively bound to one gallium cation. In the plateau region of the TG curves (550–600 K) less 2-propylamine was measured than 1-propylamine. 2-PA may be converted to a limited degree to DPA, since some ammonia evolved at somewhat lower temperatures than propene. Nevertheless, the greater part of 2-PA decomposed simultaneously to ammonia and propene in a wide temperature range (between 550 and 750 K). At higher temperature, dehydrogenation, oligomerization, and cracking occurred.

A detailed comparative study of the interaction of 1-propylamine with H-, Ga-, In-, and Cu-MFI zeolites was also presented by Kanazirev et al. in [114]. The results illustrated well the advantages of using simultaneous TG-DTA-MS-GC techniques. Identification of the evolved products by MS/GC essentially contributed to the understanding of the physical and chemical processes giving rise to the observed weight losses and heat effects.

8 Conclusions

Thermal analysis proved to be a highly effective tool for the characterization of zeolites and the investigation of intracrystalline processes accompanied by weight changes and/or heat effects (e.g. dehydration, dehydroxylation, deammoniation, phase transitions, crystal collapse, oxidation, or reduction of lattice constituents). In the case of thermal reactions associated with gas evolution it may be appropriate to analyze the volatile products by combination of TG and DTA with complementary techniques, e.g., titration methods, on-line MS, and/or on-line GC. Nature and consequences of thermochemical processes in zeolites are generally better understood when thermoanalytical techniques are complemented by, e.g., XRD, FTIR- and/or MAS NMR spectroscopy applied to investigate zeolite specimens, in which changes in composition and structure may have occurred during previous thermoanalytical experiments. Also, TG and DTA may be successfully used, generally in combination with complementary techniques, to get insight into (i) the formation of coke during catalytic processes; (ii) the decomposition behavior of organic guest molecules introduced by adsorption and template molecules occluded during the synthesis in the channel system of zeolites; (iii) interactions of alkylamines and alcohols with "bridging" hydroxyls; and (iv) the hydrophobic and organophilic properties of high-siliceous zeolite varieties.

References

1. Cammenga HK, Epple M (1995) Angew Chem Int Ed Engl 34:1171
2. Auroux A (2002) Topics in Catalysis 19:205
3. Epple M (1994) J Therm Anal 42:559
4. Gimzewski E (1991) J Therm Anal 37:1251
5. Langier-Kuzniarowa A (1993) J Therm Anal 39:1169
6. Gabelica Z, Nagy BJ, Derouane EG, Gilson JP (1984) Clay Miner 19:803
7. Li C-Y, Rees LVC (1986) Zeolites 6:217
8. Muller JCM, Hakvoort G, Jansen JC (1998) J Therm Anal Calorim 53:449
9. Nyman M, Gu BX, Wang LM, Ewing RC, Nenoff TM (2000) Microporous Mesoporous Mater 40:115
10. Ray GJ, Nerheim AG, Donohue JA (1988) Zeolites 8:458
11. Anderson MW, Klinowski J (1986) J Chem Soc, Faraday Trans I 82:1449
12. Weitkamp J, Kleinschmit P, Kiss A, Berke CH (1993) In: von Ballmoos R, Higgins JB, Treacy MMJ (eds) Proc 9th Int Conf Zeolites, Montreal, 1992. Butterworth-Heinemann, London, p 79
13. Long Y, Jiang H, Zeng H (1997) In: Chon H, Ihm S-K, Uh YS (eds) Progress in zeolite and microporous materials 1997, Proc 11th Int Zeolite Conf, Seoul, Korea, August 12–17, 1996. Elsevier, Amsterdam, Stud Surf Sci Catal 105:787
14. Yang H, Ping Z, Niu G, Jiang H, Long Y (1999) Langmuir 15(16):5382
15. Qian B, Jiang H, Sun Y, Long Y (2001) Langmuir 17:1119

16. Fajula F, Pacheco VM, Figueras F (1987) Zeolites 7:203
17. Ernst S, Kokotailo GT, Weitkamp J (1987) Zeolites 7:180
18. Lee EFT, Rees LVC (1987) Zeolites 7:545
19. Li J, Qiu J, Sun Y, Long Y (2000) Microporous Mesoporous Mater 37:365
20. Li C-Y, Rees LVC (1986) Zeolites 6:60
21. Zi G, Yi T (1988) Zeolites 8:232
22. Abdillahi MM, Gharami MS, Siddiqui MAB (1994) J Therm Anal 42:1275
23. Esenli F, Kumbasar I (1994) In: Weitkamp J, Karge HG, Pfeifer H, Hölderich W (eds) Zeolites and related microporous materials: State of the art 1994, Proc 10th Int Zeolite Conf, Garmisch-Partenkirchen, Germany, July 17–22, 1994. Elsevier, Amsterdam, Stud Surf Sci Catal 84:645
24. Braunbarth CM, Behrens P, Felsche J, van de Goor G, Wildermuth G, Engelhardt G (1996) Zeolites 16:207
25. Ullrich B, Adolphi P, Zwahr H, Schomburg J (1989) Zeolites 9:412
26. Karge HG (1991) Coke Formation on Zeolites. In: van Bekkum H, Flanigen EM, Jansen JC (eds) Introduction to Zeolite Science and Practice. Stud Surf Sci Catal 58:531
27. Chen WH, Pradhan A, Jong SJ, Lee TY, Wang I, Tsai TC, Liu SB (1996) J Catal 163:436
28. Choudhary VR, Sivadinarayana C, Devadas P, Sansare SD, Magnoux P, Guisnet M (1998) Microporous Mesoporous Mater 21:91
29. Bauer F, Geidel E, Geyer W, Peuker Ch (1999) Microporous Mesoporous Mater 29:109
30. Zhuang Y, Ng FTT (2000) Appl Catal A: General 190:137
31. Liu H, Li T, Tian B, Xu Y (2001) Appl Catal A: General 213:103
32. Chen D, Gronvold A, Rebo HP, Moljord K, Holmen A (1996) Appl Catal A: General 137:L1
33. Zhu W, van de Graaf JM, van den Broeke LJP, Kapteijn F, Moulijn JA (1998) Ind Eng Chem Res 37:1934
34. Zhu W, Kapteijn F, Moulijn JA (2000) Phys Chem Chem Phys 2:1989
35. Zhu W, Kapteijn F, Moulijn JA (2001) Microporous Mesoporous Mater 47:157
36. Zhu W, Kapteijn F, van der Linden B, Moulijn JA (2001) Phys Chem Chem Phys 3:1755
37. Zhu W, Kapteijn F, Moulijn JA, Jansen JC (2000) Phys Chem Chem Phys 2:1773
38. Alexander SM, Bibby DM, Howe RF, Meinhold RH (1993) Zeolites 13:441
39. Aiello R, Crea F, Nastro A, Subotic B, Testa F (1991) Zeolites 11:767
40. Fejes P, Kiricsi I, Tasi Gy, Hannus I, Bertóti I, Székely T (1989) Zeolites 9:392
41. Karge HG, Beyer HK (1991) In: Jacobs PA, Jaeger NI, Kubelkova L, Wichterlova B (eds) Zeolite chemistry and catalysis, Proc Int Symp, Prague, Czechoslovakia September 8–13, 1991. Elsevier, Amsterdam, Stud Surf Sci Catal 69:43
42. Kanazirev V, Price GL, Dooley KM (1991) In: Jacobs PA, Jaeger NI, Kubelkova L, Wichterlova B (eds) Zeolite chemistry and catalysis, Proc Int Symp, Prague, Czechoslovakia September 8–13, 1991. Elsevier, Amsterdam, Stud Surf Sci Catal 69:277
43. Kanazirev V, Neinska Y, Tsoncheva T, Kosova L (1993) In: von Ballmoos R, Higgins JB, Treacy MMJ (eds) Proc 9th Int Zeolite Conf, Montreal, 1992. Butterworth-Heinemann, London, p 461
44. Beyer HK, Mihályi RM, Minchev Ch, Neinska Y, Kanazirev V (1996) Microporous Materials 7:333
45. Mihályi RM, Beyer HK, Mavrodinova V, Minchev Ch, Neinska Y (1998) Microporous Mesoporous Mater 24:143
46. Ohgushi T, Niwa T, Araki H, Ichino S (1997) Microporous Materials 8:231

47. Testa F, Crea F, Nastro A, Aiello R, Nagy BJ (1991) Zeolites 11:705
48. Zhao D, Qiu S, Pang W (1993) Zeolites 13:478
49. Howden MG (1993) Zeolites 13:315
50. Franklin KR, Lowe BM (1988) Zeolites 8:495
51. Araya A, Lowe BM (1986) Zeolites 6:111
52. Zones SI (1997) In: Occelli ML, Kessler H (eds) Synthesis of porous materials, zeolites, clays and nanostructures. Marcel Dekker Inc, New York, p 93
53. De Witte B, Patarin J, Guth JL, Cholley T (1997) Microporous Materials 10:247
54. Gies H, Marler B (1992) Zeolites 12:42
55. Crea F, Nastro A, Nagy BJ, Aiello R (1988) Zeolites 8:262
56. ElHage-Al Asswad J, Dewaele N, Nagy BJ, Hubert RA, Gabelica Z, Derouane EG, Crea F, Aiello R, Nastro A (1988) Zeolites 8:221
57. Soulard M, Bilger S, Kessler H, Guth JL (1987) Zeolites 7:463
58. Patarin J, Soulard M, Kessler H, Guth JL, Baron J (1989) Zeolites 9:397
59. Delmotte L, Soulard M, Kessler H (1990) Chem Lett 1215
60. Bilger S, Soulard M, Kessler H, Guth JL (1991) Zeolites 11:784
61. Soulard M, Bilger S, Kessler H, Guth JL (1991) Zeolites 11:107
62. Tavolaro A, Mostowicz R, Crea F, Nastro A, Aiello R, Nagy BJ (1992) Zeolites 12:756
63. Kosanovic C, Cizmek A, Subotic B, Smit I, Stubicar M, Tonejc A (1995) Zeolites 15:51
64. Yi KH, Ihm SK (1993) Microporous Materials 1:115
65. Occelli ML, Innes RA, Pollack SS, Sanders JV (1987) Zeolites 7:265
66. Araya A, Blake AJ, Harrison ID, Leach HF, Lowe BM, Whan DA, Collins SP (1992) Zeolites 12:24
67. Franco MJ, Perez Pariente J, Fornes V (1991) Zeolites 11:349
68. Wang B, Gao Q, Gao J, Ji D, Wang X, Suo J (2004) Appl Catal A: General 274:167
69. Feijen EJP, Matthijs B, Grobet PJ, Martens JA, Jacobs PA (1997) In: Chon H, Ihm S-K, Uh YS (eds) Progress in zeolite and microporous materials 1997, Proc 11th Int Zeolite Conf, Seoul, Korea, August 12–17, 1996. Elsevier, Amsterdam, Stud Surf Sci Catal 105:165
70. Goossens AM, Feijen EJP, Verhoeven G, Wouters BH, Grobet PJ, Jacobs PA, Martens JA (2000) Microporous Mesoporous Mater 35–36:555
71. Thangaraj A, Rajamohan PR, Ganapathy S, Ratnasamy P (1991) Zeolites 11:69
72. Franklin KR, Lowe BM (1988) Zeolites 8:501
73. Mostowicz R, Nastro A, Crea F, Nagy BJ (1991) Zeolites 11:732
74. Borade RB, Clearfield A (1994) Zeolites 14:458
75. Fjellvåg H, Lillerud KP, Norby P, Sørby K (1989) Zeolites 9:152
76. Schreyeck L, Caullet P, Mougenel JC, Guth JL, Marler B (1996) Microporous Materials 6:259
77. Pál-Borbély G, Szegedi Á, Beyer HK (2000) Microporous Mesoporous Mater 35–36:573
78. Perez-Pariente J, Martens JA, Jacobs PA (1988) Zeolites 8:46
79. Perez-Pariente J, Martens JA, Jacobs PA (1987) Appl Catal 31:35
80. Hegde SG, Kumar R, Bhat RN, Ratnasamy P (1989) Zeolites 9:231
81. Camblor MA, Corma A, Valencia S (1998) Microporous Mesoporous Mater 25:59
82. Vaudry F, Di Renzo F, Fajula F, Schulz P (1994) In: Weitkamp J, Karge HG, Pfeifer H, Hölderich W (eds) Zeolites and related microporous materials: State of the art 1994, Proc 10th Int Zeolite Conf, Garmisch-Partenkirchen, Germany, July 17–22, 1994. Elsevier, Amsterdam, Stud Surf Sci Catal 84:163
83. Borade RB, Clearfield A (1996) Microporous Materials 5:289
84. Parker LM, Bibby DM, Patterson JE (1984) Zeolites 4:168

85. Bourgeat-Lami E, Di Renzo F, Fajula F, Mutin PH, Des Courieres T (1992) J Phys Chem 96:3807
86. Kanazirev V, Price GL (1996) J Catal 161:156
87. Kosslick H, Tuan VA, Fricke R, Jedamzik J, Lanh HD (1991) J Therm Anal 37:2631
88. Nigro E, Crea F, Testa F, Aiello R, Lentz P, Nagy BJ (1999) Microporous Mesoporous Mater 30:199
89. Borade RB (1987) Zeolites 7:398
90. Testa F, Crea F, Diodati GD, Pasqua L, Aiello R, Terwagne G, Lentz P, Nagy BJ (1999) Microporous Mesoporous Mater 30:187
91. Jacob NE, Joshi PN, Shaikh AA, Shiralkar VP (1993) Zeolites 13:430
92. Chandwadkar AJ, Abdulla RA, Hegde SG, Nagy BJ (1993) Zeolites 13:470
93. Chandwadkar A, Bhat RN, Ratnasamy P (1991) Zeolites 11:42
94. Reddy JS, Reddy KR, Kumar R, Ratnasamy P (1991) Zeolites 11:553
95. Singh AP, Reddy KR (1994) Zeolites 14:290
96. Singh AP (1992) Zeolites 12:858
97. Borade RB, Adnot A, Kaliaguine S (1991) Zeolites 11:710
98. Camblor MA, Pérez-Pariente J, Fornes V (1992) Zeolites 12:280
99. Kumar R, Thangaraj A, Bhat RN, Ratnasamy P (1990) Zeolites 10:85
100. Occelli ML, Eckert H, Wölker A, Auroux A (1999) Microporous Mesoporous Mater 30:219
101. Borade RB, Clearfield A (1994) Microporous Materials 2:167
102. Borade RB, Clearfield A (1992) J Phys Chem 96:6729
103. Camblor MA, Corma A, Perez-Pariente J (1993) Zeolites 13:82
104. Tuel A, BenTaarit Y (1993) Zeolites 13:357
105. Tuel A, BenTaarit Y (1994) Zeolites 14:272
106. Parrillo DJ, Adamo AT, Kokotailo GT, Gorte RJ (1990) Appl Catal 67:107
107. Kofke TJG, Gorte RJ, Farneth WE (1988) J Catal 114:34
108. Kofke TJG, Gorte RJ, Kokotailo GT, Farneth WE (1989) J Catal 115:265
109. Kofke TJG, Gorte RJ, Kokotailo GT (1989) Appl Catal 54:177
110. Kanazirev V, Dooley KM, Price GL (1994) J Catal 146:228
111. Price GL, Kanazirev V (1997) Zeolites 18:33
112. Biaglow AI, Parrillo DJ, Kokotailo GT, Gorte RJ (1994) J Catal 148:213
113. Biaglow AI, Parrillo DJ, Gorte RJ (1993) J Catal 144:193
114. Kanazirev VI, Price GL, Dooley KM (1994) J Catal 148:164
115. Ruiz JAC, Ruiz VSO, Airoldi C, Pastore HO (2004) Appl Catal A: General 261:87
116. Parrillo DJ, Dolenec D, Gorte RJ, McCabe RW (1993) J Catal 142:708
117. Price GL, Kanazirev V, Church DF (1995) J Phys Chem 99:864
118. Kanazirev V, Price GL (1994) In: Weitkamp J, Karge HG, Pfeifer H, Hölderich W (eds) Zeolites and related microporous materials: State of the art 1994, Proc 10th Int Zeolite Conf, Garmisch-Partenkirchen, Germany, July 17–22, 1994. Elsevier, Amsterdam, Stud Surf Sci Catal 84C:1935

Mol Sieves (2007) 5: 103–154
DOI 10.1007/3829_003

Published online: 8 April 2006

Characterization of the Pore Size of Molecular Sieves Using Molecular Probes

Yvonne Traa · Sarah Sealy · Jens Weitkamp (✉)

Institute of Chemical Technology, University of Stuttgart, 70550 Stuttgart, Germany
jens.weitkamp@itc.uni-stuttgart.de

Abstract This Chapter deals with the evaluation of the pore size of crystalline microporous solids with molecular probes. Only methods where the dimensions of the probe molecules and the pore width are similar are discussed. This means that the adsorption or the selectivities and/or conversions of the reaction depend, in an unambiguous manner, on the pore width. After the introduction, in Sect. 2, some general aspects are discussed that are important for the detailed understanding of the methods covered, namely dimensions of probe molecules and intracrystalline cavities as well as molecular sieving.

Section 3 is devoted to adsorption, i.e., the use of molecular probes without chemical reactions. This includes mainly the characterization of various zeolites in comparison with one another and the discussion of molecular probes for zeolites with different pore sizes. Finally, in Sect. 4 shape-selective catalytic reactions are reviewed which have been employed for characterizing the width of micropores. As an introduction, the basics of shape-selective catalysis in microporous materials are discussed. The main test reactions dealt with are, for monofunctional acidic molecular sieves, the competitive cracking of *n*-hexane and 3-methylpentane (Constraint Index) as well as the isomerization and disproportionation of *m*-xylene. For bifunctional molecular sieves, the isomerization and hydrocracking of long-chain *n*-alkanes (Refined or Modified Constraint Index) as well as the hydrocracking of C_{10} cycloalkanes such as butylcylohexane (Spaciousness Index) are reviewed.

Abbreviations

a	adsorbed in the pores
a/e	adsorbed very slowly and in small amounts in the pores
AFI	three-letter code of the International Zeolite Association for zeolite SSZ-24
AFR	three-letter code of the International Zeolite Association for zeolite SAPO-40
BEA	three-letter code of the International Zeolite Association for zeolite Beta
BET	Brunauer Emmett Teller
CFI	three-letter code of the International Zeolite Association for zeolite CIT-5
CHA	three-letter code of the International Zeolite Association for zeolite chabazite
CI	Constraint Index
*CI**	Refined or Modified Constraint Index
CIT	California Institute of Technology
CON	three-letter code of the International Zeolite Association for zeolite CIT-1
$C_8S.I.$	C_8 Selectivity Index
CVD	chemical vapor deposition
DDR	three-letter code of the International Zeolite Association for zeolites decadodecasil 3R, Sigma-1
DFT	density functional theory
2,2-DM-Bu	2,2-dimethylbutane
DON	three-letter code of the International Zeolite Association for zeolite UTD-1
e	excluded from the pores
EMC	Elf Mulhouse Chimie
EMT	three-letter code of the International Zeolite Association for zeolite EMC-2
ERI	three-letter code of the International Zeolite Association for zeolite erionite
EU	Edinburgh University
EUO	three-letter code of the International Zeolite Association for zeolites ZSM-50, EU-1
FAU	three-letter code of the International Zeolite Association for zeolites X and Y
FER	three-letter code of the International Zeolite Association for zeolites ZSM-35, ferrierite
FT-IR	Fourier transform infrared
n-Hx	*n*-hexane
HPLC	high-pressure liquid chromatography
IFR	three-letter code of the International Zeolite Association for zeolite ITQ-4
IR	infrared
ITQ	Instituto de Tecnologia Quimica Valencia

IZA	International Zeolite Association
k	first-order rate constants
KFI	three-letter code of the International Zeolite Association for zeolite ZK-5
LHSV	liquid hourly space velocity
LTA	three-letter code of the International Zeolite Association for zeolite A
LTL	three-letter code of the International Zeolite Association for zeolite L
m	mass
MAZ	three-letter code of the International Zeolite Association for zeolite Omega
MCM	Mobil composition of matter
M-CPn	methylcyclopentane
MEA	multiple-equilibrium analysis
MEL	three-letter code of the International Zeolite Association for zeolite ZSM-11
MFI	three-letter code of the International Zeolite Association for zeolite ZSM-5
MIN-1	minimum dimension through a molecule
MIN-2	second minimum dimension through the same molecule perpendicular to *MIN-1*
3-M-Pn	3-methylpentane
2-M-Pr	2-methylpropane
MOR	three-letter code of the International Zeolite Association for zeolite mordenite
MTG	methanol to gasoline
MTH	methanol to hydrocarbons
MTO	methanol to olefins
MTT	three-letter code of the International Zeolite Association for zeolite ZSM-23
MTW	three-letter code of the International Zeolite Association for zeolite ZSM-12
MWW	three-letter code of the International Zeolite Association for zeolites MCM-22, SSZ-25, ITQ-1
n	molar amount
$\dot{n}$	molar flow
NES	three-letter code of the International Zeolite Association for zeolite NU-87
NMR	nuclear magnetic resonance
NU	new (ICI)
p_i	partial pressure at the adsorber outlet
$p_{i,0}$	partial pressure at the adsorber inlet
r	distance
R	ratio of rate constants observed experimentally for the formation of *o*- and *p*-xylene from *m*-xylene under conditions of diffusional limitations
R_{CN}	index defined as the ratio of the adsorption capacity of cyclohexane to that of *n*-hexane
S	selectivity
SAPO	silicoaluminophosphate
SI	Spaciousness Index
SR	Selectivity Ratio
SSI	Shape Selectivity Index
SSZ	Standard Oil synthetic zeolite
$t_{0.3}$	time for o-xylene sorption of 30% of the *p*-xylene capacity
T	temperature
TIPB	1,3,5-triisopropylbenzene
TON	three-letter code of the International Zeolite Association for zeolite ZSM-22
TOS	time on stream
TPD	temperature-programmed desorption

type A β-scission	the classification of β-scissions of alkyl carbenium ions has been introduced in [120]. The salient feature of type A β-scissions is that they start from a tertiary and lead again to a tertiary carbenium ion
type B β-scission	The classification of β-scissions of alkyl carbenium ions has been introduced in [120]. The salient feature of type B β-scissions is that they start from a tertiary and lead to a secondary carbenium ion or *vice versa*
UTD	University of Texas at Dallas
UV/Vis	ultraviolet/visible
$\dot{V}$	volumetric flow rate
VFI	three-letter code of the International Zeolite Association for zeolite VPI-5
VPI	Virginia Polytechnic Institute
V_{ROA}	relative *o*-xylene adsorption velocity
X	conversion
Y	yield
ZK	zeolite Kerr
ZSM	Zeolite Socony Mobil
ϕ	zeolite Socony Mobil potential
σ	"kinetic diameter", which appears as a parameter in the Lennard-Jones potential

1 Introduction

With molecular probes, various properties of crystalline microporous solids can be explored. One example is the determination of surface acidity/basicity by adsorption and desorption of basic/acidic probe molecules (e.g., ammonia, pyridine, carbon dioxide, chloroform or deuterochloroform) (cf. this book series, vol 7, Chap. 1) and observing the sorption processes by IR spectroscopy (vibrational spectroscopy) (cf. this book series, vol 4, Chap. 1), NMR spectroscopy (cf. this book series, vol 4, Chap. 2), mass spectrometry or gas chromatography. Another possibility is the evaluation of surface hydrophobicity/hydrophilicity by sorption of mixtures of non-polar and polar substances. The assessment of micropore volume and pore size can be accomplished by adsorption of xenon monitored by ^{129}Xe NMR (cf. this book series, vol 5, Chap. 4) spectroscopy, by adsorption of nitrogen and other small molecules usually followed gravimetrically or volumetrically (cf. this book series, vol 6, Chap. 4) or by determining the heats of adsorption by means of microcalorimetry (cf. this book series, vol 6, Chap. 5).

This chapter deals with the evaluation of the pore size of crystalline microporous solids with molecular probes, but only those methods that are based on size effects will be discussed, i.e., where the dimensions of the probe molecules (or of the transition states/product molecules formed from them) and the pore width are similar. These methods include adsorption of molecules of different sizes large enough to "feel" the presence of the micropores and, therefore, allowing an assessment of the pore width [1], and test reactions in which the selectivities and/or conversions depend, in an unambiguous manner, on the pore width, i.e., shape-selective reactions. In the early days of zeolite science, these two techniques were most popular tools for collecting information on the approximate crystallographic pore size of zeolites with unknown structures. With the advent of more sophisticated and highly efficient crystallographic methods, a rapid determination even of complex new structures became feasible. Hence, the initial incentive for the application of methods using molecular probes has shifted [2]: Nowadays, these techniques are primarily used as quick tests for probing the effective pore width under catalytically relevant conditions and/or of molecular sieves manipulated and modified with post-synthesis methods such as chemical vapor deposition (CVD), deliberate or unwanted coking, isomorphous substitution of framework atoms and the like.

In Sect. 2, some general aspects will be discussed that are important for the detailed understanding of the methods covered. Section 3 will be devoted to adsorption, i.e., the use of molecular probes without chemical reactions. Finally, in Sect. 4 shape-selective catalytic reactions, which have been employed for characterizing the width of micropores, will be reviewed.

2
General Aspects

2.1
Dimensions of Probe Molecules and Intracrystalline Cavities

For the discussion of size effects, it is vital that the dimensions be defined in an appropriate way. This problem was tackled in an excellent paper by Cook and Conner [3]. These authors stress that the hard-sphere picture underlies the thinking about adsorption, i.e., adsorbate molecules and the adsorbent are generally both considered to be rigid structures composed of hard-sphere atoms or ions. However, in reality the molecules and the host matrix are in continuous vibration and often quite flexible. Therefore, if results of adsorption experiments are to be explained using the hard-sphere picture, the dimensions of the probe molecules and of the intracrystalline cavities should be brought in line with this simple picture. In other words, the model employed for interpreting the adsorption process should be consistent with the type of dimensions used for the description of the probe molecules and the intracrystalline cavities. By contrast, the "kinetic diameter" σ (from the Lennard-Jones potential, cf. Fig. 1) of the adsorbate and the pore size based on the ionic oxygen radius of 0.135 nm are generally used, and these dimensions do not satisfy this criterion of consistency [3]. In many instances, molecules do diffuse into pores that are considered to possess a width lower than the molecular diameter. An example is cyclohexane (σ = 0.60 nm, [4])

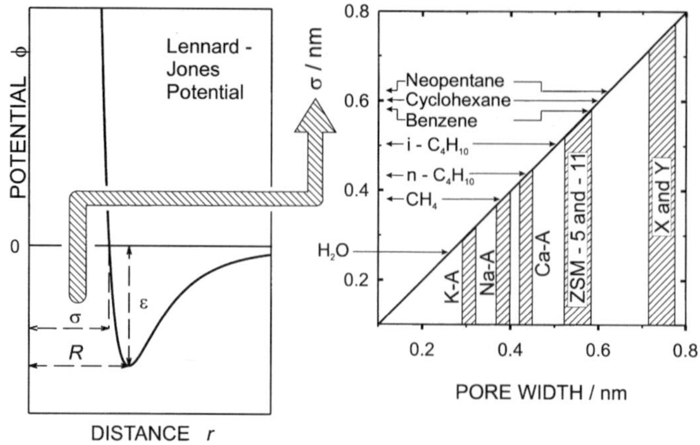

Fig. 1 Dimensions of probe molecules and intracrystalline cavities (σ is the "kinetic diameter" that appears as a parameter in the Lennard-Jones potential)

in zeolite ZSM-5 (0.53×0.56 and 0.51×0.55 nm, [5], cf. Fig. 1) [6]. For this reason, Cook and Conner [3] proposed a re-definition of framework atom sizes, based on the average physical extension of electron density distributions in zeolite frameworks. With this modification, the maximum dimension of the ZSM-5 pores would be 0.63 nm, now permitting cyclohexane to enter its pores.

Another fact that should be taken into account is that, for cylindrical pores, the consideration of just one molecular dimension is usually insufficient. Hardly any molecule has spherical symmetry. In this context, Webster et al. [7] advanced the concept of effective minimum dimensions of molecules. *MIN-1*, the minimum dimension through a molecule, and *MIN-2*, the second minimum dimension through the same molecule perpendicular to *MIN-1*, determine whether or not this molecule can enter the pores of a given material. These dimensions can be calculated with molecular orbital theory and allow a more sophisticated description of the molecular behavior inside the pores.

Furthermore, one should consider the flexibility of molecules and how this flexibility can affect the diffusion and adsorption inside the pores. Choudhary and Akolekar [8] proposed the shuttlecock-shuttlebox model to account for the fact that larger molecules than expected do diffuse into pores of a given width. Their model envisages the compression of alkyl groups of branched molecules similar to the compression of feathers of a shuttlecock in a shuttlebox.

In addition, one should keep in mind that the pore dimensions given in the "Atlas of Zeolite Framework Types" [5] are coarse data and subject to changes due to both the experimental conditions and the precise form of the porous material. For example, the effective pore dimension varies with temperature. The best way for probing this are catalytic tests performed at different temperatures [9]. This will be discussed in more detail in Sect. 4. Another parameter that can affect the pore size is the aluminum content: Framework aluminum tends to reduce the pore volume and to broaden the pore size distribution. Steam treatment has been reported to reduce the apparent pore size [10]. Furthermore, the shape of the pore aperture can change during adsorption [4], and zeolites can undergo structural rearrangement, which might alter the pore size [11]. Wu and Ma [12] showed that the adsorption capacity for various hydrocarbons on zeolite ZSM-5 decreases as the radius of the cation increases. Thus, certain parts of the channel system can be blocked by cations, and the pore aperture can be contracted. In the extreme case, molecules are excluded from the pore system. For example, uncalcined offretite with large organic cations in its channels adsorbs neither cyclohexane nor *n*-hexane, whereas the calcined zeolite does [13]. In the "Atlas of Zeolite Framework Types" [5], the pore size of the species examined first is given. In zeolite minerals, different cations are often present, which reduce the pore size, compared to the synthetic zeolites and the pure silica analogues, which undergo no such reduction. Thus, the reader should pay attention to the exact form of the species the pore size of which was determined. Another possible reason for pore blockage is amorphous material in the intracrystalline cavities [14]. Finally, preadsorption of polar molecules such as water or ammonia often affects the subsequent adsorption of other molecules by clustering around the cations, reducing the apparent pore size and adsorption capacity and eventually blocking the pores [15]. Therefore, particular attention should always be paid to the hydration state of the zeolite.

In conclusion, one should be very careful when predicting whether or not a given molecule has access to the pores of a given material merely from tabulated sizes of molecules and cavities. Only if the dimensions of the probe molecules and the dimensions of the intracrystalline cavities are chosen in such a way that they are consistent with each other and the adsorption model, meaningful predictions are possible [9].

2.2 Molecular Sieving

Molecular sieving is the selective adsorption of molecules into the intracrystalline void system of a molecular sieve and the exclusion of others due to their dimensions being above the critical size. One example is illustrated in Fig. 2: A gaseous mixture of *n*-pentane and 2-methylbutane was continuously passed over a fixed bed of calcium-exchanged zeolite A (or zeolite "5A") with an effective pore diameter of ca. 0.5 nm (5 Å). This is large enough to allow for the diffusion of the *n*-alkane molecules through the 8-membered ring windows of zeolite A, but too small for the uptake of the branched alkane. Hence, 2-methylbutane breaks through at the adsorber outlet directly after the onset of the experiment, while the smaller *n*-pentane is completely ad-

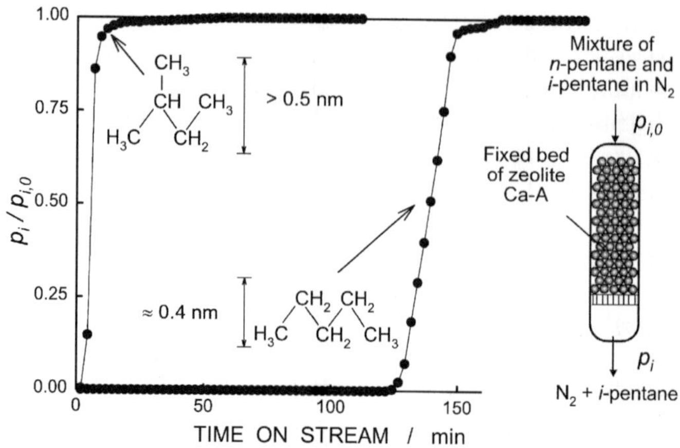

Fig. 2 Breakthrough curves for the adsorption of an *n*-pentane/2-methylbutane mixture over zeolite Ca-A in a fixed-bed flow-type adsorber

sorbed by the zeolite for about 2 h, whereupon its adsorption capacity is exhausted.

The example displayed in Fig. 2 represents the extreme case in molecular sieving, namely pore size exclusion, i.e., one molecular species is so bulky that it is completely prevented from entering the intracrystalline cavities. In many instances, however, molecules do enter the pores, but their diffusion inside the pores is very slow. In such cases, very different results for the adsorption capacity will be obtained depending on whether or not the system was given enough time to reach adsorption equilibrium. Furthermore, for the purpose of pore size characterization, one must keep in mind that adsorptive separation on zeolites can be accomplished not only by molecular sieving, but also by selective cation-adsorbate interaction, by selective sorption due to hydrophilic or hydrophobic surface properties or by selective sorption due to acidic/basic surface properties. The discussion in this chapter will essentially be restricted to molecular sieving, however, a clear-cut distinction between the different mechanisms of adsorptive separation is sometimes difficult.

Obviously, if adsorption at the external surface of a molecular sieve occurs to a significant extent, this will obscure the desired information on the pore size. Masuda and Hashimoto [16] demonstrated that up to 50% of the total amount adsorbed can be located on the external surface, if the zeolite crystals are very small. One possibility to eliminate the undesired adsorption of the probe molecule(s) on the external surface is to add another component (e.g., as a solvent in liquid-phase adsorption), the molecules of which are so bulky that they are completely hindered from entering the pores. These bulky molecules will often cover the external surface exclusively, especially if their concentration in the bulk fluid phase is high.

Another complication arises when the size of the probe molecules closely matches the size of the pore. Under these conditions, branched isomers can be more strongly adsorbed than linear ones. For example, Santilli et al. [17] reported that for zeolites of pore sizes between 0.6 and 0.7 nm, there was an increased preference for branched versus linear hexane isomers, giving rise to inverse shape selectivity. This phenomenon should be borne in mind when adsorption over a range of pore sizes is being studied. Santilli et al. [17] argued that attractive forces between the zeolite walls and hexane isomers stabilize the branched isomers relative to *n*-hexane. Schenk et al. [18], on the other hand, were of the opinion that *n*-hexane is excluded from the pores because its effective size is greater, i.e., adsorption of shorter, more compact isohexanes is preferred to that of the longer *n*-hexane. This was demonstrated using Grand Canonical Monte Carlo simulations and points to the value of using molecular simulations in understanding adsorbate-absorbent interactions, especially in relation to shape-selective effects.

Wu et al. [19] pointed out a few more pitfalls: They found that the amount of cyclohexane sorbed in large crystals of zeolite ZSM-5 was much lower after 2 h of equilibration than that sorbed in smaller crystals of the same zeolite. This observation was explained by increasing time required to reach adsorption equilibrium with increasing crystallite size [19]. Another fact that could hinder the adsorption are imperfections of the large crystals, which are invisible with conventional characterization techniques. Therefore, when interpreting adsorption results, one should always take into account the crystallite size of the samples. Another factor that can severely affect the results are impurities in the adsorptives [19]. Even at low concentrations, impurities in an adsorptive can lead to erroneous results due to selective adsorption of the impurity, especially if its molecular dimensions are small.

3 Adsorption of Probe Molecules with Different Sizes

Probing the pore width of microporous materials by adsorptives of different molecular size has been a popular method since the beginning of zeolite science. Consequently, a vast amount of results are scattered in the scientific and patent literature. No comprehensive discussion of these results is aimed at in the present review. Rather, a few selected and instructive examples will be presented.

3.1 Characterization of Various Zeolites in Comparison

Adsorption of an appropriate set of components with different molecular dimensions is a widely accepted procedure for characterizing the pore sizes of zeolites. A very thorough study on several 8-, 10- and 12-membered-ring zeolites was made by Wu et al. [19], which can be looked upon as a basis for a large part of the later work. During static adsorption at room temperature, substantial amounts of *n*-hexane were sorbed by all zeolite samples, since its critical dimensions are smaller than or equal to their crystallographic pore openings. The largest adsorptive, mesitylene (1,3,5-trimethylbenzene), was sorbed only in the pores of the 12-membered-ring zeolites. The authors concluded that the ability to sorb benzene and cyclohexane classifies zeolite ZSM-23 (MTT) (the structural code of the International Zeolite Association is given in parentheses [5]) and ZSM-48 as medium-pore zeolites. However, having only one-dimensional channels, zeolites ZSM-23 and ZSM-48 adsorbed less benzene or cyclohexane than zeolites ZSM-5 (MFI) and ZSM-11 (MEL), which have intersecting 10-membered-ring channels. The lower adsorption capacity of the 12-membered-ring zeolite ZSM-12 (MTW) as compared to zeolite Y (FAU) with its very spacious pore system was proposed to be a consequence of its denser structure with one-dimensional channels (like the large-pore zeolite mordenite (MOR); the 8-membered-ring pore channels of mordenite do not adsorb the hydrocarbons used). The lesser amount of mesitylene sorbed by ZSM-12 as compared to mordenite was said to be consistent with its smaller crystallographic pore opening, impeding the diffusion of the large mesitylene molecules into its intracrystalline voids [19].

Wu et al. [19] also applied a dynamic adsorption method using a thermogravimetric analyzer. These experiments were carried out at 373 K; equilibrium was judged to have been achieved when the weight gain was less than 17 ng/s. ZSM-23 and ZSM-48 exhibited a comparable capacity ratio of 3-methylpentane to *n*-hexane as ZSM-5 and ZSM-11, but a considerably lower ratio of uptake rate for these two hydrocarbons, reflecting again the unidimensionality of the channels in ZSM-23 and ZSM-48. However, the capacity of cyclohexane sorbed by ZSM-48 was larger and the rate of uptake faster than on ZSM-23, consistent with the greater eccentricity of the pore openings and the channels of ZSM-23.

In a further evaluation of their experimental data, Wu et al. [19] determined effective pore sizes of the zeolites by using the smaller cycloalkanes as a measure of the minor axis and the larger aromatics as a measure of the major axis of the effective pore opening. Just one example is given here for illustration: Since the uptake rate of cyclohexane (about 0.5×0.6 nm size) in ZSM-23 was very low, the minor axis of the ZSM-23 channels was concluded to be about 0.45 nm, approaching that of the smallest dimension of cyclohexane. Its ability to sorb *p*-xylene (about 0.4×0.6 nm size) and *o*-xylene less readily (about 0.4×0.7 nm size) indicates that its major axis appears to be only 0.65 nm. Table 1 gives an overview of which molecules can enter the pores of important zeolites. When using this table, the reader should always be aware that the adsorption behavior is dependent on the experimental con-

Table 1 Molecules adsorbed in (a), adsorbed very slowly and in small amounts in (a/e) or excluded from (e) the pores of various zeolites [17, 19–24]

Zeolite	*n*-Hexane (0.39 × 0.43 ×0.91) [1]	3-Methyl-pentane (0.46 × 0.58 ×0.86)	Benzene (0.34 × 0.62 ×0.69)	*p*-Xylene (0.37 × 0.62 ×0.86)	Cyclo-hexane (0.47 × 0.62 ×0.69)	2,2-Dime-thylbutane (0.59 × 0.62 ×0.67)	*o*-Xylene (0.41 × 0.69 ×0.75)	Mesitylene (0.37 × 0.78 ×0.85)
Ca,Na-A (LTA) [2]	a	e	e	e	e	e	e	e
H-ZSM-23 (MTT)	a	a	a	a	a/e	e	a/e	e
ZSM-35 (FER)	a	— [3]	— [3]	— [3]	a	— [3]	— [3]	e
H-ZSM-48 [4]	a	— [3]	a	— [3]	a	— [3]	— [3]	e
ZSM-22 (TON)	a	— [3]	— [3]	— [3]	a	— [3]	— [3]	— [3]
H-ZSM-11 (MEL)	a	a	a	a	a	a	a	e
H-ZSM-5 (MFI)	a	a	a	a	a	a	a	e
EU-1 (EUO)	a	— [3]	— [3]	— [3]	a	— [3]	— [3]	— [3]
Na-MCM-22 (MWW)	a	— [3]	— [3]	— [3]	a	— [3]	a	a
H-ZSM-12 (MTW)	a	a	a	— [3]	a	a	— [3]	a/e [5]
H-Mordenite (MOR)	a	a	a	— [3]	a	a	— [3]	a
CIT-1 (CON)	a	— [3]	— [3]	— [3]	— [3]	— [3]	— [3]	a
L (LTL)	a	a	— [3]	— [3]	a	a	— [3]	a [6]
SSZ-24 (AFI)	a	— [3]	— [3]	— [3]	a	a	— [3]	a/e [5,6]
H-Beta (BEA)	a	a	— [3]	— [3]	a	a	— [3]	a
EMT (EMT)	a	— [3]	— [3]	— [3]	a	— [3]	— [3]	— [3]
H-Y (FAU)	a	— [3]	a	— [3]	a	a	— [3]	a

[1] Molecular dimensions in nm estimated from Courtald space-filling models; the two smaller parameters define the critical dimensions [19].
[2] In parentheses, the structural code of the International Zeolite Association is given [5].
[3] This adsorption experiment was not carried out in [17, 19–24].
[4] No three-letter code has been assigned to zeolite ZSM-48 by the IZA Structure Commission yet.
[5] The virgule indicates that different results were reported by different authors.
[6] In this experiment, 1,3,5-triisopropylbenzene (kinetic diameter ≈ 0.85 nm [23]) was adsorbed instead of mesitylene.

ditions and on changes in the pore size, which can be caused by ion exchange and so forth (cf. Sect. 2.1).

Guil et al. [22] chose similar probe molecules, i.e., *n*-hexane, toluene, *m*-xylene and mesitylene, for probing the pore size of the new zeolite ITQ-4 (IFR) and comparing it to several medium- and large-pore zeolites. These authors defined so-called packing densities as the amount of adsorbate that fills up the micropores during single-component adsorption per micropore volume (as determined by nitrogen sorption). The micropore volume of zeolite ITQ-4 was found to be nearly as large as the one of zeolite Beta (BEA), suggesting the existence of wide channels and great cavities. However, its *n*-hexane packing density, which is very sensitive to the presence of cavities, was between those of zeolites SSZ-24 (AFI) and ZSM-12 (MTW), both of which possess unidirectional 12-membered-ring channels. Furthermore, the packing density of *m*-xylene in ITQ-4 was 4.55 mmol g^{-1}. From this, the authors concluded that ITQ-4 possesses 12-membered-ring channels, since in 10-membered-ring channels, into which *m*-xylene can penetrate, the packing density was much lower. Mesitylene did not have access to the micropores of ITQ-4 [22].

Santilli et al. [17] studied the competitive, dynamic adsorption of *n*-hexane, 3-methylpentane and 2,2-dimethylbutane at 130 °C on several small-, medium-, large- and extra-large-pore zeolites. Except for zeolites erionite (ERI) and ZSM-23 (MTT), the three hydrocarbons were sorbed on all zeolites used. Zeolite erionite sorbed only *n*-hexane, ZSM-23 only *n*-hexane and 3-methylpentane.

Lourenço et al. [25] recorded adsorption isotherms of cyclohexane on zeolites H-mordenite, H-ZSM-5, H-SAPO-40 (AFR) and H-Y ($n_{Si}/n_{Al} = 4.5$). They proposed that the lowest rate of cyclohexane adsorption and the low total amount adsorbed on H-mordenite were in line with the presence of the unidirectional channel system in this zeolite. This contrasted with the high adsorption rate on H-ZSM-5, which has a three-dimensional channel system that facilitates the diffusion of cyclohexane, despite the pore diameter being smaller than that of H-mordenite. On zeolite H-Y with its large cavities, cyclohexane was adsorbed fast and in large amounts. In H-SAPO-40, only the main 12-membered-ring channels were accessible to cyclohexane. The high sorption rate and capacity indicated that its structure involves a larger void volume and less constraints than that of H-mordenite.

Lobo et al. [23] used adsorption experiments to characterize the new extra-large-pore zeolite UTD-1 (DON): The adsorption capacities of *n*-hexane, 2,2-dimethylbutane and cyclohexane were similar for zeolites SSZ-24 (AFI), L (LTL) and UTD-1, but considerably larger for zeolites Y and VPI-5 (VFI). This was explained by the former having one-dimensional pore systems, the latter having large cavities or very large 18-membered-ring channels (VFI). The pore size of the different zeolites was assessed using 1,3,5-triisopropylbenzene: This large molecule (kinetic diameter $\approx$ 0.85 nm) had no access to the channels of SSZ-24 (diameter 0.73 nm), adsorbed very slowly in the pores of zeolite Y (diameter 0.74 nm), but had easy access to the pores of the 14-membered-ring zeolite UTD-1 and to VPI-5. (Strangely enough and uncommented by the authors, 1,3,5-triisopropylbenzene was adsorbed on zeolite L, in which the diameter of the 12-membered-ring channels is 0.71 nm.)

Yuen et al. [26] employed linear dialkylazodicarboxylate chromophores with different sizes of the alkyl groups as probes for pore size characterization. The uptake of the azo compounds from an isooctane solution was easily monitored using a UV/Vis spectrophotometer. The method demonstrated no uptake of the chromophore-containing molecules for small-pore zeolites, somewhat hindered uptake for medium-pore zeolites and variable uptake rates for large-pore zeolites, because the solvent could compete with the azo compound. For medium-pore zeolites, the uptake of the chromophore-containing molecules was inversely related to the crystallite size. The technique allows for a quick probing of the pore system of novel zeolite materials. In this study, the structure of zeolite SSZ-25 (MWW) was examined in comparison to the known structures of various zeolites, such as ZSM-5, ZSM-22, ZSM-23, Y, Beta and other SSZ zeolites. Water was found to compete with the azo compounds in the adsorption experiments, therefore, it should carefully be excluded from the system. As expected, the presence of coke reduced the uptake of the chromophore-containing molecules. Though a large amount of adsorption isotherms and other useful data were collected in this study, the authors themselves stressed that the analysis of the data is hardly foolproof and that additional data is needed, for example concerning crystallite size, zeolite composition and so on, before this method can possibly be used as a standard test.

A vast amount of adsorption data was collected by Breck [27]: Adsorption capacities of most of the common gases and hydrocarbon vapors on various zeolites and zeolite minerals are listed in this book. Where a marked effect of the cation on the sieving characteristic is evident (cf. Sect. 2.1), adsorption capacities for different cation-exchanged forms of the zeolites are given. For example, the potassium form of zeolite A ("3A") has a smaller, the calcium form ("5A") a larger pore size than the sodium form of zeolite A ("4A"). It is concluded that on zeolite Na-A, for example, molecules with a kinetic diameter larger than 0.36 nm are not adsorbed (cf. Table 2). Breck's book continues to be very useful as a quick reference on whether a molecule can be adsorbed on a specific zeolite or not, though it does not contain information on the more recent zeolites.

For the characterization of small-, medium- and large-pore zeolites by adsorption, Otake [21] proposed the so-called R_{CN} index, defined as the ratio of the adsorption capacity of cyclohexane to that of n-hexane, as a simple and versatile tool. Whereas catalytic test reactions (cf. Sect. 4) are only applicable to medium-pore *or* large-pore zeolites, but not to both at the same time, the

Table 2 Adsorption properties of various zeolites [27]

Zeolite	does not adsorb molecules with a kinetic diameter
Na-A (LTA)	> 0.36 nm at 77 K, > 0.40 nm at 300 K
Ca-A (LTA)	> 0.43 nm at 300 K, > 0.44 nm at 420 K
Erionite (ERI) (mineral)	> 0.43 nm at 400 K
Chabazite (CHA) (mineral)	> 0.43 nm at 400 K
Mordenite (MOR) (mineral)	> 0.48 nm at 300 K
Ca-X (FAU)	> 0.78 nm at 300 K
Na-X (FAU)	> 0.80 nm at 298 K
Na,K-L (LTL)	> 0.81 nm at 323 K

R_{CN} index was claimed to be linearly dependent on the average diameter of the largest pore over a wide range of pore sizes of zeolites with 8- to 12-membered rings. A further advantage of this index is that the absolute amount of adsorbate is immaterial. Only the selectivity of adsorption determines the index. It was demonstrated that the R_{CN} index could also detect subtle changes in the pore size due to ion exchange: The R_{CN} value decreased with increasing cation radius of cation-exchanged Beta zeolites. However, Na-Beta and Ca-Beta had different R_{CN} values, even though the cation radii of Na and Ca are the same. Similar inconsistencies appeared when comparing different zeolites. In summary, though the proposed index is an interesting tool for characterizing zeolite pore sizes, it is in an early stage of development. More data and a refinement are needed, before one can judge upon its general applicability and usefulness.

3.2 Various Methods of Pore Size Characterization by Adsorption

Besides the relatively simple methods for pore size characterization by adsorption, which were presented in Sect. 3.1, a wealth of other methods have been proposed. Some of them will be described in this section.

First, methods similar to the determination of the R_{CN} index (cf. Sect. 3.1, last paragraph) were introduced, the difference being that the time dependence of the adsorption process is taken into consideration: In a study on ZSM-5 and ZSM-11 zeolites, Harrison et al. [28] defined the so-called Pore Constraint Index as the ratio of the amount of *n*-hexane adsorbed after 15 min to the amount of 2,3-dimethylbutane (or cyclohexane) adsorbed after 15 min. This index is based on the fact that, under the experimental conditions chosen, molecules with a kinetic diameter less than the dimensions of the zeolite pore openings, such as *n*-hexane, were rapidly adsorbed and approached their sorption capacity within 15 min. In contrast, molecules such as 2,3-dimethylbutane and cyclohexane, the diameters of which are close to the channel dimensions, were only slowly adsorbed. In such cases, the amounts sorbed after 15 min are not true sorption capacities, but represent relative rates of adsorption. Such data can, however, still yield valuable information on the zeolite pore structure: A plot of sorbate uptake on ZSM-5 after 15 min versus the kinetic diameter of the sorbate molecule revealed a sharp cut-off in sorbate uptake at a molecular diameter of approximately 0.58 to 0.60 nm, which agrees well with the pore dimensions of ZSM-5. By contrast, on zeolite ZSM-11, even 2,3-dimethylbutane and cyclohexane approached their adsorption capacities within 15 min. Thus, the decreasing value of the Pore Constraint Index in going from silicalite-1 over ZSM-5 and ZSM-5/ZSM-11 intermediates to ZSM-11 was accounted for in terms of decreasing channel tortuosity, allowing for a more rapid rate of diffusion of larger molecules into the zeolite pores. Therefore, this test facilitates an estimation of pore sizes within a small range, which could also be useful for the assessment of pore size changes by CVD. In addition, the method reveals an important advantage of adsorption tests over catalytic test reactions namely, they can be used for zeolites with different n_{Si}/n_{Al} ratios, in other words, the pure silica analogue silicalite-1 can be tested as well.

A similar adsorption test was proposed in the patent literature [29]: Zeolites selective for the production of *p*-dialkylbenzenes were characterized by determining relative rates of adsorption of *p*- and *o*-xylene. Highly selective zeolites, such as ZSM-5, typically had an adsorption capacity of $m_{p\text{-xylene}}/m_{\text{zeolite}} \geq 0.01$. The equilibrium adsorption capacity was determined gravimetrically at 120 °C; *p*-xylene was preferentially chosen since it is the xylene isomer which reaches equilibrium within the shortest time, but *m*- or *o*-xylene or isomer mixtures can also be used. The time for *o*-xylene sorption of 30% of the *p*-xylene capacity was larger than 10 min with selective zeolites. This sorption time, the so-called $t_{0.3}$, could be extended considerably by precoking of the catalyst. Hence, this adsorption test is useful for characterizing pore size reduction by coking and modification with oxides of antimony, phosphorus, boron or magnesium or CVD. $t_{0.3}$ was found to be a direct measure of the diffusivity [30] and of the pore tortuosity [31]. A similar parameter, which was also claimed to be a measure of the pore tortuosity, is the relative *o*-xylene adsorption velocity, $V_{\text{ROA}} \equiv (m_{o\text{-xylene, adsorbed at180 min}})/(m_{p\text{-xylene, adsorbed at infinite time}})$ [31]. Baeck et al. [32] showed that the rate constant for *n*-hexane sorption on Mg-ZSM-22 decreased strongly with increasing deposition time of silicon alkoxide. Therefore, the rate constant is also a measure of the contraction of the pore opening. However, the adsorption capacities of *n*-hexane were unchanged by CVD, suggesting that only the pore opening at the external surface of the crystallites was reduced, whereas the channel diameters in the interior of the crystals were not.

Santilli [33] developed the so-called pore probe technique, which allows for the measurement of the absolute steady-state concentrations of molecules within the pores of zeolites even at temperatures near or at typical reaction conditions, i.e., the adsorption capacity during the reaction can be determined. To achieve this, the catalyst/adsorbent is exposed to the feed flow till the product stream reaches steady state. One then switches to a nitrogen flow, and the catalyst is cooled quickly in order to avoid further reaction. The material desorbed is collected and analyzed by gas chromatography using an external standard. Blank runs using adsorbents without pores are employed to measure the steady-state concentrations of all feed components in the dead space and on the external surface. By comparing the amount of C_6 hydrocarbons in the pores of H-ZSM-5 and Na-ZSM-5, the influence of an acid-catalyzed reaction on the adsorption capacity could be determined [33]. In addition, the change of the adsorption capacity with the temperature can be calculated.

Choudhary et al. [34] studied zeolite adsorption properties by injecting different compounds into a gas chromatographic column packed with zeolite particles at temperatures close to those employed in catalytic processes. The flowing carrier gas eluted the adsorbate, and the peak was recorded. The retention time data were corrected for the voids present in the zeolite column and connecting tubes. A sharp decrease in the retention time (or sorption) of the sorbate (during the passage of the sorbate pulse) on the zeolite was observed with an increase in the steric hindrance due to chain branching. The method appears to be useful for determining whether a molecule can enter the channels or not. Thielmann et al. [35] used a similar technique that combines thermal desorption with inverse gas chromatography. A peak during thermodesorption indicates that the molecule concerned had access to the pore system. In addition, this method seems to permit a discrimination between adsorption in micropores, mesopores and on the external surface [35]. Anunziata and Pierella [36] also performed desorption experiments, but they used a conventional TPD (temperature-programmed desorption) equipment with a flame ionization detector. During desorption of *m*-xylene or mesitylene from zeolite ZSM-11, they observed one peak at low temperature, mostly due to the desorption of *m*-xylene and mesitylene adsorbed on the external surface, and one peak at higher temperature, mostly due to molecules that partially penetrated the pores of the zeolite. Naphthalene did not have access to the channels and was desorbed from the external surface in one peak at relatively low temperatures.

Denayer et al. [37] considered the fact that most industrial zeolite processes are performed at elevated hydrocarbon pressures or in the liquid phase, but adsorption of organic molecules is commonly studied at low partial pressures in the gas phase. In the liquid phase, the adsorbent is saturated with adsorbate molecules, which approaches the industrial conditions in a much more realistic way than the zero coverage limit that is usually studied in the gas phase. In the liquid phase, non-idealities, such as adsorbate-adsorbate interactions and surface heterogeneity, become more important. Therefore, the situation is complex, and liquid-phase adsorption properties cannot be extrapolated easily from gas-phase adsorption isotherms. However, data on competitive adsorption of organic molecules with different polarities and carbon numbers at temperatures relevant to catalysis and elevated pressures are lacking in the literature. Denayer et al. [37] used a classical HPLC (high-pressure liquid chromatography) setup with pellets or crystals of the adsorbent in the HPLC column and a refractometer as detector, which allowed work at temperatures exceeding boiling temperatures of the fluids employed as mobile phases. The authors found that the competitive adsorption between *n*-alkanes strongly depends on the pore size of the zeolite. In the large-pore zeolite Y, adsorption was governed by chain segment interactions rather than interaction with complete molecules. Thus, whereas the adsorption constants strongly depend on the carbon number of the *n*-alkanes *in the gas phase*, short-chain and long-chain *n*-alkanes adsorb in a non-selective way *in the liquid phase*, as they are built of the same carbon chain segments. Selective adsorption occurred only in the presence of aromatics or polar molecules. The molecular weight of the molecules had only a minor influence. However, when the amount of adsorbed molecules decreased, long-chain *n*-alkanes started to be adsorbed preferentially over short-chain *n*-alkanes as they then had more free space to align themselves with the zeolite surface. In the smaller pore system of ZSM-5, *n*-alkanes were stretched along the pore axis and had a stronger interaction with the zeolite. Sorbate-sorbate interactions were reduced, and a stronger competitive adsorption behavior between short-chain and long-chain *n*-alkanes was observed. However, short-chain *n*-alkanes (up to *n*-heptane) had access to the two channel types of ZSM-5, whereas longer *n*-alkanes had only restricted access to the pore system. In addition, it was observed that the polarity of the zeolite determines to what extent long-chain *n*-alkanes are adsorbed preferentially over shorter *n*-alkanes. The work of Denayer et al. [37] shows in an impressive way the lack of realistic data on adsorption properties of porous materials and is an appeal to perform competitive adsorption experiments in the liquid phase and/or under elevated pressure.

An unusual method of pore size characterization was applied by Yoon et al. [38]: A series of brightly colored charge-transfer complexes was assembled in situ by intercalation of various aromatic donors with different pyridinium acceptors in a number of large-pore zeolites. Upon the deliberate introduction of water vapor into these variously colored zeolites, the diffuse reflectance spectra underwent pronounced spectral shifts of the charge-transfer bands, the magnitude of the shifts being uniquely dependent on the molecular size/shape of the donors and acceptors as well as the dimensions of the zeolite cavities. For example, the bathochromic shift of the charge-transfer band of complexes of methylviologen with 1,4-dimethoxybenzene decreased rapidly with increasing pore size from zeolite mordenite (MOR) over zeolite L (LTL) to zeolite Omega (MAZ). In zeolite Y (FAU) with its spacious supercavities, no shift at all was observed. Thus, intermolecular charge-transfer complexes provide a method for probing the pore width of large-pore zeolites, which is especially useful because the other techniques proposed up to now mostly employ molecular probes more appropriate for the characterization of medium-pore zeolites. Another useful spectroscopic method is FT-IR spectroscopy. With this technique, Trombetta et al. [39] showed, for example, that *o*-xylene and pivalonitrile do not reach the internal zeolitic OH groups of zeolite ZSM-5. However, FT-IR spectroscopy is more appropriate for the characterization of acidic and basic sites of zeolites than for their pore size and will not further be dealt with in this review article.

Another widely applied method for pore size characterization is microcalorimetry: Derouane et al. [40] found that the heats of adsorption of methane increase considerably with decreasing pore size of the zeolitic adsorbents employed. These results were extended by many researchers, but are not within the scope of this chapter.

An approach including much modelling work was adopted by Webster et al. [41]: They used smaller probe molecules (O_2, N_2, CO, Ar, CH_4) to characterize the 8-membered-ring channels and larger probe molecules or atoms (C_2H_6, C_3H_8, *n*-C_4H_{10}, *i*-C_4H_{10}, Xe, SF_6) for the 12-membered-ring channels of zeolite mordenite. All these molecular probes had access to the channel system, but by using such a broad variety, the channels could be well assessed by gradually increasing the size of the molecules. The authors defined three different interaction types between the probe molecules and the pores: (i) strong interaction with two opposing walls, (ii) strong interaction with one wall and weak interaction with the opposite wall and (iii) interaction with one wall only. Since the larger probes were excluded from the small channels, they could provide information on the large channel and the aperture of the small channels only, yielding no information on the small channels directly. By contrast, the smaller molecules could access regions that the larger molecules could not. The data sets were analyzed using the new multiple-equilibrium analysis (MEA) method, providing adsorption capacities, accessible surface areas, pore volumes as well as enthalpies and entropies of adsorption. The big advantage of this technique is its accuracy and the possibility to determine *accessible* surface areas (e.g., 265 $m^2\,g^{-1}$ for Na-MOR and 346 $m^2\,g^{-1}$ for H-MOR), whereas with the BET model values of 458 (Na^+ form) and 462 (H^+ form) $m^2\,g^{-1}$ were found.

A considerable amount of modeling work has been done regarding the adsorption and diffusion of molecular probes in crystalline microporous solids. Molecular modeling is a useful adjunct in gaining insight into adsorption processes over molecular sieves and determining whether a given molecule will enter the pores. It thus can also be used in understanding experimental data. Deka and Vetrivel [42] demonstrated, for instance, the efficiency of the force field energy minimization technique for the study of adsorption of large molecules inside the micropores of zeolites. Molecular modeling will only be briefly mentioned here, and the reader is referred to Pikunic et al. [43], who explain in detail the various types of modeling techniques currently available.

The main methods used at present are density functional theory (DFT) and molecular simulation techniques such as Grand Canonical Monte Carlo and Gibbs Ensemble Monte Carlo [43]. DFT belongs to the group of molecular modeling techniques called statistical thermodynamic theories. DFT is best suited to simple pore systems and spherical molecules, and for these systems it is computationally faster and as accurate as the molecular simulation techniques. However, it becomes unwieldy for more complex pore systems and non-spherical molecules. In such cases the more rigorous approach of molecular simulation is preferred.

3.3 Molecular Probes for Zeolites with Different Pore Sizes

3.3.1 Small-Pore Zeolites

For the characterization of small-pore zeolites, most of the molecular probes discussed so far cannot be employed because they are too large to enter the pores. Kerr [44] used *n*-hexane, water and cyclohexane for characterizing zeolite ZK-5 (KFI). Water and *n*-hexane were adsorbed in large amounts, whereas only a very small amount of cyclohexane was adsorbed. Eder and Lercher [45] reported that, whilst propane, *n*-butane, *n*-pentane and *n*-hexane had access to the pores of ZK-5, the isoalkanes had none.

Den Exter et al. [46] studied the adsorption of several small molecules on aluminum-free deca-dodecasil 3R (DDR) and also observed that linear alkanes and alkenes could enter the pores, whereas isobutane was only adsorbed in mesopores or on the external surface. Stewart et al. [47] reported, however, that even *n*-butane was excluded (at 273 K) from the pores of zeolite Sigma-1 (DDR), the aluminosilicate analogue of deca-dodecasil 3R. This could be due to pore size narrowing caused by the cations balancing the negative framework charges.

No doubt, the most thoroughly characterized small-pore zeolite is zeolite A (LTA) [9, 14, 27, 48, 49]: This zeolite can be used for the separation of, for instance, *n*-butanol from iso-butanol [48], *n*-butane from isobutane [14, 48] and oxygen from nitrogen [27, 48, 49]. With nitrogen ($\sigma = 0.364$ nm) and oxygen ($\sigma = 0.346$ nm), subtle changes of the pore size of zeolite A were probed: On zeolite K-A with its small pore size ("3A") neither oxygen nor nitrogen were adsorbed at – 196 °C, whereas on Ca-A with its considerably larger pore size ("5A") both molecules were adsorbed at this low temperature [48]. The pore size of Na-A ("4A") lies in between the pore sizes of K-A and Ca-A, and on this zeolite the sieving characteristics changed with temperature: At – 196 °C, oxygen but no substantial amount of nitrogen was adsorbed. At

– 75 °C or even higher temperatures, nitrogen was adsorbed in larger quantities than oxygen [48]. This preferential adsorption is, however, not due to the molecular size, because both molecules have access to the channel system at this temperature, but due to chemical properties of the zeolite and the probe molecules. If the pore size of zeolite Na-A was slightly decreased by CVD, nitrogen was no longer adsorbed [49].

3.3.2
Medium-Pore Zeolites

As an excellent catalyst for the skeletal isomerization of butenes, zeolite H-ferrierite (FER) has attracted significant interest. Adsorption was used to characterize its pore system containing two perpendicularly intersecting channel systems, one consisting of 10-membered rings, the other of 8-membered rings. Van Well et al. [50] found that the pore volumes occupied by *n*-heptane and *n*-hexane were significantly lower than the ones occupied by *n*-pentane, *n*-butane and propane. From this, the authors supposed that the complete pore structure of FER is not accessible to *n*-heptane and *n*-hexane. By comparing ^{13}C NMR spectra of molecules adsorbed on H-ferrierite and Na-ZSM-22 (TON), a zeolite with one-dimensional 10-membered-ring channels, they concluded that *n*-hexane adsorbs exclusively in the 10-membered-ring channels of H-ferrierite. Similar conclusions were drawn by Eder and Lercher [45]: As detected by IR spectroscopy, only about 90% of the acid sites of zeolite H-FER were accessible to *n*-hexane. Isobutane and isopentane were adsorbed only slowly in small amounts. The authors pointed out that there is an additional interaction between the acid site and the sorbed hydrocarbon. A point to be borne in mind is that these interactions can lead to synergistic and antagonistic effects, depending on whether or not probe molecules are sited in positions different from those preferred in the absence of acid sites.

The largest amount of work in this field was devoted to ZSM-5 (MFI), the most important medium-pore zeolite. ZSM-5 is the cheapest medium-pore zeolite and therefore with regard to proposed applications of this zeolite, it is important to know whether certain molecules can access its pore system or not. This subject was first broached in Table 1, and Table 3 is designed to extend the picture for zeolite ZSM-5. When using this table, the reader should always be aware that it depends, *inter alia*, on the experimental conditions whether or not molecules have access to the pores.

Multicomponent adsorption experiments on zeolite ZSM-5 in the gas phase were performed by Klemm and Emig [61]: During adsorption of the three xylene isomers on the zeolite in a recycle adsorber, an excess of *m*-xylene and a lack of *p*-xylene were observed at the adsorber outlet in the initial stage of the experiment. These effects disappeared when approaching steady state and were explained by a higher diffusion coefficient of

Table 3 Adsorption properties of zeolite MFI [16, 19, 51–63]

Compound	Adsorption in MFI zeolites	Refs.
n-Hexane	Yes	[19, 53, 54, 57, 59, 62]
3-Methylpentane	Yes	[19, 53, 54, 62]
Benzene	Yes	[16, 19, 51, 53, 57, 58]
Toluene	Yes	[16, 54, 58]
p-Xylene	Yes	[16, 19, 51, 52, 54, 56–61]
Cyclohexane	Yes	[19, 53]
Ethylbenzene	Yes	[55, 57]
1-Methylbutylbenzene	Yes	[51]
Methylnonanes	Yes	[52]
1,4-Diisopropylbenzene	Yes	[56]
2,3-Dimethylbutane	Yes	[54]
1,2,4-Trimethylbenzene	Yes	[19, 59]
Cresol isomers	Yes	[56]
2-Methylnaphthalene	Yes	[63]
m-Xylene	Yes [1]	[52, 53, 55–57, 59]
Neopentane	Yes [2]/no [3]	[51, 52]
2,2-Dimethylbutane	Yes [4]/no [5]	[19, 53, 62]
o-Xylene	Yes [2,4]/no [3,6]	[19, 51, 52, 57, 59, 60]
1,3-Diisopropylbenzene	No	[56]
1-Methylnaphthalene	No	[63]
cis-Decalin	No	[51]
trans-Decalin	No	[51]
Tetramethylsilane	No	[51]
Cyclooctane	No	[51]
1-Ethylpropylbenzene	No	[51]
Mesitylene	No	[16, 19, 51, 54]

[1] Not adsorbed during competitive adsorption of all xylene isomers at 283 K in the liquid phase [56]
[2] Adsorption at room temperature in the liquid phase [51]
[3] Adsorption at 333 K in the gas phase [52]
[4] Adsorption at 373 K in the gas phase [19]
[5] Adsorption at room temperature [53] or at 323 K [62] in the gas phase
[6] Adsorption at 303 K in the liquid phase [60]

p-xylene. Namba et al. [56] performed competitive adsorption experiments of xylene isomers in a batch adsorber in the liquid phase using as solvent 1,3,5-triisopropylbenzene, the molecular dimension of which was too large to enter the pores of the zeolite. This technique may eliminate the effect of the external surface, because the solvent, the concentration of which is much higher than that of the adsorptives, covers the external zeolite surface. At 283 K, only *p*-xylene was adsorbed on zeolite H-ZSM-5, whereas at higher temperatures all three isomers were adsorbed, even though the *p*-xylene adsorption capacities were much higher than those of *o*- and *m*-xylene. Dessau [51] obtained similar results with xylenes at the same experimental conditions. During competitive adsorption of *n*-alkanes and aromatics on ZSM-5, he observed a marked preference for the *n*-alkanes, in distinct contrast to adsorption studies on faujasites. In counterdiffusion studies, in which *p*-xylene was adsorbed initially, it was rapidly displaced upon addition of *n*-nonane, because this hydrocarbon has a higher molecular weight. The author argued that this behavior of ZSM-5 might be due to the fact that, unlike A and Y zeolites, it contains no large cavities in which *n*-alkanes can coil around themselves. The coiling of high molecular weight alkanes inside A and Y zeolites should result in an additional entropy loss upon sorption, thereby reversing the normal order of preferential adsorption of the higher molecular weight component.

Weitkamp et al. [62, 63] performed competitive adsorption experiments in a flow-type fixed-bed adsorber in the gas phase. During adsorption of an *n*-hexane/3-methylpentane/2,2-dimethylbutane mixture over Na-ZSM-5, 2,2-dimethylbutane broke through immediately (cf. Fig. 3), i.e., this hexane isomer could not enter the zeolite pores. 3-Methylpentane and *n*-hexane had access to the pore system, but *n*-hexane was adsorbed preferentially. Therefore, 3-methylpentane reached partial pressures p_i at the adsorber outlet, which were higher than the partial pressure $p_{i,0}$ at the adsorber inlet, because it was displaced from the zeolite pores by *n*-hexane [62]. During adsorption of a 2-methylnaphthalene/1-methylnaphthalene mixture on zeolite H-ZSM-5, the bulkier 1-methylnaphthalene was not adsorbed at 100 °C, whereas

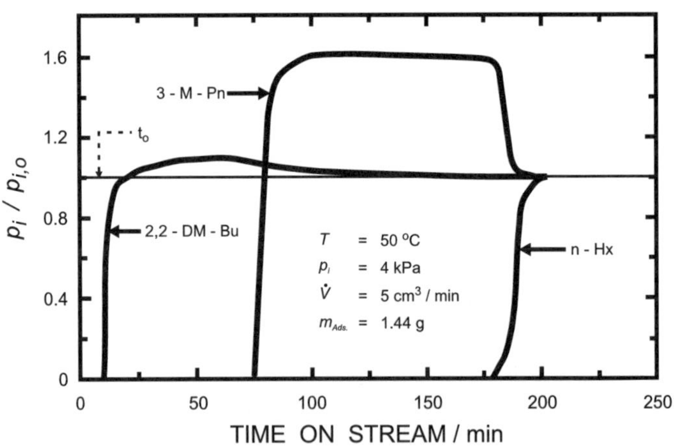

Fig. 3 Breakthrough curves for the adsorption of an *n*-hexane/3-methylpentane/2,2-dimethylbutane mixture over zeolite Na-ZSM-5 (*n*-Hx = *n*-hexane, 3-M-Pn = 3-methylpentane, 2,2-DM-Bu = 2,2-dimethylbutane)

2-methylnaphthalene could enter the zeolite pores [63]. On zeolite H-ZSM-12, both isomers were adsorbed, the larger isomer breaking through first. On zeolite Na-Y, the interaction between the zeolite and 1-methylnaphthalene was stronger than that with 2-methylnaphthalene. Therefore, the smaller 2-methylnaphthalene was displaced by 1-methylnaphthalene.

Another medium-pore zeolite that was characterized by adsorption is zeolite MCM-22 (MWW) [20, 24] and its pure silica analogue ITQ-1 (MWW) [64]: Corma et al. [24] determined adsorption isotherms of different probe molecules in the gas phase. The uptake of *m*-xylene was about half the value for toluene. The value for *o*-xylene was much lower and approximately equal to that of 1,2,4-trimethylbenzene. These results prompted the authors to suggest the existence of micropores and cavities of two different sizes, i.e., narrower micropores to which only toluene has access and wider micropores or cavities which are penetrated by toluene as well as *m*-xylene. The much lower uptake of *o*-xylene and 1,2,4-trimethylbenzene was ascribed to adsorption at the entrances of the micropores and cavities. In another series of experiments, toluene adsorption isotherms were recorded after preadsorption of other adsorptives and an intermittent outgassing. The toluene uptake was lower when preadsorption had been performed, the effect being less important for preadsorbate molecules of larger size [24]. Ravishankar et al. [20] found that the adsorption capacities of *n*-hexane, cyclohexane, *m*-xylene and mesitylene of zeolite MCM-22 were virtually independent of the aluminum content. MCM-22 adsorbed moderate amounts of mesitylene, even though this hydrocarbon should not be able to enter the 10-membered-ring pore openings. Therefore, the authors suggested that some of the large 12-membered-ring cavities should be accessible from the external surface, either through defect centers or through the presence of some 12-membered-ring cavities at the external surface [20].

3.3.3
Large-Pore and Extra-Large-Pore Zeolites

For large-pore zeolites, a suitable choice of probe molecules for pore size characterization are tertiary alkylamines with alkyl groups of varying bulkiness or with perfluorinated alkyl groups [65]. Adsorption of the critically-sized molecule perfluorotriethylamine, $(C_2F_5)_3N$, showed that the effective aperture of the dehydrated mineral faujasite is substantially smaller than that of synthetic zeolite Y, since its adsorption capacity was much smaller. Perfluorotriethylamine was not adsorbed at all on Ca-X and Ca-Y zeolite. Li-X, Na-X and Cs-X adsorbed $(C_2F_5)_2NC_3F_7$ ($\sigma = 0.77$ nm), $(C_3H_7)_3N$ ($\sigma = 0.81$ nm) and $(C_4H_9)_3N$ ($\sigma = 0.81$ nm), but not $(C_4F_9)_3N$ ($\sigma = 1.02$ nm), whereas Ca-X and Ba-X only adsorbed $(C_2F_5)_2NC_3F_7$. From this, the authors concluded that the effective pore diameter of Na-X is about 0.9 to 1.0 nm and that of Ba-X or Ca-X 0.8 to 0.9 nm [65].

Davis et al. [66] distinguished extra-large-pore zeolites from large-pore zeolites by adsorption experiments with 1,3,5-triisopropylbenzene; this molecule was too bulky to enter the pore system of faujasite with reasonable uptake rates, but did have easy access to the channels of VPI-5.

4 Catalytic Test Reactions

Catalytic test reactions for probing the pore width of porous materials have much in common with adsorption tests with the same objective. Adsorption, of course, does occur during these catalytic tests as well, but in addition, a chemical reaction takes place in a shape-selective manner. In order to design and understand test reactions for probing pore dimensions, the fundamentals of shape-selective catalysis should be addressed.

4.1 Shape-Selective Catalysis in Microporous Materials

Various definitions have been used for shape-selective catalysis. Of these, the most straightforward and useful one is as follows [2]: Shape-selective catalysis encompasses all effects in which the selectivity of a catalytic reaction depends, in an unambiguous manner, on the pore width or pore architecture of the porous solid. A comprehensive review of the fundamentals of shape-selective catalysis has been published recently [67]. The shape selectivity effects observed experimentally are usually classified into three types, namely reactant, product and restricted transition state shape selectivity [68].

Both reactant and product shape selectivity have their origins in mass transfer limitations, i.e., in the hindered diffusion of reactant and product molecules, respectively, in the pores of the zeolite catalyst. Therefore, we refer to those two shape selectivity effects as mass transfer shape selectivity. One example of the reactant type of mass transfer shape selectivity, the competitive cracking of *n*-octane and 2,2,4-trimethylpentane, is depicted in Fig. 4. This situation can be understood as molecular sieving combined with catalytic conversion: 2,2,4-trimethylpentane is too bulky to enter the pores of the zeolite and is, therefore, hindered from reaching the catalytically active sites inside the pores of the zeolite. This molecule can only be converted at catalytic sites located on the external surface of the zeolite crystallites, or it can leave the reactor without being converted. By contrast, the slender *n*-octane does have access to the pores of the zeolite, where it is readily converted. The net effect, which can be detected at the reactor outlet, is the selective cracking of *n*-octane into, e.g., an *n*-butene and *n*-butane, which are slender molecules that do not experience diffusional hindrance on their way out of the zeolite pores.

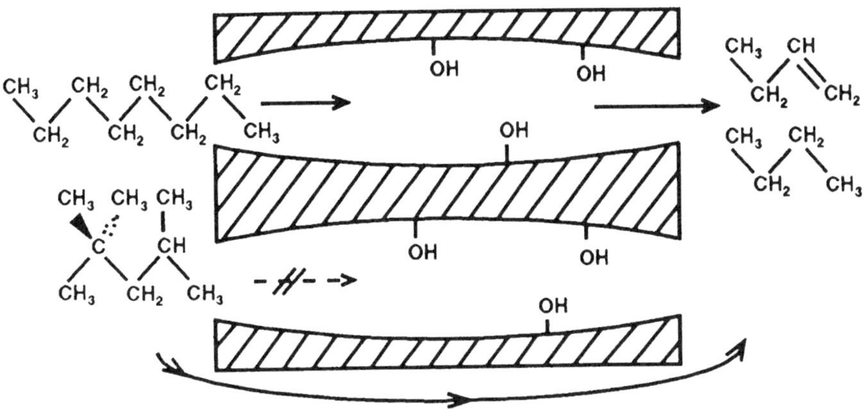

Fig. 4 Selective cracking of *n*-octane in the presence of 2,2,4-trimethylpentane as an example of reactant shape selectivity

As an example of the product type of mass transfer shape selectivity, the acid-catalyzed ethylation of toluene has been chosen, and is shown in Fig. 5. Product shape selectivity is the reverse of reactant shape selectivity: Both reactants are small enough to enter the zeolite pores, but of the potential products (*o*-, *m*- and *p*-ethyltoluene), only the slim *p*-ethyltoluene is small enough to leave the pore system. The two bulkier ethyltoluene isomers, even though they may form in relatively spacious intracrystalline cavities or at channel intersections, are unable to escape from the pores and do not occur in the reactor effluent. Ultimately, these entrapped product molecules may be catalytically transformed into more slender ones (e.g., isomerized into *p*-ethyltoluene), which are able to leave the pores, or into coke, which is deposited inside the pores. Webster et al. [9] distinguished two types of product shape selectivity: Preferential-diffusion product shape selectivity occurs, if two or more reaction

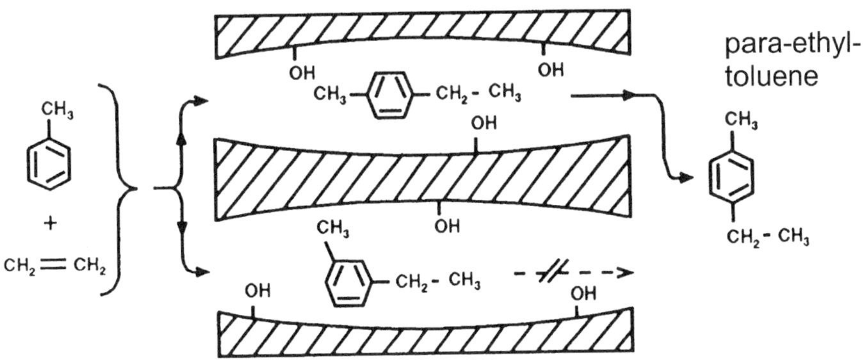

Fig. 5 Selective formation of *p*-ethyltoluene in the alkylation of toluene with ethylene as an example of product shape selectivity

products formed within the confines of the structure have effective diffusivities that sufficiently differ from each other, thus allowing one product to preferentially diffuse out of the structure. Often, the lower diffusivity of the bulkier isomer allows for isomerization to the diffusion-preferred (smaller) product. Size-exclusion product shape selectivity arises from an actual confinement of a reaction product within the structure of the solid.

A typical example of restricted transition state shape selectivity is depicted in Fig. 6. *m*-Xylene can undergo acid-catalyzed isomerization into *p*-xylene (and *o*-xylene, which is omitted in Fig. 6 for clarity) and transalkylation into toluene and one of the trimethylbenzene isomers. It is evident that transalkylation is a bimolecular reaction and as such it necessarily proceeds via bulkier transition states and intermediates than the monomolecular isomerization. In a zeolite with the appropriate pore width, there will be just enough space for the accommodation of the transition states and intermediates for the monomolecular reaction, but no room for the formation of the bulky transition states and intermediates of the bimolecular reaction, the net effect being a complete suppression of the latter reaction. As opposed to mass transfer shape selectivity, restricted transition state shape selectivity is due to intrinsic chemical effects, which emerge from the limited space around the intracrystalline active sites.

As long as one assumes that there is no catalytic contribution from the external surface, the influence of the crystallite size is as follows: If mass transfer shape selectivity is operative, the length of the intracrystalline diffusion path and hence the measurable selectivity effects, will decrease with decreasing crystallite size. If restricted transition state shape selectivity is operative, the measurable selectivity will be independent of the crystallite size. If, on the other hand, there is a significant contribution from the non-selective external surface, the influence of the crystallite size on the measurable catalytic

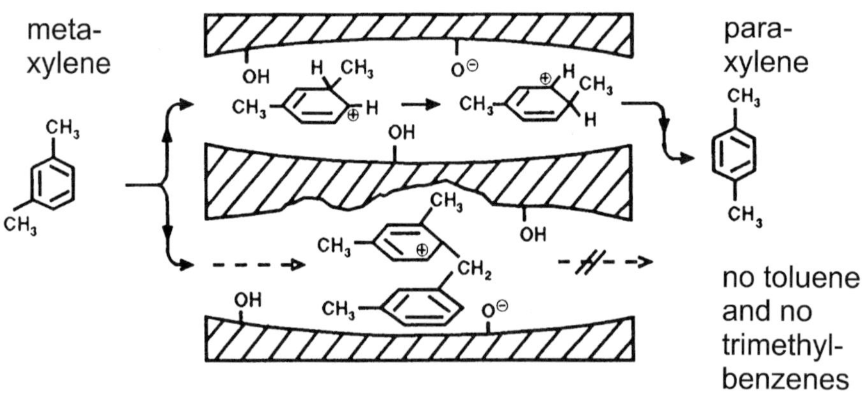

Fig. 6 Suppression of trimethylbenzene plus toluene formation during *m*-xylene isomerization as an example of restricted transition state shape selectivity

selectivity will be more complex. The selectivity will decrease with decreasing crystallite size irrespective of whether the reaction is controlled by mass transfer or restricted transition state shape selectivity. There will only be a gradual difference, which will be difficult to assess.

Several catalytic test reactions for the characterization of the effective pore widths of zeolites have been proposed so far; the results of the best known test reactions were expressed in terms of the quantitative criterions Constraint Index (*CI*), Refined or Modified Constraint Index (CI^*) and Spaciousness Index (*SI*). Extended review articles on these indices and other test reactions have previously been published [2, 69–71].

4.2
Test Reactions for Monofunctional Acidic Molecular Sieves

4.2.1
Competitive Cracking of *n*-Hexane and 3-Methylpentane – The Constraint Index, *CI*

The method of characterizing the effective pore width of zeolites by catalytic tests was first employed by researchers at Mobil Oil Corp. They introduced the Constraint Index [72], which is based on the competitive cracking of an equimolar mixture of *n*-hexane and 3-methylpentane on the monofunctional acidic form of the zeolite. As long as the catalyst pores are sufficiently spacious, branched alkanes are cracked at higher rates than their unbranched isomers. The opposite holds for medium-pore zeolites, such as H-ZSM-5. Based on this shape selectivity effect, the Constraint Index was defined at Mobil as the ratio of first order rate constants (k) of the cracking of *n*-hexane and 3-methylpentane:

$$CI \equiv \frac{k_{n\text{-Hx}}}{k_{3\text{-M-Pn}}} \,. \qquad (1)$$

The Constraint Index has been routinely used in Mobil's patents for about two decades. In the scientific literature, precise experimental conditions for its determination were given (reaction temperature between 290 and 510 °C, liquid hourly space velocity (LHSV) between 0.1 and 1 h^{-1}, 10 vol.-% of each reactant in helium as carrier gas, mass of catalyst around 1 g, overall conversion 10 to 60%, fixed-bed reactor at atmospheric pressure) [72]. The reactor effluent is to be analyzed after 20 min time on stream. With the performance equation for integral fixed-bed reactors [73], Eq. 1 can be rewritten as

$$CI \equiv \frac{\log(1 - X_{n\text{-Hx}})}{\log(1 - X_{3\text{-M-Pn}})} \qquad (2)$$

$(X = \text{conversion})$.

According to Frilette et al. [72], the Constraint Index allows for a classification of zeolites into large-pore (12-membered-ring), medium-pore (10-

membered-ring) and small-pore (8-membered-ring) molecular sieves:

$CI < 1$: large-pore materials;

$1 \leq CI \leq 12$: medium-pore materials;

$12 < CI$: small-pore materials.

Constraint Indices taken from the literature are summarized in Table 4. By and large, a correct classification of 8-, 10- and 12-membered-ring zeolites can be achieved. Exceptions are zeolites ZSM-12 (MTW) and MCM-22 (MWW). ZSM-12 possesses 12-membered-ring pores which are, however, strongly puckered and hence, narrowed, and on the basis of its Constraint Index it would be classified as a material with 10-membered-ring pores. Zeolite MCM-22 possesses large intracrystalline cavities, which are only accessible via 10-membered-ring windows. In this case, the Constraint Index of 1.5 can be looked upon as an averaged value characterizing the mean space available in the 10-membered-ring pores and in the large cavities.

Values for the Constraint Indices of more recent zeolite structures are given in [78], revealing even more shortcomings of the Constraint Index test. One example is that zeolites with 14-membered rings and pore openings greater than 0.8 nm cannot be distinguished from large-pore zeolites, simply because the degree of absence of spatial constraints cannot vary. Another example is that some zeolites with large internal cavities and pores composed of 8- or 9-membered rings do not generate Constraint Indices higher than zeolites with 10-membered-ring pores, probably again because the Constraint Index test averages the space in the pores and the cavities.

Table 4 Constraint Indices for selected zeolites. Data from the patent literature [74, 75], values in parentheses from the open literature [76, 77]. The *dashed line* is to indicate that zeolite ZSM-12 belongs to medium-pore zeolites if classified after CI even though it is defined as a large-pore zeolite by its ring size

Zeolite	*CI*	Classification after *CI*	True ring size
Eronite (ERI)	38	small pores	8-ring
ZSM-23 (MTT)	9.1		10-ring
ZSM-22 (TON)	7.3 (7.4)		10-ring
ZSM-5 (MFI)	6–8.3 (4.6)		10-ring
ZSM-11 (MEL)	5–8.7	medium pores	10-ring
ZSM-50 (EUO)	2.1		10-ring
MCM-22 (MWW)	1.5		10-ring
ZSM-12 (MTW)	2.3		12-ring
Mordenite (MOR)	0.5 (1.0)		12-ring
Beta (BEA)	0.6–2.0	large pores	12-ring
X or Y (FAU)	0.4 (0.2)		12-ring

Other advantages and disadvantages of the Constraint Index have been discussed in detail in [2]. The reaction can suffer from a relatively fast catalyst deactivation, if large-pore zeolites are used, which makes a reliable determination of the *CI* difficult, and a (sometimes pronounced) temperature dependence for medium-pore zeolites. The reason for the latter has been investigated in much detail by Haag et al. [79, 80]. These authors showed, in a convincing manner, that two basically different cracking mechanisms can be operative, namely the classical bimolecular chain-type mechanism involving tri-coordinated alkylcarbenium ions and a monomolecular mechanism via non-classical penta-coordinated alkylcarbonium ions. The latter has a higher activation energy, hence its contribution increases with increasing reaction temperature, and due to its requiring less space, the Constraint Index for a given zeolitic material decreases with increasing temperature. However, Macedonia and Maginn [81] recently used Monte Carlo integration methods employing a classical molecular mechanics force field to predict values for the Constraint Indices of 12 different zeolites. These authors concluded that it is not necessary to invoke such a change in mechanism to explain decreasing Constraint Indices with increasing temperature. Their calculations indicated that the impact of confinement on the bimolecular transition state decreases with increasing temperature, effecting a decrease of the Constraint Index. This was said to be caused by a competition between energetic confinement effects that dominate at lower temperatures and entropic effects that become dominant at high temperature.

As to the nature of the shape selectivity effects, Haag et al. [82] demonstrated in a most impressive study with H-ZSM-5 samples of equal concentration of acidic sites but different crystal sizes (0.05 to 2.7 μm) that neither the measurable rate of cracking of *n*-hexane nor that of the bulkier 3-methylpentane depend on the length of the intracrystalline diffusion paths. From this finding, the selectivity effects encountered in the competitive cracking of *n*-hexane and 3-methylpentane have to be interpreted in terms of intrinsic chemical effects (i.e., restricted transition state shape selectivity) rather than by mass transport effects. Haag et al. [82] suggested that the rate-controlling step in the chain-type mechanism of acid-catalyzed alkane cracking via carbocations is the chain-propagating hydride transfer between a cracked alkylcarbenium ion and a feed molecule, requiring significantly more space for the transition state if the feed alkane is branched, the net effect being a significant inhibition of cracking of 3-methylpentane. The simulations performed by Macedonia and Maginn [81] indicated, however, that for zeolites, the pores of which are too small to accommodate the bimolecular transition state, such as ZSM-23 (MTT) and ferrierite (FER), the monomolecular mechanism dominates, with the measured Constraint Index attributed to reactant shape selectivity. Only for zeolites, the pores of which are large enough for the bimolecular transition state but small enough for confinement effects, the bimolecular reaction was predominant, and the selectivity was based on restricted transition state shape selectivity.

Recently, Baeck et al. [32] demonstrated that the Constraint Index is useful for probing subtle changes of the pore size as effected by CVD. The Constraint Index determined on zeolite Mg-ZSM-22 (TON) increased from about 9.9 to about 13.3 after 3 h of deposition of tetraethyl orthosilicate, reflecting the decrease of the pore opening by CVD.

In conclusion, with the Constraint Index, the idea of probing the pore width by catalytic test reactions was introduced into zeolite science. It no doubt fostered the search for alternative and improved catalytic test reactions. In spite of its shortcomings enumerated above, the Constraint Index test has been widely used, and ample data is available in the literature.

4.2.2
Isomerization and Disproportionation of *m*-Xylene

On acidic catalysts, *m*-xylene can undergo isomerization into *o*- and *p*-xylene and disproportionation (or transalkylation) into toluene and 1,2,3-, 1,2,4- or 1,3,5-trimethylbenzene (cf. Fig. 7). While it is obvious that disproportionation is necessarily a reaction involving a bimolecular transition state, there is some ambiguity as to whether the acid-catalyzed isomerization of xylenes proceeds via a monomolecular or a bimolecular pathway [83, 84].

The use of *m*-xylene conversion for the characterization of the effective pore width of zeolites was first proposed by Gnep et al. [85]. These authors identified three selectivity criteria that may furnish valuable information on the effective pore width: (i) the relative rates of formation of *o*- and *p*-xylene, (ii) the ratio of rates of disproportionation and isomerization and (iii) the distribution of the trimethylbenzene isomers formed in the disproportionation reaction.

Criterion (i) is based on the finding that, in the absence of shape selectivity, *o*- and *p*-xylene are formed at virtually the same rate. With decreasing pore width, however, the formation of *p*-xylene is increasingly favored over the formation of the bulkier *o*-xylene [85]. This effect is best interpreted in terms of product shape selectivity, i.e., progressively hindered diffusion of *o*-xylene molecules, as the pores become narrower.

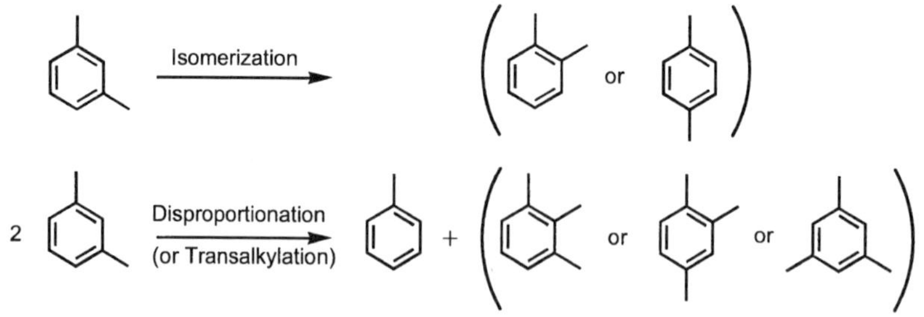

Fig. 7 Principal reaction pathways of *m*-xylene over acidic catalysts

Criterion (ii) is a quantitative expression of the observation that, with decreasing pore width, isomerization of *m*-xylene is more and more favored over its disproportionation. At 350 °C, the bimolecular disproportionation is completely suppressed in the medium-pore zeolite ZSM-5, whereas isomerization (which is likely to proceed via the monomolecular mechanism at these high temperatures) does not experience significant hindrance [85]. Later, Olson and Haag [30] demonstrated that, if zeolite catalysts with different pore systems are used, a linear relationship results between the ratio of rate constants of disproportionation and isomerization and the dimensions of the largest zeolite cavities. One of the conclusions of this study was that the suppression of disproportionation is due to restricted transition state shape selectivity rather than to mass transfer effects.

Finally, criterion (iii) was deduced from the observation that the disproportionation of *m*-xylene in the spacious cavities of zeolite Y and in the much narrower channels of mordenite (MOR) resulted in significantly different distributions of the trimethylbenzene isomers: Of the product obtained from H-mordenite, 94% of the trimethylbenzenes consisted of the 1,2,4-isomer. On the other hand, 74% of the 1,2,4-isomer and 24% of the 1,3,5-isomer were found in the product from H-Y. (On both zeolites, only negligible amounts of 1,2,3-trimethylbenzene were formed, and this was attributed to its low concentration in thermodynamic equilibrium [85]). To interpret the hindered formation of 1,3,5-trimethylbenzene from *m*-xylene in mordenite, the authors mainly invoked restricted transition state shape selectivity.

m-Xylene isomerization as a catalytic test reaction was later adopted by a number of other groups [2, 67], and the criteria proposed by Gnep et al. [85] were refined and applied to a broader structural variety of molecular sieves. Dewing [86] introduced the so-called *R* value, defined as the ratio of rate constants observed experimentally for the formation of *o*- and *p*-xylene from *m*-xylene under conditions of diffusional limitations. Joensen et al. [87] coined the term Shape Selectivity Index (*SSI*), which is defined as the yield ratio of *p*- and *o*-xylene observed on a shape-selective zeolite and extrapolated to zero conversion minus the same ratio obtained under identical experimental conditions on a catalyst with sufficiently large pores, i.e., in the complete absence of shape selectivity in *m*-xylene isomerization. Neither Dewing's *R* value nor the Shape Selectivity Index by Joensen et al. received much attention by others.

A critical evaluation of the published data revealed [2] that a comparison of the rates of isomerization and disproportionation, if applied quantitatively to a larger number of zeolites, does not allow for a reliable ranking of large-pore molecular sieves. No correlation at all exists between criterion (ii) and the window or pore size [2]. Therefore, criterion (ii) will not further be dealt with.

According to Martens et al. [88] there seems to be, however, very valuable information on the pore architecture in the distribution of the trimethylbenzenes formed from *m*-xylene in large-pore zeolites. If disproportionation occurs at all in 10-membered-ring zeolites, it gives 1,2,4-trimethylbenzene exclusively, probably because the diphenylmethane-type carbocation intermediate leading to this particular isomer can be best accommodated in their narrow pores. Exceptions are zeolites ZSM-50 (EUO) and NU-87 (NES) [89] in which 1,2,3- and 1,3,5-trimethylbenzene are formed as well. This can be rationalized if one assumes that the disproportionation takes place, at least in part, in the spacious side pockets running perpendicular to the 10-membered-ring channels (EUO) or in the 12-membered-ring channels (NES). For the zeolites with the largest pores or cavities, the distribution of the trimethylbenzenes is close to equilibrium. Hence, mechanistic, i.e., kinetic conclusions from these data should be drawn with great care, and therefore, criterion (iii) is not further considered here.

With respect to criterion (i), it has been concluded [2] that it allows for a safe discrimination between 10-($S_{p\text{-xylene}}/S_{o\text{-xylene}} \geq 2$, S = selectivity) and 12-membered-ring ($S_{p\text{-xylene}}/S_{o\text{-xylene}} \approx 1.0$ to 1.5) zeolites. However, later work revealed that the situation is more complex if the experimental conditions for the test are not exactly defined and if one has non-uniform pore systems, as in the case of zeolite MCM-22 (MWW) (cf. Table 5 and cited references). This zeolite has two-dimensional sinusoidal channels (0.41 × 0.51 nm), large supercavities (0.71 nm diameter, 1.82 nm height) accessible through 10-membered ring openings (0.40 nm × 0.55 nm) and large hemicavities on the external surface (0.71 nm diameter, 0.70 nm depth).

As demonstrated by Laforge et al. [90], each of these pore systems affect the $S_{p\text{-xylene}}/S_{o\text{-xylene}}$ ratio entirely differently. These authors successively deactivated or poisoned the acid sites within each pore system to examine the role they played during *m*-xylene transformation. Using this method, it was shown that the external hemicavities gave an $S_{p\text{-xylene}}/S_{o\text{-xylene}}$ ratio of 1.1 and a low selectivity to disproportionation. The supercavities, on the other hand, gave an $S_{p\text{-xylene}}/S_{o\text{-xylene}}$ ratio of 3.5. Furthermore, 98.5% of the disproportionation reactions and 35% of the isomerization reactions occurred in the supercavities. In the sinusoidal pores, disproportionation was sterically hindered and only isomerization resulted, giving an $S_{p\text{-xylene}}/S_{o\text{-xylene}}$ ratio of 42.5. Of the *m*-xylene transformation that took place over MCM-22, 42% occurred in the supercavities, 36% in the sinusoidal channels and 22% in the external hemicavities. Thus, the final product distribution is a combination of the contribution of all three pore systems, and one cannot speak of a single pore size. In addition, the acid sites in the different pore locations deactivated at different rates, altering the product distribution over time.

Thus, not surprisingly, the selectivity ratios $S_{p\text{-xylene}}/S_{o\text{-xylene}}$ for zeolite MCM-22 reported in the literature differ greatly (cf. Table 5). Whereas Corma et al. [92–94] and Adair et al. [95] determined their experimental data dependent on time on stream and extrapolated to zero time in order to account for the deactivation of the catalysts due to coking and their different decay rates, Ravishankar et al. [91] obtained the selectivity ratio after 2 h time on

Table 5 Results from *m*-xylene reactions over zeolite H-MCM-22 (MWW) [90–95] (*TOS* = time on stream, X = conversion, *S* = selectivity)

Crystallite size	n_{Si}/n_{Al}	*TOS*/h	*T*/K	$\frac{\dot{n}_{H_2}}{\dot{n}_{m\text{-xylene}}}$	$X_{m\text{-xylene}}$	$\frac{S_{p\text{-xylene}}}{S_{o\text{-xylene}}}$	Refs.
1–2 μm	14	2	573	4.0	58.1%	1.03	[91]
1–2 μm	14	2	573	4.0	39.7%	1.02	[91]
1–2 μm	14	2	573	4.0	20.2%	1.06	[91]
1–2 μm	14	2	573	4.0	9.5%	1.07	[91]
1–2 μm	14	2	573	4.0	4.1%	0.87	[91]
0.5 μm	11	0	623	0	not given	2.0	[92–94]
submicrometer	14	0	590	0	not given	2.6	[95]
not given	10	0.03	623	0	13.3%	3.8	[90]
not given	10	24	623	0	7.7%	4.0	[90]

stream. However, since in the latter experiments hydrogen was present in the feed (as opposed to the former), deactivation might still be absent after 2 h time on stream. Possibly, the presence of hydrogen can account for the low $S_{p\text{-xylene}}/S_{o\text{-xylene}}$ ratio observed in [91].

Regarding the group of zeolites that provide access to the intrazeolitic pore system only via 10-membered-ring channels, but have significant space inside the pores due to large cavities (MWW), large side pockets (EUO) or 12-membered-ring bridges (NES), MCM-22 (MWW) and EU-1 (EUO) have selectivity ratios in the range of medium-pore zeolites (2.6 and 3.9, respectively), but the ratio of NU-87 (NES) is considerably lower (1.3 to 1.6, typical of large-pore zeolites) [89, 95] – which was explained by the high residence time of the isomers in the 12-membered-ring channels [89]. Corma et al. [96] also reported values typical of 12-membered-ring zeolites for zeolite NU-87. However, zeolite CIT-1 (CON), having 10- and 12-membered-ring pore openings, was classified as medium-pore zeolite by Corma et al. [96], but as large-pore zeolite by Adair et al. [95]. More experimental data on the characteristics of more recently discovered zeolites in the *m*-xylene test reaction can be found in [25, 95–98].

The data discussed so far may lead one to conclude that the *m*-xylene test reaction is of limited use for probing the pore width of zeolites. However, Jones et al. [97] reported that the reactions of *m*-xylene could give, under proper experimental conditions, information enabling the characterization of medium-, large- *and* extra-large-pore zeolites with one and the same test reaction, whereas all other test reactions proposed so far are only appropriate for medium- *or* large-pore zeolites. These authors found that large- and extra-large-pore zeolites with one-dimensional channel systems, such as CIT-5 (CFI), SSZ-24 (AFI), SSZ-31 and UTD-1 (DON), have an $S_{p\text{-xylene}}/S_{o\text{-xylene}}$ ratio < 1, whereas multidimensional zeolites such as zeolites Y (FAU) and Beta (BEA) display ratios > 1 (note, however, that zeolite L (LTL) with unidimensional pores has a selectivity ratio > 1 as well). In addition, the authors stress that it is imperative that all zeolites be compared at exactly the same experimental conditions, for example, similar conversion levels and even the same flow rate (sufficiently high to ensure the absence of significant external diffusion limitations on reaction selectivity) are important. In addition, all zeolites should have roughly the same crystal size.

In conclusion, the *m*-xylene test reaction has found widespread application for probing the pore width of zeolites, probably because it is easy to handle (the number of products which have to be considered is small). However, this reaction cannot be applied blindly, especially if the zeolite possesses a non-uniform pore system. Used intelligently, perhaps together with selective poisoning, this reaction can provide valuable information on the role of various pore systems within a zeolite, played during a catalytic reaction. Furthermore, this could open a new perspective for the application of catalytic test reactions in modern catalysis research.

If properly handled, the test reaction might also allow for a meaningful classification of several classes of zeolite materials. However, a safe ranking of zeolites with respect to their pore width within these classes is certainly difficult and less safe.

4.2.3
Other Test Reactions

Csicsery [99, 100] recognized pronounced shape selectivity effects during the reaction of ortho-ethyltoluene in acidic zeolites. A wealth of information on the pore size of the catalyst can be deduced from the product distributions. Of course, all three criteria advanced for *m*-xylene conversion (cf. previous section) are applicable to Csicsery's test as well. The latter furnishes, however, additional information, *inter alia* because two types of alkyl groups of different bulkiness can be transferred. Because of the larger size of ortho-ethyltoluene and the intermediates, transition states and products derived from it, this test might be much more appropriate for probing the pores of 12-membered-ring zeolites than the *m*-xylene test.

The disproportionation of ethylbenzene into benzene and the isomeric diethylbenzenes was originally proposed by Karge et al. [101] as a test reaction to collect rapid information on the number of strong Brønsted acid sites in zeolites. Later, the reaction was found suitable for probing the pore width as well. From comparative catalytic experiments with a variety of 10- and 12-membered-ring zeolites [102, 103], the following criteria were established: (i) 12-membered-ring zeolites exhibit an induction period whilst 10-membered-ring zeolites do not; (ii) after the induction period, there is very little or no deactivation in 12-membered-ring zeolites, whereas in all 10-membered-ring zeolites, there is considerable deactivation from the very beginning of the catalytic experiment; (iii) on 12-membered-ring zeolites, the distribution of the diethylbenzenes is ca. 5 mol-% ortho-, 62 mol-% meta- and 33 mol-% para-isomer; on 10-membered-ring zeolites, the ortho-isomer is often completely absent or it appears as a very minor product, especially at the onset of the catalytic experiment. The combined application of the above criteria allows for a safe discrimination between medium- and large-pore zeolites (for the structures examined so far), but not yet for a ranking of the zeolites according to their effective pore width. However, Das et al. [104] demonstrated that the *p*-diethylbenzene selectivity increased from 33.4% on H-ZSM-5 to 99.4% on silylated H-ZSM-5 and attributed this effect to the pore size regulation achieved by vapor deposition of the bulky tetraethyl orthosilicate followed by its calcination at 813 K. In a later study [105], the authors employed reaction mixtures containing 80% *m*-xylene and 20% ethylbenzene. With this mixture, essentially two reactions occurred on ZSM-5 zeolites at 678 K, namely *m*-xylene isomerization and ethylbenzene dealkylation to benzene and ethylene. With increasing silylation time, the *m*-xylene conversion decreased much more strongly than the ethylbenzene conversion. After a silylation period of 210 min, the conversion of *m*-xylene was zero, whereas the ethylbenzene conversion was still about 20%. This example shows that such test reactions can be useful for probing subtle differences in the pore size. However, they need to be designed, defined and refined for a small pore size range.

Kim et al. [106] proposed a test reaction particularly suitable for probing the effective pore width of large-pore zeolites. They chose a very bulky reactant, namely *m*-diisopropylbenzene, which has no access to 10-membered-ring pores, for the conversion with propene. Under the reaction conditions chosen, isomerization of *m*-diisopropylbenzene to its para-isomer and alkylation to the three isomeric triisopropylbenzenes occurred. It is the rationale in the design of this catalytic test reaction that 1,3,5-triisopropylbenzene is a bulkier product than its 1,2,4-isomer. A problem associated with this test reaction is the deactivation of the catalyst due to its direct exposure to an alkene under a substantial partial pressure. Nonetheless, with a sophisticated sampling strategy, from the selectivity ratio of the 1,3,5- to the 1,2,4-isomer, conclusions could be drawn on the effective pore width of the investigated materials that are in general agreement with their Spaciousness Indices (cf. Sect. 4.3.2). To minimize coke deposition in the catalyst pores, Singh et al. [107] proposed a modification of the test, i.e., the use of isopropanol instead of propene as alkylating agent and of a large excess of *m*-diisopropylbenzene and a solvent. Unfortunately, the authors restricted their studies to zeolites Y, mordenite and Beta, so that the complete and critical evaluation of this test reaction for probing extra-large-pore zeolites is still lacking.

The alkylation of biphenyl with propene has been suggested as a test reaction for examining the pore size of pillared clays, large-pore zeolites and related microporous materials [108]. The major selectivity criterion discussed is the content of *o*-isopropylbiphenyl in the monoalkylated product fraction. The reaction could have a much broader potential for characterizing large- and even extra-large-pore molecular sieves, but more systematic work on the influence of the pore size and geometry of the catalyst on the distribution of the mono-, di- and, desirably, trialkylated products is needed.

An index analogous to the Constraint Index, yet designed for large-pore zeolites, was proposed by Miller [109]. He introduced the C_8 Selectivity Index ($C_8S.I.$), defined as

$$C_8S.I. \equiv \frac{\ln(1 - X_{n\text{-Oc}})}{\ln(1 - X_{2,2,4\text{-TM-Pn}})}, \tag{3}$$

as determined during competitive cracking of a 1 : 1 mixture of *n*-octane (*n*-Oc) and 2,2,4-trimethylpentane (2,2,4-TM-Pn) [109]. The rationale of this test is that *n*-octane can easily enter the pores, whereas 2,2,4-trimethylpentane cannot. This test is also suitable for bifunctional zeolites, since the effect of the metal on the numerical value of the index has been demonstrated to be small [109].

Doblin et al. [110] proposed a more general reaction constraint index, namely the Selectivity Ratio (*SR*), defined as

$$SR \equiv \frac{\log(1 - X_{\text{less branched isomer}})}{\log(1 - X_{\text{more highly branched isomer}})}, \tag{4}$$

as determined during competitive cracking of a 1 : 1 mixture of two alkanes with the same number of carbon atoms. With this definition, Mobil's Constraint Index (Sect. 4.2.1) becomes a special case of the Selectivity Ratio. (Note that, in the original papers of Miller [109] and Doblin et al. [110], the indices were formally defined in a slightly different manner which, from a chemical reaction engineering viewpoint, needed some re-definition.) Compared to the Constraint Index, the Selectivity Ratio is a more flexible index, since probe molecules can be chosen from a variety of alkanes to more closely match the catalyst pore size to be characterized. This is an interesting approach, but more experimental data collected on a much broader variety of zeolites including extra-large-pore materials is needed, before the usefulness of *SR* can be assessed.

Flego and Perego [111] recently proposed the aldol condensation of acetone as a more unusual test for characterizing both the acid site density and the pore dimension of small-, medium- and large-pore zeolites in their acidic forms. The products can be readily analyzed by UV/Vis spectroscopy. Medium-pore zeolites were shown to give phorone as the final product, whereas large-pore zeolites tend to favor the formation of isophorone. However, no quantitative criterion was coined, hence this test reaction appears to need more refinement, before it can be recommended for a more general application.

The last test reaction for acidic zeolites that has to be discussed is the "methanol conversion test": The acid-catalyzed Methanol-to-Hydrocarbons (MTH) reaction is known to proceed in a stepwise manner [112, 113]: With increasing severity, dimethyl ether, olefins and a mixture of aromatics plus alkanes are successively formed. By suitable adjustment of the reaction conditions, either the yield of light olefins or that of aromatics plus alkanes can be maximized, the corresponding process variants being referred to as Methanol-to-Olefins (MTO) and Methanol-to-Gasoline (MTG), respectively. Large amounts of water are necessarily formed in the MTH reaction, and the risk exists that this brings about an undesired steaming of the zeolites with a concomitant framework dealumination. To minimize these effects, the zeolites to be tested are to be used in a form with a sufficiently high n_{Si}/n_{Al} ratio, yet a minimum amount of aluminum must be present in the framework since it generates the Brønsted acidity and, hence, the catalytic activity.

The "methanol conversion test" relies on the finding that the selectivity of the MTH reaction is strongly influenced by the zeolite pore geometry. Zeolite structures with 8-membered-ring pore apertures, such as erionite (ERI) or chabazite (CHA) are capable of converting methanol selectively to light olefins. Aromatics, if formed at the catalytic sites, would be trapped inside the large cavities existing in most of these small-pore zeolites. On medium-pore zeolites, such as ZSM-5, aromatics do occur in the product, but due to the space limitations inside the pores, the aromatics distribution terminates at around C_{10} (durene is widely considered as the bulkiest aromatic hydrocarbon occurring in an MTG product) with a maximum at around C_8. In zeolite mordenite (MOR), the bulky polymethylbenzenes tend to be the main products, whereas in ZSM-12 (MTW) with its smaller effective pore diameter, a broader aromatics distribution has been observed [112]. These and other selectivity effects in the MTH reaction were exploited by Yuen et al. [114] for the characterization of AFI and CHA molecular sieves. Webster et al. [9] used the results from the test performed at different temperatures with zeolite H-ZSM-5 to assess the change of the effective pore dimension with temperature. The authors concluded that the effective channel size of H-ZSM-5 is between 0.662 and 0.727 nm at 300 °C, the *MIN-2* (cf. Sect. 2.1) dimensions of the reaction products *p*-xylene (which is formed) and *o*-xylene (which is not formed), and at least 0.764 nm at 370 °C, the *MIN-2* dimension of the reaction product 1,2,3-trimethylbenzene. Upon increasing the temperature, the dimensions of the channel intersections increase as well, namely from 0.817 by 0.888 nm to 0.908 by 0.909 nm [9].

Bendoraitis et al. [115] used a test reaction similar to methanol conversion (i.e., based on mass transfer shape selectivity) but this focused on reactant shape selectivity instead of product shape selectivity: These authors determined the catalytic pore sizes of ZSM-5, ZSM-23 (MTT) and mordenite on the basis of the sizes of molecules converted during the dewaxing of waxy distillate feeds to 0.55 by 0.70, 0.45 by 0.65 and 0.90 by 1.0 nm, respectively. As these examples show, this kind of test reaction can provide much useful information. However, the evaluation of the data is troublesome, since many products have to be considered and no criterion for an easy quantitative analysis has been defined. In addition, the crystal size of the materials examined should be kept constant, which is often difficult.

Mesoporous aluminosilicates assembled from microporous materials present a special challenge to the researcher because they have not one but two pore systems. How does one conduct catalytic test reactions on such catalysts in order to unequivocally determine the pore size? One way is to inactivate the acid sites in one of the pore systems, using an appropriate basic molecule, before studying the pores in the remaining system.

Sun et al. [116] used this method to study the microporosity in a series of ordered mesoporous aluminosilicates. First, the samples were treated using a bulky, basic organic molecule, namely 5,7-dibromo-8-hydroxy-quinoline. This molecule deactivated the acid sites in the mesopores but left those in the micropores untouched. Once the mesopores have been deactivated, the pore size of the micropores can be established using shape-selective, catalytic test reactions in much the same way as described in the foregoing sections. Using carefully selected test reactions, Sun et al. [116] were able to demonstrate that the micropores of the mesoporous aluminosilicates are distinguishable from one another. Specifically, it was shown that the base-treated material Al-SBA-15 was catalytically active for 2-propanol dehydration but not for the cracking of cumene and 1,3,5-triisopropylbenzene (TIPB); the treated MAS-9 (assembled from ZSM-5 nanoclusters) was catalytically active for 2-propanol dehydration and cumene cracking but not for TIPB cracking, and the treated MAS-7 (Beta nanoclusters) was catalytically active for 2-propanol dehydration, cracking of cumene and cracking of TIPB. Thus using this method, it could be established that Al-SBA-15 contains micropores of 0.34 to 0.50 nm, MAS-9 micropores of 0.50 to 0.74 nm and MAS-7 micropores of 0.74 to 0.87 nm.

4.3
Test Reactions for Bifunctional Molecular Sieves

In typical bifunctional molecular sieves, the Brønsted acid form of a zeolite is modified by a small amount, i.e., typically 0.1 to 1 wt.- %, of a hydrogenation/dehydrogenation component, usually platinum or palladium. Since the early 1980s, the potential of catalytic reactions occurring on bifunctional forms of zeolites for characterizing the pore size and pore architecture of zeolites has been explored. Particular emphasis was placed on the isomerization and hydrocracking of long-chain *n*-alkanes, such as *n*-decane (cf. Sect. 4.3.1), and to hydrocracking of cycloalkanes, for example, butylcyclohexane (cf. Sect. 4.3.2). A factor these reactions have in common is that they are conducted under hydrogen, which is activated by the metal component of the catalyst. One very favorable repercussion of the presence of hydrogen is that coke formation and the concomitant catalyst deactivation are absent or very slow. This not only makes the experiments much easier (neither the conversion nor the selectivities vary with time on stream), but also eliminates the risk of progressively narrowing the pores by the deposition of carbonaceous residues formed in the test reaction.

The mechanisms of hydrocarbon conversion over bifunctional zeolite catalysts have been extensively discussed (cf., e.g., [2, 117–120]) and are beyond the scope of this chapter. In brief, the saturated hydrocarbon reactant is first dehydrogenated on the noble metal to the corresponding olefins which, in turn, are protonated at Brønsted acid sites. The resulting carbenium ions, while adsorbed at acid sites, undergo skeletal rearrangements and β-scissions. Finally, the product carbocations are desorbed from the acid sites as alkenes, which are hydrogenated on the noble metal. At low conversion, for example, monobranched cations, formed from *n*-alkanes in the first rearrangement step, will usually be readily desorbed and hydrogenated to isoalkanes, and these appear as the sole products. Upon raising the conversion, e.g., by increasing the residence time or the reaction temperature, the monobranched alkylcarbenium ions undergo another rearrangement. The resulting dibranched carbenium ions desorb and appear in the product as dibranched isoalkanes. Upon further increasing the conversion, the dibranched cations rearrange once more into tribranched ones. These can undergo the very rapid type A β-scission [120]. Its salient feature is that β-scission starts from a tertiary and leads again to a tertiary carbenium ion. Tertiary carbenium ions are much more stable than secondary or primary carbenium ions. Indeed, in the absence of shape selectivity, i.e., using a zeolite or any other catalyst with sufficiently spacious pores, the mechanism of hydrocracking of long-chain *n*-alkanes proceeds via this route: Several skeletal rearrangements are followed by type A β-scission; besides, there is a smaller contribution of type B β-scission [120] of dibranched carbenium ions, involving tertiary and the less stable secondary carbenium ions.

4.3.1 Isomerization and Hydrocracking of Long-Chain *n*-Alkanes and Light (C_7) Cycloalkanes – The Refined or Modified Constraint Index, *CI**

Whereas hydrocracking of *n*-alkanes proceeds via highly branched and, hence, relatively bulky alkylcarbenium ions in large- and extra-large-pore zeolites, the tribranched precursor ions of the favorable type A β-scission cannot form under the steric constraints imposed in medium-pore zeolites. The system is then forced into alternative routes via less bulky intermediates, i.e., the narrower the pores, the higher will be the contributions of β-scissions involving less stable secondary or even primary carbenium ions. This shift in the hydrocracking mechanism with decreasing pore width brings about a large number of selectivity changes in the hydrocracked product.

Striking shape selectivity effects do not only occur in hydrocracking, but also in skeletal isomerization, which precedes hydrocracking. If *n*-decane is converted on a bifunctional catalyst with sufficiently spacious pores, such as zeolite Pt/Ca-Y, the product mixture obtained at low conversion consists of all possible isodecanes with one branch, i.e., 2-, 3-, 4- and 5-methylnonane, 3- and 4-ethyloctane as well as 4-propylheptane [121]. As the catalyst pores become narrower, the bulkier isomers cannot be formed any more inside the pores or cannot escape from there. On Pt/H-ZSM-5, for example, neither 4-propylheptane nor the two ethyloctane isomers are formed [120, 122].

An even more subtle shape selectivity effect concerns the formation of the four isomeric methylnonanes. In the absence of shape selectivity and at low conversions, i.e., in the kinetic regime, 2- and 5-methylnonane form about half as fast as 3- and 4-methylnonane. This can be readily understood in terms of the branching mechanism *via* protonated cyclopropanes, cf. [121]. 2-Methylnonane is the kinetically preferred isomer in 10-membered-ring zeolites. Based on the observation that, with decreasing pore width of the zeolite, the amount of 2-methylnonane formed from *n*-decane at low conversions increases relative to the other methylnonanes, the Refined or Modified Constraint Index, *CI**, was defined as

$$CI^* \equiv Y_{\text{2-methylnonane}}/Y_{\text{5-methylnonane}} \qquad (\text{at}\,Y_{\text{isodecanes}} \approx 5\%) \qquad (5)$$
$$(Y = \text{yield})$$

in the isomerization of *n*-decane [123]. It should be noted that the reaction pathways, on which Mobil's original Constraint Index (cf. Sect. 4.2.1) and the Modified Constraint Index are based, are entirely different. The only feature both indices have in common is that their numerical values increase with decreasing pore size of the zeolite [67].

*CI** values taken from the literature are presented in Fig. 8. It is evident that the *CI** values for 10-membered-ring zeolites extend over a relatively broad range, namely from ca. 2.7 to 15; hence, this is the range where the Modified Constraint Index is particularly useful. On the other hand, only a very narrow range, namely from ca. 1 to 2.3, is available for 12-membered-ring zeolites. It should, however, be kept in mind when using this or other indices, that the accuracy of indices determined from catalytic tests is limited. For example, for ZSM-5, *CI** values of 4.5 and 7.0 have been determined [2].

While the Modified Constraint Index is now widely employed for characterizing 10-membered-ring molecular sieves, the precise origin and nature of the shape selectivity effects on which it is based have not yet been fully elucidated. From the finding that the diffusion coefficients of the four isomeric methylnonanes in H-ZSM-5 and H-ZSM-11 at 80 °C, i.e., at a *sub-catalytic temperature*, were practically identical [122], the strongly preferred formation of 2-methylnonane was attributed to restricted transition state shape selectivity rather than to mass transfer shape selectivity. Essentially the same conclusion was drawn by Martens and Jacobs [126] from *n*-decane isomerization studies with Pt/H-ZSM-5 of different crystal size. Increasing the crystal size from 1 to ca. 15 μm brought about only a minor increase in *CI** from 7.5 to 9.4, and this was considered to be essentially consistent with the earlier interpretation of the favored formation of 2-methylnonane in 10-membered-ring zeolites.

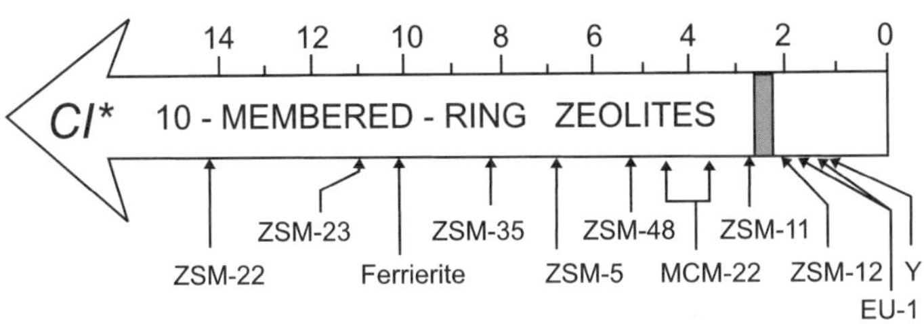

Fig. 8 Refined or Modified Constraint Indices for various zeolites. Data from [93, 123, 124]. The *numbers above the arrow* denote the numerical value of *CI**

Webb III and Grest [127] recently used molecular dynamics simulations to estimate the self-diffusion coefficients of *n*-decane and the methylnonane isomers at what they considered a *catalytically relevant temperature* of 327 °C (actually, catalytic studies on the isomerization of long-chain *n*-alkanes such as *n*-decane are often performed at temperatures around 250 °C). These authors arrived at the conclusion that the diffusion coefficients of the methylnonane isomers were clearly different in all medium-pore zeolites studied. The diffusion coefficients were minimal for different isomers in the various zeolites. This was explained by effects related to the correspondence between the molecular structure and the voids present at intersections of the main channels. Molecular motion of the monobranched nonanes was regarded as being dominated by competition amongst CH_3 groups to place themselves at energetically favorable positions along the channel, and diffusion down the channel was accomplished by jumps between successive pocket mouths. This means that molecules that are incommensurate with the void system diffuse faster, because these molecules are less efficiently locked to intersection positions. The relevant result was that, for five of the seven zeolites included in the study, the order of CI^* values was the same as their order of the ratios of diffusion coefficients for 2-methylnonane and 5-methylnonane. From this, Webb III and Grest concluded that the shape selectivity effect underlying the Modified Constraint Index CI^* does have to do, at least in part, with different mass transfer rates of the isomeric methylbranched isomers [127].

In summary, the literature on diffusion coefficients of isomeric alkanes in zeolites is very scant. The available information on diffusion coefficients of methylnonanes is contradictory. We expect a clarification in this respect by future research work.

The Modified Constraint Index exploits the selectivity occurring in the first step of the *n*-alkane reaction network. Numerous additional shape-selectivity effects are encountered in the consecutive reactions, namely the formation of dibranched isomers and hydrocracking. After a careful inspection of all these effects, Martens et al. [123, 126] defined a total of eight quantitative criteria. For a thorough discussion of these criteria and their usefulness for characterizing the pore width of microporous materials, the reader is referred to the pertinent literature [2, 123, 126]. While the Modified Constraint Index has been well accepted for characterizing the pore width of medium-pore zeolites, the seven other criteria did not reach any significant popularity.

In certain cases, for instance, if zeolites with very large pore sizes or void dimensions are to be characterized, it can be advantageous to use a probe molecule with more than ten carbon atoms. For example, an attempt was undertaken to detect differences between the pore systems of zeolite Y (FAU) and zeolite ZSM-20 (FAU/EMT), which is an intergrowth of zeolites FAU and EMT [127]. For this purpose, the isomerization of *n*-tetradecane was chosen as a catalytic test reaction with the rationale that there are three isomers with a propyl side chain (4-, 5- and 6-propylundecane) and even one with a butyl side chain (5-butyldecane) among the monobranched product isomers and that such bulky isomers may be suitable for detecting subtle differences in the pore systems. Indeed, such differences were found: 5-butyldecane was completely lacking in the product obtained on Pd/H-ZSM-20 at low conversion ($X_{n\text{-tetradecane}} \approx 2\%$), while it did appear at the same conversion on zeolite Pt/Ca-Y. Moreover, the three isomeric propylundecanes appeared in significantly different distributions on both bifunctional catalysts [128].

Martens and co-workers [129, 130] went one step further and suggested the use of *n*-heptadecane as probe molecule for the characterization of large-pore zeolites. In this case, however, the information was acquired from the product selectivity of hydrocracking, rather than from the distribution of isomerization products of varying bulkiness, as in the test with *n*-tetradecane proposed by Weitkamp et al. [128].

Very recently, the isomerization of methylcyclohexane was proposed as a test reaction to simultaneously probe the acid site density, the acid site strength and the pore size [131]. The reaction observed is the ring contraction of methylcyclohexane to ethylcyclopentane and the five isomeric dimethylcyclopentanes. The authors claim that the selectivity ratio of the bulkier *trans*-1,2-dimethylcyclopentane to the less bulky *trans*-1,3-dimethylcyclopentane is a measure for the pore size due to product shape selectivity effects. The ratio was found to vary in the range of 0.1 to 0.5 for medium-pore zeolites and from 1.2 to 2.4 for large-pore zeolites and amorphous solid acids, the equilibrium value being 1.8 to 1.9.

At higher reaction temperatures, Raichle et al. [132] demonstrated that the selectivity for ethane, propane and *n*-butane (the mixture of these alkanes is a high-quality feedstock to steamcrackers for the manufacture of ethene and propene) during the hydroconversion of methylcyclohexane on acidic zeolites increases significantly with increasing spatial constraints inside the zeolite pores and correlates well with the Modified Constraint Index. This example shows that the Modified Constraint Index data are capable of providing valuable quantitative information concerning the shape-selective performance of zeolites with 10-membered-ring pores in a completely different reaction. This is particularly noteworthy in view of the fact that the CI^* data have been collected on *bifunctional* forms of the zeolites, whereas Raichle et al. [132] employed their zeolite catalysts in their *monofunctional* acidic forms.

In conclusion, shape-selective test reactions based on the isomerization and hydrocracking of long-chain *n*-alkanes are now well established for characterizing the pore width of medium-pore zeolites. While a variety of quantitative criteria have been proposed, the most popular one by far is the Modified Constraint Index CI^*, which is based on the selectivity of *n*-decane isomerization at low conversion on a bifunctional form of the zeolite. Among the features that render the determination of CI^* so straightforward is the lack of catalyst deactivation. Expectedly, CI^* is particularly sensitive in a certain range of pore widths only, i.e., where the pore width strongly influences the selectivity of isomerization and hydrocracking. As clearly shown in Fig. 8, this is the region of 10-membered-ring zeolites. By contrast, CI^* is of little use in the range of 12-membered-ring zeolites. Fortunately, there is another index based on a different test reaction, which is ideally complementary to CI^*. This is the Spaciousness Index, *SI*.

4.3.2 Hydrocracking of C_{10} Cycloalkanes such as Butylcyclohexane – The Spaciousness Index, *SI*

Isomerization and hydrocracking of cycloalkanes follow essentially the same mechanistic rules as the reactions of alkanes. The cycloalkane undergoes several steps of skeletal isomerization until a structure is reached from which the favorable type A β-scission can start. In comparison with alkanes, however, the carbon-carbon bond rupture *inside* the cycloalkane ring proceeds more sluggishly, which is best interpreted in terms of an unfavorable orbital orientation in the transition state of β-scission of cyclic carbenium ions [133, 134]. Therefore, hydrocracking of C_{10} cycloalkanes over bifunctional catalysts gives entirely different carbon number distributions than hydrocracking of a C_{10} alkane. The products are independent of the feed hydrocarbon. Pentylcyclopentane, butylcyclohexane, diethylcyclohexanes or tetramethylcyclohexanes can be used. Preferentially, butylcyclohexane is employed. Pentylcyclopentane is, from a chemical viewpoint, equally well or even better suited, but it is usually not readily available.

It has been shown that, during hydrocracking of C_{10} cycloalkanes, almost exclusively methylcyclopentane and isobutane are formed *in the absence of spatial constraints* [135, 136]. This can be readily accounted for by the following sequence of steps: Starting from, for example, butylcyclohexane and excluding *endocyclic* carbon-carbon bond rupture, several rearrangement, ring contraction and ring enlargement steps take place until one of three possible carbenium ions displayed in Fig. 9 is reached which can undergo *exocyclic* type A β-scission. Type A β-scission of all these three carbenium ions happens to lead to methylcyclopentane and isobutane as hydrocracked products (Fig. 9). Small amounts of by-products (propane, *n*-butane, pentanes and cycloalkanes other than methylcyclopentane) are in line with some contribution of *exocyclic* type B β-scissions [2].

Again, the tribranched intermediates for the type A β-scission are rather bulky. Therefore, it has been predicted on the basis of carbocation chemistry that, as the pores become narrower, hydrocracking of C_{10} cycloalkanes proceeds much less selectively and a broader variety of hydrocracked products is formed, and this has indeed been found experimentally [137]. A thorough investigation of all features of shape-selective hydrocracking of C_{10} cycloalkanes revealed that the yield ratio of isobutane and *n*-butane in the hydrocracked products is a most valuable indicator for the effective pore width

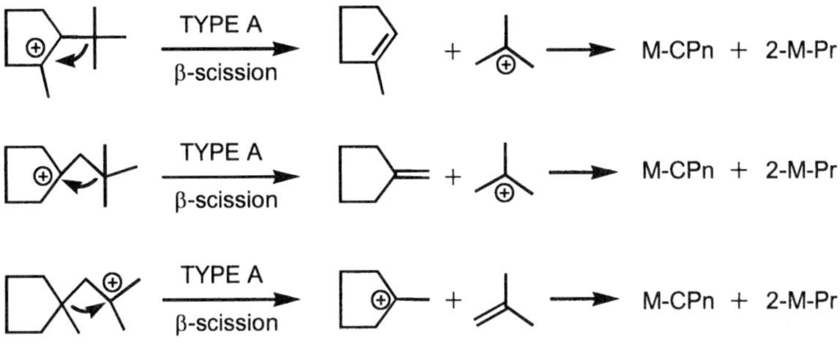

Fig. 9 All possible exocyclic type A β-scissions of C_{10}-cycloalkylcarbenium ions give methylcyclopentane (M-CPn) and isobutane (2-methylpropane, 2-M-Pr)

of the zeolite. Since this ratio increases with increasing space inside the pores, it was named the Spaciousness Index, *SI* [137, 138].

SI values for a number of microporous materials are given in Fig. 10. It is evident that the Spaciousness Index is particularly valuable for ranking 12-membered-ring molecular sieves for which it covers a wide range from ca. 3 (ZSM-12) to more than 20 (faujasite). By contrast, all 10-membered-ring zeolites have essentially the same Spaciousness Index around 1. Thus, the index is of little or no use in this range of pore widths, nor does it seem to be sensitive for extra-large-pore zeolites, since the *SI* value for amorphous $SiO_2 - Al_2O_3$, which is probably mesoporous or even macroporous, has been found to be essentially equal to that for zeolite Y [67].

In the routine application, the Spaciousness Index offers several advantages: (i) As in all hydrocarbon reactions carried out on bifunctional catalysts, there is no deactivation. (ii) The analysis for isobutane and *n*-butane is very easy and can be done quickly. (iii) The *SI* value is in a very broad range independent of the butylcyclohexane conversion, the yield of hydrocracked products, the reaction temperature and other parameters. Hence, there is no prescribed conversion or yield, and a tedious search for the appropriate experimental conditions is superfluous.

Evidence has been obtained [2] that the Spaciousness Index is based on restricted transition state shape selectivity rather than on mass transfer effects,

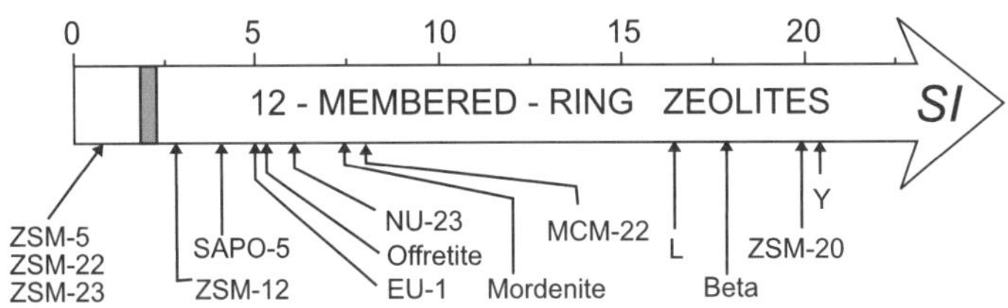

Fig. 10 Spaciousness Indices for various zeolites. Data from [137]. The *numbers above the arrow* denote the numerical value of *SI*

at least for 12-membered-ring zeolites including ZSM-12 (MTW). By using ball-shaped ZSM-12 crystallites with a diameter of ca. 0.5 μm and rod-like crystallites of the same zeolite with 11 μm by 1.5 μm, no significant influence on the *SI* could be detected [138]. Nor was there any significant influence on the *SI* when the n_{Si}/n_{Al} ratio was varied from 70 to 300. The independence of *SI* of the conversion implies that there is no interconversion of isobutane and *n*-butane under the conditions applied. This was confirmed experimentally for two zeolites, namely Pd/H-ZSM-5 (MFI) and Pd/H-L (LTL) [139]. It has furthermore been demonstrated for zeolite EU-1 (EUO) that the amount of noble metal introduced into the pores does not influence the selectivity of the shape-selective hydrocracking of butylcyclohexane [139], at least not within a reasonable range.

In conclusion, for the characterization of large-pore molecular sieves, the Spaciousness Index is the method of choice. The carbocation intermediates, which govern the selectivity of hydrocracking of C_{10} cycloalkanes, seem to be ideally suited for exploring the space available in the whole range of 12-membered-ring materials [2].

4.4
Are Monofunctional or Bifunctional Forms of Molecular Sieve Catalysts More Suitable?

To make a molecular sieve catalyst bifunctional, a strong hydrogenation/dehydrogenation component has to be incorporated into the pores of its acidic form. In the vast majority of cases, either platinum or palladium have been used for this purpose. The catalytic experiments have to be conducted in a hydrogen atmosphere. Obvious questions that have to be asked, are:

a) Does the noble metal under the conditions of the test reaction develop an undesired catalytic activity of its own, viz. for metal-catalyzed skeletal isomerization and hydrogenolysis? Such side reactions could falsify the selectivities of the reactions via carbocations described in Sect. 4.3.
b) Does the noble metal introduced into the pores influence their effective width, i.e., are the zeolite pores slightly narrowed by the metal clusters generated?

As a matter of experience, the answer to question a) is almost always favorable: Isomerization (cf. the Modified Constraint Index, Sect. 4.3.1) and hydrocracking (cf. the Spaciousness Index, Sect. 4.3.2) via the bifunctional route are usually significantly faster than the reactions on the noble metal. This is particularly so for hydrocarbons with as many as ten carbon atoms like *n*-decane (CI^*) and butylcyclohexane (*SI*). The typical temperature range of their conversion on bifunctional zeolite catalysts is around 250 °C or below, whereas their skeletal isomerization and hydrogenolysis on the noble metal begin to occur at or even above ca. 300 °C. It is, hence, generally easy to determine CI^* and SI without disturbances by reactions on the metal. Since the risk of undesired reactions on the metal is lower for palladium than for platinum (this holds especially for skeletal isomerization and to a lesser extent for hydrogenolysis), the use of palladium, typically with a loading of 0.3 wt.-% on the dry zeolite, as hydrogenation/dehydrogenation component inside the pores has been recommended [2].

An answer to question b) can immediately be obtained from a quantitative estimation of the metal cluster density inside the pores of zeolites, which are typically used as catalysts, i.e., faujasite (FAU) and ZSM-5 (MFI). If introduced into the pores via ion exchange with the tetraamine complex and after an adequate activation of the ion-exchanged zeolite (see vol 3 of this book series, Chap. 4 "Preparation of Metal Clusters in Zeolites"), the metal will reside in its zero-valent state in the form of small clusters located inside supercavities of faujasite or pore intersections of ZSM-5. An estimation for palladium gives cluster sizes of ca. 60 Pd atoms and 25 Pd atoms inside a supercavity of faujasite and a pore intersection of ZSM-5, respectively. For a typical palladium loading of 0.3 wt.-% it is estimated that one out of ca. 1550 supercavities in faujasite and one out of ca. 625 pore intersections in ZSM-5 are filled with a palladium cluster. It is obvious from these figures that, in typical bifunctional zeolites, with their low metal loadings, the vast majority of cavities and pore intersections are totally unaltered and free from metal atoms. Hence the risk of narrowing the zeolite pores by incorporating the noble metal can be considered as negligible.

Among the important advantages of hydrocarbon reactions on bifunctional catalysts in a hydrogen atmosphere is the lack of self-poisoning by coke formation and the absence of catalyst deactivation. By contrast, the formation of carbonaceous deposits is very frequently encountered during hydrocarbon conversion in monofunctional acidic zeolites. While the exact location of these deposits (e.g., inside cavities, at intersections, inside pores, at the pore mouths or on the external surface) is often difficult to determine and largely unknown, it is common experience that the build-up of carbonaceous deposits brings about a gradual selectivity change in favor of the more slender product molecules. This effect is generally referred to as "coke selectivation".

The general conclusion is, therefore, that the risk of modifying the effective pore width by test reactions using bifunctional zeolites is *much lower* than in reactions catalyzed by monofunctional acidic zeolites.

5 Conclusions

Among the salient features of zeolites and zeolite analogues are their strictly uniform pore shapes and pore widths in the same range as molecular dimensions. Obviously, these properties can be exploited for characterizing zeolitic pores, especially their effective widths, by means of molecular probes. The available techniques rely on adsorption or shape-selective catalysis. Instructive examples for both methods have been presented in this review with emphasis on their respective advantages, drawbacks and pitfalls.

Acknowledgements The authors thank Dr. Rainer Albert Rakoczy for helpful discussions. Financial support by Deutsche Forschungsgemeinschaft, Fonds der Chemischen Industrie and Max Buchner-Forschungsstiftung is gratefully acknowledged. Y.T. thanks the Ministerium für Wissenschaft, Forschung und Kunst Baden-Württemberg and the Landesstiftung Baden-Württemberg gGmbH for special funding.

References

1. Grubner O, Jírů P, Rálek M (1968) Molekularsiebe. VEB Deutscher Verlag der Wissenschaften, Berlin, pp 1–176
2. Weitkamp J, Ernst S (1994) Catal Today 19:107
3. Cook M, Conner WC (1999) In: Treacy MMJ, Marcus BK, Bisher ME, Higgins JB (eds) Proc 12th Int Zeolite Conf, vol 1. Mater Res Soc, Warrendale, Pennsylvania, p 409
4. Breck DW (1984) Zeolite Molecular Sieves, 2nd Edn. Robert E. Krieger Publishing Company, Malabar, Florida, p 633
5. Baerlocher C, Meier WM, Olson DH (2001) Atlas of Zeolite Framework Types, 5th Revised Edn. Elsevier, Amsterdam, pp 1–302
6. Harrison ID, Leach HF, Whan DA (1984) In: Olson D, Bisio A (eds) Proc Sixth Int Zeolite Conf. Butterworths, Guildford, p 479
7. Webster CE, Drago RS, Zerner MC (1998) J Am Chem Soc 120:5509
8. Choudhary VR, Akolekar DB (1989) J Catal 117:542
9. Webster CE, Drago RS, Zerner MC (1999) J Phys Chem B 103:1242
10. Snyder RAJ (1965) US Patent 3,224,167 (assigned to Union Carbide Corporation)
11. Conner WC, Vincent R, Man P, Fraissard J (1990) Catal Lett 4:75
12. Wu P, Ma YH (1984) In: Olson D, Bisio A (eds) Proc Sixth Int Zeolite Conf. Butterworths, Guildford, p 251
13. Aiello R, Barrer RM, Davies JA, Kerr IS (1970) Trans Faraday Soc 66:1610
14. Nishimura Y, Takahashi H (1971) Kolloid Z Z Polym 245:415
15. Breck DW, Eversole WG, Milton RM, Reed TB, Thomas TL (1956) J Am Chem Soc 78:5963
16. Masuda T, Hashimoto K (1994) In: Hattori T, Yashima T (eds) Zeolites and Microporous Crystals, Studies in Surface Science and Catalysis, vol 83. Kodansha, Tokyo, Elsevier, Amsterdam, p 225
17. Santilli DS, Harris TV, Zones SI (1993) Microporous Mater 1:329
18. Schenk M, Calero S, Maesen TLM, van Benthem LL, Verbeek MG, Smit B (2002) Angew Chem Int Ed 41:2499
19. Wu EL, Landolt GR, Chester AW (1986) In: Murakami Y, Iijima A, Ward JW (eds) New Developments in Zeolite Science and Technology, Studies in Surface Science and Catalysis, vol 28. Kodansha, Tokyo, Elsevier, Amsterdam, p 547
20. Ravishankar R, Joshi PN, Tamhankar SS, Sivasanker S, Shiralkar VP (1998) Ads Sci Technol 16:607
21. Otake M (1993) J Catal 142:303

22. Guil JM, Guil-López R, Perdigón-Melón JA (2000) In: Corma A, Melo FV, Mendioroz S, Fierro JLG (eds) 12th Int Congr Catalysis, Studies in Surface Science and Catalysis, vol 130, Part C. Elsevier, Amsterdam, p 2927
23. Lobo RF, Tsapatsis M, Freyhardt CC, Khodabandeh S, Wagner P, Chen C-Y, Balkus KJ Jr, Zones SI, Davis ME (1997) J Am Chem Soc 119:8474
24. Corma A, Corell C, Pérez-Pariente J, Guil JM, Guil-López R, Nicolopoulos S, Gonzalez Calbet J, Vallet-Regi M (1996) Zeolites 16:7
25. Lourenço JP, Ribeiro MF, Ramôa Ribeiro F, Rocha J, Gabelica Z (1996) Appl Catal A: General 148:167
26. Yuen L-T, Geilfuss JS, Zones SI (1997) Microporous Mater 12:229
27. Breck DW (1984) Zeolite Molecular Sieves, 2nd edn. Robert E. Krieger Publishing Company, Malabar, Florida, p 596
28. Harrison ID, Leach HF, Whau DA (1984) In: Olson D, Bisio A (eds) Proc Sixth Int Zeolite Conf. Butterworths, Guildford, p 479
29. Haag WO, Olson DH (1978) US Patent 4,117,026 (assigned to Mobil Oil Corporation)
30. Olson DH, Haag WO (1984) In: Whyte TE Jr, Dalla Betta RA, Derouane EG, Baker RTK (eds) Catalytic Materials: Relationship Between Structure and Reactivity, ACS Symposium Series, vol 248. Am Chem Soc, Washington, DC, p 275
31. Namba S, Kim J-H, Yashima T (1994) In: Hattori T, Yashima T (eds) Zeolites and Microporous Crystals, Studies in Surface Science and Catalysis, vol 83. Kodansha, Tokyo, Elsevier, Amsterdam, p 279
32. Baeck SH, Lee KM, Lee WY (1998) Catal Lett 52:221
33. Santilli DS (1986) J Catal 99:335
34. Choudhary VR, Singh AP, Kumar R (1991) J Catal 129:293
35. Thielmann F, Butler DA, Williams DR, Baumgarten E (2000) In: Sayari A, Jaroniec M, Pinnavaia TJ (eds) Nanoporous Materials II, Studies in Surface Science and Catalysis, vol 129. Elsevier, Amsterdam, p 633
36. Anunziata OA, Pierella LB (1995) In: Beyer HK, Karge HG, Kiricsi I, Nagy JB (eds) Catalysis by Microporous Materials, Studies in Surface Science and Catalysis, vol 94. Elsevier, Amsterdam, p 574
37. Denayer JF, Bouyermaouen A, Baron GV (1998) Ind Eng Chem Res 37:3691
38. Yoon KB, Huh TJ, Kochi JK (1995) J Phys Chem 99:7042
39. Trombetta M, Armaroli T, Alejandre AG, Solis JR, Busca G (2000) Appl Catal A: General 192:125
40. Derouane EG, Andre J-M, Lucas AA (1988) J Catal 110:58
41. Webster CE, Cottone A III, Drago RS (1999) J Am Chem Soc 121:12127
42. Deka RC, Vetrivel R (1998) J Catal 174:88
43. Pikunic J, Lastoskie CM, Gubbins KE (2002) In: Schüth F, Sing KSW, Weitkamp J (eds) Handbook of Porous Solids, vol 1. Wiley, Weinheim, p 182
44. Kerr GT (1966) Inorg Chem 5:1539
45. Eder F, Lercher JA (1997) J Phys Chem B 101:1273
46. den Exter MJ, Jansen JC, van Bekkum H (1994) In: Weitkamp J, Karge HG, Pfeifer H, Hölderich W (eds) Zeolites and Related Microporous Materials: State of the Art 1994, Studies in Surface Science and Catalysis, vol 84, Part B. Elsevier, Amsterdam, p 1159
47. Stewart A, Johnson DW, Shannon MD (1988) In: Grobet PJ, Mortier WJ, Vansant EF, Schulz-Ekloff G (eds) Innovation in Zeolite Materials Science, Studies in Surface Science and Catalysis, vol 37. Elsevier, Amsterdam, p 57
48. Milton RM (1959) US Patent 2,882,243 (assigned to Union Carbide Corporation)
49. Niwa M, Yamazaki K, Murakami Y (1989) Chem Lett 441

50. van Well WJM, Cottin X, de Haan JW, Smit B, van Hooff JHC, van Santen RA (1999) In: Treacy MMJ, Marcus BK, Bisher ME, Higgins JB (eds) Proc 12th Int Zeolite Conf, vol 1. Mater Res Soc, Warrendale, Pennsylvania, p 91
51. Dessau RM (1980) In: Flank WH (ed) Adsorption and Ion Exchange with Synthetic Zeolites, ACS Symp Series, vol 135. Am Chem Soc, Washington, DC, p 123
52. Jacobs PA, Beyer HK, Valyon J (1981) Zeolites 1:161
53. Foger K, Sanders JV, Seddon D (1984) Zeolites 4:337
54. Anderson JR, Foger K, Mole T, Rajadhyaska RA, Sanders JV (1979) J Catal 58:114
55. Wu P, Debebe A, Ma YH (1983) Zeolites 3:118
56. Namba S, Kanai Y, Shoji H, Yashima T (1984) Zeolites 4:77
57. Karsli H, Çulfaz A, Yücel H (1992) Zeolites 12:728
58. Talu O, Guo C-J, Hayhurst DT (1989) J Phys Chem 93:7294
59. Ma YH, Tang TD, Sand LB, Hou LY (1986) In: Murakami Y, Iijima A, Ward JW (eds) New Developments in Zeolite Science and Technology, Studies in Surface Science and Catalysis, vol 28. Kodansha, Tokyo, Elsevier, Amsterdam, p 531
60. Kurganov A, Marmé S, Unger K (1994) In: Weitkamp J, Karge HG, Pfeifer H, Hölderich W (eds) Zeolites and Related Microporous Materials: State of the Art 1994, studies in surface science and catalysis, vol 84, Part B. Elsevier, Amsterdam, p 1299
61. Klemm E, Emig G (1999) In: Treacy MMJ, Marcus BK, Bisher ME, Higgins JB (eds) Proc 12th Int Zeolite Conf, vol 1. Mater Res Soc, Warrendale, Pennsylvania, p 235
62. Weitkamp J, Ernst S, Schwark M, unpublished results
63. Weitkamp J, Schwark M, Ernst S (1989) Chem Ing Tech 61:887
64. Du H, Kalyanaraman M, Camblor MA, Olson DH (2000) Microporous Mesoporous Mater 40:305
65. Breck DW, Flanigen EM (1968) In: Molecular Sieves. Soc Chem Ind, London, p 47
66. Davis ME, Saldarriaga C, Montes C, Carces J, Crowder C (1988) Nature 331:698
67. Weitkamp J, Ernst S, Puppe L (1999) In: Weitkamp J, Puppe L (eds) Catalysis and Zeolites. Springer, Berlin Heidelberg New York, p 327
68. Csicsery SM (1976) In: Rabo JA (ed) Zeolite Chemistry and Catalysis, ACS Monograph, vol 171. Am Chem Soc, Washington, DC, p 680
69. Forni L (1998) Catal Today 41:221
70. Weitkamp J, Ernst S (1988) Catal Today 3:451
71. Traa Y, Weitkamp J (2002) In: Schüth F, Sing KSW, Weitkamp J (eds) Handbook of Porous Solids, vol 2. Wiley, Weinheim, p 1015
72. Frilette VJ, Haag WO, Lago RM (1981) J Catal 67:218
73. Levenspiel O (1972) Chemical Reaction Engineering, 2nd edn. Wiley, New York, p 484
74. Morrison RA (1987) US Patent 4,686,316 (assigned to Mobil Oil Corporation)
75. Apelian MR, Degnan TF, Fung AS (1993) US Patent 5,234,872 (assigned to Mobil Oil Corporation)
76. Ernst S, Kumar R, Neuber M, Weitkamp J (1988) In: Unger KK, Rouquerol J, Sing KSW, Kral H (eds) Characterization of Porous Solids, Studies in Surface Science and Catalysis, vol 39. Elsevier, Amsterdam, p 531
77. Ernst S, Weitkamp J, Martens JA, Jacobs PA (1989) Appl Catal 48:137
78. Zones SI, Harris TV (2000) Microporous Mesoporous Mater 35–36:31
79. Haag WO, Dessau RM (1984) In: Proc 8th Int Cong Catalysis, vol 2. Verlag Chemie, Weinheim, p 305
80. Haag WO, Dessau RM, Lago RM (1991) In: Inui T, Namba S, Tatsumi T (eds) Chemistry in Microporous Crystals, Studies in Surface Science and Catalysis, vol 60. Elsevier, Amsterdam, p 255

81. Macedonia MD, Maginn EJ (2000) AIChE J 46:2504
82. Haag WO, Lago RM, Weisz PB (1982) Faraday Discuss Chem Soc 72:317
83. Csicsery SM (1969) J Org Chem 34:3338
84. Corma A, Sastre E (1991) J Catal 129:177
85. Gnep NS, Tejada J, Guisnet M (1982) Bull Soc Chim Fr I-5
86. Dewing J (1984) J Mol Catal 27:25
87. Joensen F, Blom N, Tapp NJ, Derouane EG, Fernandez C (1989) In: Jacobs PA, van Santen RA (eds) Zeolites: Facts, Figures, Future, Studies in Surface Science and Catalysis, vol 49, Part B. Elsevier, Amsterdam, p 1131
88. Martens JA, Perez-Pariente J, Sastre E, Corma A, Jacobs PA (1988) Appl Catal 45:85
89. Llopis FJ, Sastre G, Corma A (2004) J Catal 227:227
90. Laforge S, Martin D, Guisnet M (2004) Microporous Mesoporous Mater 67:235
91. Ravishankar R, Bhattacharya D, Jacob NE, Sivasanker S (1995) Microporous Mater 4:83
92. Corma A (1998) Microporous Mesoporous Mater 21:487
93. Corma A, Corell C, Martínez A, Pérez-Pariente J (1994) In: Weitkamp J, Karge HG, Pfeifer H, Hölderich W (eds) Zeolites and Related Microporous Materials: State of the Art 1994, Studies in Surface Science and Catalysis, vol 84, Part A. Elsevier, Amsterdam, p 859
94. Corma A, Corell C, Llopis F, Martínez A, Pérez-Pariente J (1994) Appl Catal A: General 115:121
95. Adair B, Chen C-Y, Wan KT, Davis ME (1996) Microporous Mater 7:261
96. Corma A, Chica A, Guil JM, Llopis FJ, Mabilon G, Perdigón-Melón JA, Valencia S (2000) J Catal 189:382
97. Jones CW, Zones SI, Davis ME (1999) Appl Catal A: General 181:289
98. Jones CW, Zones SI, Davis ME (1999) Microporous Mesoporous Mater 28:471
99. Csicsery SM (1970) J Catal 19:394
100. Csicsery SM (1987) J Catal 108:433
101. Karge HG, Ladebeck J, Sarbak Z, Hatada K (1982) Zeolites 2:94
102. Weitkamp J, Ernst S, Jacobs PA, Karge HG (1986) Erdöl Kohle Erdgas Petrochem 39:13
103. Karge HG, Wada Y, Weitkamp J, Ernst S, Girrbach U, Beyer HK (1984) In: Kaliaguine S, Mahay A (eds) Catalysis on the Energy Scene, Studies in Surface Science and Catalysis, vol 19. Elsevier, Amsterdam, p 101
104. Das J, Bhat YS, Halgeri AB (1993) Ind Eng Chem Res 32:2525
105. Halgeri AB, Bhat YS (1994) In: Hattori T, Yashima T (eds) Zeolites and Microporous Crystals, Studies in Surface Science and Catalysis, vol 83. Kodansha, Tokyo, Elsevier, Amsterdam, p 163
106. Kim MH, Chen CY, Davis ME (1993) In: Davis ME, Suib SL (eds) Selectivity in Catalysis, ACS Symp Series, vol 517. Am Chem Soc, Washington, DC, p 222
107. Singh PS, Shaikh RA, Bandyopadhyay R, Rao BS (1998) In: Rao TSRP, Dhar GM (eds) Recent Advances in Basic and Applied Aspects of Industrial Catalysis, Studies in Surface Science and Catalysis, vol 113. Elsevier, Amsterdam, p 473
108. Butruille JR, Pinnavaia TJ (1992) Catal Lett 12:187
109. Miller SJ (1994) Microporous Mater 2:439
110. Doblin C, Mathews JF, Turney TW (1994) Catal Lett 23:151
111. Flego C, Perego C (2000) Appl Catal A: General 192:317
112. Chang CD (2000) In: Song C, Garcés JM, Sugi Y (eds) Shape-Selective Catalysis, ACS Symp Series, vol 738. Am Chem Soc, Washington, DC, p 96
113. Stöcker M (1999) Microporous Mesoporous Mater 29:3

114. Yuen L-T, Zones SI, Harris TV, Gallegos EJ, Auroux A (1994) Microporous Mater 2:105
115. Bendoraitis JG, Chester AW, Dwyer FG, Garwood WE (1986) In: Murakami Y, Iijima A, Ward JW (eds) New Developments in Zeolite Science and Technology, Studies in Surface Science and Catalysis, vol 28. Kodansha, Tokyo, Elsevier, Amsterdam, p 669
116. Sun Y, Han Y, Yuan L, Ma S, Jiang D, Xiao F (2003) J Phys Chem B 107:1853
117. Weitkamp J (1975) In: Ward JW, Qader SA (eds) Hydrocracking and Hydrotreating, ACS Symp Series, vol 20. Am Chem Soc, Washington, DC, p 1
118. Weitkamp J (1978) Erdöl Kohle Erdgas Petrochem 31:13
119. Weisz PB (1962) In: Eley DD, Selwood PW, Weisz PB (eds) Advances in Catalysis, vol 13. Academic, New York, p 137
120. Weitkamp J, Jacobs PA, Martens JA (1983) Appl Catal 8:123
121. Weitkamp J (1982) Ind Eng Chem Prod Res Dev 21:550
122. Jacobs PA, Martens JA, Weitkamp J, Beyer HK (1982) Faraday Discuss Chem Soc 72:353
123. Martens JA, Tielen M, Jacobs PA, Weitkamp J (1984) Zeolites 4:98
124. Jacobs PA, Martens JA (1986) Pure Appl Chem 58:1329
125. Souverijns W, Rombouts L, Martens JA, Jacobs PA (1995) Microporous Mater 4:123
126. Martens JA, Jacobs PA (1986) Zeolites 6:334
127. Webb EB III, Grest GS (1998) Catal Lett 56:95
128. Weitkamp J, Ernst S, Cortés-Corberán V, Kokotailo GT (1986) In: 7th Int Zeolite Conf, Preprints of Poster Papers. Jpn Assoc Zeolite, Tokyo, p 239
129. Martens JA, Vanbutsele G, Jacobs PA (1993) In: von Ballmoos R, Higgins JB, Treacy MMJ (eds) Proc 9th Int Zeolite Conf, vol 2. Butterworth-Heinemann, Stoneham, p 355
130. Feijen EJP, Martens JA, Jacobs PA (1996) In: Hightower JW, Delgass WN, Iglesia E, Bell AT (eds) 11th Int Congr Catalysis – 40th Anniversary, Studies in Surface Science and Catalysis, vol 101, Part B. Elsevier, Amsterdam, p 721
131. McVicker GB, Feeley OC, Ziemiak JJ, Vaughan DEW, Strohmaier KC, Kliewer WR, Leta DP (2005) J Phys Chem B 109:2222
132. Raichle A, Scharl H, Traa Y, Weitkamp J (2001) In: Galarneau A, Di Renzo F, Fajula F, Védrine J (eds) Zeolites and Mesoporous Materials at the Dawn of the 21st Century, Studies in Surface Science and Catalysis, vol 135. Elsevier, Amsterdam, p 302 and full paper No. 26-P-10 on accompanying CD
133. Brouwer DM, Hogeveen H (1970) Rec Trav Chim 89:211
134. Weitkamp J, Ernst S (1990) In: Barthomeuf D, Derouane EG, Hölderich W (eds) Guidelines for Mastering the Properties of Molecular Sieves, NATO ASI Series B, vol 221. Plenum Press, New York, p 343
135. Weitkamp J, Ernst S, Karge HG (1984) Erdöl Kohle Erdgas Petrochem 37:457
136. Ernst S, Weitkamp J (1985) In: Proc Int Symp Zeolite Catalysis, Siófok, Hungary, May 13–16. Acta Phys Chem, p 457
137. Weitkamp J, Ernst S, Kumar R (1986) Appl Catal 27:207
138. Weitkamp J, Ernst S, Chen CY (1989) In: Jacobs PA, van Santen RA (eds) Zeolites: Facts, Figures, Future, Studies in Surface Science and Catalysis, vol 49, Part B. Elsevier, Amsterdam, p 1115
139. Weitkamp J, Chen CY, Ernst S (1988) In: Inui T (ed) Successful Design of Catalysts, Studies in Surface Science and Catalysis, vol 44. Elsevier, Amsterdam, p 343

Mol Sieves (2007) 5: 155–248
DOI 10.1007/3829_004

Published online: 10 February 2006

NMR of Physisorbed ^{129}Xe Used as a Probe to Investigate Molecular Sieves

J.-L. Bonardet[1] · A. Gédéon[1] · M.-A. Springuel-Huet[1] · J. Fraissard[2] (✉)

[1]Laboratoire S.I.E.N., University P. and M. Curie, 4 place Jussieu, 75252 Paris Cedex 05, France

[2]Laboratoire PMMH, University P. and M. Curie, ESPCI, 10 rue Vauquelin, 75231 Paris Cedex 05, France
jfr@ccr.jussieu.fr

Abstract Xenon is an inert gas with a very large electron environment that makes it sensitive to any interaction, even physical. In the case of ^{129}Xe isotope (spin 1/2), the resulting electronic perturbation is directly transmitted to the nucleus and, therefore, affects the nuclear magnetic resonance chemical shift. In this chapter, we report some applications of this technique in both fundamental and applied research in the field of microporous and mesoporous solids.

Keywords ^{129}Xe-NMR · Optically polarized xenon · Zeolites · Microporous solids · Mesoporous solids · Clays · Carbon · Adsorption · Porosity · Metal catalysts

Abbreviations

a_{Xe}	Xe atom radius (0.22 nm)
A	internal surface area
A_n	preexponential factor of k_n
b	k_a/k_d ratio
C	molecular constant characterizing the polarizability (Sect. 2.2.3), relative water concentration in zeolite (Sect. 2.4.3)
d	shortest distance from Xe to the shell wall in potential calculation (Sect. 2.2)
D	Xe atom diameter (0.44 nm), diffusion coefficient (Sect. 2.6)
D_0	pre-exponential factor of diffusion coefficient
$E_{\mathrm{a}}(n)$	activation energy relative to k_n
g	magnetic-field gradient in pulsed-field gradient NMR
I_{a}, I_{b}, I_{c}	intensity of line a, b, c, respectively
k	Boltzmann constant
k_{a}	rate constant of adsorption

k_d	rate constant of desorption
k_n	rate coefficient for a given Xe atom to leave a n-cage
$\overline{\ell}$	mean free path of a Xe atom adsorbed in a microporous solid
L	pore width (Sect. 2.2.4)
$M(\omega^{(n)}, \omega^{(m)})$	joint probability of finding a Xe atom in a n-cage before t_{mix} and in a m-cage afterwards
n-cage	Na – A cavity containing n (m) atoms
n_a	number of adsorption sites in a unit volume
n_{EM}	number of metal particles determined by electronic microscopy
n_{H_2}	number of chemisorbed H_2 molecules
n_p	number of metal particles
N	number of adsorbed Xe atoms per gram of dehydrated solid
N_A	number of Xe atoms located on surface sites of the internal pore wall
N_V	number of Xe atoms situated in the internal pore volume
N_S	number of adsorbed Xe per cavity
t_{mix}	mixing time in 2D NMR experiment
T	absolute temperature (Kelvin)
p	probability of a xenon atom being at the cavity wall (Sect. 2.2), number of differentiated particles (Sect. 4)
P	gaseous pressure
P(Xe)	equilibrium Xe pressure
$P(n)$	probabilty of finding a n-cage among all of them
$\langle r^2(\Delta)\rangle_{intra}$	mean square displacement of molecules diffusing inside the crystallites
$\langle r^2(\Delta)\rangle_{inter}$	mean square displacement of molecules diffusing outside the crystallites
R	radius of the spherical shell representing the cage wall (Sect. 2.2.4), radius of adsorbent particle or gas constant
$R_{n,m}$	overall rate of Xe atoms going from a n-cage to a m-cage
$S_{i,j\ldots}$	surface of an $i(j\ldots)$-type particle
$V_{i,j\ldots}$	part of the volume of zeolite containing $i(j\ldots)$-type particles
W	van der Waals interaction energy (Sect. 2.2.3), concentration of Na – Y or ZSM-5 zeolite in a mixture with Na – A (Sect. 2.6.3)
α_i	probability of Xe – S_i collision
α_{Na-Y}	probability of Xe – Na – Y zeolite wall collision
γ	gyromagnetic ratio
$\gamma'(\Delta)$	relative amount of molecules diffusing in intercrystallite space
ΔH	NMR linewidth
δ_w	solvent effects on chemical shift due to van der Waals interactions
δ_a	chemical shift of a Xe atom located on surface sites of the internal pore wall, chemical shift of line a (Sect. 2.2)
δ_b, δ_c	chemical shift of line b and c, respectively (Sect. 2.5.2)
δ_i	chemical shift of Xe atoms interacting with an i-type particle
δ_{Na-Y}	chemical shift of Xe atoms interacting with the Na – Y pore wall
δ_V	chemical shift of a Xe atom situated in the internal pore volume
$\langle\delta_V\rangle$	average value of δ_V
δ_S	chemical shift of adsorbed Xe due solely to interactions with the internal pore wall, δ_S is a weighted average of δ_a and δ_V
δ_{Xe}	contribution of Xe – Xe interactions to the chemical shift
δ_{SAS}	contribution of Xe-strong adsorption site interactions to the chemical shift
δ_E	contribution of electric field due to the cations to the chemical shift
δ_M	contribution of cation paramagnetism to the chemical shift

$\delta_{N\to 0}$	chemical shift extrapolated to zero concentration
δ_0	chemical shift of an isolated Xe atom
δ_{ref}	same meaning as δ_0
$\delta_1, \delta_2, \delta_3$	characterize the Xe – Xe interactions in the virial expansion of the xenon density
δ'	pulse width in pulsed-field gradient NMR
Δ	time separation between two identical field-gradient pulses in pulsed-field gradient NMR
ε	energy coefficient in Lennard-Jones potential function
λ	fitting parameter in δ_S empirical expression (Sect. 2.6), degree of cationic exchange in faujasite (Sect. 3.3)
λ_i	probability of Xe collisions with i-type particles in a volume V_i
μ	fitting parameter in δ_S empirical expression
μ_i	probability of Xe collisions with Na – Y in a volume V_i
θ	average number of benzene molecules per cage
ρ	xenon density
ρ_{Xe}	xenon density in binary mixture
ρ_{A}	A-gas density in binary mixture (Xe-A)
ρ_{a}	xenon density of atoms located on surface sites of the internal pore wall
ρ_V	xenon density of atoms situated in the internal pore volume
ρ_g	concentration of xenon located in the free volume of faujasite supercages, similar to ρ_V (Cheung model)
σ_{d}	diamagnetic contribution to the shielding constant
σ_{p}	paramagnetic contribution to the shielding constant
τ_{intra}	intracrystallite mean lifetime
$\tau_{\mathrm{intra}}^{\mathrm{diff}}$	minimum intracrystallite mean lifetime
ξ_g	analogous to δ_1 in the Cheung model
ψ	intensity of the spin-echo signal in pulsed-field gradient NMR

1 General Introduction

1.1 Introduction

The central idea of the pioneers [1] of this technique was to find a molecule which was nonreactive and particularly sensitive to its environment, to physical interactions with other chemical species and to the nature of adsorption sites, and which could, therefore, be used as a probe to determine in a new way solid properties difficult to detect by classical physicochemical techniques. In addition, this probe needed to be detectable by NMR, since this technique is particularly suitable for investigating electron perturbations in rapidly moving molecules.

The isotope ^{129}Xe, which the authors proposed, is this ideal probe. It has given rise in the last 25 years to a great number of applications both in fundamental and applied research. Fragmentary reviews have already been published [2–7]. We present here some examples of the applications of the Xe-NMR technique to the study of microporous and mesoporous materials.

1.2 NMR Properties of Xenon

Xenon was discovered by Ramsay and Travers in 1898. The element has a Z value of 54. Among the nine stable xenon isotopes, only two, ^{129}Xe and ^{131}Xe, have nonzero spin and are therefore of interest for NMR studies. Table 1 gives their natural abundance, their magnetic properties [8–10] and their sensitivity to detection (Table 1).

At a field of 2.3488 Tesla, the resonance frequencies of ^{1}H, ^{13}C and ^{129}Xe are 100, 25.144 and 27.600 MHz, respectively. The resonance of ^{129}Xe occurs, therefore, at a frequency about 10% higher than that of ^{13}C at the same magnetic field strength. This makes it possible to use a spectrometer designed for ^{13}C detection with a slight change of the circuit tuning. According to the natural abundance of the various isotopes, the sensitivity of detection of ^{129}Xe (the isotope most widely studied) with respect to ^{1}H and ^{13}C is 5.60×10^{-3} and 31.8, respectively. Thus, ^{129}Xe is easier to observe than ^{13}C, neglecting differences in relaxation times. Of course, pure monoatomic xenon is expected to have a long spin relaxation time T_1, but relaxation is enhanced by interaction with paramagnetic species, even with few impurities. For example, ^{129}Xe is easily detectable in xenon gas at 0.5 MPa and at 27 MHz if 1% of O_2 is added. We will see later that the T_1 of ^{129}Xe adsorbed in microporous solids is rather short, typically in the range of 10 ms to a few seconds, even at low temperature. By using optical polarization techniques (see Sect. 6) [11–13] the sensitivity of detection can be increased by several orders of magnitude, which widens the field of applications of this technique. The use of a continuous flow approach [13–17] for the production of hyperpolarized xenon (denoted HP Xe in this text) is particularly beneficial for several applications (porous materials [15], flames [16], microimaging [17], etc.).

Table 1 Natural abundance and magnetic properties of ^{129}Xe and ^{131}Xe

Isotope	Natural abundance in xenon (%)	Spin quantum number	Magnetic moment (μ_N)	Electric quadrupole moment ($\times 10^{-28}$ m^2)	Sensitivity to detection
^{129}Xe	26.4	1/2	– 0.77247		5.60×10^{-3}
^{131}Xe	21.2	3/2	0.68697	– 0.12	5.61×10^{-4}

In theory, the nuclear relaxation time T_1 of adsorbed xenon should provide interesting information about local xenon structure and dynamics. Unfortunately, this is only true for extremely pure solids. The importance of paramagnetic impurities for relaxation is well known. Now, real catalysts, which are the most studied of all microporous solids, always contain relatively high concentrations of all sorts of impurities. For this reason we shall not discuss relaxation very much in what follows. The most important NMR characteristic of ^{129}Xe is its chemical shift which ranges over 7500 ppm for all the xenon compounds and over 1500 ppm if only adsorption is considered. This is a very wide range compared to ^{1}H (20 ppm) and ^{13}C (300 ppm) for diamagnetic systems. The chemistry of xenon began in 1962 [18]. Since that time many xenon compounds have been synthesized and ^{129}Xe NMR has proved to be the best technique for determining their structure [19, 20]. General information about these xenon derivatives can be found in [21] and [22]. However, ^{129}Xe NMR appeared to be of general interest in many fields: in physics, to study the interactions of xenon with solvents and with organic or bioorganic ligands, etc. General information on these applications can be found in the review by Reisse [23]. As discussed by Ramsey [24] the shielding constant consists of two parts:

- a diamagnetic contribution, σ_d, which depends only on the fundamental state of the electronic system and can be related to the charge on the atom considered;
- a paramagnetic contribution, σ_p, which depends on the excited states and above all on the symmetry of the valence orbitals. The shielding constant of the isolated xenon atom (gas pressure extrapolated to zero) is due to σ_d only. The paramagnetic contribution, which expresses the deviation from spherical symmetry, is zero.

In most xenon compounds the electron distribution is not spherically symmetrical and the σ_p values show a large variation, which depends on the nature of the atom to which the xenon is bonded, and the bond type. Except in a very small number of cases the chemical shifts, δ, measured relative to the isolated atom, are positive (downfield).

In this paper we will focus our attention on ^{129}Xe NMR as a means of probing the properties of inorganic microporous materials such as zeolites.The very large and extremely polarizable electron cloud of xenon makes this atom particularly sensitive to its close environment. Small variations in the physical interactions with the latter cause marked perturbations of the electron cloud which are transmitted directly to the xenon nucleus and greatly affect the NMR chemical shift.

Results on liquid media amply demonstrate the sensitivity of the chemical shift of xenon to physical interactions. Figure 1 shows that ^{129}Xe gas to solution shifts are linearly related to the ^{13}C gas to solution shifts of methane in the same solvents. Since the shifts are attributed exclusively to van der Waals

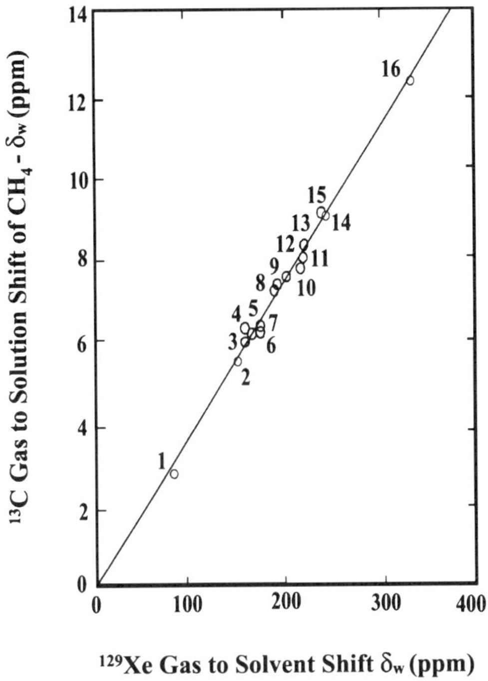

Fig. 1 Gas to solution shifts of ^{13}C in $^{13}CH_4$ in various solvents (corrected for susceptibility) vs. gas to solution shifts of ^{129}Xe in the same solvents. **1**, C_6F_6; **2**, CH_3NO_2; **3**, cyclo-C_5H_{10}; **4**, TMS; **5**, cyclo-C_6H_{12}; **6**, CH_3CN; **7**, CH_3COCH_3; **8**, $C_6H_5NO_2$; **9**, C_6H_6; **10**, C_6H_5Cl; **11**, $CHCl_3$; **12**, CCl_4; **13**, CS_2; **14**, $(CH_3)_2SO$; **15**, CH_3I; **16**, CH_2I_2. (From [25], Copyright Wiley (1991). Reproduced with permission)

effects, this variation proves that ^{129}Xe is 27.1 times more sensitive to these effects than ^{13}C [25]. Generally speaking, in the observed solvent effects on chemical shifts the part due to van der Waals interactions, δ_W, is the most important for heavy atoms. δ_W does not depend only on the solvent (Fig. 1) but sometimes also on the nuclear site in a given solute molecule. The effect of δ_W on the observed chemical shifts in homogeneous media has been the subject of much theoretical work which has been very well reviewed by Rummens [26].

Finally, it should be remembered that for every chemical shift measurement with an external reference a correction of the magnetic susceptibility related to the volume should be made; this was first mentioned by Dickinson [27], then developed mainly by Bothner-By and Glick [28] for homogeneous media. Bonardet and Fraissard have described the means of performing this correction in the measurement of shifts of phases adsorbed on solids [29]. However, in the present study, given the very large shifts of adsorbed xenon, this correction is negligible.

1.3 ^{129}Xe-NMR Chemical Shift in Xenon Gas

Before reviewing the NMR spectroscopy of ^{129}Xe adsorbed in microporous structures, it seems necessary to summarize some results relative to the ^{129}Xe chemical shift in xenon gas or in mixtures of xenon and other gases.

After the first detection of gaseous xenon [30] the subsequent studies by Carr et al. [31–33] and by Brinkmann [34] showed that over a wide range of density the shift can be expressed quite well by a virial expansion of the xenon density ρ:

$$\delta(T,r) = \delta_0 + \delta_1(T)\rho + \delta_2(T)\rho^2 + \delta_3(T)\rho^3 \quad (1)$$

where $\delta(T,\rho)$ is the resonance shift at a temperature T and a density ρ. δ_0 is the position of the isolated atom. Experimentally this corresponds to virtually zero pressure (in what follows we shall always use this reference). $\delta_i(T)$ are the virial coefficients of the shift as a function of density.

The most exact values were obtained by Jameson et al. [35]. At 298.15 K:[1]

$$\delta_1 = 0.548 \pm 0.004 \text{ ppm/amagat}^1 \quad (2a)$$

$$\delta_2 = (0.169 \pm 0.02)\ 10^{-3} \text{ ppm/amagat}^2 \quad (2b)$$

$$\delta_3 = (-\ 0.163 \pm 0.01)\ 10^{-5} \text{ ppm/amagat}^3 \quad (2c)$$

For densities below about 100 amagat, terms of an order higher than one are negligible. In this case, the linear variation yields a value of 0.55 ppm chemical shift per amagat. δ_1 results from binary collisions. Higher-order terms are an indication that three or more body collisions become important at higher density. Various authors have calculated the value of δ_1 from the interaction of two xenon atoms at the moment of collision [35, 36].

Adrian [36] has demonstrated that only the exchange interactions of the wave functions of the colliding atoms make a significant contribution to the chemical shift. In his opinion the van der Waals interactions were of negligible importance. Later, Jameson et al. showed that Adrian underestimated the role played by repulsive interactions on the pressure dependence of δ; these interactions must be taken into account [35].

The coefficients δ_i in Eq. 1 are temperature-dependent [37, 38]. Fitting $\delta_1(T)$ to a polynomial of degree 4, Jameson et al. give [38]:

$$\delta_1(T) = 0.536 - 0.135 \times 10^{-2}\tau + 0.132 \times 10^{-4}\tau^2 - 0.598 \times 10^{-7}\tau^3 + 0.663 \times 10^{-10}\tau^4 \text{ ppm/amagat}, \quad (3)$$

where $\tau = T - 300$ K.

[1] One amagat is the density (2.689×10^{19} atoms/cm^3) of an ideal gas at standard temperature (273.15 K) and pressure (1 atmosphere).

Dybowski et al. [6] have proposed a strictly empirical function to describe the temperature dependence of δ_1:

$$\delta_1(T) = 0.467 + 19.29 \exp(-\,0.0187T). \tag{4}$$

It must be remembered that the shift δ_0 of an isolated xenon atom is strictly temperature-independent [38].

Finally, the paper of Cowgill and Norberg [39] shows that the effect of the xenon pressure on the chemical shift is of the same type for all three phases, gas, liquid, and solid. This means that "the repulsive exchange forces may also have a role in the liquid and solid states" [39]. This result is particularly interesting for the interpretation of the δ values of adsorbed xenon.

1.4 ^{129}Xe NMR Chemical Shift in Mixtures of Xenon and Other Gases

In a mixture of xenon with another gas A, the shift of xenon depends on the nature of A and the collision frequency with the latter. When the xenon density is so low that only terms for binary collisions need to be considered, one may generalize Eq. 1 to a multi-dimensional virial expansion of the shift [35, 38, 40]:

$$\delta(T) = \delta_0 + \delta_{1,\mathrm{Xe}}(T).\rho_{\mathrm{Xe}} + \delta_{1,\mathrm{A}}(T).\rho_{\mathrm{A}}, \tag{5}$$

where ρ_{Xe} and ρ_{A} are the densities of Xe and A, respectively. Since $\delta_{1,\mathrm{Xe}}$ has previously been determined, $\delta_{1,\mathrm{A}}(T)$ can be calculated from data (Table 2).

Table 2 Measured values of second virial coefficients $\delta_{1,\mathrm{A}}$ ($T = 298$ K) of the ^{129}Xe NMR shift for interactions with various gases A (from [35] and [40])

A	$\delta_{1,\mathrm{A}}$ (ppm/mol cm^{-3})	$\delta_{1,\mathrm{A}}$ (ppm/amagat)
Ar	3119 ± 216	$0.139 \pm 10 \times 10^{-3}$
CO_2	3823 ± 151	$0.171 \pm 7 \times 10^{-3}$
CF_4	4327 ± 121	$0.193 \pm 5 \times 10^{-3}$
CHF_2	4287 ± 198	$0.191 \pm 9 \times 10^{-3}$
CH_2F_2	4965 ± 239	$0.222 \pm 10 \times 10^{-3}$
CH_3F	4314 ± 170	$0.192 \pm 9 \times 10^{-3}$
CH_4	6201 ± 125	$0.277 \pm 12 \times 10^{-3}$
Kr	6070 ± 375	$0.271 \pm 12 \times 10^{-3}$
HCl	7678 ± 132	$0.343 \pm 15 \times 10^{-3}$
Xe	12283 ± 90	$0.548 \pm 24 \times 10^{-3}$
NO	17769 ± 256	$0.793 \pm 11 \times 10^{-3}$
O_2	21037 ± 216	$0.939 \pm 10 \times 10^{-3}$

$\delta_{1,A}(T)$ is also temperature-dependent. The effect on the xenon shift of xenon-other gas interactions is smaller than that of interactions between xenon atoms, except for O_2 and NO. The large shift induced by gaseous O_2 and NO arises from a "contact overlap" mechanism. Overlap of the Xe (5s) and the O_2 (π_g) or NO (π^*) orbitals is the major contributor to the spin density at the Xe nucleus [41–43]. It should be remarked that this is the first example of a contact shift of this type.

Previously, the contact-shift was only observed when the unpaired electron was on the same molecule as the nucleus under observation. This result will also be important in the case of xenon adsorbed on paramagnetic or metallic solids.

Finally, for certain studies mentioned below, where the xenon is coadsorbed with another phase at high concentration, we felt it useful to recall the studies of Stengle et al. [44, 45] and Muller [46] concerning xenon dissolved in various hydrocarbon solvents, as well as the critical review already cited [23]. The former have shown that in the calculation of the shift of xenon, the reaction field model is valid within a group of closely related solvents, but it fails when extended to collections of randomly selected solvents. According to Muller, in calculating the van der Waals interactions one must take into account what he calls "the solvent cohesive energy density", which corresponds to the idea of the "volume of mixing effect" [47].

2 Zeolites

2.1 ^{129}Xe-NMR of Adsorbed Xenon: Generalities

As in the gas phase the main information has been obtained from the analysis of the variation of the chemical shift against the xenon concentration, generally at 300 K. The amount of adsorbed xenon is expressed as the number, N, of atoms per gram of anhydrous zeolite or the number of atoms, N_s, per cage (zeolites Y, A, erionite, etc.).

Fraissard et al. have shown that the chemical shift of adsorbed xenon is the sum of several terms corresponding to the various perturbations it suffers [2–4, 48].

$$\delta = \delta_{\mathrm{ref}} + \delta_{\mathrm{S}} + \delta_{\mathrm{Xe}} + \delta_{\mathrm{SAS}} + \delta_{\mathrm{E}} + \delta_{\mathrm{M}}. \quad (6)$$

δ_{ref} is the reference (gaseous xenon at zero pressure). δ_{S} arises from interactions between xenon and the surface of the zeolitic pores, provided that the solid does not contain any electrical charges. In this case it depends only on the dimensions of the cages or channels and on the ease of xenon diffusion. $\delta_{\mathrm{Xe}} = \delta_{\mathrm{Xe\text{-}Xe}} . \rho_{\mathrm{Xe}}$ corresponds to Xe – Xe interactions; it increases with the local density of adsorbed xenon and becomes predominant at high xenon pressure. When the Xe – Xe collisions are isotropically distributed (large spherical cage) the relationship $\delta = f(N)$ is a straight line (Fig. 2, line 1). The slope, $d\delta/dN$, is proportional to the local xenon density and, therefore, inversely proportional to the "void volume".

If the Xe – Xe collisions are anisotropically distributed (narrow channels) the slope of this function increases with N (Fig. 2, line 2).

When there are strong adsorption sites (SAS) in the void space interacting with xenon much more than the cage or channel walls, each xenon spends a relatively long time on these SAS, particularly at low xenon concentrations. The corresponding chemical shift δ will be greater than in the case of a noncharged structure (influence of δ_{SAS}, Fig. 2, line 3). When N increases, δ must decrease if there is a fast exchange of the atoms adsorbed on SAS with those adsorbed on the other sites. When N is high enough, the effect of Xe – Xe interactions again becomes the most important and the dependence of δ on N is then similar to that of Fig. 2. In this case $\delta_{N\to 0}$, which is the chemical shift extrapolated to zero concentration, depends on the nature and number of these strong adsorption sites. Often these SAS are more or less charged, and sometimes paramagnetic cations. The theoretical curve, is then displaced downfield (Fig. 2, line 3). The difference expresses the effect, δ_E, of the electrical field and, if it exists, δ_M due to the magnetic field created by these cations. If strong adsorption sites of different types (for example various highly charged cations or metal particles) are distributed in solid pores (for example zeolite supercages), Xe interacts with them and the chemical shift of xenon is different depending on the presence or absence of these targets and their surface. If the exchange between these various

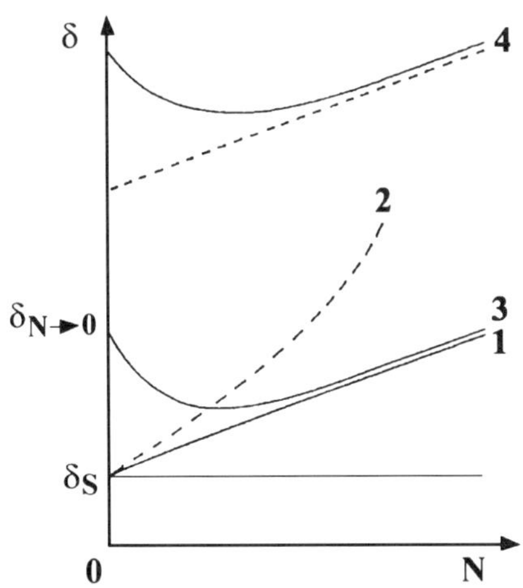

Fig. 2 $\delta = f(N)$ variations (see text). (Reprinted from [48] with permission from Elsevier Science)

sites is fast with respect to the NMR time-scale, an average signal is observed, while a multiple signal spectrum corresponds to a situation of slow exchange [2–4].

Finally, there is nothing to prevent us from recording the Xe NMR spectra of the previous samples in the presence of an active gas (for example, hydrogen on Pt-zeolite). In this complex case, too, the xenon NMR spectrum gives information about the distribution of this active gas in the sample [2–4].

2.2 Models for the Interpretation of Chemical Shifts Due to the Porosity

2.2.1 Introduction

The δ_S term—Eq. 6—characterizing the Xe-surface interactions, has been related to the dimensions and the shape of the pores, or more precisely to the mean free path, $\overline{\ell}$, of a Xe atom inside the pore volume. The mean free path is defined as the average distance travelled by a Xe atom between two successive collisions against the pore walls.

To explain the hyperbolic relation obtained between δ_S and $\overline{\ell}$, for various zeolites, a simple model of fast site exchange has been used [49, 50]. Other physisorption models have been proposed such as:

- the model described by Derouane et al. which is based on the calculation of the van der Waals energy as a function of the surface curvature [51, 52].
- models based on Lennard-Jones potential curve calculations [53, 54].

2.2.2 Fast Site Exchange

For simplicity, exchange between adsorbed and gaseous phases is assumed to be negligible, which is reasonable at room temperature at least for zeolites. Chen and Fraissard have shown that the chemical shift varies by about 5% between powder and highly compressed ZSM-5 and Na – Y samples [55].

At 300 K, which is the usual recording temperature, the experimental chemical shift, δ, is the average value of the shift of xenon in rapid exchange between a position A on the pore surface (defined by δ_a) and a position in the volume V of the cavity or channel (defined by δ_v) (Fig. 3).

$$\delta = \frac{N_a\delta_a + N_V\delta_V}{1} gN_a + N_V \qquad (7)$$

where N_a and N_V are the numbers of xenon atoms in each state. This equation is valid whatever the xenon concentration; $\delta = \delta_s$ for $N = 0$. In this case N_a and N_V are the probabilities of there being a Xe atom at the surface (A) or in the volume (V).

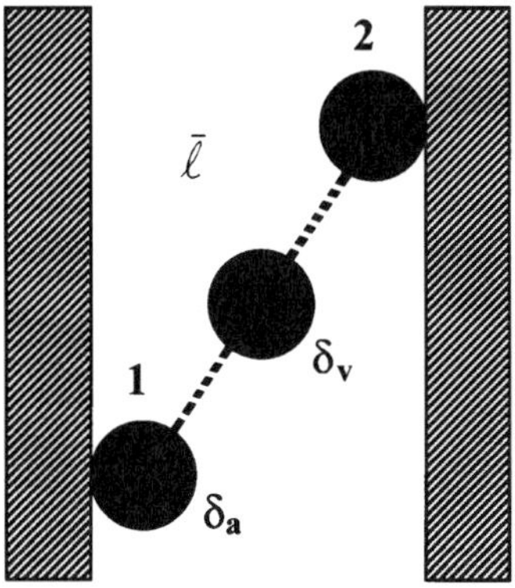

Fig. 3 Scheme of fast site exchange. (Reprinted from [48], with permission from Elsevier Science)

δ_a depends on Xe-surface interactions [56]. δ_V is a function of δ_a and the distance ℓ. In fact, $\delta_V = \delta_a$ when the xenon atom leaves the surface; then δ_V decreases during the journey between sites 1 and 2, hence the need to determine a mean value $\langle\delta_V\rangle = f(\delta_a, \bar{\ell})$ where $\bar{\ell}$ is the average value of ℓ and is called the mean free path imposed by the structure [49, 56].

Considering a random distribution of the Xe atoms in the pore, that means there are not any preferential states, N_a and N_V can be expressed in terms of the xenon diameter D, the surface area A, the void space $V - AD$ and the density of xenon either adsorbed on the surface (ρ_a) or in the void volume (ρ_V), with ρ_a proportional to ρ_V at low concentration ($\rho_a = k\rho_V$). When $\bar{\ell}$ is not too small, $\langle\delta_V\rangle \approx 0$.

$$\delta_S = \frac{AD\rho_a}{AD\rho_a + (V_A D)\rho_V}\delta_a = \frac{Dk}{D(k-1) + \frac{V}{A}}\delta_a = \frac{\alpha}{\beta + \frac{V}{A}}\delta_a, \tag{8}$$

with $\alpha = kD$ and $\beta = D(k-1)$. $\frac{V}{A}$ is proportional to $\bar{\ell}$. We shall see later the application of $\bar{\ell}$ in Sect. 2.6.

More simply, Johnson and Griffiths wrote:

$$\delta_S = \delta_a p + \delta_V(1-p), \tag{9}$$

where the probability p of a xenon atom being at the cavity wall is proportional to the surface area to volume ratio of the cavity. They found a fairly good linear relationship between δ_S and $\frac{A}{V}$ for a certain number of zeolite structures [57].

Cheung et al. [50] also used the fast site exchange model, even at 144 K, and expressed the chemical shift of xenon adsorbed in Y zeolites at temperature T and concentration ρ as:

$$\delta(T,\rho) = \frac{bn_a\delta_a + (1 + b\rho_g)\xi_g\rho_g}{1 + bn_a + b\rho_g}, \tag{10}$$

where $b = k_a/k_d$ is the ratio of the rate constants for adsorption and desorption of a xenon atom on a site; n_a is the number of adsorption sites in a unit volume; ρ_g is the concentration of xenon located in the free volume of faujasite supercages and ξ_g corresponds to δ_1 of Jamesons' expression; δ_a has the same meaning as above.

At zero concentration ($\rho_g = 0$), Eq. 10 reduces to:

$$\delta(T,\rho) = \delta_S = \frac{bn_a}{1 + bn_a}\delta_a, \tag{11}$$

which is similar to Eq. 8.

Cheung et al. then distinguished two cases:

- ***adsorption is weak***, i.e. $b\rho_g \ll 1$, even when ρ_g is large; Eq. 10 becomes

$$\delta(T,\rho) = \delta_S + \frac{(\xi_g - b\delta_S)}{(1 + bn_a)^2}\rho. \tag{12}$$

  Eq. 12 shows that δ is a linear function of ρ, the total Xe concentration in the supercage, which is true for Y zeolites and in general for zeolite structures with rather large pore volumes (A, beta zeolites ...) where the distribution of Xe – Xe interactions is isotropic. One can notice that the slope of this linear function is smaller than ξ_g ($= \delta_1$) which is consistent with the experimental observation: 0.40 ppm/amagat for Y zeolite compared to 0.548 for the gas phase.
- ***adsorption is strong***, i.e. $b\rho_g \gg 1$; Eq. 10 reduces to:

$$\delta(T,\rho) = \frac{n_a\delta_a + \xi_g n_a^2}{\rho} - 2n_a\xi_g + \rho\xi_g. \tag{13}$$

  It is evident that there is an initially rapid decrease of $\delta(T,\rho)$ with increasing coverage (term $1/\rho$) before the $\rho\xi_g$ term becomes predominant. At higher concentration, the gradient of chemical shift against density should approach that of the free gas.
  Eq. 13 accounts for the behavior of the ^{129}Xe NMR chemical shift in the presence of multivalent cations or supported metals [58].

2.2.3 Influence of Surface Curvature

This model factorizes and scales the physical interaction of a sorbed atom or molecule with the curved surface environment of micropores, the atom or molecule being represented by its polarizability and the solid (adsorbent) by a frequency-dependent dielectric function [51, 59]. It can be shown that the van der Waals interaction between the trapped species and the curved wall of a sphere can be factorized into the product of two terms accounting for the intrinsic molecular property (polarizability) and extrinsic surface curvature effects, respectively:

$$W = -\frac{C}{4d^3}\left(1 - \frac{d}{2a}\right)^{-3}, \tag{14}$$

with C a molecular constant, d, the distance from the point molecule to the pore wall (i.e. its van der Waals radius, r_m, in the physisorbed state) and, a, the pore radius. At temperatures such that $kT < W$, spherical molecules or atoms stick and oscillate with respect to the pore wall and hop as a whole over their barriers rather than collide with the wall and be reflected through the cavity or pore as is generally assumed. The pore or cage center never corresponds to an equilibrium position.

Therefore, these authors wrote that the chemical shift δ_S induced by interaction with the pore wall is directly proportional to the molecular interaction energy with the surface, W (K being a molecular constant):

$$\delta_S = KW. \tag{15}$$

For xenon adsorbed in a certain number of zeolites a fairly linear correlation between δ_S and W, the molecular interaction energy with the surface, has been obtained [52] (Fig. 4).

Even if these two models, fast site exchange and surface curvature effect are different, the relation between δ_S and W is not surprising since W corresponds to the chemical shift δ_a, (Xe atom interacting with the surface) and δ_S depends itself on δ_a.

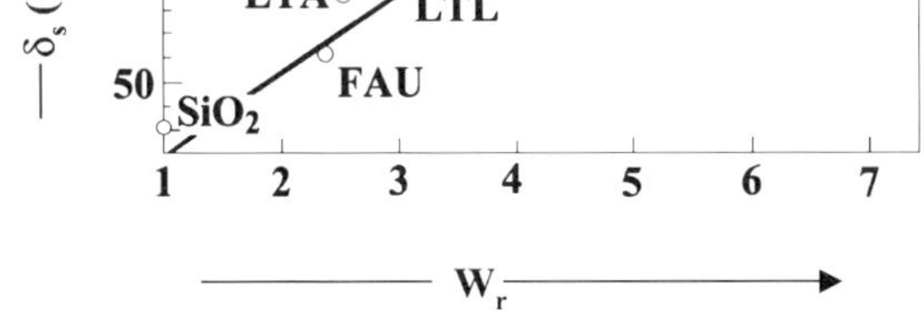

Fig. 4 The surface-induced NMR chemical shift δ, as a function of the relative van der Waals energy W_r for ^{129}Xe trapped in a variety of zeolites. W_r is the ratio $W/W(0)$, $W(0)$ corresponding to the flat surface ($d/a = 0$). Zeolite structures are designed by their three-letter code. (Reprinted from [52], with permission from Elsevier Science)

These two models are limiting cases. The real situation should depend on the temperature and the adsorption energy which depends itself on the pore size (assuming identical chemical nature of the surface).

In order to define exactly the influence of these factors, some other models used a Lennard-Jones potential to describe the physical interaction between a Xe atom and a surface consisting of oxygen atoms.

2.2.4
Lennard-Jones Potential Curves

Ripmeester and Ratcliffe considered the potential energy between a xenon atom and a spherical shell representing the cage wall. Oxygen atoms are smeared over this shell with a density similar to that of the zeolite A cage. The Xe – O interaction is described by a Lennard-Jones potential function which can be integrated over the whole sphere:

$$U_{\mathrm{tot}}(d) = 4\varepsilon \frac{2\Pi R}{R-d} \left\{ \frac{\alpha^{12}}{10} \left(\frac{1}{d^{10}} - \frac{1}{(2R-d)^{10}} \right) - \frac{\alpha^6}{4} \left(\frac{1}{d^4} - \frac{1}{(2R-d)^4} \right) \right\}, \tag{16}$$

where R is the spherical shell radius, d is the shortest distance from Xe to the shell wall; $\alpha = 0.349$ nm and $\varepsilon = 0.8975$ kJ/mol [53]. Plots of this function for different values of R are shown in Fig. 5.

Several points appear:

- The potential energy goes through an absolute minimum for $R = (2.5)^{1/6}\alpha$, i.e. in this case, $R = 0.407$ nm (which corresponds to $\bar{\ell} = 0.047$ nm). Therefore, the strength of the binding site is intimately linked to its ability to make efficient contact with xenon atoms. For sphere radii both larger and smaller than 0.407 nm, the Xe-surface interaction is weaker.
- On the other hand, when $R \approx 0.45$ nm ($\bar{\ell} \approx 0.09$ nm), the potential profile becomes quite flat and for larger R values a central hump appears. This means that for $R > 0.45$ nm the Xe atom no longer samples the cage uniformly but spends more time near the cage wall than in the center of the cage.

From these calculations, Ripmeester and Ratcliffe concluded that for small cages, δ_S values can be expected to reflect the true void space. However, for large cages, the δ_S value is a complicated function of sorption energy, void space and temperature, the δ_S variations with temperature being a means of distinguishing the two types of behavior, as Chen and Fraissard have shown [60].

To determine the chemical shift of Xe adsorbed in such a cage, Ripmeester and Ratcliffe chose a potential curve with a hump which they approximated by a rectangular potential. In this simple picture Xe atoms are either in the potential well (with the chemical shift δ_a) or in the central part of the cage

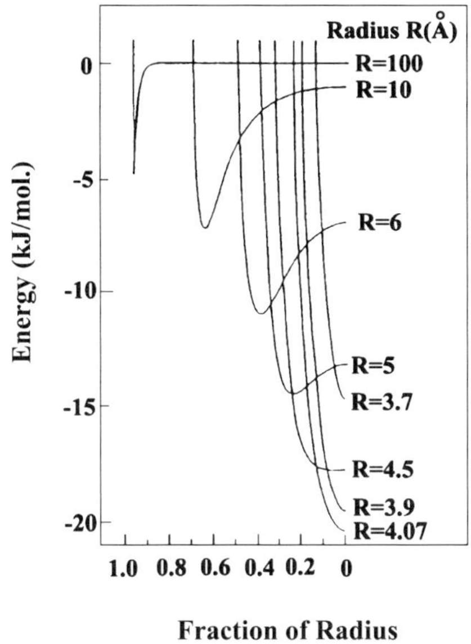

Fig. 5 Calculated Lennard-Jones potential curves for a xenon atom in a spherical cage shown as a function of cage radius. (Reprinted with permission from [53]. Copyright (1990) American Chemical Society)

(with the chemical shift $\langle\delta_V\rangle$). Writing the chemical shift as the average of δ_a and $\langle\delta_V\rangle$ weighted by the fractional population above (p_a) and below (p_b) the energy barrier they obtained:

$$\delta = [1 - p_a(r^3/R^3)]\delta_a + p_a(r^3/R^3)\langle\delta_V\rangle, \qquad (17)$$

where R and r are the radius of the pore and the hump respectively.

If it is assumed that $\langle\delta_V\rangle = 0$ and at the zero concentration limit, Eq. 17 reduces to:

$$\delta_S = \{1 - [r^3 \exp(-\Delta E/RT)]/(R^3 - r^3 + R^3 \exp(-\Delta E/RT)\}\delta_a, \qquad (18)$$

ΔE being the potential difference between the hump and the well.

When T tends to zero, δ_S tends to δ_a, i.e. the Xe atom sticks to the surface; while when T tends to infinite, δ_S tends to $2/[1 + R^3/(R^3 - r^3)]\delta_a$. Consequently, only the limiting high-temperature δ_S value should reflect the true void space, as it depends only on R/r and δ_a. At any T value, the chemical shift takes an intermediate value which depends on the void space as well as the adsorption energy of the Xe atom on the surface, hence the need to perform variable temperature experiments.

Cheung made calculations similar to those of Ripmeester for a xenon atom trapped between two infinite parallel layers, and obtained similar results: the appearance of a central hump when the distance between the layers increases [54]. He considered that the chemical shift of a xenon atom is directly proportional to the van der Waals interaction energy (W_r), being itself represented by the first term (attractive part) in the Lennard-Jones potential, U.

When U has a single minimum at the center (micropore of about the same size as the xenon atom), and assuming that the potential near the minimum can be described by a parabola, Cheung showed that the chemical shift increases linearly with increasing temperature. Conversely, he obtained a decrease in the chemical shift with increasing temperature when the pore is much larger than the xenon atom. This result is completely consistent with experimental observations [60, 61].

Modelling the potential by a rectangular potential, Cheung also calculated the chemical shift and obtained a general equation:

$$\frac{1}{\delta_S(T)} = A(T) + B(T)\left(\frac{L - 2a_{Xe}}{m}\right), \tag{19}$$

where $A(T)$ and $B(T)$ depend on the well width, the well depth and the temperature; L and a_{Xe} are the pore width and the Xe atom radius, respectively; m is equal to 1, 2 or 3 if the pore is layer-like (1), cylindrical (2) or spherical (3). The expression, $(L - 2a_{Xe})/m$, is exactly that used by Fraissard's group to calculate the mean free path, $\bar{\ell}$, of a Xe atom in pores, which can be considered to fit these simple pore models. Once again, Eq. 19 is quite consistent with the empirical relationship obtained by Springuel-Huet et al. [49].

Molecular dynamics simulations, in which the interaction energy is calculated with a Lennard-Jones potential, was used to correlate the chemical shift with the number of binary collisions [62].

2.2.5 Discussion

In fact, the first two models are the two limiting cases of the latter. Everything depends on the calculated values of the xenon-surface interaction and the temperature. In the opinion of Derouane et al. [51, 52], the xenon atom sticks to the surface and never goes through the cage center, whereas Fraissard et al. [49] and Cheung et al. [50] consider that the xenon atom samples the void space freely. The energy W, calculated by Derouane, corresponds to the chemical shift, δ_a (Eq. 7), of the fast site exchange. It is, therefore, not surprising to obtain a relationship between W and δ_S, since in Fraissard's model δ_S depends on δ_a and $\langle\delta_V\rangle$, $\langle\delta_V\rangle$ being itself a function of δ_a.

It must be noted that Cheung et al. have used the fast site exchange model although they performed NMR experiments at 144 K and correlated the chemical shifts with the pore size of zeolites and the type of cations [63, 64]. These authors were able to assume that the xenon atoms can sample the void space more or less uniformly because very high xenon loadings were used.

Derouane et al. calculated the van der Waals energy for closed spheres and used a correction coefficient for channels [51, 52]. Ripmeester et al. also chose closed spheres of different radii [53]. Vigné-Maeder used more realistic structural models but the zeolites are still supposed to be cation free and the frameworks rigid [62]. Their calculations are certainly worthwhile, but in zeolite structures cages are almost never closed (except, for example, in Na – A zeolite where, due to the small apertures, they can be considered as closed), the framework is more or less flexible and more often cations are present. Since all these approximations change the potential curves calculated in these simple ideal cases, these calculations are not always easy to use for determining the pore size. Nevertheless, these models are attempts, the only ones to date, to interpret the chemical shift.

We would mention here that the xenon mobility is a very important parameter which must be taken into account when interpreting the data. Pulsed-field gradient measurements give Xe diffusion rate constants in zeolites (A, Y and ZSM-5) of the order of 10^{-8}–10^{-9} m^2 s^{-1} [65–67]. This means that, in these materials, at room temperature the movement of the xenon averages its interactions over a great number of zeolite cages and channels on the NMR time scale. When intercrystallite exchange occurs, the number of lines, chemical shifts, linewidths and relaxation times in the NMR spectra may be dramatically changed. Intercrystallite exchange depends of course on the interparticle diffusion barriers, hence on temperature, on the crystallite size and the xenon concentration. Many studies have clearly shown the influence of these parameters [55, 60, 68–73]. We have seen above the influence of xenon mobility on the different models.

The rapid movement of xenon atoms through large-pore zeolites at room temperature usually limits resolvable structural details to macroscopic dimensions. Nevertheless, Chmelka et al. have measured macroscopic adsorbate distribution heterogeneities in Na – Y zeolite which, when coupled with a crystallite size scaling analysis, make it possible to estimate diffusivities of coadsorbed organic molecules [74]. In such cases, other complementary experimental methods are often needed to check interpretations of ^{129}Xe NMR results [75–78].

It is obvious that, when the temperature is decreased, xenon diffusion is limited and information becomes more localized. The greater the mean free path, $\bar{\ell}$, the larger the influence of temperature. Chen and Fraissard have studied the effect of temperature on the chemical shift of xenon adsorbed in Na – Y zeolite [60]. Figure 6 shows the decrease of δ and the changes in the slopes of the $\delta = f(N)$ curves for temperatures ranging from 173 to 373 K. The variation of the slopes, due to Xe – Xe interactions, directly reflects the diffusion. At low temperature and concentration ($N_s < 0.5$ Xe atom/cage) δ is constant, which means there is no Xe – Xe interaction. The slope then increases rapidly for $0.5 < N_s < 1$ Xe atom/cage. For $N_s > 1$, the slope becomes constant with a maximum for $N_s \approx 1$. The relative variation of the slopes for $N_s \approx 1$ with respect to temperature fits well the relative temperature dependence of the second virial coefficient of chemical shielding of xenon

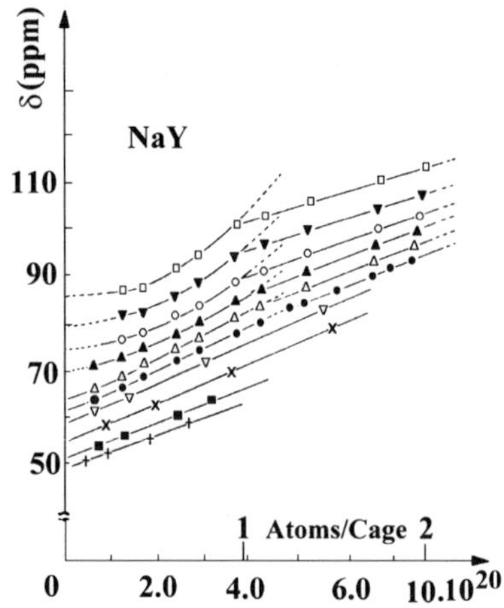

Fig. 6 $\delta = f(N)$ variations for Y zeolite at temperatures T (K): 173; 193; 213; 233; 253; 273; 300; 323; 353; 373 (from the *top* to *bottom*). (Reprinted with permission from [60] Copyright (1992) American Chemical Society)

gas determined by Jameson et al. [79], showing that for this concentration ($N_s \approx 1$) xenon diffusion in Na – Y zeolite is similar to that in xenon gas. For $N_s > 1$ the slope is constant and lower than for $N_s < 1$. This lower value of the slope can be explained by the occurrence of three-body collisions whose probability is no longer negligible.

It is possible to avoid intercrystalline diffusion by coating the external surface. For example octamethylcyclotetrasiloxane has been used in the case of faujasite structure [80, 81].

Na – A zeolite has received special attention because diffusion of xenon between α-cavities is very slow due to the blockage of the apertures by Na cations. In such systems, xenon can be introduced at elevated temperature and pressure. The sensitivity of the ^{129}Xe chemical shift to Xe – Xe interactions means that signals corresponding to different Xe concentrations in the α-cages can be resolved (Fig. 7) [82]. Under these conditions, studies of xenon distribution statistics, assuming different models with binomial or hypergeometric distribution, have been carried out [83, 84], sometimes in the presence of other adsorbed molecules [85, 86]. Jameson et al. found experimental deviations from these statistical models, explained by attractive Xe – Xe interactions which favor clustering at low to medium loadings [84]. The observed shifts and their temperature dependence are interpreted by using the results of Grand Canonical Monte Carlo simulation [87].

Cheung has derived a cell theory of an interacting lattice gas from the basic principle of statistical mechanics that the equilibrium distribution of X particles among Y cells (when X and Y are very large) is given by the most probable distribution. He could calculate the probability distribution of the

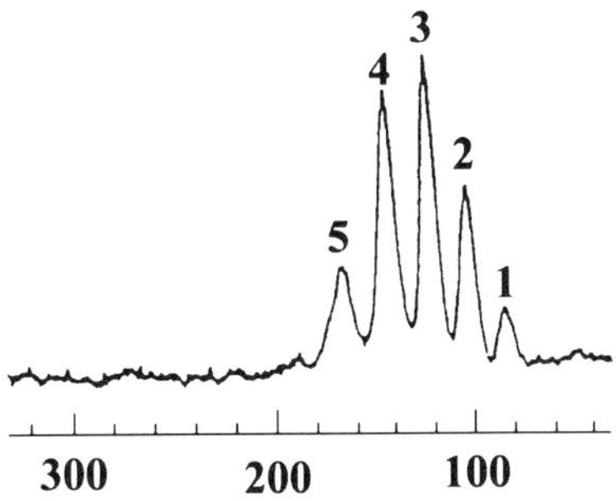

Fig. 7 Room-temperature spectrum of xenon adsorbed in Na – A zeolite at 523 K and xenon pressure of 4 MPa. Each peak corresponds to a definite number of xenon atoms (1, 2, 3, 4 and 5) in the α-cages. (Reprinted with permission from [82]. Copyright (1988) American Chemical Society)

gas particles among the cells [88]. To describe ^{129}Xe NMR data for xenon trapped in the α-cages of the Na – A zeolite, it is necessary that, in addition to the attractive interactions between the particles, the repulsive interactions should also be included when the xenon atoms begin to fill the α-cages of the zeolite.

With more sophisticated NMR experiments (multisite magnetization transfer experiments or two-dimensional exchange NMR) and simulations, the dynamics of xenon movement between α-cages can be investigated [86, 89]. Larsen et al. give an activation energy of about 60 kJ/mole for cage-to-cage migration [86].

2.2.6
Porosity Studies

In order to obtain from $\delta(^{129}\text{Xe})$ precise data about the void space of a zeolite of unknown structure or on the dimensions of structural defects, the term δ_S has been related to the size and shape of the internal volume by means of the mean free path $\overline{\ell}$ of adsorbed xenon imposed by the structure (Eq. 8 and Fig. 3). In the fast exchange model, for a sufficiently high temperature, a xenon atom moves freely inside the void space. This random movement is the basis of the determination of the mean free path, either calculated for a closed sphere or an infinite channel, or using a computer in the general case [56].

The computer is given the analytical expressions of the modeled structures of zeolites. A xenon atom is then randomly circulated within the space. The distance, ℓ, travelled between two successive collisions is calculated. The mean free path, $\overline{\ell}$, is obtained by averaging these distances (Table 3). The plot

Table 3 Values of δ_S and mean free path $\overline{\ell}$ for various molecular sieves. The case of infinite cylinder and sphere can be rigourously solved by calculation: infinite cylinder: $\overline{\ell} = D_C - D_{Xe}$; sphere: $\overline{\ell} = 0.5(D_S - D_{Xe})$ where D_C, D_S and D_{Xe} are the diameters of the cylinder, the sphere and the xenon atom, respectively. The values of pore size given above are from [90]

Molecular sieves	δ_S (ppm)	$\overline{\ell}$ (nm)	Characteristics of the void spaces accessible to xenon atom
A, ZK4	87	0.37	sphere, diameter 1.14 nm with six 8-ring openings of 0.4–0.5 nm, depending on the cation
L	90	0.31	one dimensional barrel-shaped channels: 12-ring openings of 0.71 nm, max. diameter 0.74 nm
Ferrierite	227	0.01	c-channel, 10-ring: 0.54 × 0.42 nm
	157	0.04	b-channel, pseudo-sphere, diameter: ≈ 0.7 nm with 2 8-ring openings: 0.48 × 0.35 nm
Mordenite	115	0.245	one dimensional channel 12-ring: 0.67 × 0.7 nm
	250	0.005	side-pocket: 0.57 × 0.26 × 0.48 nm
Offretite	108	0.2	one dimensional channel 12-ring: 0.67 × 0.68 nm
ZSM-12	90	0.13	one dimensional channel, non planar 12-ring 0.55 × 0.59 nm
ZSM-20	58	0.56	intergrowth of cubic and hexagonal faujasite: normal supercage with four 12-ring apertures of 0.74 nm diameter, "maxi" supercage with five 12-ring openings, 0.71 × 0.71 nm (for two) and 0.74 × 0.65 nm (for three) and "mini" supercage with three 12-ring openings 0.74 × 0.65 nm
ZSM-23	114	0.045	one dimensional channel, 10-ring: 0.52 × 0.45 nm
ZSM-48	96	0.11	one dimensional channel, 10-ring: 0.53 × 0.56 nm
EU-1	94	0.17	one dimensional channel, 10-ring: 0.41 × 0.57 nm "side pockets": 0.68 × 0.58 nm, 0.81 nm deep
Ω	75	0.30	one dimensional channel, 12-ring, 0.74 nm
Theta-1, Nu-10, ZSM-22	130	0.05	one dimensional channel, 10-ring: 4.4 × 5.5 nm
Erionite	99	0.17	cavity 0.63 × 0.63 × 1.51 nm with six openings: 0.36 × 0.52 nm
$AlPO_4$-5, SAPO-5	56	0.29	one dimensional channel, 12-ring: 0.73 nm
$AlPO_4$-8	58	0.39	one dimensional channel, 14-ring: 0.79 × 0.87 nm
$AlPO_4$-11, SAPO11	120	0.056	one dimensional channel, 10-ring: 0.39 × 0.63 nm
$AlPO_4$-17	72	0.17	erionite structure
SAPO-31	85	0.09	one dimensional channel, non planar 12-ring: 5.3 nm
SAPO-34	84	0.18	chabazite structure: cavity 0.67 × 0.67 × 1 nm with six 8-ring openings 0.38 nm
MAPO-36	74	0.26	one dimensional channel, 12-ring: 0.65 × 0.75 nm
Y, SAPO-37	58	0.56	faujasite structure
SAPO-41	108	0.11	one dimensional channel, 10-ring: 0.70 × 0.43 nm
VPI-5	49	0.77	one dimensional channel, 18-ring: 1.21 nm

of $\delta_S = f(\bar{\ell})$ for a few classical zeolites (A, Y, L, MOR, Rho ...), with usual Si/Al ratios, has hyperbolic shape, curve I in Fig. 8, and Fraissard et al. first wrote:

$$\delta_s = \frac{1}{\lambda + \mu\bar{\ell}}\delta_a, \tag{20}$$

which is consistent with Eq. 8, λ and μ are fitting parameters. This expression has to be modified in the light of more recent work: δ_a cannot be constant for all structures. It was said above that δ_a is related to the molecular interaction energy (W) of Derouane's model, which depends on the surface curvature, hence on $\bar{\ell}$. Moreover, $\langle\delta_V\rangle$ has been assumed to be equal to zero; this is certainly true for large void volumes (compared to the xenon diameter), such as in faujasite; but for zeolites whose channels are small compared to the xenon diameter, $\langle\delta_V\rangle$ is not zero and depends on δ_a and $\bar{\ell}$, as we have seen in Sect. 2.2.

On the other hand, Chen and Fraissard have shown that the fast site exchange hypothesis is valid in the case of Y zeolite for $T >$ ca. 303 K, which is not far from the experimental temperature [60]. This is certainly not true for zeolites with smaller void volumes.

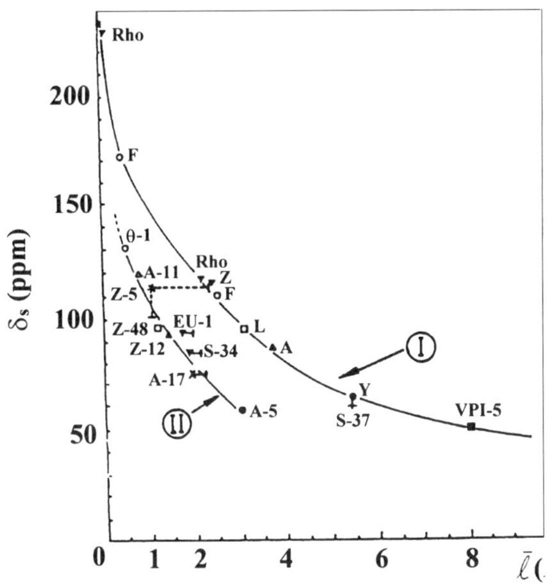

Fig. 8 $\delta_S = f(\bar{\ell})$. See Table 1 for the coordinates of all molecular sieves.
Curve I (classical zeolites): Rho (cavity and prism), Ferrierite: F (channels b and c), Mordenite: Z (main channel), L, A, Y.
Curve II (new molecular sieves): Theta-1: Θ-1, $AlPO_4$-11: A-11, ZSM-5: Z-5, ZSM-48: Z-48, EU-1, ZSM-12: Z-12, SAPO-34: S-34, $AlPO_4$-17: A-17, $AlPO_4$-5: A-5, SAPO-37: S-37, VPI-5. (Reprinted from [2–4] with permission from Springer-Verlag)

Another important point is the dependence of the chemical shift on the chemical composition of the zeolite surface, in particular upon the number of charges, i.e. on the Si/Al ratio. This dependence is more pronounced when the temperature is low and the pore diameter is small. In the case of faujasite, either in the Na^+ form or decationized, the δ_S values decrease monotonically by 4 ppm when Si/Al increases from 1.28 to 54 [2–4]. Such variations have been also observed by Chao et al. [90]. These authors showed that the reduction of the δ_S value with decreasing Al content of faujasite type zeolites is not linear and depends on the pore size and extra-framework species formed by dealumination. For ZSM-5 and ZSM-11 (narrow channels) the dependence of δ_S on the aluminum concentration is greater than for faujasite (large cavities) [91–93]. For instance, δ_S increases linearly with the Al content of the framework on ZSM-5 and ZSM-11 zeolites [93]. In addition, this variation shows a break at about [Al] = 2 atoms per unit cell (Fig. 9) which demonstrates that the xenon-wall interactions change at this concentration. This can be explained by a change of the Al distribution in the framework. Finally, we mention the difference between the $\delta_S = f(N)$ relationships for ZSM-5 and ZSM-11 for [Al] $\geq$ 2 Al atoms per unit cell (Fig. 9). Despite the similarity of the structure of these zeolites, the sensitivity of the ^{129}Xe NMR technique is sufficient to differentiate between them. The variation of the δ_S value with framework composition has been also reported for SAPO-37 [94], $AlPO_4$-5 and SAPO-5 [95] molecular sieves.

These various deviations from the simple model are responsible for the fact that many newly studied structures, mainly structures with very low cation content, do not fit the first $\delta_S = f(\bar{\ell})$ curve; they form a cloud of points limited by a second curve for $0.05 < \bar{\ell} < 0.4$ nm, curve II in Fig. 8 [48].

Although the models are essentially qualitative this technique has been widely used to study microporous solids.

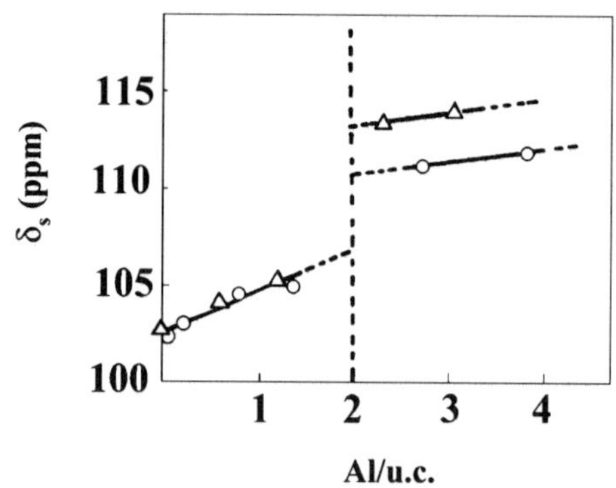

Fig. 9 δ_S variation versus Al content of the framework: ○ NaZSM-5; △ NaZSM-11. (Reprinted with permission from [93]. Copyright (1992) American Chemical Society)

2.2.6.1
Zeolite with a Single Type of Void Volume

When the zeolite structure contains only one type of pore (more or less cylindrical channels or spherical cavities) accessible to xenon, the xenon spectrum consists of a single line characteristic of the adsorption zone.

The main information is generally obtained by analyzing the dependence of the chemical shift of this signal on the xenon concentration. The amount of xenon adsorbed, e.g., at 300 K is expressed as the number, N, of atoms per gram of anhydrous zeolite or the number, N_S, per cage or part of the channel. At 300 K, the $\delta = f(N)$ variation is a straight line when the distribution of Xe – Xe collisions is isotropic, i.e. in the case of large cavities (Y, ZK4 ...). The slope, $d\delta/dN$, is proportional to the local Xe density and therefore inversely proportional to the internal free volume. If the distribution of Xe – Xe collisions is anisotropic (narrow channels, diameters between 4 and 7.5 Å), the slope of this curve increases with N [48, 96] (Fig. 10).

The chemical shift, δ_S, at zero concentration is clearly related to the structure: the smaller the channels or the cavities, or the more restricted the diffusion, the greater δ_S. For example, δ_S (ZK4) is greater than δ_S (Y) not only because the α-cage diameter (1.14 nm) is slightly smaller than the supercage one (1.3 nm) but, also and especially, because the diffusion of xenon in ZK4 zeolite is more difficult than in Y. The dimensions of the openings allowing passage from one α-cage to another are similar to those of xenon atoms, whereas those of the supercages are twice as big as Xe.

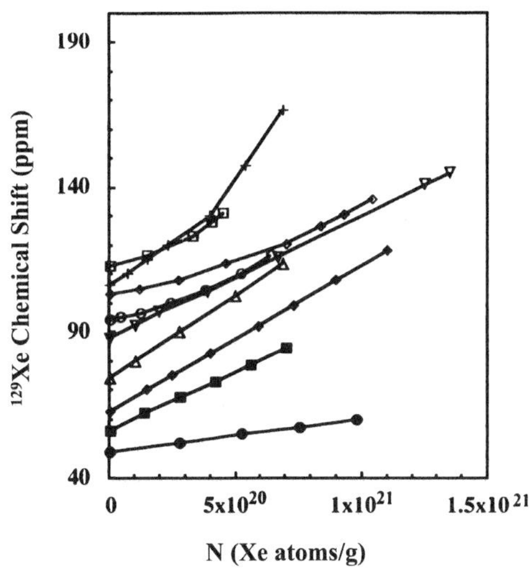

Fig. 10 $\delta = f(N)$ variations for different molecular sieves: □ ZSM-23, + SAPO-41, ◇ silicalite-1, ○ EU-1, ▽ ZK-4, △ MAPO-36, ◇ Beta, ■ LZY-52, (•) VPI-5. (Reprinted from [2–4], with permission from Springer-Verlag)

Many papers deal with studies of Y zeolites [97–100], where it is often a question if structural changes arise after dealumination. Conclusions are frequently based on variations of the internal volumes [90, 101, 102]. Some other less common zeolites or related materials have been investigated, such as ZSM [92, 93], L [103], Theta [104], VPI-5 [105, 106], SAPO-34 [107], SAPO-37 [95, 108], SAPO and $AlPO_4-5$ [95], SAPO and $AlPO_4-11$ [109], etc.

The chemical shifts, δ_S, of Na – Y and SAPO-37, which is isostructural with Na – Y, are very similar, showing that to a first approximation the fast exchange hypothesis is valid for such an open structure [94, 95]; the chemical variation of the surface does neither modify the residence time of the adsorbed xenon on the surface nor the δ_a value. Note that Davis et al. found chemical shift differences for these two structures, but this was due to a partial destruction of the SAPO-37 structure, which is fragile in the presence of humidity [110].

For very small pore diameters, the Si/Al ratio is relatively more important and for a given mean free path, $\bar{\ell}$, the chemical shift varies from curve I (say Si/Al < 6 or 7) to curve II (higher Si/Al ratios), Fig. 8. The influence of the chemical composition of the surface is highlighted by the existence of these two curves. The main reason for this chemical shift variation is that the residence time of a xenon atom on the surface, τ_a, is not negligible at room temperature.

Nevertheless, the empirical and experimental relationship $\delta_S = f(\bar{\ell})$—Eq. 20—was used with success for the approximate determination of the internal void volume space of beta zeolite when the structure was still unknown [111].

2.2.6.2
Zeolites with Several Types of Void Volume

Except in the case of Na – A, already discussed, ^{129}Xe-NMR spectra have as many components as there are different types of void volume in the zeolites, at least if exchange between xenon adsorbed in the different zones is slow on the NMR time scale.

This is the case of ferrierite whose spectra have two lines corresponding to the two types of channels [112]. This example is particularly interesting: the diffusion from one channel to another is slow and into one type of channels only one Xe atom can enter. Therefore, the corresponding signal is independent of the xenon pressure. Only the line intensity increases with the filling degree.

It is also the case of Rho zeolite, but here the spectrum depends on the nature of the cation and on the temperature. For H – Rho, there is a rapid exchange between cavities and prisms. The two characteristic lines are then obtained only at low temperature [113]. On the contrary, for Cs – Rho there is only one line, since the Cs cations are located in the prisms and prevent Xe atoms from being there. The ^{129}Xe NMR study of this structure was later extended more thoroughly to a set of H – Cs Rho samples with variable Cs content [114] and for Cd Rho [115]. It must be noted that in these two papers the attributions of the lines have been changed: the high-frequency signal previously attributed to xenon adsorbed in the prisms (or at least xenon exchanging between cavity and prism, depending on the temperature) is assigned to the cavities, and the low-frequency signal previously attributed to xenon adsorbed in the cavities is assigned to xenon adsorbed at the external surface. The flexibility of this structure as a function of the temperature has also been studied [116].

Mordenite, whose porous structure consists of one-dimensional channels connected to side pockets in the perpendicular direction, has also been studied by ^{129}Xe NMR [61, 117]. The spectra have two signals corresponding to channels and side-pockets, respectively, at least if the temperature is low enough to prevent exchange between these two sites. The temperature of coalescence depends on the nature of the cations. It is about 273 K for H^+ and about 370 K for Na^+. If the cation is Cs^+, the side-pockets are no longer accessible to xenon, and there is only one line. Moudrakovski et al., using 2D NMR could determine rate constants for xenon exchange between the main channels and side pockets [118].

Cloverite is a zeolite with two types of void spaces for xenon adsorption. But in the only published paper, to our knowledge, its ^{129}Xe-NMR gives a single-line spectrum corresponding to the large supercages [119]. The authors conclude that the apertures of the second type are blocked by residual carbonaceous products arising from template decomposition.

When there are several signals corresponding to various types of void volumes, the line intensities can be used to study the distribution of adsorbed xenon in these different spaces for a given total xenon concentration as a function of the temperature and the location of cations. In this way one can obtain interesting information about the structures of zeolites, e.g. zeolite intergrowths and crystallinity.

2.2.6.3
Zeolite Intergrowths and Crystallinity

In view of what has been said about intercrystallite diffusion, if there is a mixture of zeolites or a structure intergrowth, each zeolite component will give rise to its own NMR lines in the spectra, providing that the diffusion of Xe between monocrystalline domains is not too fast and prevents the averaging of Xe-zeolite interactions. If the latter condition is satisfied and since NMR spectroscopy is quantitative, the measurement of peak areas allows the determination of the zeolite composition. Of course, the adsorption isotherms of standard components must be known. This has been checked with a synthetic mixture of Ca-A and Na – Y. The composition was easily and precisely (to ± 1%) found [120].

In the same way, the composition of a ferrierite-mordenite intergrowth, which was very poorly detected by X-ray diffraction, was determined [112]. It should be pointed out that the zeolite structures which give intergrowths are usually very similar, which means that the void volumes and the corresponding lines are very close. The case of ferrierite-mordenite is particularly favorable, because mordenite gives a high-frequency line (about 250 ppm, due to side-pockets) well separated from the others. Moreover, after line decomposition, a signal coming from ferrierite can be used to obtain additional information about the composition of the mixture. On the contrary, it has so far been impossible to resolve offretite-erionite intergrowths [121].

Chen et al. obtained a line corresponding to the $AlPO_4-8$ structure in a VPI-5 sample; $AlPO_4-8$ is the degradation product of VPI-5 [105]. A ZSM-5-ZSM-11 intergrowth in a ZSM-8 sample could also be seen [121, 122].

The problem of crystallinity is of the same type; there are several approaches:

- At a given pressure, comparing the intensity of the "well crystallized" lines with that of the standard zeolite gives a measure of crystallinity [120].
- It has been said that the slope of the $\delta = f(N)$ curve, expressed in Xe atoms/g, depends on N. If the amount of xenon adsorbed by the amorphous phase is negligible at room temperature, then the ratio of the slopes of the two $\delta = f(N)$ curves for one sample and the standard is also a measure of the crystallinity.
- Finally, in the most favorable case, treatment of the zeolite can create secondary porosity. This situation reduces to that of zeolites with several zones and has been observed for Y zeolite (Fig. 11), mordenite, or with the dealumination of ZSM5 during the methanol to gasoline conversion [121].

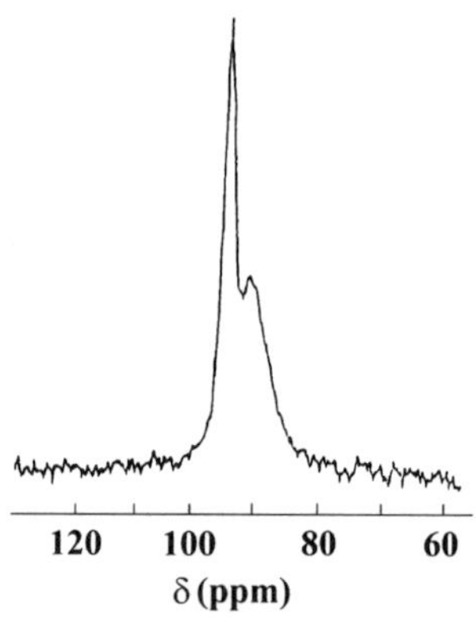

Fig. 11 Room temperature spectrum of xenon adsorbed in dealuminated Y zeolite at a pressure of 0.2 MPa. The smaller signal corresponds to secondary porosity. (Reprinted from [2–4], with permission from Springer-Verlag)

2.3
Influence of Cations

2.3.1
Introduction

It has been explained in Sect. 2.1 that, when xenon adsorbs more strongly on some adsorption sites (SAS), a term δ_{SAS}, characteristic of this interaction, must be considered. Indeed, these sites can be saturated while the adsorption on the other sites continues to increase monotonically with [Xe]. At low Xe pressure, the term δ_{SAS} predominates. Therefore, when the pressure increases, adsorption occurs on the weaker sites. The result is a decrease of the observed chemical shift which is a weighted average of δ_{Xe} and δ_{SAS}. Afterwards the chemical shift again increases due to important Xe – Xe interactions at higher pressures. These characteristic $\delta = f(N)$ plots have been observed for X and Y zeolites containing cations like Mg^{2+}, Ca^{2+}, Zn^{2+}, Cd^{2+}, rare earth cations Y^{3+}, La^{3+}, Ce^{3+} and even paramagnetic cations Ni^{2+}, Co^{2+}, Ru^{3+}. Most of these studies have been carried out as a function of the cation exchange degree or the temperature of thermal treatment (i.e. the extent of dehydration of the zeolite). Since Xe atoms can only interact with cations located in the supercages and not with those in the sodalite cages or prisms, the location of cations can be deduced. The migration of cations inside the crystals, between different sites, have been studied in terms of their hydration state.

2.3.2
Case of H and Alkali-Metal Ions

It has been shown that for X and Y zeolites the influence of the H^+ and Na^+ cations on the ^{129}Xe NMR is negligible at room temperature [1–4, 97, 99, 123, 124]. The chemical shift is then given by the terms δ_S and δ_{Xe} of Eq. 6 and is roughly independent of the value of the Si/Al ratio, and therefore, of the number of H^+ or Na^+ ions (Fig. 12). These results prove that in X and Y supercages, the time-averaged field $\langle\delta_E\rangle$ due to these cations is negligible at 300 K. At very low N, the motion of each atom is disturbed only by cage walls. Consequently, the chemical shift, δ_S (58 ± 2 ppm) obtained by extrapolation of the line $\delta = f(N)$ to $N = 0$ can be considered as characteristic of the zeolite structure. The increase of δ with N results from mutual interactions between Xe atoms. However, for analogous zeolites containing other monovalent cations, several ^{129}Xe NMR studies have shown that the chemical shifts vary linearly with xenon loading [99, 123, 124]. The plots $\delta = f(N)$ depend on the type of cation; the heavier the cation, the greater the shift. For example, Ito et al. [99] have shown that $\delta_{N\to 0}$ is 78 and 99 ppm for KY and RbY, respectively.

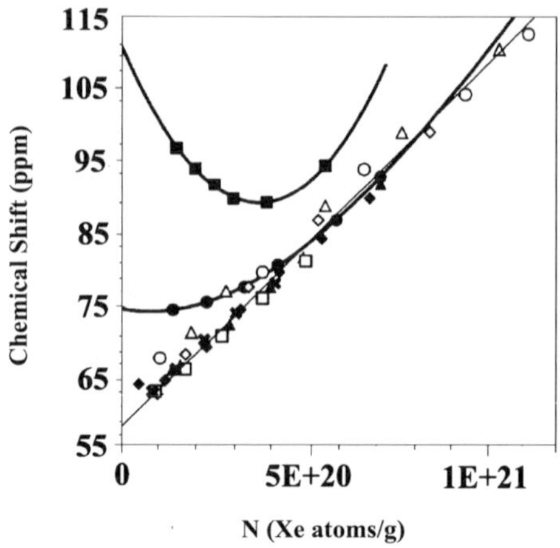

Fig. 12 $\delta = f(N)$ variation for □ HY, NaY with different Si/Al ratios: △ 1.35, ◇ 2.42, (×) 54.2, ○ 1.28 and for MgY zeolite for various magnesium contents: ◆ 47%, ▲ 53%, ● 62%, ■ 71%

2.3.3
Influence of Divalent Cations: Diamagnetic and Paramagnetic Ions

Fraissard et al. [97–99] have studied xenon adsorption on $Mg_\lambda - Y$ zeolites, under vacuum at 773 K, where λ denotes the degree of cation exchange with Na^+ (Fig. 12). When Mg^{2+} cations are in the sodalite cages or the hexagonal prisms without any contact with xenon ($\lambda < 53\%$), δ is a linear function of the xenon concentration (as for Na – Y). When some Mg^{2+} cations are situated within the supercages ($\lambda > 53\%$), one observes values of δ (compared with Na – Y) that are greatest when λ is high, especially at low xenon concentrations. For each value of λ the curve of $\delta = f(N)$ passes through a minimum; the higher λ is, the more pronounced and further shifted to higher concentration is the minimum. According to Ito et al. [99], the experimental value of $\delta_{N \to 0}$ for $N = 0$ is roughly proportional to the square of the electric field at the nuclei of xenon atoms adsorbed on Mg^{2+} cations.

The large positive shift and the parabolic behavior of the $\delta = f(N)$ curves in the case of divalent cations was attributed first by Fraissard et al. [2–4] to the high polarizability of xenon and the distortion of the xenon electron cloud by the strong electric fields created by the 2+ cations. According to Cheung et al. [50], the high value of δ is not solely due to that. They suggest the formation of a partial bond between these two species formed by the donation of a xenon 5p electron to the empty s orbital of the divalent cation. A similar model concerning electron transfer from xenon to platinum was proposed by Ito et al. to explain the large shift in δ for platinum supported on Na – Y (see Sect. 2.5 [182]).

The effect of the dehydration and rehydration of zeolites on the influence of Ca^{2+} cations can also be studied by ^{129}Xe NMR [99]. For example, the chemical shift δ and the signal width $\Delta\omega$ of xenon adsorbed on $Mg_{70}-Y$ increase with increasing dehydration of the solid. The values of δ and $\Delta\omega$ are greatest when dehydration is complete. Conversely, the spectra evolve with rehydration: for a given xenon pressure, the line due to xenon in the supercages containing only bare Mg^{2+} decreases in favor of the line corresponding to Mg^{2+} surrounded by OH ligands and water molecules yielding hydrated $(MgOH)^+$. ^{129}Xe NMR, therefore, makes it possible to follow the diffusion of an adsorbate in a zeolite crystallite.

The problem is more difficult in the case of paramagnetic cations (Ni^{2+}, Co^{2+}). One must consider the term δ_M which influences the chemical shift. This term may be very large and leads to δ values of several thousand of parts per million [125]. In this case, the location and the oxidation state of the cations can be studied, in particular during reduction, oxidation or dehydration of such samples. For example, Scharpf et al. [126] have succeeded, using this technique, in following the reduction and reoxidation of the $Ni_{7.5}Na-Y$ zeolite dehydrated under vacuum at 623 K. When the sample was reduced at lower temperatures (about 373 K), two types of environments for the xenon atoms were evident: one corresponding to xenon in the nickel-exchanged material, and the other corresponding to xenon in contact with Na – Y or HY. Reduction at higher temperatures produced a low-frequency shift of the first resonance, indicating that this environment becomes more like the environment of xenon in the Na – Y zeolite. At the highest temperature (643 K) studied, only the line corresponding to xenon in Na – Y was detected. These results prove that nickel ions are removed from the supercages upon reduction. Using this method, Scharpf et al. showed that reoxidation under 80 kPa of oxygen at various temperatures does not reverse the process of reduction. The distribution of the Ni^{2+} cations in the different sites of the zeolites has been studied by Bansal et al. for zeolites containing a low extent of exchange-cations [127] and by Gédéon et al. [128, 129] in the case of Ni – Na – Y zeolites with a high Ni^{2+} concentration. By following the variation of the chemical shift with the number of atoms adsorbed, the evolution of the environment of these cations can be studied as a function of the pretreatment temperature, for each level of exchange, and their migration from the supercages.

In the case of Co – Na – Y, with 15% of Na^+ cations exchanged, Bonardet et al. [130] have studied the dependence of the δ_M term on the pretreatment temperature T_t (Fig. 13). For $300 \le T_t \le 423$ K the chemical shift extrapolated to zero pressure, $\delta_{N\to 0}$, and the slopes of the $\delta = f(N)$ plots decrease. This corresponds to the departure of water molecules freeing the pores. The small shift difference compared to Na – Y is due to the presence of $Co(H_2O)_6^{2+}$ partially blocking the pores. When T_t is higher than 423 K, the dramatic increase in chemical shift and the change in the shape of the $\delta = f(N)$ plots arising from paramagnetic $Xe-Co^{2+}$ interactions prove that, despite the migration

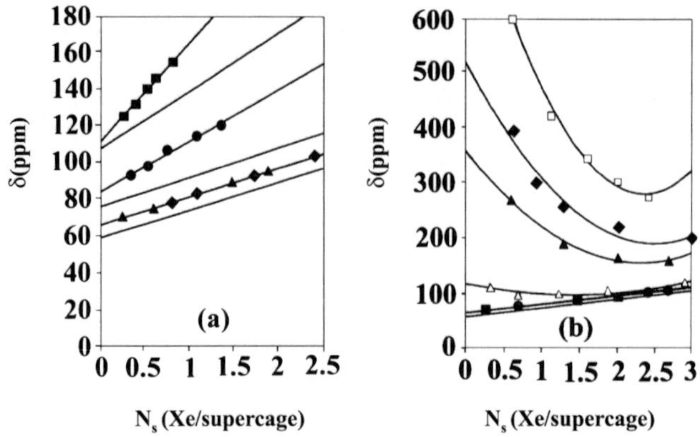

Fig. 13 $\delta = f(N_S)$ variation for CoNaY (15% of Na exchanged): (a) T_t (K) ■ 300, • 323, ▲ 373, ♦ 423; (b) T_t (K) ■ 373, • 423, △ 523, ▲ 573, ♦ 623, □ 773. *Full line:* $\delta = f(N_S)$ curves for pure NaY zeolites treated at the same T_t as CoNaY. (Reprinted from [130], with permission from Elsevier Science)

of Co^{2+} out of supercages as shown by X-ray diffraction [131], $Xe-Co^{2+}$ interactions are still detectable when water molecules are progressively eliminated from the coordination sphere of Co^{2+} cations. It has been proposed that Co^{2+} cations are located in $S_{II'}$ sites of the sodalite cages.

2.3.4 Case of Trivalent Cations

The ^{129}Xe NMR technique has also been used to study the dehydration processes of La^{3+} and Ce^{3+} exchanged Y zeolites [132, 133]. In these studies, the authors have confirmed that all the cations La^{3+} and Ce^{3+} migrate to the hexagonal prisms and the sodalite cages when the zeolites are completely dehydrated. Later, Fraissard and coworkers [134] have extended the ^{129}Xe NMR investigation to other faujasite-type zeolites with trivalent cations, i.e., Y, Na – Y, La, Na – Y, Ce, Na – X and Y, Na – X. The chemical shifts of adsorbed xenon follow a concave plot against the adsorbed xenon concentration, very similar to the previous ^{129}Xe NMR study of Mg^{2+}, Ca^{2+}, Zn^{2+}, Cd^{2+}, Ni^{2+} ... in X and Y zeolites. This result suggests strong adsorption of xenon on the Y^{3+} ions located inside the zeolite supercage. In contrast, the ^{129}Xe NMR chemical shift for Ce – X, La – X, Ce – Y and La – Y zeolites [132, 134] indicates that Ce^{3+} and La^{3+} cations are located in hexagonal prisms or sodalite cages inaccessible to xenon atoms. The example of Y^{3+} cations proves that not all trivalent cations migrate out of the supercages during thermal treatment of the solid.

The location of Ru^{3+} ions in Y zeolites and their redox behavior have been studied by Shoemaker et al. [69]. The increase in shifts in the down-field direction was attributed to the interaction of xenon atoms with the paramagnetic Ru^{3+} centers. From chemical shift and line-width variations, these authors have shown that after reduction at room temperature, the ruthenium particles are highly dispersed and located in the zeolite cages. However, migration of the metal occurs upon evacuation.

2.3.5
Effect of the Electronic Structure (nd^{10}): Ag^+, Cu^+, Zn^{2+}, and Cd^{2+} Cations

A particularly interesting case is that of faujasite-type zeolites containing Ag^+ or Cu^+ cations, the external electron structure of which is nd^{10}. In experiments on silver-exchanged Na – Y and Na – X zeolites, Fraissard and coworkers found an unusual and unexpected up-field (i.e. low-frequency) ^{129}Xe chemical shift of about – 40 to – 50 ppm (relative to xenon gas at zero pressure) for fully silver-exchanged zeolites in the limit $N \rightarrow 0$ [135–139]. This is in contrast to all other ^{129}Xe NMR studies on cation-exchanged X and Y zeolites which all show high-frequency shifts. Figure 14 shows the ^{129}Xe NMR isotropic chemical shifts of xenon in Na – X and in the fully silver-exchanged zeolite Ag – X. The shifts in dehydrated and oxidized Ag – X are distinctly lower than that for Na – X over the range of concentration studied. Most remarkably, the shifts decrease with Xe concentration, exhibiting negative values in the range – 40 to – 50 ppm at low xenon concentration. In contrast to these results, the samples reduced at 373 K, and 673 K show high-frequency shifts with respect to Na – X. After reduction at 373 K, δ increases steadily with the number of xenon atoms per supercage (N_s) from + 100 ppm to about

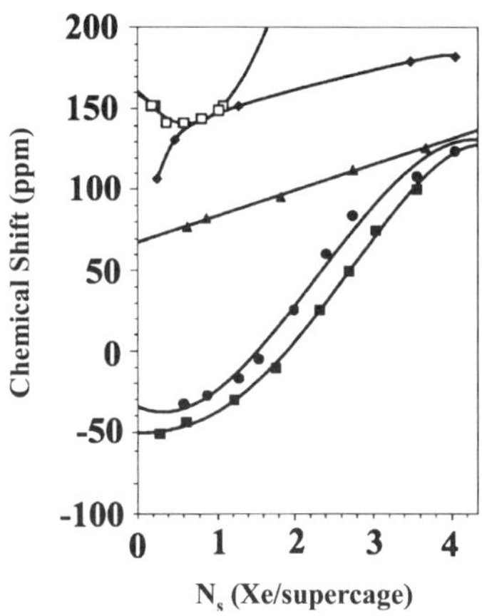

Fig. 14 $\delta = f(N_S)$ variation for the zeolites: ▲ Na – X, • Ag – X (dehydrated at 673 K), ■ Ag – X (oxidized at 723 K), ◆ Ag – X (reduced at 373 K); □ Ag – X (reduced at 673 K). (Reprinted from [135], with permission from Elsevier Science)

+ 170 ppm for about $0.3 < N_s < 4$, whereas after reduction at 673 K the shift values are between + 140 and + 160 ppm for $0.1 < N_s < 1$, with a shallow minimum at about $N_s = 0.7$. In the latter case, the plot of $\delta = f(N)$ has the classical shape of zeolite-supported metals. This unusual low-frequency shift in the dehydrated and oxidized samples has been attributed to specific $Ag^+ - Xe$ interactions which are caused by the existence of a fully occupied d-shell in this cation. In order to explain the up-field shift, the authors postulate a short-lived $Ag^+ - Xe$ complex and a $4d^{10}$-$5d^0$ donation from the transition metal cation to xenon [135]. Similar phenomena have also been observed for the ^{77}Se resonance in the presence of Ag^+ cations [140].

Later experiments showed that fully exchanged CuY [141, 142] and CuX [143] zeolites exhibit shifts which are much lower than expected. The parabolic form of the $\delta = f(N)$ curves, which was expected because of the presence of paramagnetic Cu^{2+} is not observed (Fig. 15). In order to explain this behavior, Gédéon et al. [143] attributed this down-field shift to a $3d^{10}$-$5d^0$ donation mechanism between Cu^+ ion and Xe. Boddenberg et al. have attributed this to the presence of sites bereft of cations in the supercages [144]. Indeed, during dehydration, there is at the same time migration of Cu^{2+} from the supercages and partial autoreduction of Cu^{2+} to Cu^+inside the supercages. These easily accessible Cu^+ cations are able to behave like Ag^+ and participate in $3d_\pi$-$5d_\pi$ electron donation. This behavior is not observed for Zn^{2+} and Cd^{2+} in X-type zeolite and seems not general among cations of nd^{10} configuration. It has been proposed that, after dehydration, Zn^{2+} and Cd^{2+} interact strongly with the matrix and are situated in S_{II}sites, a position which

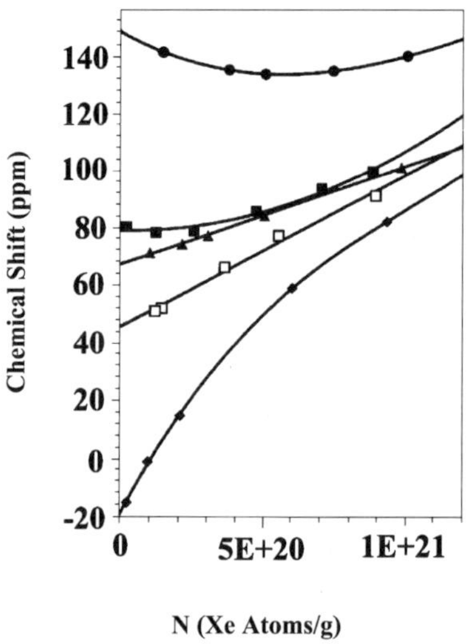

Fig. 15 $\delta = f(N)$ variation for the zeolites: (▲) NaX, (◇) AgX, (•) CdX, (■) ZnX, (*o*) CuX. (Reprinted from [143], with permission from Elsevier Science)

prevents such d_{π} donation. The higher charge of the cation, responsible for a greater polarization of the electronic cloud leading to a high chemical shift, can also be put forward and may compete with the electron donation.

Gédéon et al. [145] have also studied the oxidation state as well as the location of copper in CuY zeolites at various levels of exchange and hydration, and at different stages of redox treatment. Information about the nature of $Xe-Cu^+$ or $Xe-Cu^{2+}$ interactions has been obtained by a combination of ESR and ^{129}Xe NMR. Similar studies have been done by Liu et al. [146]. The combined ESR and ^{129}Xe NMR studies enabled these authors to investigate the reduction/oxidation processes of the copper ions.

More recently, quantum chemical ab-initio calculations have been performed by Freitag et al. [147] for the interaction between xenon atoms and the cations Li^+, Na^+, K^+, Cu^+ and Ag^+. The ^{129}Xe NMR chemical shifts have been calculated. For the alkali ions a small down-field shift is obtained which is attributed to the polarization of the Xe wave function by the charge of the cation. For the Cu^+ and Ag^+ an additional up-field shift is found which is caused by a mixing of the 5p, 4p, 3p orbitals at Xe with the 3d or 4d orbitals at Cu^+ or Ag^+. This leads to an increased magnetic shielding and consequently to low chemical shifts.

2.3.6
Location and Number of Cations: Quantitative ^{129}Xe NMR

The concept of quantitative ^{129}Xe NMR spectroscopy concerns the analysis of experimentally detected isotropic chemical shifts in terms of site specific shifts and of the concentrations of various cation sites as well as the exploration of the information content of the widths of the resonance lines.

A systematic investigation of the effect of the cation on the chemical shift of ^{129}Xe adsorbed on zeolites has been carried out by Fraissard and coworkers. We have seen that these authors suggested an explanation of the ^{129}Xe NMR shift in terms of different contributions from Xe – Xe interactions and Xe-zeolite interactions. The magnetic and electric contribution has been evaluated in the case of nickel-exchanged Y zeolite [127, 128].

Cheung et al. has offered a quantitative interpretation by the use of a model in which the xenon atoms adsorbed on the wall of the supercages are in rapid exchange with those in the gaseous phase [50]. Later, Liu and coworkers [148] have carried out a detailed investigation of the effect of ion exchange in Na – Y zeolite on the adsorption strength toward xenon atoms and on the change in the ^{129}Xe NMR chemical shifts. They found that the dependence of the ^{129}Xe NMR chemical shift on the amount of adsorbed xenon can be fitted with a second order polynomial in all cases. This is interpreted by treating the adsorbed xenon as a two dimensional gas with a virial expansion model similar to the case of xenon in the normal gaseous state. According to this model, the major force for the adsorption and various kinds of interaction of xenon is the van der Waals force, but the coulombic force of the cations also plays a role.

In a series of papers [137, 149–155], a model of xenon adsorption was introduced that is capable of explaining quantitatively the dependence of the chemical shift on the concentration of sorbed xenon in Y-zeolites exchanged with transition metal ions such as silver, copper, zinc and cadmium. According to this model, the adsorption isotherms of Xe in transition metal ion-exchanged zeolites can be quantitatively described by the isotherm equation:

$$N_S = n_1 \frac{k_1 P}{1 + k_1 P} + n_2 \frac{k_2 P}{1 + k_2 P} + k_3 P, \tag{21}$$

where N_S is the concentration of Xe sorbed in the supercages at the pressure P; n_1 and n_2 are the concentrations of the transition metal cation sites which were identified with ions residing on the crystallographic S_{III} (index 1) and S_{II} (index 2) positions; k_1 and k_2 are the Langmuir adsorption constants on the respective sites; k_3 is a Henry adsorption constant which describes the interaction of Xe with the residual supercage surface not occupied by the transition metal cations. Figure 16 shows, as an example, the xenon adsorption isotherm for the oxidized AgY zeolite as well as its components. The concentrations of the cations, n_i, and their adsorption constants k_i are obtained from the curve fit.

Using this model with zeolites containing different levels of silver cations, it is possible to monitor the number of the silver cations (species 1 and 2) in the supercages as a function of the overall silver content (Fig. 17).

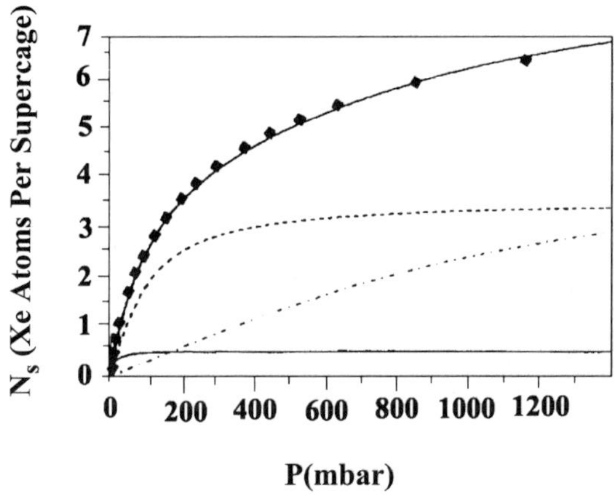

Fig. 16 Overall and composite adsorption isotherms of xenon in the zeolite AgY(ox) according to the model discussed in the text. The composite isotherms belong to — type 1, - - - type 2, and -·- type 3 sites in the supercages. (Reprinted from [137], with permission from Elsevier Science)

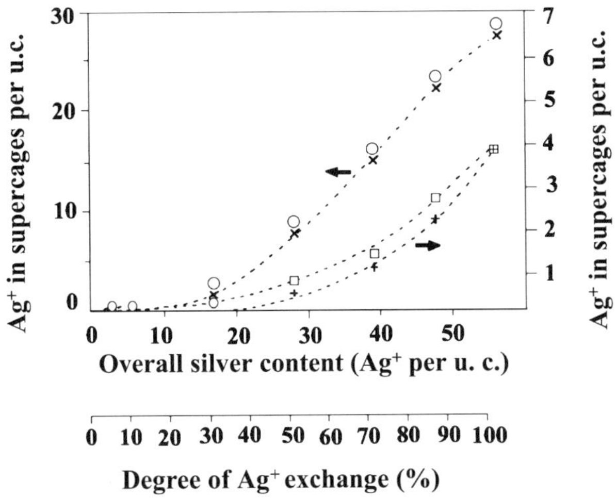

Fig. 17 Concentrations of supercage silver cations in dehydrated (673 K) and oxidized (673 K) AgNaY zeolites as a function of the overall silver content. Type 1: Ag^+ in (+) dehydrated and □ oxidized zeolites. Type 2: Ag^+ in × dehydrated and ○ oxidized zeolites. (Reprinted from [137], with permission from Elsevier Science)

The experimentally observed ^{129}Xe NMR chemical shifts, $\delta = f(P)$, can be quantitatively reproduced with the aid of the equation:

$$\delta = \frac{1}{N_s}\left(\frac{n_1 k_1 P}{1+k_1 P}\delta_1 + \frac{n_2 k_2 P}{1+k_2 P}\delta_2 + k_3 P \delta_3\right) + FN_S. \tag{22}$$

The individual chemical shifts δ_i of xenon interacting with the metal cation sites ($i = 1, 2$) and the residual surface ($i = 3$) can be determined. The term FN_S takes into account the mutual Xe – Xe interactions to a first-order approximation [38].

Boddenberg and coworkers [155] have demonstrated that the observed values of the linewidths as well as their quite different dependence on pressure can be explained on the basis of a theoretical approach which is capable of explaining semi-quantitatively the ^{129}Xe NMR line widths versus xenon concentration curves for a series of zinc- and cadmium-exchanged Y zeolites. This theoretical model traces the origin of line broadening back to inhomogeneous spatial distributions of transition metal cation sites interacting strongly with the encaged xenon atoms. Using Eqs. 21 and 22, these authors determined the distribution of Zn^{2+} [156].

2.3.7
Cation Exchange Between Different Zeolites

^{129}Xe NMR of adsorbed xenon can be used to follow the exchange of cations in the zones accessible to xenon in a given crystallite, and to visualize the exchange of cations between different zeolites. Chen et al. have studied the cation exchange process between RbX and Na – Y zeolites. The two peaks detected at 144 and 89 ppm correspond to xenon adsorbed in dehydrated Rb – Na – X and Na – Y zeolites, respectively [73]. Upon mechanical mixing at 300 K, the intensities of the two signals decrease and a third broad signal at about 141 ppm appears. After further treatment of the mixture at 673 K and 0.01 Pa, there is a further larger decrease in the intensities of the Rb – Na – X and Na – Y signals, the peak at 141 ppm disappears and two additional broad signals are seen at 128 and 109 ppm.

The two xenon resonance signals corresponding to the two pure zeolites can be restored upon cooling the sample to 200 K, which proves that the third signal at 141 ppm is a coalescence of two signals. The mixture obtained directly in the tube by shaking is very inhomogeneous with regard to the distribution of the crystallites of the two zeolites. There are three types of zones: pure Rb,Na – X, pure Na – Y, and a Rb,Na – X,Na – Y mixture. Because of slow diffusion, the three corresponding lines can be distinguished at 300 K. Consequently, the information obtained by ^{129}Xe NMR at 300 K is characteristic of macroscopic zones, i.e. zones containing several crystallites. The extent of these zones depends on the mean lifetime of the xenon atom in the crystallites compared with the NMR timescale. These zones can be reduced by lowering the temperature of the NMR experiment in order to obtain more information about localization. The rate of exchange of xenon atoms between one crystallite and another must also depend on the crystallite size and, above all, on the barrier for diffusion to the external surface; all other things being equal, this barrier is related to the size of the windows of the cavities and channels.

The low-temperature ^{129}Xe NMR spectra of xenon adsorbed on the zeolite mixture show that, down to 200 K, there are still four well-resolved peaks, two of them corresponding to the original Rb,Na – X and Na – Y zeolites. The other two signals located between those of RbNa – X and Na – Y should correspond to the $Rb_{58-x}Na_{23+x}$ X and Rb_xNa_{56-x}Y zeolites, that result from ion exchange between the two original solids.

However, in an investigation of a mixture of dehydrated Na – Y and Ca – Y, Ryoo et al. [70] pointed out that the single ^{129}Xe NMR peak observed after thorough mixing of the samples could result from fast xenon exchange between the crystallites, rather than from ion exchange. This is supported by the observation of two ^{129}Xe NMR peaks on cooling the sample, when the exchange rate of xenon between particles is reduced [71]. If solid-state ion exchange had occurred, the two signals would be seen if the temperature was reduced sufficiently to retard the exchange of xenon between cages, which would only be expected at much lower temperatures, while in fact two peaks are observed at temperatures as high as 263 K.

2.3.8 Effect of Nonframework Aluminum

It is well known that nonframework aluminum, Al_{NF}, plays an important role in the catalytic properties of zeolites. The presence of these species is usually checked by ^{27}Al NMR spectroscopy but the nature of these species is still unclear. Chen et al. [157, 158] have studied the nature of the Al_{NF} in MFI-type zeolite by ^{129}Xe NMR and ^{27}Al NMR. ^{27}Al NMR showed that the reference sample (R) did not contain any nonframework aluminum (Table 4). Conversely, the samples prepared in fluoride medium, A, B and C always contained some Al_{NF}. The quantity of Al_{NF} in the samples depends on the synthesis conditions.

The four samples studied by ^{129}Xe NMR have different chemical shift variations with N (Fig. 18). For sample R, δ increases monotonically with N. For the samples synthesized in the fluoride medium, δ first decreases and then increases with N. This indicates the presence of some strong adsorption sites inside the channels of the samples. These sites can only be more or less

Table 4 Aluminum concentration per unit cell

Sample	Total aluminum	Framework Al (Al_F)	Nonframework Al (Al_{NF})
R	4.0	4.0	0
A	8.0	4.0	4.0
B	3.4	2.9	0.5
C	2.2	1.6	0.6

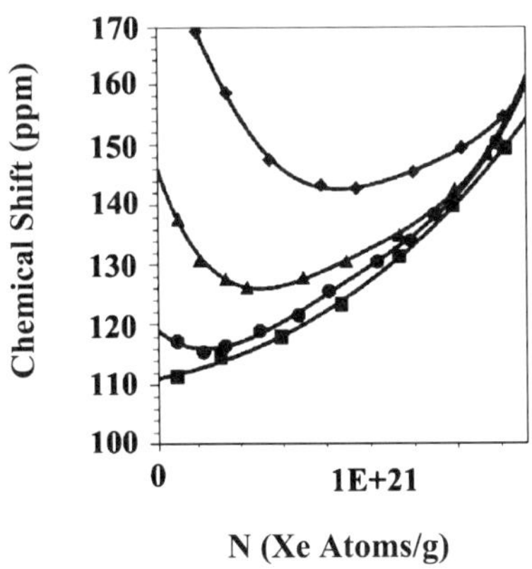

Fig. 18 $\delta = f(N)$ variations: • A; ▲ B; ♦ C; ■ R. See Table 4. (Reprinted from [157], with permission from Elsevier Science)

charged Al_{NF} species. Comparison of the number of Al_{NF} and the $\delta_{N\to 0}$ value shows that the charge on Al_{NF} increases in the order A < B < C. The average charge on each Al_{NF} atom depends on the amount of Al_{NF} and on the Al_{NF}/Al_F ratio inside the zeolite.

Similar upward curvatures in xenon chemical shift plots have been reported by Bonardet et al. [159] for xenon adsorbed in steam-treated ultrastable NH_4Y zeolites, and attributed by these authors to specific interactions of Xe with extra-framework aluminum species. Likewise, Bradley et al. [160] observed highly curved chemical shift versus coverage plots for xenon in aluminosilicate and gallosilicate MFI zeolites and proposed the existence of charged extra-framework aluminum species. Theses authors showed that ^{129}Xe NMR can be used to indicate changes in the nature of the extra-framework Al during the steam treatment.

To detect the Al_{NF} species in Y zeolites, it is necessary to decrease the experiment temperature, i.e., the xenon mobility. This is justified by the fact that in Y zeolites, where the pores are bigger than in ZSM-5, the influence of the pore surface on the chemical shift is smaller at a given temperature.

2.4 Bulk and Distribution of Adsorbed Phases

2.4.1 Introduction

The use of NMR techniques such as relaxation time measurements, pulse-field gradient or line-shape analyses to study the molecular dynamics and consequently the location of adsorbed molecules (water, light hydrocarbons ...) in the micropores of zeolites does not allow us to easily obtain pertinent information about the adsorbate distribution. The simple ^{129}Xe NMR method, sometimes coupled with multiquantum NMR measurements, has been successfully used to follow the encumbering or the blocking of the internal volume of molecular sieves during adsorption or coking.

2.4.2 Study of Adsorbed Organic Molecules

The first study of the bulk and distribution of organic molecules trapped in a zeolite was performed by Pines and coworkers [161, 162]. They examined the distribution of hexamethylbenzene (HMB) in an Na – Y zeolite dehydrated at 673 K under vacuum (10^{-4} Pa). They observed two signals, the relative intensities of which depended on the number of HMB molecules in the supercages. When the loading corresponded to an average of 0.5 molecule per supercage, the two signals had the same intensity (Fig. 19). When the sample was heated to 573 K, the spectra displayed a single line, suggesting

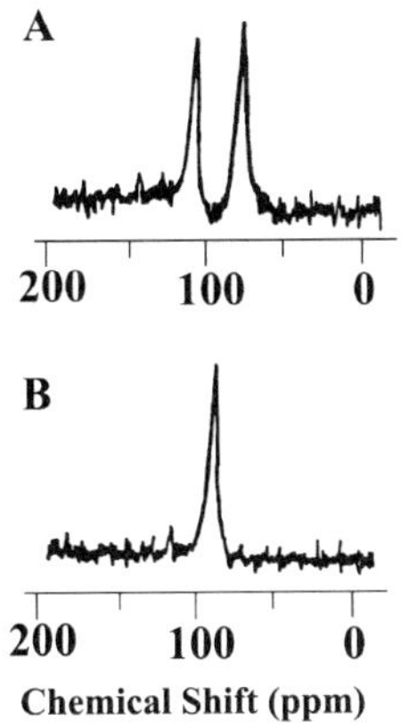

Fig. 19 ^{129}Xe NMR spectra of Xe ($P_{Xe} = 40$ kPa) adsorbed in NaY zeolite with an average loading corresponding to 0.5 HMB molecule/supercage. A: before heating; B: after heating at 573 K for 2 h. (Reprinted with permission from [161]. Copyright (1987) American Chemical Society)

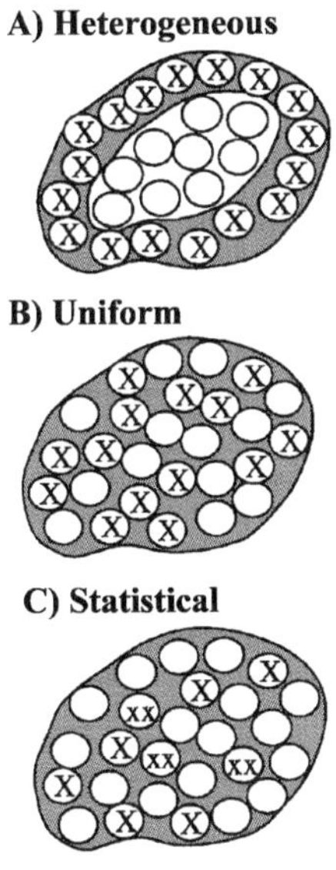

Fig. 20 Possible dispersion of guest molecules among cavities of a zeolite particle when the average coverage is less than one molecule per cavity. (Reprinted with permission from [161]. Copyright (1987) American Chemical Society)

that the guest molecules were then homogeneously distributed among the zeolite cavities.

This is consistent with the distributions shown in Fig. 20: 50% of supercages are occupied by one HMB molecule and 50% are free of HMB (B) or there is a statistical distribution (C). Both are macroscopically uniform in the zeolite, and both would yield only one average xenon NMR line. Unfortunately, because of the high xenon mobility the resonance line is averaged and the two distributions cannot be distinguished by the ^{129}Xe NMR technique alone. Nevertheless, with additional information from multiple-quantum NMR spectroscopy, it is possible to conclude unambiguously that the distribution inside the supercages is homogeneous.

The same authors [71, 163] extended their study to other molecules: *n*-hexane, benzene and 1,3,5-trimethylbenzene (TMB). Table 5 presents values of δ for $P_{Xe} = 0$ and the slope of the $\delta = f(N)$ curve when the xenon concentration is 0, 1 or 2 molecules per supercage. The samples with adsorbed benzene and TMB show similar values of δ at all concentrations of guest species and with *n*-hexane the largest values of the chemical shift. Certain conclusions can be drawn regarding the location of the guest molecules inside the supercages. As suggested by Lechert et al. [164] and by de Mallmann and Barthomeuf [165], benzene is probably "attached" to the walls of the supercages near cation sites. Since the geometry of TMB is similar to that of benzene and has the same chemical shift at zero xenon pressure, it may be attached in a similar manner. The much larger interaction between xenon and *n*-hexane is attributed to the fact that the molecules have no preference for being attached parallel or perpendicular to the supercage surface and, therefore, protrude into the center of the cage allowing a greater interaction. Figure 21 resumes the different situations allowed for adsorbed molecules.

The variation of the slope of the $\delta = f(N)$ variation is related either to the void volume or to the channel blockage. For the three benzene samples, the similarity of the slopes indicates that benzene does not interfere with xenon

Table 5 $\delta_{N\to 0}$ and slope of the $\delta = f(N)$ variations for different guest molecules at various loadings

Guest molecules	Molecules/cage	δ at $N = 0$ (ppm)	Slope $\delta = f(N)$ (ppm/10^{20} Xe g^{-1} of NaY)
	2	89.7 ± 0.1	3.8 ± 0.1
benzene	1	75.1 ± 1.1	3.9 ± 0.2
	0.5	68.6 ± 0.1	4.0 ± 0.1
	2	89.0 ± 1.3	5.4 ± 0.3
trimethylbenzene	1	75.9 ± 0.4	4.5 ± 0.1
	0.5	71.2 ± 0.7	3.6 ± 0.1
	2	$1\,07.7 \pm 1.2$	4.4 ± 0.3
n-hexane	1	86.7 ± 0.8	3.4 ± 0.2
	0.5	74.7 ± 0.5	3.5 ± 0.1
none (this work)		58.9 ± 0.5	4.4 ± 0.1
none (ref. 1)		58 ± 4	4.6 ± 0.2

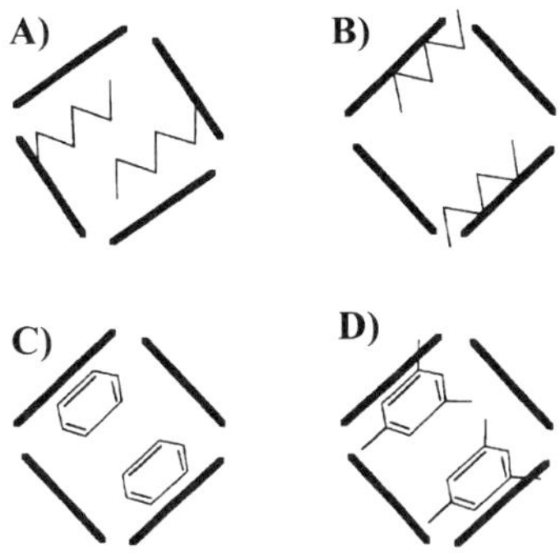

Fig. 21 Schematic illustration of the various guest species adsorbed inside the NaY zeolite supercages. Two *n*-hexane molecules adsorbed: normal to the supercage (A); parallel to the supercage (B); two benzene molecules (C); two TMB molecules (D). (Reprinted with permission from [163]. Copyright (1990) American Chemical Society)

by taking up volume in the middle of the supercages or by blocking the channels. For TMB, the situation is not so clear. If TMB adsorbs like benzene, it would block the channels of the zeolite because of its larger kinetic diameter (0.75 nm) and then interfere with xenon atoms. On the other hand, if it did not attach to the supercage walls, it could occupy a large fraction (up to 40% for 2 molecules per supercage) of the void volume, but in both cases the slope would increase with the xenon concentration. It is difficult to explain the *n*-hexane data: one reason for the larger slope, observed only at a concentration of two *n*-hexane molecules, may be that the probability of having the 2 molecules normal to the surface is greater than when the molecules are attached like benzene and TMB, near the S_{II} sites (Fig. 21A). In this case, interaction with xenon is enhanced.

Liu and coworkers [166, 167] investigated in more detail the distribution of benzene in Na – X and Na – Y zeolites. First of all, they showed that benzene diffuses very slowly at room temperature in the zeolite crystallites. Even after 120 days, adsorption equilibrium is not reached and it is necessary to heat the sample to 523 K for 10 h to obtain a homogeneous distribution.

Moreover, the variation of the ^{129}Xe NMR linewidth, ΔH, with the average number of benzene molecules per cage, θ, shows an unexpected marked increase for $2.5 < \theta < 3.5$ (Fig. 22). The small linear increase in ΔH with θ (for $\theta < 2.5$) is in agreement with the decrease in the mean free path of xenon proposed by Demarquay and Fraissard [56], and is due either to a decrease in the void space available to the xenon atoms or to an increase in site-hopping motion of the adsorbed benzene molecules. The fact that the linewidths are nearly independent of the xenon concentration proves that xenon atoms can move rapidly between supercages and that the effects of Xe – Xe interactions are reduced. For $2.5 < \theta < 3.5$, the $\Delta H = f(\theta)$ curves pass through a marked maximum ($\theta = 3$), whose position depends on the xenon loading.

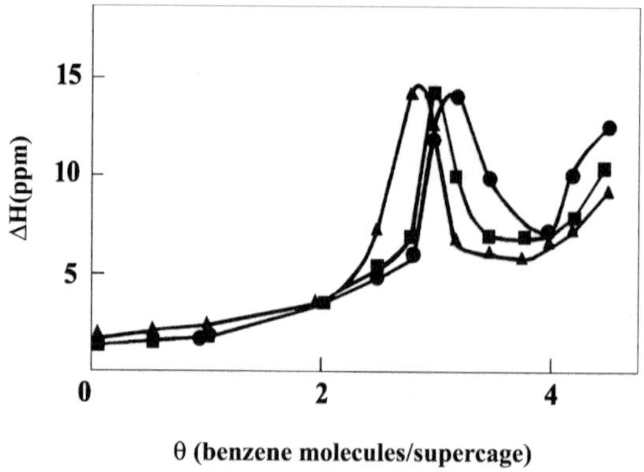

Fig. 22 The variation of ^{129}Xe linewidth spectra with respect to benzene coverage, θ (average number of benzene molecules per supercage); • $P_{Xe} = 6.7$ kPa; ■ $P_{Xe} = 40$ kPa; ▲ $P_{Xe} = 90$ kPa. (Reprinted from [166], with permission from Elsevier Science)

These results are interpreted by the authors as reflecting a mutual hindrance between xenon atoms and benzene molecules, which rearrange by cooperative migration of benzene-Na^+ complexes towards the center of the supercages. These results seem to agree with those obtained by Lechert et al. [164] from ^{1}H NMR measurements, by Renouprez et al. [168, 169] from neutron diffraction studies and by de Mallmann and Barthomeuf [165] from IR results. These authors found that benzene molecules start to aggregate at about $\theta = 2.5$. For $\theta > 4$, ΔH increases again with increasing θ, which is consistent with a cooperative interaction of benzene within the limited void space of the supercage. For $\theta > 4.5$, benzene molecules are packed in the supercage into which xenon can no longer enter.

2.4.3
Study of the H_2O/Na–Y System

Gédéon et al. [170] have shown that it is possible, in a Na – Y zeolite, to differentiate between water adsorbed in regions accessible or not to xenon, to measure the volume of water in pores and to determine the blockage of windows by water molecules.

Figure 23 shows the variations of δ_S (δ value for $N = 0$) and the slope $d\delta/dN$ with the water concentration, C. The decrease in δ_S with C proves that right from the beginning of dehydration, the number of water molecules per supercage decreases. Furthermore, δ_S stops decreasing and remains constant when $C < 0.15$.

This result shows that for $0 < C < 0.15$ the mean free path ($\overline{\ell}$) of xenon at zero concentration is constant. It may be deduced that for $C < 0.15$ the supercages and the windows between supercages are completely dehydrated. In other words, the water molecules are all in the sodalite cages and the prisms.

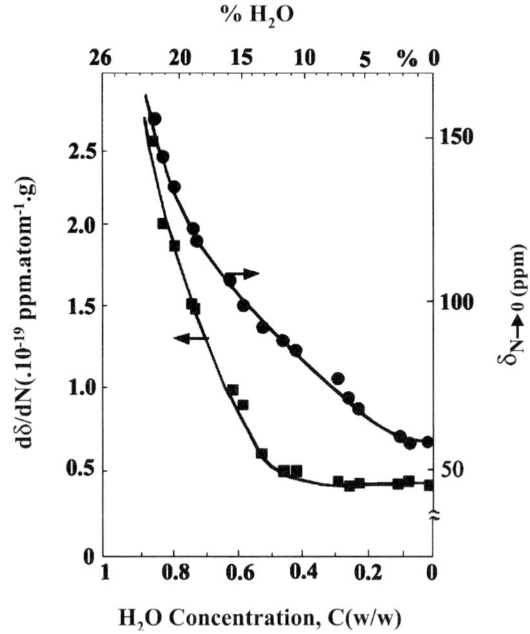

Fig. 23 Effect of the water concentration C [*upper* (w/w of anhydrous solid); *lower* (relative concentration)] on shift $\delta_{N\to 0}$ (•) and the slope of the $\delta = f(N)$ variations (■). (Reprinted from [170], with permission from Elsevier Science)

This result is compatible with the fact that each sodalite cage can contain a maximum of four water molecules corresponding to $C = 0.15$. As for the slope, it remains constant for $C < 0.4$, whereas δ_S continues to decrease down to $C = 0.15$. This means that the void space available to xenon atoms stays about the same ($d\delta/dN$ constant) but that the mean free path increases (δ_S decreases) because xenon atoms diffuse more easily between the supercages. It is easy to understand from these results that for $0.15 < C < 0.4$ the water molecules are in the vicinity of the 8 Å apertures, which they obstruct more or less depending on the value of C. Dehydration at these windows does not change the pore volume of the zeolites very much but leads to an easier diffusion of xenon between supercages and therefore greatly affects $\overline{\ell}$ and δ_S.

2.4.4
Study of Coke Deposition in Zeolites

The deactivation of zeolites by carbonaceous deposits (coke) is a crucial problem in industrial cracking, and many studies have shown that deactivation depends on the nature and the structure of the zeolites. ^{129}Xe NMR spectroscopy has proved to be an efficient tool for locating coke inside the cage or on the external surface of the zeolite. This technique is also able to show the role of extra-framework aluminum in the cracking activity of zeolites.

The study by Ito et al. [171] concerns coke deposition formed by *n*-hexane or propylene cracking on HY zeolites. The single ^{129}Xe NMR line is narrow and almost symmetrical for samples coked with *n*-hexane; it becomes much broader and asymmetrical for samples coked with propylene, indicating that xenon atoms are much less mobile in the cages. For a sample highly loaded with coke (33% w/w) formed from propylene cracking, the spectrum consists of a broad asymmetrical highly shifted component (157 ppm) and a narrow, almost symmetrical line at ca. 10 ppm. The first line is characteristic of strongly adsorbed xenon atoms in the modified supercages, while the other resonance line may be attributed to xenon atoms adsorbed in mesopores or macrocavities formed by the coke between the crystallites. The $\delta = f(N)$ variations show nearly identical slopes for noncoked and 5% coked samples, whereas slopes are much larger for samples coked by propylene to 15 and 33% w/w. These results suggest that the available free volume is slightly reduced when the coke level is low but is seriously affected at high coke loading. By comparing the remaining free volume obtained from the slope and that calculated from the average density of the coke (with the assumption that all the coke is deposited in supercages), the authors show that part of the coke blocks the access to other supercages; in this case, the coke formed must be essentially aromatic or polyaromatic.

Barrage et al. [172, 173] have studied in detail the coke distribution and the influence of dealumination in dealuminated HY and HZSM-5 zeolites. The variations of δ versus [Xe] for dealuminated HY zeolites, which have undergone various hydrothermal treatments, are similar to that described by Ito and Fraissard [98] for Ca – Y or Mg – Y zeolites and by Gédéon et al. [129] for NiY. In all cases, there are adsorption centers stronger than H^+ or Na^+. The curvature, observed at low xenon concentration, is therefore attributed to the electric field created by nonframework aluminum species equivalent to more or less complexed Al^{3+} cations (see Sect. 2.3.8). In the case of coked samples, even at low loading, this curvature vanishes (Fig. 24) and δ varies linearly over the entire concentration range. This linearity shows that the coke is deposited firstly on the nonframework aluminum preventing the Al – Xe interactions.

Examination of the $\delta = f(N)$ curves for 3 and 10.5% coked samples shows that the slopes increase markedly (by a factor of 2.4) and $\delta_{N\to 0}$ somewhat less. These results mean that, after covering the strong adsorption sites (i.e. extra-framework Al), coke forms more homogeneously, affecting both the windows between the α-cages (δ_S increases) and the volume of the supercages ($\mathrm{d}\delta/\mathrm{d}N$ increases). When the coke content reaches 15%, the slope is almost identical to that of the 10.5% coked sample but the strong increase of δ_S shows that the mean free path of xenon is very restricted. Moreover, a second Xe NMR signal appears (45 ppm) independent of the pressure; this is attributed to xenon

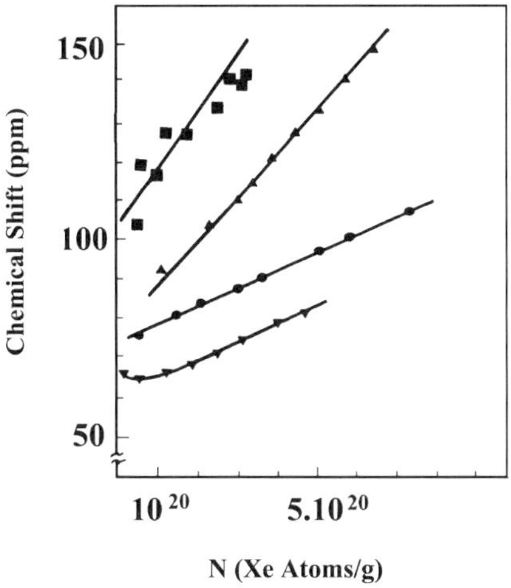

Fig. 24 $\delta = f(N)$ variation at 300 K for: ▼ HY-reference sample, • HY 3%-coked sample, ▲ HY 10.5%-coked sample, ■ HY 15%-coked sample. (Reprinted from [172], with permission from Baltzer Science)

adsorbed in micro- or meso-cavities of coke formed at the external surface of the crystallites.

A variable-temperature NMR study of adsorption shows that, for the 3% coked sample, δ_S and $d\delta/dN$ are temperature-dependent (Fig. 25). For each temperature the slope of the $\delta = f(N)$ plot is close to that of the straight section of the plot for the reference sample, while the value of δ_S is higher. The free volume is, therefore, hardly affected but xenon diffusion is restricted. The authors conclude that the coke is located initially at the windows between the supercages. At higher coke content (10.5%), δ_S and $d\delta/dN$ become independent of the adsorption temperature. This result is similar to that obtained for zeolites with narrow channels. It must be assumed that the high coke content considerably reduces the diameters of the cages and windows, and consequently the mean free path of xenon. The size of the cavities still accessible to the xenon becomes of the same order of magnitude as that of the xenon atoms.

Similar studies have been performed by Tsiao et al. [174] and Liu et al. [175]. The former studied HZSM-5 coking after exposure to 2-butene at 823 K. They found a linear relationship between the chemical shift and the xenon uptake whatever the coke loading. The values of δ_S are the same for three different coked samples but the slope increases with the amount of coke deposited. They concluded that deactivation results in channel blockage by coke deposition at the openings or at channel intersections. The latter authors investigated the formation of coke in the disproportionation of *n*-propylbenzene on various zeolites. For HY zeolites their results agree with ours [172]. In the case of HZSM-5 zeolites, they found δ variations analogous

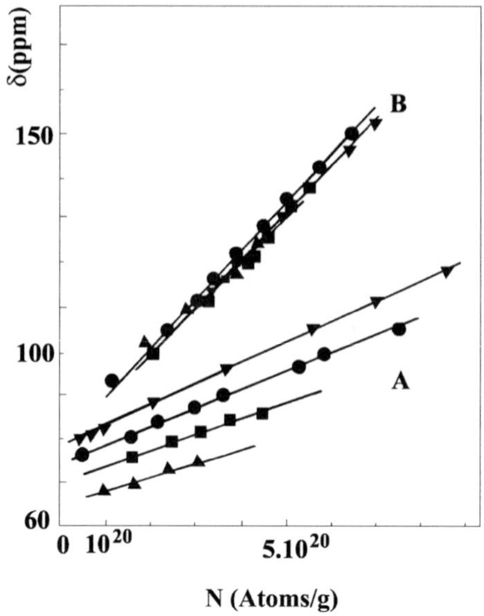

Fig. 25 $\delta = f(N)$ variation on dealuminated HY zeolites. A (3%-coked samples), B (10.5% coked) at different adsorption temperatures: (τ) 273 K, (•) 300 K, (■) 319 K, (σ) 338 K. (Adapted from [172], with permission from Baltzer Science)

to those observed previously by Bonardet et al. [176] after coking by acetone conversion. At low coke loading, they observed two signals which they attributed to two adsorption sites: one in the opening of the pores or mid-channel positions, the other at the channel intersections. In our case we also observed, at low coke content, two xenon NMR lines for adsorbed xenon, but we attributed the second signal (which disappears when the coke level increases) to extra-framework aluminum adsorption sites, in agreement with the ^{27}Al NMR spectrum, which shows the presence of octahedrally coordinated aluminum.

It has been also possible to study the reoxidation of coked zeolites. For example, Bonardet et al. [176, 177] have shown that total reoxidation of the carbonaceous residues, even under mild conditions induces structure defects (additional dealumination) and/or partial amorphization leading to a significant loss of internal microporous volume of the catalyst. In the case of highly dealuminated HZSM-5 zeolites (Si/Al > 80) [173] the values of δ_S obtained are similar for fresh and samples coked during conversion of methanol. Thus, at low concentrations xenon is mainly adsorbed in zones which are not influenced by coke deposits, i.e. which do not contain coke. This result shows that the coke distribution is heterogeneous. A greater slope is observed for the coked samples than for the fresh catalyst, but this slope is the same for both coked samples (7 and 14% w/w). This proves that the available free volume is seriously reduced either by blocking of channels or by deposits in the channel intersections. On the other hand, the identical curves for samples with 7 and 14% of coke show that a high proportion of the coke must be located outside the crystallites (at least the part of coke above 7%).

Miller et al. [178] studied the coking of HY zeolites by propylene cracking, but they combined ^{129}Xe NMR with argon sorption measurements. They found that the increase in xenon chemical shift (in comparison with a fresh sample) depends on the aromaticity of the coke. Moreover, argon sorption showed a bimodal distribution in the pore apertures: one, free of coke (0.72 nm) and the other greatly restricted by the coke deposits (0.58 nm). ^{129}Xe NMR by itself cannot detect the two environments, because of the rapid exchange of xenon atoms between the two sites.

Finally, comparison of ^{129}Xe and ^{1}H pulsed-field gradient NMR spectroscopy [179] made it possible to locate the carbonaceous residues on Na – A zeolites and to distinguish unambiguously deposits in the intracrystalline pore system and on the external surface of crystallites. The small decrease in the intracrystalline diffusion coefficient of methane adsorbed in noncoked and coked A zeolites (10^{-9} to 0.6×10^{-9} m^2 s^{-1}) and the marked effect observed on the intracrystalline mean lifetimes (τ_{intra} goes from 3 ms to 100 ms) showed that the carbonaceous compounds are mainly deposited as a layer on the external surface, acting as a barrier to molecular mass transfer between the crystallites.

2.5 Study of Supported Metals and Chemisorption Thereon

2.5.1 Introduction

In what follows we shall reason on the basis of faujasite structure zeolites, but the results obtained are easily generalized to other types of solids.

Let us assume now that in various Na – Y zeolite supercages there are distributed solid particles with chemically different surfaces S_i. The index, i, denotes a particle type of population n_i, among the p types that exist ($0 < i < p$). Corresponding to each of these populations, there is a shift δ_i characteristic of the Xe – S_i collisions, being the product of two terms:

$$\delta_i = \begin{pmatrix} \text{Term for the} \\ \text{chemical nature} \\ \text{of } S_i \end{pmatrix} \begin{pmatrix} \text{Term for the} \\ \text{frequency of Xe–}S_i \\ \text{collisions} \end{pmatrix} \tag{23}$$

The spectrum depends on the lifetime of Xe on each adsorption site:

1. If the lifetime of xenon on each surface S_i is long (from an NMR point of view), the spectrum of the adsorbed xenon should theoretically contain as many components δ_i as there are target types, the intensity of each

being proportional to the number, n_i, of targets i, of the same type, the total being $p + 1$, since the component (shift $\delta_{\text{Na-Y}}$) due to xenon striking the supercage surfaces and the other Xe must also be taken into account. This situation could be obtained by recording the spectrum at sufficiently low temperature, but the components would be very broad and almost undetectable by classical NMR methods.

2. In the opposite case, where the xenon has a very short lifetime at each adsorption site and can, moreover, diffuse rapidly across several supercages contained in the same crystallite of the zeolite, all the above signals coalesce. The spectrum consists of only one component whose position depends on the values of δ_i, each weighted by the probability α_i of Xe – S_i collision:

$$\delta = \sum \alpha_i \delta_i + \alpha_{\text{Na-Y}} \delta_{\text{Na-Y}} \quad \text{with} \quad \sum \alpha_i + \alpha_{\text{Na-Y}} = 1. \tag{24}$$

In this case, it is, of course, very difficult to get much information about this system from this single line. Fortunately, the spectrum depends also on the distribution of the different particles i, j... inside the zeolite crystallites.

Among the different models considered [180, 181] we should note the following case, which is fundamental in the study of chemisorption on metal-loaded zeolites (Fig. 26). Let us suppose that in each zeolite crystallite the particles i, j... (with surfaces S_i, S_j...) are grouped according to their nature in volumes V_i, V_j... Unless the concentration of one of the species is really very small, the quantity of xenon exchanging between two adjacent zones V_i and V_j is negligible compared to the quantities of xenon in each zone. The system appears then as though there were several independent samples. The NMR spectrum must therefore contain p components corresponding to each type of target, with shift δ_i, such that:

$$\delta_i = \lambda_i \delta_i + \mu_i \delta_{\text{Na-Y}}. \tag{25}$$

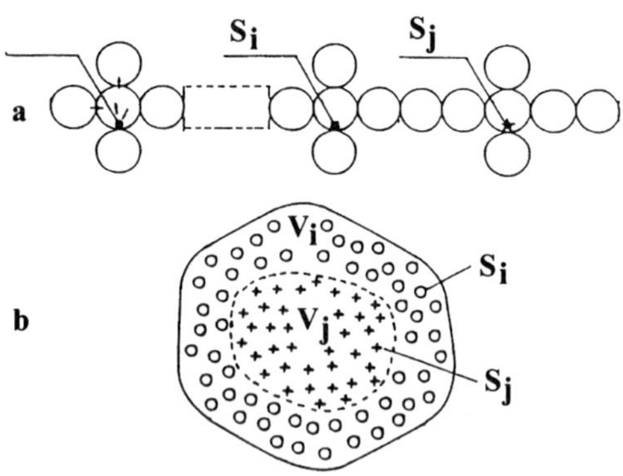

Fig. 26 Model of distribution of S_i and Sj particles grouped according to their nature in volumes V_i and V_j

λ_i and μ_i are the probabilities of xenon collisions with S_i and Na – Y, respectively.

In general, such a situation can arise whenever there are, in the zeolite crystallites, zones that are clearly differentiated either by the nature of the particles or by the distribution of the particles that can, in this case, be of the same type.

Let us consider now a catalyst M*x*– βG – Na – Y containing *n* metal particles per gram of sample (each containing on average *x* atoms of metal M) with a total amount β of adsorbed gas G. There is a distribution of atoms G (or molecules) over the totality of the metal, the number, *i*, of G atoms (or molecules) per particle being possibly different from one particle to another. Now, the chemical nature of the surface of the metal particle will be indicated by changes in the number, *i*, of G atoms (or molecules) per particle. For each number, *i*, there is therefore a corresponding term $\delta_i = \delta_{\mathrm{M}x+i\mathrm{G}}$ characteristic of the collision between Xe and the particles M*x* + *i*G.

As we have seen above, the form of the spectrum of xenon adsorbed on such a sample will depend on the Xe–(M*x* + *i*G) interaction, on the numbers *i*, *j*... of atoms or molecules of G chemisorbed on the various particles and on the distribution of M*x* + *i*G, M*x* + *j*G ... particles within a zeolite crystallite.

The main applications of this technique concern the study of the distribution of phases chemisorbed on supported metal particles and the determination of the particle size.

2.5.2
Chemisorption of Hydrogen

The spectrum of xenon adsorbed at 300 K on a Pt_x– Na – Y sample consists of a single line (denoted *a*) whose chemical shift $\delta(a)$ is always much greater than $\delta_{\mathrm{Na-Y}}$ whatever the xenon pressure [58] (Fig. 27). This signal is due to the coalescence of the line of high chemical shift, δ_{Pt}, due to xenon adsorbed on the platinum particles, and that of shift $\delta_{\mathrm{Na-Y}}$ corresponding to Xe atoms colliding with the walls of the supercages or with other Xe atoms. δ_{Pt} is very high, about 1000 ppm, and practically independent of the particle size [182]. For the sake of simplicity, we shall say that line a is characteristic of Xe – Pt collisions.

After chemisorption of a very small amount of hydrogen (the number, $n_{\mathrm{H_2}}$, of hydrogen molecules being much smaller than the number of metal particles determined by electron microscopy, n_{EM}) a second signal (denoted b) is detected; its shift, δ(b), is between that of δ(a) and $\delta_{\mathrm{Na-Y}}$, and it corresponds to Xe atoms adsorbed on particles that have chemisorbed hydrogen.

The existence of these two signals proves that the distribution of H atoms on the particles divides the zeolite crystallite into two zones. At ambient temperature, hydrogen is chemisorbed by the first particles encountered when it penetrates the zeolite crystallites, thus defining two zones: a central one (a)

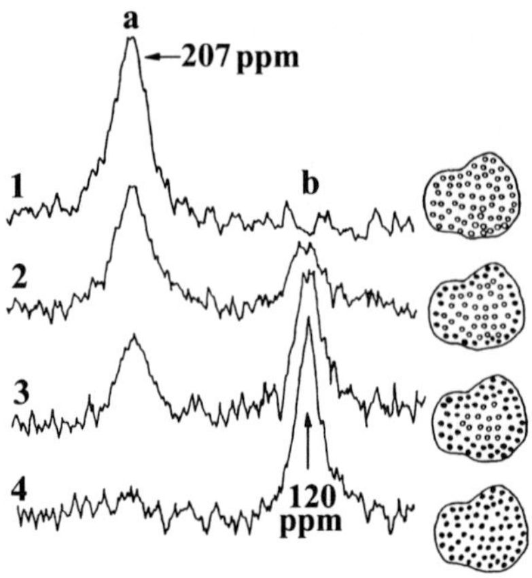

Fig. 27 Spectra of xenon adsorbed on Pt-βH-NaY at different numbers, n_{H_2}, of chemisorbed hydrogen molecules; $0 \leq n_{H_2} \leq n_{part.}$; ∘∘∘ bare Pt particles; (•••) Pt particles with chemisorbed H_2 (see text). (Reprinted from [2–4], with permission from Elsevier Science)

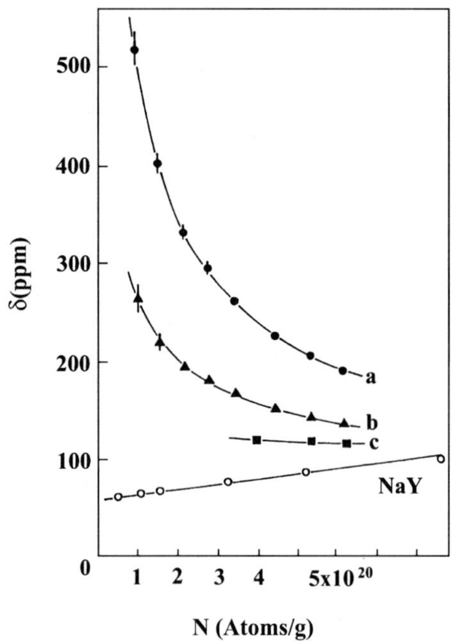

Fig. 28 $\delta = f(N)$ variation for Pt – Y zeolites. **a** line a, **b** line b, **c** line c. See text. (Reprinted from [180], with permission from P.A. Jacobs)

corresponding to bare particles and a peripheral one (b) containing particles that have chemisorbed hydrogen (Fig. 28).

Furthermore, when n_{H_2} increases while remaining very small, line b increases at the expense of a, but without any change in their NMR characteristics (such as chemical shift and line width), showing that each particle in b bears the same amount of hydrogen, or at least that the average distribution of the chemisorbed hydrogen is constant in zone b, regardless of the size of this zone. The intensity, I_a, of the signal falls to zero when that of b, I_b, is a maximum. Considering the literature data concerning the variation of the heat of chemisorption with the coverage and the evolution of δ with the chemisorbed H_2 concentration, the first authors [58] concluded that the amount chemisorbed at the beginning was 2H per particle. In this case, I_b is maximum when the total number of chemisorbed H_2 molecules, n_{H_2}, is equal to the real number of particles, n_p, which one can determine exactly from n_{H_2}. One finds that n_p is always much greater than n_{EM} determined by electron microscopy, confirming thus that, except in very rare cases, this latter technique is unable to detect very small particles. When n_{H_2} becomes greater than n_p, a signal c appears ($\delta(c) < \delta(b)$), generally poorly resolved from b and corresponding to particles that have more than 2H on their surface.

Boudart et al. have extended this study to various Pt/Na – Y samples. They observed that the number of Pt atoms per metal particle determined by ^{129}Xe NMR was always smaller than that obtained by other techniques: X-ray diffraction, EXAFS, WAXS and TEM [183, 184], if it is assumed, as in [58], that the clusters are sufficiently small for Xe colliding with them to "see" the first two hydrogen atoms adsorbed on the clusters no matter where it strikes. They propose a second explanation, namely, that the Pt clusters are large so that they almost fill the supercages of Y-zeolite. Then Xe will interact only with Pt atoms exposed in each of the four windows of the supercages. If, in addition, the first two H atoms adsorbed on Pt facing a window stay at that window, the endpoint indicates the point at which there is one H_2 adsorbed per window of the supercage. Accordingly, there are four times as many Pt atoms per cluster as obtained by the first assumption. The second assumption gives results which are compatible with the results given by other physical techniques.

In fact, ^{1}H NMR has shown, at least in the case of Pt on Al_2O_3 [185] or SiO_2 [186] that H_2 is chemisorbed at 300 K by the first particles encountered, with a degree of coverage at equilibrium of about 0.5. It is only beyond this coverage that H_2 diffuses easily in the sample. I_b therefore reaches a maximum when the coverage of each particle is close to 0.5. Another solution consists of chemisorbing H_2 at high temperature or, which would be equivalent, homogenizing the phase initially chemisorbed at 300 K over the whole sample, by raising the temperature (if possible to 673 K in order to overcome spillover, at least partially). As an example, Fig. 29, line 1 shows the spectrum of a Pt,Na – Y sample which has adsorbed a small amount of H_2 at 300 K [181]. When the sealed sample (therefore at constant total H_2 concentration) is heated , there is a parallel decrease in I_a and δ_a of line a, and an increase in I_b and δ_b of line b. This shows that zone b extends at the expense of a, as a result of desorption from b-type particles and diffusion of H_2 throughout the entire sample. When the temperature and the heating time are sufficient, the H_2 is homogeneously distributed throughout the sample and a single line is obtained (Fig. 29, line 4). If the total H_2 concentration is very low, the probability that there are more than 2H per particle is negligible. The sample consists then of particles which are either bare or carry 2H. In this case δ_b is linearly dependent on the H_2 concentration and goes through a ***critical*** point where all the particles carry 2H, from which the number of particles can be determined [181, 184, 187].

^{129}Xe NMR is therefore an interesting technique for determining the average number of atoms per particle. Moreover, this technique gives detailed information about the cluster distribution within the zeolite crystallites. A narrow Lorentzian signal of xenon adsorbed on samples without H_2 indicates a homogeneous distribution of clusters in the Y crystallites. In the opposite case, a broad peak and sometimes even a second peak is detected. Finally, if the situation arises ^{129}Xe NMR can be used to determine the concentrations of Pt located inside and outside the crystallites [183].

Chmelka et al. have shown that ^{129}Xe NMR can be used to monitor the location of metal clusters and cluster precursors as a function of calcination conditions for Na – Y zeolite-supported platinum catalysts [75, 188]. Their results indicate that for the reduction conditions imposed, the formation of highly dispersed platinum clusters within the Y-zeolite matrix is best achieved by employing a calcination temperature close to 673 K. Incomplete

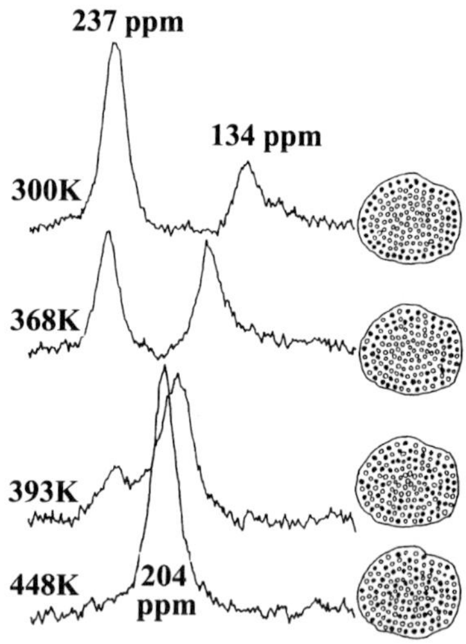

Fig. 29 Spectrum of xenon adsorbed on Pt_x – βH-NaY after heat treatment at (1) 300 K, (2) 368 K, (3) 393 K, (4) 448 K ($n_{H_2} < n_{part}$), (○○○) bare Pt particles, (●●●) Pt particles with chemisorbed H_2 (see text). (Reprinted from [2–4], with permission from Elsevier Science)

decomposition of the ion-exchanged $Pt(NH_3)_4^{2+}$ complex during calcination at 473 K results in migration of nearly all platinum to the exterior surface of the zeolite crystallite during reduction. Calcination temperatures significantly above 673 K induce decomposition of the shielded precursor species and subsequent migration of the metal into the sodalite cavities. A substantial amount of the platinum confined within the sodalite cavities migrates back into the supercages during reduction at 673 K [75, 188]. A similar study was carried out by Yang et al. [189].

Studying the surface of platinum supported on alumina, Boudart et al. have shown that metals supported on nonmicroporous materials can be probed by Xe NMR as previously demonstrated for zeolites. In particular, the amounts of adsorbed hydrogen and oxygen, as well as the reaction of adsorbed hydrogen with dioxygen, can be followed by this technique [190].

2.5.3 Chemisorption of other Gases (G)

The distribution of the first molecules of a gas G chemisorbed on Pt particles depends above all on the nature of this gas. For example, at 300 K, oxygen behaves very similarly to hydrogen.

In the same way, at 300 K, carbon monoxide half saturates the first Pt particles encountered when entering the zeolite crystallite (apparent stoichiometry 1 CO/2 Pt) [191]. In contrast, at 673 K one can obtain at low coverage a homogeneous distribution corresponding to one CO molecule per particle.

Thus, by means of this technique, it is possible *in all cases* to determine quantitatively the distribution of gases chemisorbed on metal particles and the distribution within the Y crystallites of Pt particles distinguished by the amount chemisorbed.

Now, in fundamental research it is important to know the particle coverage. For example, by means of the results obtained by ^{129}Xe NMR spectroscopy for the local distribution of CO chemisorbed on the very small platinum particles (six atoms on average) supported on Na – Y zeolite, it has been possible to determine precisely the effects of back-donation from the metal to CO and of dipole-dipole coupling between chemisorbed CO on the variation of the stretching frequency, ν_{CO} , with surface coverage [192].

There is a further point of interest concerning the chemisorption of CO at 300 K. Under the experimental conditions employed, the xenon technique can detect only changes occurring inside the supercages. It is, therefore, insensitive to the chemisorption of CO on the Pt particles located on the external surface of the Y crystallites. If very thin layers of the solid are used, the xenon will detect only the first CO molecules chemisorbed on the internal metal particles when the outer particles are saturated. Assuming a stoichiometry of one CO per 2 Pt atoms it is then possible to determine the number of Pt atoms located on the surface of the external Pt particles [180].

Finally, this technique can also be used to determine the distribution of several gases chemisorbed on zeolite-supported metal particles [181].

2.5.4 Metals other than Platinum

^{129}Xe NMR has been applied to the study of catalysts based on metals other than Pt. For example, by means of this technique, Coddington et al. [193] have shown that adsorption of $Mo(CO)_6$ in zeolite Na – Y followed by decomposition at 473 K produces uniformly dispersed Mo_2 clusters in the zeolite supercages; heating to 673 K causes sintering to an average cluster size of three or four Mo atoms.

Finally, ^{129}Xe NMR seems to be particularly useful for studying bimetallic catalysts, amongst other things, to determine the size of the particles and their distribution in the Y crystallites, as well as to estimate their electron deficiency (metal-support interaction), as has been shown by Ichikawa et al. on $Rh_{6-x}Ir_x$/Na – Y bimetals with $x = 0.6$ [194].

2.6 Xenon Diffusion

2.6.1 Introduction

The study of the diffusion of xenon in a microporous system is particularly interesting for the characterization of the solid itself and for the study of the diffusion of other adsorbates. It can be performed by comparing chemical shifts with various standards or by means of the now-classical pulse-field gradient NMR (PFG NMR) technique.

We think it is worthwhile recalling first of all the characteristics of pure xenon in order to better understand the effect of the adsorbent. Reference [195] is the first systematic investigation of self-diffusion in a monoatomic gas over a large density and temperature range. Within experimental error the results are the same with ^{129}Xe and ^{131}Xe. For example, at 298 K and 25 amagat, $D = 2.054 \times 10^{-7}$ m^2 s^{-1}. D is inversely proportional to the density and increases with the temperature.

2.6.2 Self-Diffusion of Xenon by the PFG NMR Technique

Pulse-field gradient NMR is a versatile tool for studying molecular transfer in zeolitic adsorbent-adsorbate systems. In particular, it allows the direct measurement of the translational molecular mobility inside the crystallites, represented by the self-diffusion coefficient D. This technique is based on the application of radio frequency pulse sequences for generating spin echos [196, 197], where during two time intervals two inhomogeneous magnetic fields with the field gradient, of intensity g, are superimposed on the constant magnetic field.

In general, under the influence of these field gradient pulses, the magnitude of the NMR signal of adsorbed species (the spin-echo intensity) is reduced by the factor

$$\psi(g,\delta',\Delta) = [1-\gamma'(\Delta)]\exp\left[-\gamma^2\delta'^2 g^2\frac{\langle r^2(\Delta)\rangle_{\text{intra}}}{6}\right] + \gamma'(\Delta)\exp\left[-\gamma^2\delta'^2 g^2\frac{\langle r^2(\Delta)\rangle_{\text{inter}}}{6}\right], \quad (26)$$

with Δ denoting the time separation between the two identical field gradient pulses, chosen to be much larger than the pulse width δ'. γ is the gyromagnetic ratio of the considered nuclei. $\langle r^2(\Delta)\rangle_{\text{intra}}$ and $\langle r^2(\Delta)\rangle_{\text{inter}}$ represent the mean square displacements of the two following subgroups: those diffusants that remain inside the individual crystallites over the whole observation time, and those that may leave their crystallites and that are able, therefore, to cover large diffusion paths through the intercrystalline space. $\gamma'(\Delta)$ denotes the relative amount of the latter subgroup.

For Na – X and ZSM-5 zeolites the $\psi(\delta',\Delta) = f(\delta'^2)$ relationship is represented by the first term of Eq. 26 and all the information is contained in the mean square displacement $\langle r^2(\Delta)\rangle_{\text{intra}}$ (Fig. 30a and b) [198]. For the A zeolite the observed echo attenuation is represented by the superposition of the two terms of Eq. 26 which contains two types of information: $\langle r^2(\Delta)\rangle_{\text{intra}}$ obtained from the shape of the second more slowly decaying part of the Ln ψ vs. δ'^2 plot and the quantity $\gamma'(\Delta)$ (Fig. 30c).

According to the Einstein relation

$$\langle r^2(\Delta)\rangle = 6D\Delta, \quad (27)$$

the self-diffusion coefficient may be determined from the slope of the mean square displacement versus the observation time, provided that the root of $\langle r^2(\Delta)\rangle$ is still smaller than the root of the mean square radius, $\langle R^2\rangle$, of the adsorbent particle.

Table 6 gives the coefficient D determined for sufficiently short observation times. Also are included the intracrystalline mean lifetime τ_{intra} calculated from $\gamma'(\Delta)$ and $\tau_{\text{intra}}^{\text{diff}}$ defined by [199]:

$$\tau_{\text{intra}}^{\text{diff}} = \langle R^2\rangle/15D, \quad (28)$$

Direct information about the existence of surface barriers is provided by comparing τ_{intra} and $\tau_{\text{intra}}^{\text{diff}}$ which represent the minimum possible lifetime calculated by Eq. 26 from the intracrystalline diffusivities. For methane these two quantities are in reasonable agreement [198], proving that molecular ex-

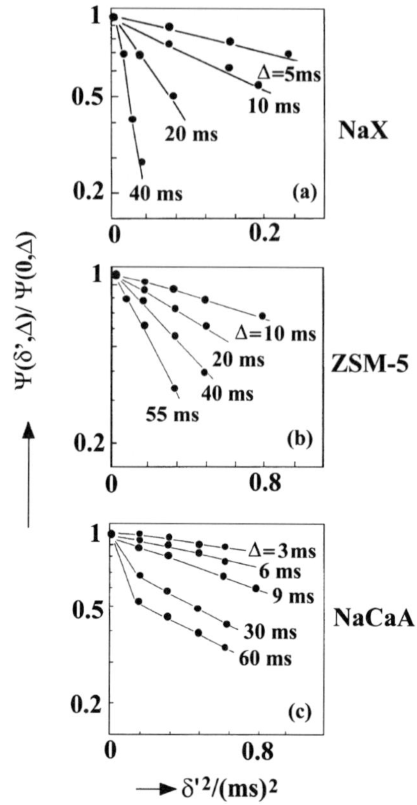

Fig. 30 Plots of the ^{129}Xe NMR signal intensity (spin echo amplitude) versus the square of the width of the gradient pulses for different observation times Δ, for zeolites NaX (crystallite diameter $= 50\,\mu m$), ZSM-5 (crystallite dimensions $100 \times 30 \times 30\,\mu m^3$), and $Na_{6.6}Ca_{2.7}A$ (crystallite diameter $= 13\,\mu m$). (Reprinted from [198], with permission from Elsevier Science)

change is controlled mainly by intracrystalline diffusion. On the contrary, in all NaCa – A specimens as well as in the considered specimen with the smaller crystallites, xenon desorption is significantly retarded by factors other than intracrystalline diffusion, i.e. by additional transport resistances on the external surface or in the surface layer of the crystallites. This result may be related to the kinetic diameter of xenon atoms, which is greater than that of CH_4, making them suitable for probing the surface permeability of adsorbents with limiting free diameter of this order of magnitude (0.4–0.5 nm for NaCa-A, 0.55 nm for ZSM-5).

For Na – X zeolite no surface resistance is observed, since the diameter of xenon is much lower than that of the windows of supercages (~ 0.75 nm). This result has been confirmed by ^{129}Xe PFG NMR tracer desorption measurements [200].

In Na – X and ZSM-5 zeolites the self-diffusion coefficients of xenon decrease with increasing concentration while for NaCa – A zeolites they are essentially constant [65]. In the case of $Na_{4.4}Ca_{3.8}A$, the observed diffusivity (1.5×10^{-9} m^2 s^{-1}) is several orders of magnitude larger than literature data based on uptake experiments (1.2×10^{-11} and 1.10×10^{-14} in [201] and [202], respectively) or computer simulation (3.3×10^{-12} in [203]). Computer simulations of xenon diffusion in ZSM-5, however, are found to be in satisfactory agreement [65].

Figure 31 shows the Arrhenius plots of the self-diffusion coefficients of Xe in Na – X, silicalite and $Na_{4.4}Ca_{3.8}A$. The diffusivities follow the simple exponential dependence [204]

$$D = D_0 \exp(-E/RT) \tag{29}$$

with D_0 and E denoting the preexponential factor and the activation energy of D, respectively. E is a measure of the energetic barriers that the molecules have to overcome on their diffusivity path in the intracrystalline space.

Table 6 indicates the influence of extra-framework cations on the magnitude of E. D_0 is the self-diffusivity that would be attained at infinitely high temperatures, i.e. under conditions where energetic influences are not any longer of importance and the diffusivity is entirely determined by the geometrical restriction of the adsorbed particles within the zeolite. Then D_0 decreases with the free diameter of the pores (X zeolites > silicalite ~ A zeolite) (Table 6).

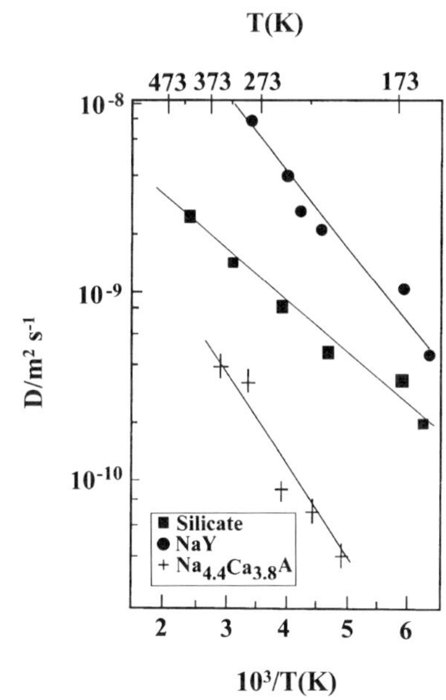

Fig. 31 Arrhenius plots of the intracrystalline self-diffusion coefficient D of xenon in: • NaX; ■ silicalite; + $Na_{4.4}Ca_{3.8}A$.

Table 6 (adapted from [198] and [204], with permission): Self-diffusion coefficients, *D*, and intracrystallite mean lifetimes τ (see text) of adsorbed xenon. Activation energy, *E*, and pre-exponential factor of *D*

Zeolite	Adsorbent: Limiting free diameter (nm)	Adsorbent: Mean crystallite diameter (μm)	Adsorbed Xenon: D $(10^{-9}\ m^2\ s^{-1})$	τ_{intra}^{diff} (ms)	τ_{intra} (ms)	D_0 $(10^{-8}\ m^2\ s^{-1})$	E $(kJ\ mol^{-1})$
NaCaA	0.4–0.5						
Ca^{2+} %							
45%		13	1 ± 0.3	3 ± 0.1	80 ± 20		
63%		20	1.5 ± 0.4	4 ± 1	45 ± 10	0.9 ± 0.3	8 ± 3
80%		5	1.5 ± 0.4	0.4 ± 0.2	25 ± 8		
NaX	0.75	50	5 ± 1.5	8 ± 2	15 ± 10	8 ± 3	6 ± 1.5
		20	5 ± 1.5	1.5 ± 0.5	5 ± 3		
Silicalite (a)	0.55	25	0.9 ± 0.3	11 ± 3	>40		
Silicalite (b)	"	$100\times 30\times 30$*	0.9 ± 0.3	160 ± 40	>70	0.9 ± 0.2	5 ± 1

* mean crystallite dimensions (μm^3)

In the case of silicalite, as a consequence of diffusion anisotropy [197], the correct dependence of the spin-echo attenuation on the gradient intensity deviates from the pattern provided by the first term of Eq. 26. However, because of the low signal/noise ratio the diffusivities, D_x, D_y and D_z, in the three principal directions were not determined, whereas this had been possible for hydrocarbons adsorbed in oriented crystallites [205]. The principal elements of the diffusion tensor of xenon adsorbed in silicalite have been determined by molecular dynamics simulation [206] (1.3, 4 and $0.28\times10^{-9}\ m^2\ s^{-1}$ for D_x, D_y and D_z, respectively).

Finally, in the case of xenon adsorbed in 5A zeolite there is a good agreement between the limiting (zero concentration) transport diffusivities measured by the ZLC technique and by PFG NMR [207]. This agreement is sufficiently rare to be worth pointing out.

2.6.3 Dynamics of Adsorbed Xenon and ^{129}Xe Chemical Shift

2.6.3.1 Encumbering of Zeolite Pores: Location of Transport Resistance

The use of the ^{129}Xe chemical shift generally gives qualitative information about xenon diffusion and especially its dependence on various parameters.

We have seen that the variation of δ with the water concentration can be used to locate water molecules outside the supercages when the concentration is low (2 H_2O molecules per 1/8 unit cell), then, at higher concentrations, at the windows between the supercages, thus reducing the diffusion of xenon between the supercages [170]. Finally, the increase of δ_S and of the slope of the δ vs. $[H_2O]$ plots indicate an increase in the frequency of molecular encounters of xenon atoms between themselves or with water molecules, i.e. a perceptible reduction of the intracrystalline free space and of course, again, a decrease in the diffusion.

With the help of a few examples it was also possible to show that there is a correlation between the ^{129}Xe shift of adsorbed xenon and ^{1}H PFG NMR diffusion of hydrocarbons used as probe molecules [179, 208].

The second parallel between ^{129}Xe NMR and ^{1}H PFG NMR concerns the effect of coking. We have seen from [172] that the variation of the $\delta = f(N)$ curve with the degree of coking makes it possible to locate the coke, to deduce its distribution outside and inside the zeolite crystallites and, in the latter case at the windows or the surface of the cages or channels, whether it is homogeneous or not. The PFG NMR technique in combination with the NMR tracer desorption technique gives the same conclusions ([179] and [209] for 5A zeolite, [210] and [211] for ZSM-5 and HY zeolites).

2.6.3.2
Industrial Zeolite Catalysts

Zeolite + Binder

Although qualitative, such studies make it possible to demonstrate the effects of various factors such as the dilution of the zeolite by another solid, the compression, the experiment temperature, etc. on the $\delta = f(N)$ curves (or $\delta = f(P_{Xe})$ and therefore on the diffusion of xenon from one crystallite to another. The information obtained is vitally important from the point of view of the application of the Xe technique to the study of industrial zeolite catalysts. Chen et al. [60] have studied the effects of these factors on Na – Y and ZSM-5 either pure or mixed or associated with Na – A, a zeolite which does not adsorb xenon under normal conditions and which therefore serves simply as a diluent, the samples being in the form of powder or compressed at pressures P_C between 0 and 314 MPa.

For a pure zeolite, at a given xenon pressure, δ increases very little with the compression. For example in the case of Na – Y (diameter of windows ~ 7.5 Å), $\delta_{N\to 0}$ increases from 57 to 60 ppm when P_C goes from 0 to 314 MPa. The variation decreases further with the pore diameter of solids.

In the case of a mixture with Na – A, where W is the concentration of the Na – Y or ZSM-5 zeolite:

$$\delta_{N\to 0} = \frac{W}{W + C}\delta_S, \tag{30}$$

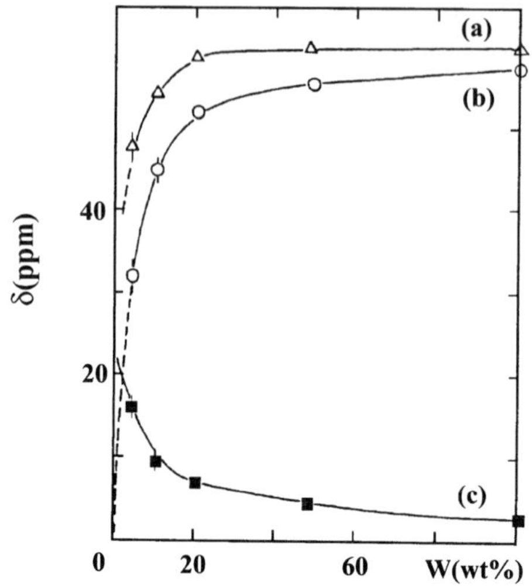

Fig. 32 $\delta_{P\to 0}$ variations with NaY concentration W, in :NaY-NaA mixture: (**a**) 314 MPa compressed; (**b**) powder; (**c**) difference between line **a** and **b**. (Reprinted with permission from [55] Copyright (1992) American Chemical Society)

C is a constant characteristic of the mixture. At low dilution, $\delta_{N\to 0}$ does not change much with dilution and compression (Fig. 32). Therefore, for routine application of the xenon NMR technique to a pure zeolite or samples containing less than 50% of nonporous impurity the result should be acceptable. At higher dilution, $\delta_{N\to 0}$ decreases rapidly with dilution. This tells us that special attention must be paid to the application of the technique to industrial catalysts which sometimes contain more than 50% of binder. For a mixture, the increase of linewidth with xenon pressure can be also a good indication of the presence of a second phase in the sample.

Equation 30 can be written:

$$1/\delta_{N\to 0} = 1/\delta_S + (C/\delta_S)W^{-1}. \tag{31}$$

Hence the $1/\delta_{N\to 0} = f(1/W)$ plot should be a straight line whose intercept gives the value of $1/\delta_S$. This allows us to check whether the zeolite has been perturbed by its environment, either during the formation of the pellet or during a catalytic reaction. It is, to the best of our knowledge, the only technique by which the zeolite component alone in a mixture can be studied. It should be however noted that, when the binder is also porous, it is better to record the spectra at low temperature. It is known that δ_S (corresponding to the pure zeolite), decreases with a temperature increase. $\delta_{N\to 0}$ shows the same variation, but the influence of the temperature increases with the degree of dilution.

Finally, the form of the xenon NMR signal depends markedly on the homogeneity of mixtures. This can be used as a tool for checking the homogeneity of industrial samples.

Zeolite-Supported Metals

The theoretical analysis of a spectrum containing several components due to the inhomogeneity of the sample can also provide information about the mobility of xenon. For example, we have seen in paragraph Sect. 2.5.2 that at the beginning of hydrogen chemisorption on a Pt/Na – Y catalyst, two Xe-NMR lines are observed; these are attributed to xenon atoms in contact either with pure platinum particles or with particles bearing hydrogen. This interpretation clearly implies that as long as not all particles are in contact with H_2, the mean lifetime of the Xe atoms in these two regions must not be much less than the inverse of the frequency difference between the two lines, since otherwise the two lines should coalesce. By a rigorous determination of the NMR line shape of a two-region system with finite exchange times, the mean lifetime of the xenon atoms in the inner core of the Pt – H/Na – Y crystallites proves to be $\tau = 1.25$ ms, and the self-diffusion coefficient of xenon in the uncovered Pt – Na – Y part is found to be equal to $D \sim 10^{-11}$ m^2 s^{-1} [211].

2.6.3.3 Na–A/Xe system

There have been many experimental and theoretical studies on the distribution, the calculation of the chemical shift and the dynamics of xenon adsorbed in the α-cages of Na – A, essentially because Xe exchange between α-cages is very slow and because one has also a good model of $(Xe)_n$ clusters with a well-defined number of atoms, n, in the cages. Indeed, ^{129}Xe NMR can be used to determine the distribution of xenon atoms among the α-cages of Na – A (Fig. 33a) [82–84].

The xenon is generally adsorbed at high temperature (523 K) and pressure (0.5 to 20 MPa) because the window openings of the α-cages are roughly 4.2 Å and in addition are blocked by the presence of Na^+. However, this does not entirely rule out intercage movement of the xenon, since pore sizes determined from adsorption studies have almost always turned out to be greater than their crystallographically computed counterparts, for two reasons [212, 213]:

1. Neither the guest molecule nor the host lattice is rigid, in that both the molecule and the oxygen framework are polarizable (i.e. capable of distortion);
2. Both guest and host are in a continuous state of vibration as the bonds holding them together bend under the influence of temperature.

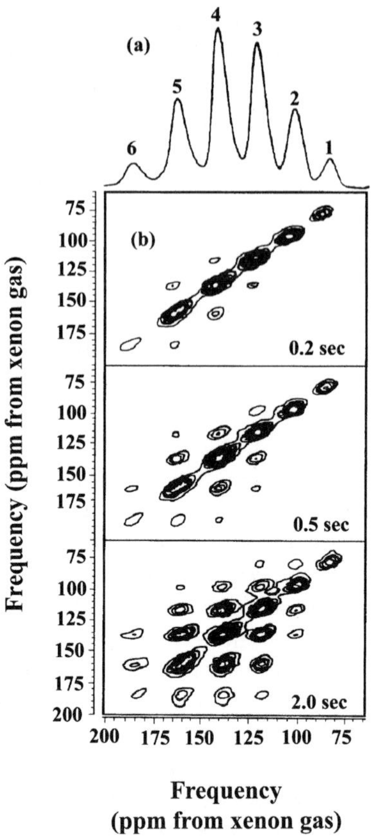

Fig. 33 **a** ^{129}Xe 1D NMR spectrum and **b** ^{129}Xe 2D exchange NMR spectra of xenon adsorbed on NaA zeolite at 523 K (P_{Xe} = 3 MPa). 2D spectra were recorded using mixing times of 0.2, 0.5, and 2.0 s. The diagonal peaks correspond to ^{129}Xe resonances from xenon in α-cages containing different numbers of occluded atoms as indicated above by the respective peaks of the 1D spectrum. The *cross peaks* are a result of intercage motion of xenon during the mixing time of the 2D experiment. (Reprinted from [86], with permission from Elsevier Science)

Nivarthi and McCormick [214] were particularly interested in the movement of xenon within the cage. They have demonstrated how ^{129}Xe NMR relaxation of adsorbed molecules can provide valuable information about sorbate location and dynamics. The decoalescence of the ^{129}Xe NMR peak observed with a loading of 1 Xe/α-cage indicates the presence of distinct adsorption sites in this cage (in front of the 4-oxygen ring windows) and a very rapid intracage motion of the Xe atoms at room temperature. The magnitude of the longitudinal relaxation time, T_1, observed for the peak corresponding to the high loading case suggests the existence of polyhedral configurations of the xenon atoms inside the α-cage with enhanced stability.

The frequency of the intracage exchange (hopping from one adsorption site to another) is, at a maximum, of the order of 3450 Hz and the corresponding activation energy, 32.0 kJ mol^{-1}, of the same order as that observed for intracrystalline diffusion. Nivarthi et al. [214] conclude that the rate-limiting step for diffusion through a zeolite may not be configurational diffusion through small windows but might be the hop between adsorption sites within a single cage, especially at high loadings, since motion is in that case restricted by a "molecular traffic jam". This conclusion should apply generally to all zeolite systems at high sorbate loadings.

Li et al. [215] extended the methods used for rare gas clusters in free space to investigate the relation between the thermodynamic and dynamic properties of occluded clusters in terms of the adsorption site of a single $(Xe)_n$ cluster inside the α-cavity. They reported how the sodium cations, primarily the type III Na^+, and the Xe – Xe interactions influence aspects of Xe adsorption, particularly the site occupancy, the site-to-site exchange rate, and the way the number of Xe atoms inside the α-cavity influences the ^{129}Xe chemical shift. They have shown that there is a locally stable site for one adsorbed xenon atom in each eight-membered and four-membered ring; the effective number of potential energy minima is 11, the number of minima found by Jameson et al. [216].

Even though the Xe-wall interaction is much larger than Xe – Xe interaction, the energy differences between the adsorption sites are not much larger than the Xe – Xe repulsions. Atoms in smaller clusters, which have low-energy sites available for all the Xe atoms, can move around without encountering the hindrance of large short-range Xe – Xe repulsive forces; but in larger clusters either congestion forces Xe atoms to occupy higher-energy sites, which makes them mobile, or some of the Xe atoms occupy the low-energy sites and experience large Xe – Xe nearest neighbor interactions. Both cases bring the Xe – Xe repulsive forces into play in the dynamics. Nonetheless, even though the Xe – Xe interaction seems to play the dominant role in the dynamics of a cluster of xenon atoms inside the Na – A α-cavity, it is the balance of this interaction against the Xe-cavity interaction that governs the size and temperature dependence of the chemical shift.

The other studies are mainly concerned with xenon diffusion from one cage to another. The well-separated peaks of the one-dimensional spectra and their relatively small line-widths indicate that intercage exchange frequency is smaller than the frequency differences between the resolved peaks. However, intercage xenon exchange is inferred from the observation of changes in the one-dimensional ^{129}Xe spectra over the time as samples equilibrate [84]. Larsen et al. [86] demonstrated that this slow intercage motion can be monitored directly by ^{129}Xe 2D exchange NMR in which magnetization transport during a time interval, t_{mix}, is measured. In Fig. 33b the ^{129}Xe frequencies before and after the mixing time are correlated.

Denoting the frequency of a n-cage (i.e. a cage containing $\underline{n}$ xenon atoms) as $\omega^{(n)}$, the normalized spectral intensity $M(\omega^{(n)}, \omega^{(m)})$ is the joint probability of finding a xenon in a n-cage before the mixing time and in a m-cage afterwards. In the experimental spectra of Fig. 33b, for the shortest mixing time (0.2 s), most of the spectral intensity is confined to the diagonal, indicating negligible intercage motion during this mixing time. For $t_{\text{mix}} = 0.5$ s, one-off diagonal cross peaks $M(\omega^{(n)}, \omega^{(n\pm1)})$, i.e. cross-peaks adjacent to the diagonal, at frequencies corresponding to cages differing in occupancy by one xenon, have significant intensity. This indicates that the dominant change in xenon frequencies with time corresponds to an increase or decrease of the cage occupancy by one xenon. When $t_{\text{mix}} = 2.0$ s, exchange is observed between all but the least populated cages. This progression of spectral features with mixing time is characteristic of mass transport in this system.

The overall rate, $R_{n,m}$, of xenon atoms going from any n-cage to any m-cage is proportional to the probability $P(n) \times P(m)$ of finding a n-cage next to a m-cage, and to the rate coefficient k_n of a given xenon leaving a n-cage

$$R_{n,m} = nP(n)P(m)k_n, \tag{32}$$

where k_n exhibits an Arrhenius form

$$k_n = A_n \exp(-E_a(n)/kT). \tag{33}$$

From the simulation of ^{129}Xe 2D spectra, Larsen et al. [86] have determined k_n as a function of cage occupancy n and the diffusivity $D = 10^{-19}\ \text{m}^2\ \text{s}^{-1}$ at ambient temperature, which is much smaller than the values accessible by PFG NMR [198].

Another possibility for studying the chemical exchange by monitoring the transfer of polarization is a one-dimensional experiment in which one selectively inverts the magnetization at site A and monitors the recovery both of this resonance and that at sites B, C, etc., with which it is exchanging [217]. The use of the DANTE (Delays Alternating with Nutations for Tailored Excitation) sequences for selective inversion, has the advantage that the total flip angle and selectivity may be controlled independently, without changing the pulse power.

By using these techniques, Jameson et al. [217] proposed and confirmed experimentally the relationships between the set of microscopic rate constants k_{mn} and the phenomenological rate constant K_{mn}. These relationships are analogous to those proposed by Larsen et al. [86]. (It should be noted, however, that the notations are different). k_{mn} is associated with the rate transfer of a single Xe atom from a cage containing the cluster $(\text{Xe})_n$ into a neighboring cage containing the cluster $(\text{Xe})_{m-1}$, thereby making the new cluster $(\text{Xe})_m$. These authors have also shown that the rate constant associated with a single Xe atom leaving a particular $(\text{Xe})_n$ is relatively independent of the destination, except when the destination cage is already highly populated.

2.6.4
^{129}Xe NMR as a Tool for Probing Intracrystalline Concentration Profiles and Transport Diffusion

In opposition to the self-diffusion measurements (PFG NMR, quasi-elastic neutron scattering) which allow the microscopic observation of molecular mobilities, the measurement of transport diffusion can only be carried out from a macroscopic point of view, i.e. by monitoring the time dependence of the molecular concentration in the gas phase or the bulk phase of the bed of zeolites. "Microscopic" observation of intracrystalline transport diffusion by ^{129}Xe NMR spectrum evolution allows us to go back to the time dependence of the intracrystalline concentration profile.

The principle depends on the influence of an adsorbate on the spectrum of xenon coadsorbed as a probe. Bansal and Dybowski [218] have studied the diffusivity of H_2O at 373 K between two layers of NiNa – Y zeolite (3.4 wt % nickel), one well dehydrated and the other very little (simply pumped at 298 K). According to these authors, if δ_1 and δ_2 correspond to the chemical shift of xenon in the two regions of H_2O concentrations, C_1 and C_2, to a first approximation it holds:

$$\delta_1 - \delta_2 \propto \langle C_1 \rangle - \langle C_2 \rangle \tag{34}$$

The diffusion of H_2O from one region to another changes the concentrations and the δ values. The evolution of shifts with time can be used to measure the diffusivity D. In particular, when the time is long, the following equation:

$$\delta_1(t) - \delta_2(t) \sim K \exp(-\pi^2 Dt/L^2), \tag{35}$$

where $2L$ is the height of the sample, makes it possible to estimate the diffusivity of water: $D(373\,\mathrm{K}) = (2 \pm 1) \times 10^{-10}\,\mathrm{m^2\,s^{-1}}$.

Kärger [219] expressed some doubts about the comparison of this coefficient with the constant for the self-diffusion of water in 13 X zeolite: $(4 \pm 1) \times 10^{-10}\,\mathrm{m^2\,s^{-1}}$ [220]. He and his co-workers have looked at this type of study in more detail. For this purpose, the adsorption/desorption process of benzene in ZSM-5 zeolite has been followed under xenon atmosphere by intimately mixing activated and loaded zeolite crystallites [221]. Figure 34 a (full line) shows the evolution of the spectrum of xenon with time, for an initial concentration of 6 and 0 molecules of benzene per u.c. and an average xenon concentration of 16 atoms/u.c. The signal of xenon situated in the supercages containing benzene (high δ) is broader and weaker than that of xenon in the benzene-free cages (small δ), because of the reduction of the adsorption capacity of xenon in the presence of benzene.

Corresponding to the two cases of intracrystalline limitation (i.e. limitation by intracrystalline diffusion) and extracrystalline limitation (i.e. limitation by external resistance, e.g. surface barriers), two different time dependences of the intracrystalline concentration profiles may be expected,

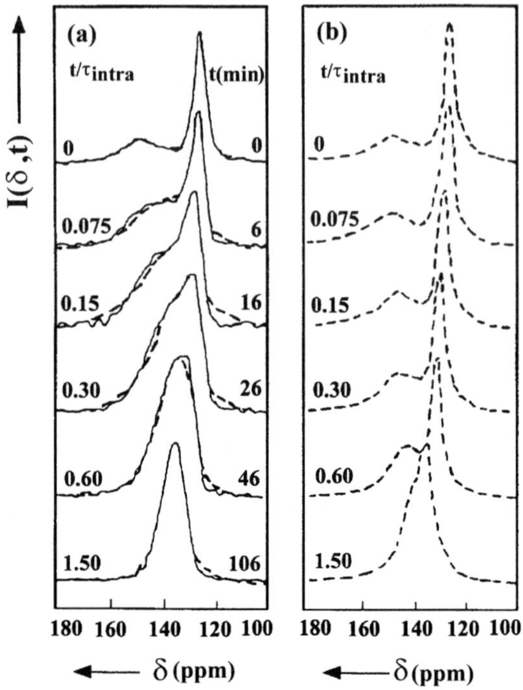

Fig. 34 ^{129}Xe NMR spectra (*full line*) during the sorption of benzene on crystallites of ZSM-5-type zeolite at 293 K with an initial concentration of 6 and 0 benzene molecules per unit cell and a mean xenon concentration of 16 atoms per unit cell. Comparison with the simulated spectra (*dashed lines*) for the limiting cases of diffusion (**a**) and barrier-controlled sorption (**b**). (Reprinted with permission from [221]. Copyright (1992) American Chemical Society)

corresponding to two distinctly different ^{129}Xe spectra. Due to the fact that for barrier-controlled adsorption/desorption the intracrystalline concentration may assume only two values, the distinction between the two lines is preserved over nearly the whole process (Fig. 34b), while for diffusion-controlled adsorption/desorption the wide range of intracrystalline concentrations leads to rapid coalescence of the two lines (Fig. 34a). These simulations indicate that the adsorption/desorption process of benzene in ZSM-5 zeolite is controlled by intracrystalline diffusion. The intracrystalline transport diffusivity deduced from these simulations is found to be $(1.3 \pm 0.3) \times 10^{-14}\ m^2\ s^{-1}$. This value is in satisfactory agreement with the results of previous uptake experiments [222–224]. However, by the present analysis it has been shown unambiguously for the first time that the adsorption/desorption process under such experimental conditions (intimately mixed activated and loaded crystallites) is controlled by intracrystalline diffusion, so that the values obtained are in fact real intracrystalline transport diffusivities.

The proposed method is limited to slow adsorption/desorption processes, since it is only applicable if the time constant of these processes is large in comparison with the time necessary for the measurement of the ^{129}Xe NMR spectra. Moreover, the use of sufficiently large zeolite crystallites is inevitable so that the displacement of the xenon atoms during the reciprocal of the difference of the chemical shifts is still much smaller than the crystallite diameter.

Springuel-Huet et al. have also studied the diffusion of benzene in H-ZSM-5 [225], varying the experimental conditions so as to approach those of applications (fluidized or fixed bed reactors, gas phase chromatography, etc.):

1. Sorbate equilibrium between loaded and unloaded well-mixed crystallites, as in the previous procedure. These authors verified that during this experiment the adsorption/desorption process is controlled by intracrystalline diffusion. In this case 1.5 hour after the onset of the experiment, the ^{129}Xe NMR spectrum has already attained its final shape.
2. Sorbate equilibrium between loaded and unloaded beds (each 5 mm deep) of zeolite crystallites. The adsorption/desorption process is found to be significantly slowed down in this procedure. In this case the rate of attaining macroscopic equilibrium is controlled by the rate of molecular propagation through the bed; the equilibrium is achieved in about 7.5 hours. The intercrystallite diffusivity, D_{inter}, and the diffusion coefficient through the bed, D, are found to be 2×10^{-5} m^2 s^{-1} and 1.3×10^{-9} m^2 s^{-1}, respectively.
3. Adsorption from the gas phase, corresponding to 6 C_6H_6 molecules/u.c. at adsorption equilibrium. In this case, the retardation of uptake in comparison with the previous procedure (about 30 minutes) is a consequence of the spatial expansion of the gas volume over the bed of zeolite crystallites.

In conclusion, ^{129}Xe NMR demonstrates the dramatic effect of the arrangement of crystallites on the rate of intercrytalline exchange of guest molecules in zeolite beds. The exchange rate of benzene molecules in a bed of well-mixed crystallites of ^{129}H-ZSM-5 is at least one order of magnitude larger than the corresponding molecular exchange between two separate layers of crystallites, even for bed depths as small as 10 mm.

We can mention a more recent application of ^{129}Xe NMR spectroscopy to study gaseous hydrocarbon diffusion in a fixed bed of ZSM5 zeolite [226].

3 Other Microporous Solids

3.1 Pillared Clays

Pillaring of clays by large cationic complexes (PILC) increases the thermal stability of materials which may have a potential catalytic importance, and generates microporosity between the layers, allowing reactive molecules to penetrate. The first study of PILC using ^{129}Xe spectroscopy was undertaken by Fetter et al. [227] in 1990. These authors observed practically no dependence of the chemical shift of xenon adsorbed at room temperature on the xenon pressure (7 ppm as P_{Xe} increases from 0 to 1.33×10^5 Pa). This suggests there is a rapid exchange with the gas phase and no Xe – Xe interactions or electric field effects. The value of the chemical shift extrapolated to zero pressure is 85 ± 2 ppm. This value is intermediate between those of Na – A (80 ppm) and L (90 ppm) zeolites. Assuming that xenon-lattice interactions are the same in zeolites and PILC and using the Demarquay–Fraissard relationship (Table 3) they obtained an average interlayer space of 1 nm, in good agreement with XRD and porosimetry measurements.

A more complete study was performed at variable temperature ($110 < T <$ 298 K) by Barrie et al. [228]. The starting material was a gelwhite L containing Ca^{2+} and Na^+ cations. Pillaring by treatment with aluminum chlorohydrate at 353 K, followed by calcination in air at 773 K, leads to the insertion of polymeric $[Al_{13}O_4(OH)_{24}]^{7+}$ cations which decompose to form alumina pillars after calcination. The interlayer spacing is 0.81 nm. Like Fetter, the authors did not find, whatever the adsorption temperature, an initial decrease in the chemical shift with xenon coverage, indicating that pillars do not act as strong adsorption sites; the value of the chemical shift at zero pressure is 96 ± 5 ppm. The Demarquay–Fraissard equation leads to values of the average interlayer space calculated for cylindrical or spherical pores (0.77 and 1.1 nm, respectively) close to the *d*-spacing obtained by XRD measurements.

From a detailed study of the effect of xenon concentration and lowering of the temperature, these authors concluded that in pillared clays xenon does not condense onto other xenon atoms but spreads out over the surface. These results show that the xenon NMR technique can be a useful tool to detect phases both in the interlayed spaces generated by pillaring and on the outer surface.

Yamanaka et al. [229] studied two kinds of $SiO_2 - TiO_2$ sol pillared clays; one prepared by air-drying (AD samples), the other by drying with CO_2 under supercritical conditions (SCD samples). At room temperature, as observed by previous authors, the chemical shift does not depend on the xenon pressure, suggesting that the pores in pillared clays are isolated with no exchange of xenon atoms between the pores, at least at the NMR timescale. By extrapolation to zero xenon pressure, values of 98 and 103 ppm are obtained for AD samples calcined at 473 K and 773 K, respectively. The average pore diameters (for spherical cavities) extracted from $\bar{\ell}$ (Eq. 20 and Table 3) were estimated to be 1.04 and 1 nm, respectively. In the case of SCD samples, which have a dual mode of porosity (micro as well as mesopores), the authors expected two distinct signals, but only one broad, slightly shifted (30 ppm) resonance line was observed. The Xe atoms must exchange rapidly between the two types of pores, because SCD samples have an open porous structure.

3.2 Heteropolyoxometalate Salts (HPOM)

Another class of catalysts, heteropolyoxometalate salts (HPOM), which have been much used in recent years as selective oxidation catalysts, has been also investigated by ^{129}Xe NMR spectroscopy. The first study on these compounds was performed by Bonardet et al. [230]. Such compounds are ionic solids with a discrete arrangement of anions and cations forming a variety of cage-like structures similar to that of zeolites. The most common are those whose anions have the Keggin structure: the basic unit consists of 12 MO_6 octahedra ($M = W$ or Mo) which, by sharing edges, form four M_3O_{13} groups assembled through common oxygen atoms, leading to a tetrahedral cavity at the center of which a nonmetallic atom (P, Si, As) is located. This anion has a charge of 3- or 4- compensated by cations. The general formulas can be written $(X^+)_3$ [P (W or Mo)$_{12}$ $O_{40}]^{3-}$ or $(X^+)_4$ Si[W or MO)$_{12}$ $O_{40}]^{4-}$with $X^+ = NH_4{}^+$, Cs^+ or K^+ for the samples studied by the authors. Generally a nonpolar gas cannot penetrate either into the Keggin anion or into the interionic region. Nevertheless, analysis of porosity by nitrogen adsorption-desorption isotherms by Moffat [231] has shown that certain of the monovalent cation salts have relatively high surface areas and a microporous structure. However, the average pore sizes ($1.9 < d < 2.9$ nm) obtained by porosimetry seemed too large in comparison with the lattice parameter of the salts (1.17 nm); for this reason the ^{129}Xe NMR technique was tested on these samples. Surprisingly and despite the small amount of xenon adsorbed, the NMR signal is easy to detect and the resonance lines are narrow ($2 < \Delta H < 7$ ppm at a resonance frequency of 24.9 MHz).

As found with zeolite Y, the $\delta = f(N)$ plots are linear over the entire range of xenon concentrations studied ($10^{19} < N < 10^{21}$ atoms/g) for all the samples. These results strongly support a narrow micropore size distribution with micropores isotropically distributed in the solid. The similarities of the xenon NMR results for zeolites and HPOM allowed the authors to apply the Demarquay–Fraissard model. The $\delta_{N\to 0}$ for three salts with the same cation (Fig. 35) are identical (72 ± 2 ppm) and lead to a mean free path of 0.46 nm. Assuming that $NH_4{}^+$ has no electrical effect, as in zeolites, and that the microporous structure consists of infinitely long cylindrical or spherical cavities, the pore diameter is 0.9 or 1.36 nm, respectively. In comparison with the lattice parameters of ammonium salts (1.1–1.3 nm), these pore sizes appear to be more reasonable than those obtained from porosimetry measurements. This microporosity could result from a translation and/or a rotation of the Keggin anion in the crystal when the size of the cation increases. This motion could create interconnections between the interstitial void volumes, with the appearance of channels into which small molecules can enter.

Later, Terskikh et al. [232] studied the structure of 12-tungstophosphoric heteropolyacids (HPA) (H_3 PW_{12} O_{40}) supported on silica. At low HPA load-

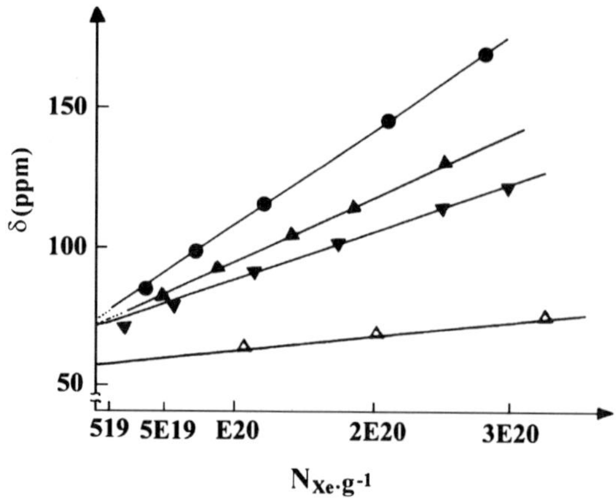

Fig. 35 $\delta = f(N)$ curves at 300 K for NH_4^+ salts: $PW_{12}O_{40}$ ▲; $PMO_{12}O_{40}$ ♦; $SiW_{12}O_{40}$ •; dehydrated NaY zeolite △

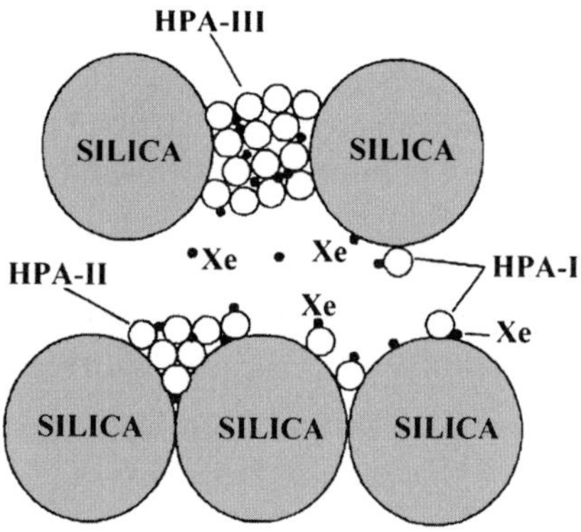

Fig. 36 Schematic representation of the sorbed and interporous xenon atoms in the pore structure of HPA/SiO_2 catalysts

ing (< 20 wt %) the surface HPA species are mainly isolated molecules, denoted HPA-I (Fig. 36). The volume and pore distribution do not change considerably compared to pure silica, therefore, the difference observed in chemical shift (≈ 20 ppm) is due to interactions of xenon with a limited number of HPA-I species acting as strong adsorption sites.

Under conditions of fast exchange, the chemical shift can be written:

$$\delta = \delta_{SiO_2} p_{SiO_2} + \delta_{HPA\text{-}I} p_{HPA\text{-}I} \quad (36)$$

where p_i represents the relative xenon population adsorbed on the surface $\underline{i}$.

The contributions of the silica and HPA-I (measured at low xenon pressure) are 69 and 95 ppm, respectively. At higher loading (40 wt %), c.a. 5 nm HPA clusters appears (denoted HPA-II in Fig. 36). The chemical shift is then written:

$$\delta = \delta_{SiO_2} p_{SiO_2} + \delta_{HPA\text{-}II} p_{HPA\text{-}II} \tag{37}$$

where $\delta_{HPA\text{-}II}$ denotes the chemical shift of xenon adsorbed in the open microporosity of the clusters. This microporosity has been confirmed by N_2 adsorption measurements. At the same time a broad downfield line of unknown origin appears at about 140 ppm. At the highest HPA loading (60 wt %) the spectrum consists of three overlapping lines at 106 (A), 117 (B) and 140 (C) ppm. The chemical shift of line A decreases when xenon pressure P increases, as for the 20 and 40% loaded samples, whereas the chemical shift of line B increases with P, as in zeolites: it is assigned to xenon adsorbed in the more closed porosity of larger ($\approx$ 50 nm) HPA clusters.

This work shows that the high sensitivity of the ^{129}Xe NMR technique makes it possible to study silica-supported HPA and to follow the clustering of HPA molecules "step-by-step".

3.3 Activated Carbon

Porous carbons and activated charcoal are widely used both as catalysts and in separation techniques. They have a microporous structure with more or less extended graphitic zones. Whereas the adsorption of xenon on graphite has been studied extensively, there has been little work with ^{129}Xe NMR. Generally the signals are very broad, as is shown in [233–236] (revue 1999).

Contrary to previous authors, Suh et al. [237] obtained better resolved and relatively narrow signals for activated carbon at ambient temperature. For different samples with high specific area ($500 < S < 1000\ m^2/g$) they obtain at relatively low xenon pressure ($P < 1000$ torrs) a linear relationship between the chemical shift and the surface xenon density ρ, $\delta = \delta_0 + \rho\delta_1$, analogous to that previously proposed by Ito and Fraissard [1]. Like these authors, they attribute the δ_0 term to the Xe-surface interaction which is sensitive to the nature of the surface. After nitric acid treatment to introduce acid sites on the surface of activated carbons, the values of δ_0 are shifted 20 ppm downfield without change in the slope (Fig. 37) despite significant variations in the specific area and the mean pore size. This could be due to the fact that the Xe – Xe interactions in the gas phase inside the pores have little effect on the value of δ_1, as compared to those at the internal surface of the pores. The latest study published on this subject was by Bansal et al. [238] in 1992. By combining microporosity measurements treated by the method of Horvath–Kawazoe [239] and extrapolating Eq. 20 to activated carbons, they obtain the dimensions of the graphitic microzones. The values (1–3 nm) are quite consistent with those measured by small angle X-rays (1.5–6 nm).

There was a recent 129Xe NMR study of a series of catalytic filamentous carbons (CFC) obtained from the gas-phase reaction on iron subgroup metal

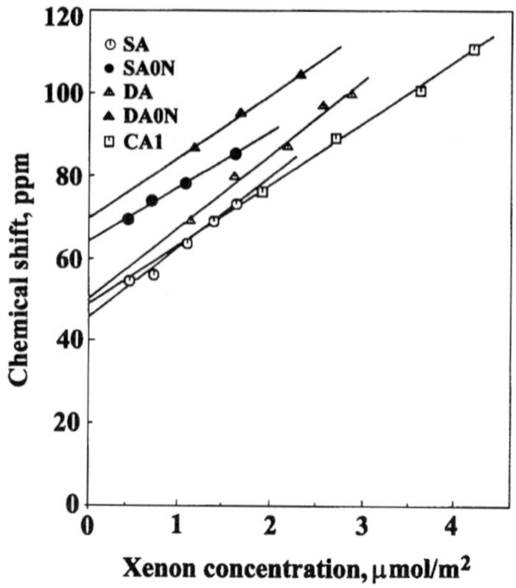

Fig. 37 Dependence of the ^{129}Xe NMR chemical shift on the concentration of adsorbed xenon for different amorphous carbons (SA0N and DA0N are samples SA and DA activated with nitric acid)

catalysts.A connection between ^{129}Xe NMR and structural, textural and paramagnetic properties of CFC was found [240]. The chemical shift δ was shown to depend on the structure of the CFC surface formed by edge, basal or both (edge and basal) graphite faces. This dependence follows a general trend of the chemical shift to increase with adsorption potential of a surface. The term describing Xe – Xe interaction ($\delta_{(Xe-Xe)}$) in confined space decreased with the average pore size of CFC granule. For hollow multi-carbon nanotubes two ^{129}Xe NMR signals were attributed to voids inside nanotubes and to interstices between the interlaced nanotubes.

4 Mesoporous Solids

4.1 Amorphous Oxides

Amorphous solids such as silica, silica-alumina, etc. have not been studied extensively by the ^{129}Xe NMR technique because their amorphous structure and widely opened porosity lead generally to broad signals. Nevertheless, we can report some studies of such materials. However, it is useful to consider first some specifics of Xe NMR spectroscopy on mesoporous systems: Very large and easily accessible pores, plus fast diffusion of xenon, cause exchange to have pronounced effects on the observed spectra.

The first study was performed in 1989 by Conner et al. [241] who studied the pore structure created by compressing, at 17.5 tons/m^2, nonporous aerosil silica spheres 12 and 7 nm in diameter. Compression of these materials created interparticle voids. Although nitrogen adsorption measurements give no evidence for micro or small mesopores, the chemical shift of adsorbed xenon, which depends on the compression (70–90 ppm) but which is not pressure-dependent, was much higher than the authors expected (Fig. 38). However, there appears to be a difference in the nature of the chemical shift of xenon adsorbed in aerosils as compared to that in the micropores of zeolites.

The absence of xenon pressure dependence, the large increase in the chemical shift with decreasing adsorption temperature and the narrowness of the NMR signal must be interpreted as the result of a fast exchange between adsorbed xenon atoms and free xenon located in the interporous space created by compressing the silica spheres. In this case, adsorbed xenon can be located in a defect of the silica surface or in small spaces at the point of contact between particles, the dimension of which must be very small (i.e. < 1 nm).

The authors showed that the experimental chemical shift is given by the equation:

$$\delta_{\text{exp}} = \frac{KP\delta_{\text{ads}} + (NV_{\text{p}}P/RT)\delta_{\text{pores}}}{KP + NV_{\text{p}}P/R}, \tag{38}$$

where K is the Henry's law coefficient for xenon adsorption, P and T are the pressure and the temperature of the experiment, N is the Avogadro number, V_{p} the pore volume of the silica and R the gas constant. If the concentration

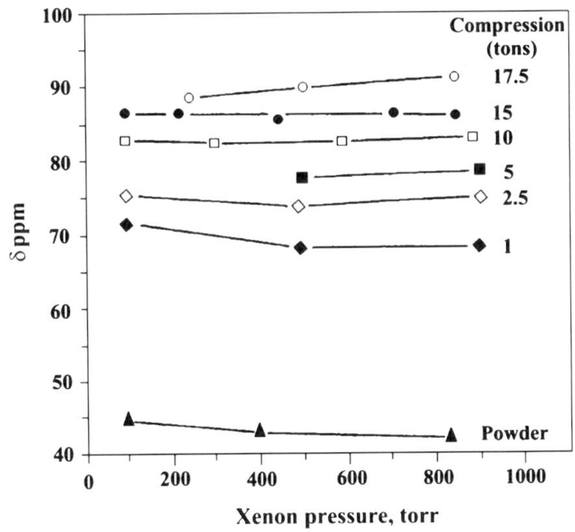

Fig. 38 Effect of compression and Xe pressure on the chemical shift of xenon adsorbed on aerosil samples

of both the adsorbate and interporous xenon have first-order dependence on the xenon pressure, the chemical shift is then not pressure dependent.

In 1990 Cheung [242] studied xenon adsorption at low temperature (144 K) on a wide range of amorphous oxides. He found a nonlinear dependence on xenon concentration; a parabolic-like curvature is observed for alumina and silica-alumina, similar to that observed in Y zeolites with 2 + cations (Ca^{2+}, Mg^{2+}, Ni^{2+}, Co^{2+} etc.). His explanation is based on the distribution of micropore sizes: at very low xenon concentration, the chemical shift is due to xenon atoms adsorbed in the smallest micropores. As the loading increases, xenon is adsorbed in larger micropores and the chemical shift, which is the exchange average of the shifts in the large and small micropores, is therefore smaller. At higher xenon pressure, Xe – Xe interactions become preponderant and the chemical shift increases again. More recently, Terskikh et al. [243] characterized the porous structure of a number of silica gels by ^{129}Xe NMR. On the basis of experimental data for samples with well defined structure they proposed an empirical correlation between the chemical shift and the parameters of the pore structure.

$$\delta = \frac{\delta_S}{1 + D/\eta KRT} = \frac{115}{1 + 0.0076D}, \qquad (39a)$$

where δ_S is the chemical shift extrapolated at zero pressure, η a factor depending on the more or less interconnecting pore system and its geometry ($\eta = 4$ for an interconnecting cylinder, 2 for slit-like non interconnecting pores), K is Henry's constant of the adsorption isotherm and D the mean pore size. According to the authors this relationship should be valid in the range $2 < D < 40$ nm. These authors proposed also this equation in another form:

$$\delta = \delta_S/(1 + V_g/KSRT), \qquad (39b)$$

where S is the specific surface area and V_g the free volume inside the adsorbent. This equation shows that the observed shift is expected to be independent of xenon pressure, which is commonly observed, at least at low pressures when Xe – Xe interactions are insignificant and the adsorption follows Henry's law. Equation 39b shows that δ depends strongly on S/V_g, the surface-to-volume ratio of the pores occupied by Xe atoms involved in fast exchange. It is therefore expected that the spectra of materials with a distribution of S/V_g will reflect this feature either by the appearance of more than one line, or by the presence of a broad line presenting a minimum in the variation of δ against Xe concentration, as shown by Cheung [242].

The presence of small particles (less then about 10 μm) in the material studied will also result in broad lines [244, 245]. Specifically, the effects of xenon exchange and bulk properties of porous materials on the spectra have been demonstrated and discussed in terms of xenon diffusion path length using 40 Å Vycor controlled-pore glass of different particle sizes [245]. According to Eq. 39b the mean pore size D related to the Vg/S as $D = \eta Vg/S$

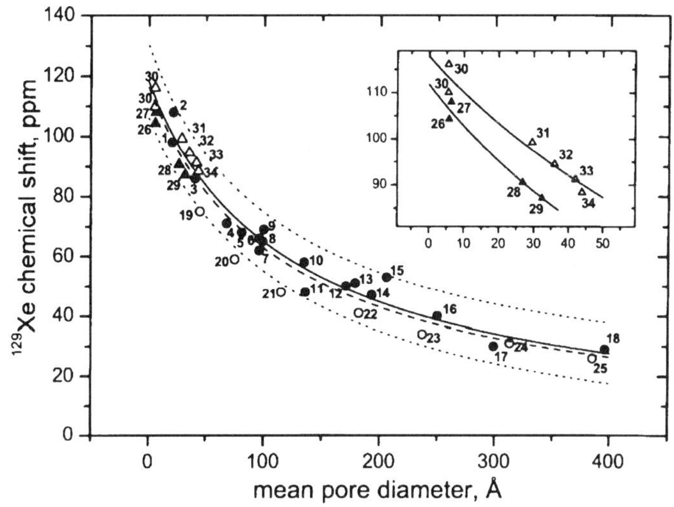

Fig. 39 ^{129}Xe chemical shifts vs. mean pore diameters for porous silica-based materials: • silica gels; ○ Vycor/CPG; ▲ and △ porous organo-silicates of two different origins. The *solid curve* is the nonlinear least-squares fit for samples 1–18, with prediction bands given at a confidence level of 95% shown as the *dotted curves*. The *dashed curve* is the fit for all samples. *Inset*: Fits for two subsets of porous organo-silicates

(η is a parameter dependent on the pore shape), can be found from the NMR experiment provided that K is known from adsorption experiments. The correlation of the observed chemical shift with D (Fig. 39) as obtained by conventional adsorption methods has uncovered a general correlation between δ and D of the form $\delta = \delta_S/(1 + D/b)$ [244, 246] that is similar to those found for zeolites [56].

For 34 silica-based materials with pores in the range of 0.5–40 nm, the parameters of the equation obtained from the fit are $\delta_S = 116 \pm 3$ ppm and $b = 117 \pm 8$ Å [246]. With some caution, the correlation can be used for the characterization of silica samples with unknown pore structure. One needs to note that even within this general correlation, subsets of materials of similar origin display yet finer correlations that indicate an accurate sensitivity of the method to details of the pore structure.

Temperature-dependent chemical shift data can be used to extract the physical parameters related to the adsorption properties of materials. In the fast exchange approximation with weak adsorption, as described by Henry's law, the temperature dependence of the observed xenon chemical shift δ for arbitrary pores can be expressed as [244]:

$$\delta = \delta_s \left(1 + \frac{V_g}{K_0 S R T^{1/2}} \mathrm{e}^{-\frac{\Delta H_{\mathrm{ads}}}{RT}} \right)^{-1} . \tag{40}$$

The heat of adsorption ΔH_{ads} can be found by fitting the experimental temperature dependence of the chemical shift and, for Xe physically adsorbed on silica-based surfaces, can range from just a few kJ/mol to about 20 kJ/mol [244]. If the pre-exponent of the Henry's constant is known from independent adsorption measurements, then the V_g/S ratio can be found as well. Application of Eq. 41 necessitates working at very low xenon concentrations to reduce the effects of Xe – Xe interactions and to prevent Xe condensation in the pores at low temperature.

Supported V_2O_5 is used as a catalyst for the selective oxidation of hydrocarbons. ^{129}Xe NMR was used to investigate structural features of V_2O_5/TiO_2 (anatase) and V_2O_5/TiO_2 catalysts [247]. The chemical shift of xenon adsorbed either on the support TiO_2 or on the oxidation catalyst V_2O_5 varies linearly with the xenon coverage, showing a lack of strong adsorption sites. Extrapolated values at low adsorption temperature, characteristic of Xe-surface interactions, are 109 ± 3 ppm for TiO_2 and 93 ± 5 ppm for V_2O_5. On the contrary, for V_2O_5/TiO_2-supported catalysts and particularly for those with a high proportion of monomeric vanadyl sites (low % of V_2O_5) a curvature in the δ-plots appears at low coverage. The authors concluded that monomeric vanadyl sites act as strong adsorption sites, excluding the existence of large vanadia domains; then, for low V_2O_5 loading, the vanadia units are well dispersed on the support. As the V_2O_5 loading increases the curvature becomes less pronounced, due to a decrease in the fraction of monomeric species and an increase in the fraction of polymeric species; this result can be compared to that for MFI zeolites with extra-framework alumina species [183]. Two-dimensional ^{129}Xe spectroscopy (Fig. 40) reveals that xenon diffuses between two distinct environments: the exchange rate shows that these environments are in close proximity.

In 1994 Mansfeld and Veeman [248] studied the sintering of Al_2O_3 and $Al_2O_3 - ZrO_2$ fibers using high xenon pressure ($4 < P < 10$ bar). For a non-sintered sample, the spectrum at 200 K exhibits two peaks, one at 228 ppm assigned to liquid xenon (considering the temperature and the pressure) and the other, at 231 ppm, which disappears after sintering, corresponding to xenon adsorbed in the pores of the fibers. Nearly the same observations can be made for the $Al_2O_3 - ZrO_2$ sample. To explain the temperature dependence of the chemical shift the authors proposed a simple theoretical model: for the low temperature region, xenon atoms can move in a potential well, as already described by Ripmeester et al. [53] and Cheung [54], due to interaction with an oxygen atom of the surface; for the high temperature region, the temperature and pressure dependence of δ are described by a Langmuir model. Finally, Oepen and Gunther [249] studied the condensation of surface hydroxyl groups of silica after high temperature treatment. For the samples pretreated at 773 and 973 K, cross-peaks between signals of inter- and intraparticle xenon are observed in 2D EXSY experiments, showing an increased rate of exchange. This is not the case for silica with a high concentration of su-

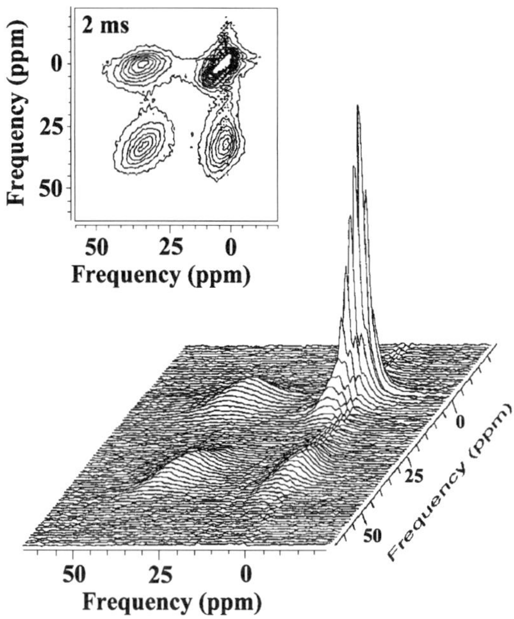

Fig. 40 ^{129}Xe 2D exchange NMR spectrum, at 11.7 T (138 MHz), of xenon adsorbed on V_2O_5 at 290 K and at xenon pressure of approximately 2 atm. The spectrum was acquired using a mixing time of 2 ms and a recycle delay of 300 ms. The *contour lines* represent 1–20% of the maximum intensity

perficial hydroxyl groups which increase the polarity of the silica surface and hinder exchange.

Different types of aerogels represent an important class of open-pore mesoporous materials with extremely low framework density, reaching values as low as 0.05 g/cm^3 and possessing great potential for various industrial applications. A recent ^{129}Xe NMR study of aerogels [250], performed as a combination of spectroscopic and spatially resolved NMR spectroscopy, has proved to be a powerful approach for characterizing the average pore structure and steady-state spatial distribution of xenon atoms in different physicochemical environments. The method offers unique information and insights into the microscopic morphology of aerogels, the dynamical behavior of adsorbates, and provides spatially resolved information on the nature of the defect regions found in these materials [250]. The extremely low density of aerogels, however, required very long accumulations. The problem of long experimental data acquisition times was addressed in another study by employing the Xe-131 isotope and a very high density of xenon [251].

A more practical approach to obtaining the Xe NMR spectra of aerogels is to employ the high sensitivity of HP Xe. Application of NMR microimaging using continuous flow HP Xe resulted in a visual picture of the dynamics of gases inside the particles of aerogels [252]. The "polarization-weighted" images of gas transport in aerogel fragments are correlated to the diffusion coefficient of xenon obtained from NMR pulsed-field gradient experiments. In another diffusion-related study [253] the ingression of HP Xe in Vycor porous glass was followed by 1D NMR imaging and by observing the intensity of the NMR signal. The resulting diffusion coefficients compared well with the results obtained from the pulsed-field gradient measurements with thermally polarized Xe.

We can also mention the study of some other mesoporous solids, such as soils [254] or of wood cellular structures [255].

4.2
New Mesoporous Solids (MCM, SBA, etc.)

The development of a novel group of mesoporous materials, such as MCM-41, MCM-48, SBA-15, etc., and related periodic porous oxides [256] has generated a great deal of interest due to their highly uniform pore sizes. They have great potential as catalysts, chromatographic supports, separation materials, photonic crystals, and in electronic devices [257]. The large pore size (diameter of 20–500 Å), surface area, and easy functionalization of the silica wall provide further applications as supports for chemical and biological reactions [258].

Often such materials are also either amorphous or poorly crystalline, which limits the applications of diffraction-based methods. After many successful applications of Xe NMR to microporous materials (zeolites, pore diameters ≤ 20 Å), as has been shown in previous paragraphs, naturally there is a growing interest in the application of this technique to such porous materials, especially with the use of hyperpolarized xenon [259–267]. Indeed these silica-based materials can be prepared with pores in a very broad range of sizes and are a good testing-ground for the chemical shift-pore size correlations.

The first studies demonstrated [259, 260], however, that the observed shifts could fall well outside the range predicted by the general correlations obtained for silicates [261, 262]. In most cases the chemical shift of adsorbed xenon is practically independent of the xenon pressure, and the observed values are almost always *below* those estimated from the empirical correlations. Since the materials are prepared as very fine powders with the particle size rarely exceeding 10 μm, the exchange between adsorbed and gas phases is expected to contribute heavily to the observed shifts. Indeed, compression of the samples [260] produced significant downfield shifts similar to those in compressed aerosils [261] or in Vycor porous glass with different particle sizes [262].

In a recent study [263] the NMR of continuously circulating hyperpolarized Xe has been used to characterize purely siliceous and ordered Al-containing mesoporous MCM-41 and SBA-15. The effect of compression on the mesopore structure has also been studied. The NMR spectra obtained can be interpreted in terms of exchange between adsorbed xenon in the pores and gaseous Xe atoms in the interparticle spaces. The greatly increased sensitivity arising from hyperpolarized Xe has also been used to detect the spectra of xenon adsorbed on very small quantities of mesoporous silica thin films [264].

Additional information about the internal surface of the pores can be obtained from the data on co-adsorption of Xe with some other small molecules [259–261, 265]. The co-adsorption studies can be particularly useful for screening intra-wall micropores, surface defects and inhomogeneities in porosity. In certain cases, using guest molecules of different sizes, it is possible to distinguish and estimate the sizes of micropores inside the mesopore walls [265].

The assessment of the distribution and accessibility of moieties attached to the walls of mesoporous silicates is another area of ^{129}Xe NMR applications.

The effects of hydrocarbon chains attached to the silica walls has been the subject of an extensive hyperpolarized ^{129}Xe NMR study [266] The variable temperature measurements revealed a nonuniform porosity and irregular pore structure, and allowed one to follow the changes in the adsorption properties of xenon due to modification of the mesopore voids [266]]. In another study [267] the presence of Pt in the ordered mesoporous silicates results in a notable chemical shift of adsorbed xenon. It was concluded that the Pt clusters are situated inside the pores of the mesoporous molecular sieve rather than on the external surface, in a similar way as shown in paragraph 2.5 with zeolites.

4.3 Xe NMR Cryoporometry

Although there are still few applications we will mention this technique. Xenon porometry is a novel method for the determination of pore dimensions. It has been developed on the basis of NMR cryoporometry [268]. In this method, a porous material is immersed in an organic substance, and the freezing and melting behaviors of the substance are explored by means of the ^{129}Xe NMR spectroscopy of xenon dissolved in the sample [269]. An example of the spectra measured at different temperatures is shown in Fig. 41. The signals have been labeled in the figure.

The origins of the components are the following: *Signal A* arises from the inner xenon thermometer, which is a capillary tube in the middle of a 10 mm sample tube containing ethylbromide and xenon gas [270]. *Signals B and C* originate from xenon dissolved in the liquid substance. *Signal B* arises from bulk substance located in the spaces between particles of porous material and on the top of porous material, and vanishes below the freezing temperature of the substance (227 K in the case of acetonitrile), as the transition from li-

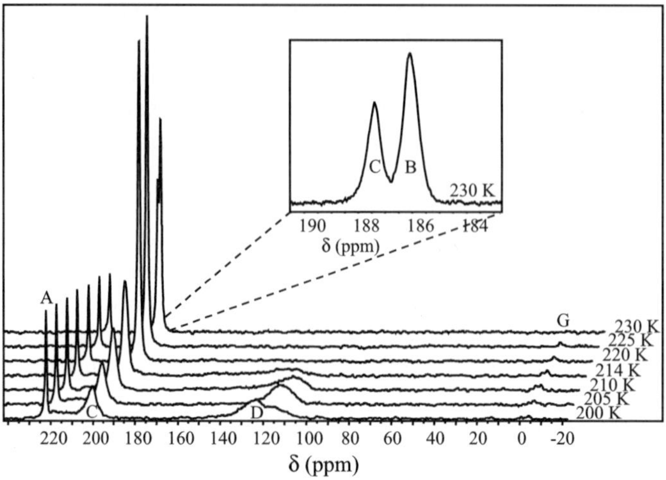

Fig. 41 The ^{129}Xe NMR spectra of the sample containing Silica gel 100, acetonitrile and xenon at different temperatures. The measurement temperatures are shown beside the spectra. The chemical shift range of the signals B and C at 230 K has been expanded in the *inset* of the figure

quid to solid is accompanied by an abrupt reduction of gas solubility. *Signal C* originates from liquid inside the pores, and gradually vanishes at lower temperatures, as the confined substance freezes. If the density of the substance increases substantially (as occurs in the case of acetonitrile), *signal D* appears at lower temperatures. This signal arises from xenon in very small gas bubbles appearing inside the pores. The bubbles build up during the freezing of confined substance, as this contracts. The signal vanishes, as the confined substance melts. As *bulk* substance freezes, bubbles also build up in between the particles of porous materials due to the density change, and *signal G* originates from xenon in this environment. Since the bubbles are much bigger than those inside the pores the chemical shift of signal G is close to that of bulk gas.

The method provides three novel possibilities for determining pore sizes:

1. As, on one hand, the melting point of the bulk substance can be deduced from the emergence of signal B and from the disappearance of signal G, and, on the other hand, the disappearance of signal D and the changes of the intensity of signal C reveal the lowered melting point of the confined substance, the melting point depression can be determined by measuring the NMR spectra at variable temperature, and the average pore size of the material can be calculated by the Gibbs–Thompson equation in the same way as in the usual NMR cryoporometry.
2. The chemical shift difference of signals C and B is dependent on pore size. By determining the correlation for a certain substance at a certain

temperature using known mesoporous materials, the pore size of an unknown sample material can be measured. This is very easy to do, because the signals can be recorded by a single scan measurement at any temperature above the bulk melting point. On the other hand, the correlation is quite inaccurate because the line-width of the signals is large compared to the distance between them. The correlation is best suited for pore size determination of smaller pore sizes in the mesoporous range.
3. As the size of the gas bubbles formed inside the pores during the freezing of the confined substance depends on the size of the pores, the chemical shift of xenon atoms inside the bubbles also depends on the pore size. These studies prove that the large majority of bubbles are isolated (i.e. they are not connected with each other) giving resonance signals characteristic of each pore size. As signal D is made up of these components, the shape of the signal represents the pore size distribution. The correlation between the chemical shift of signal D and the pore size can be determined using known reference samples. After that, the pore size distribution of an unknown material can be determined by measuring its NMR spectrum [269].

5 Conclusions

Since the first applications on Xe-NMR of adsorbed xenon used as a probe, this technique has shown its interest in many applications.

In the case of zeolites it is possible:

- to determine the dimensions and the form of their internal free volumes without knowing their structures;
- to reveal structure defects, produced for example by dealumination, and to determine their characteristics;
- to calculate the short-range crystallinity, as opposed to that determined by X-rays;
- to follow the mechanism of the synthesis and crystallization of zeolites during their preparation;
- to locate cations in the zeolite structure and to follow their migration or change in their environment depending on various factors;
- to locate any “encumbering” species. For example, adsorbed molecules, extra-framework species, coke formed during catalytic cracking reactions, etc.;
- to determine the dispersion of metals (particularly when the particles are too small to be seen by electron microscopy) and the distribution of the molecules adsorbed on the particles (as opposed to the mean coverage);
- to compare the diffusion of species as a function of the size of the zeolite “windows”.

The porosity of new mesoporous solids (MCM, SBA, etc) is relatively well defined by means of nitrogen adsorption-desorption techniques. Xe-NMR is however an excellent complement, in particular to demonstrate the roughness of the pore surface or, more recently, to study systems with a zeolite monolayer on the surface. It is also useful for various applications similar to the previous ones for zeolites. This list of applications is certainly not exhaustive. This technique can also be applied to any microporous or mesoporous solids such as clays, amorphous oxides, activated carbons, solid polymers, etc. provided one work at low temperature.

This wide field of applications has been considerably extended by the advent of xenon hyperpolarization which increases the sensitivity of NMR detection by 3 or 4 orders of magnitude.

6 Appendix: Hyperpolarized Xenon

Optical pumping is a procedure whereby the population of the spin states of the valence electron of an alkali metal atom can be changed. In this way a population distribution very different from that of thermodynamic equilibrium is achieved [271]. By irradiating an alkali metal vapor with circularly polarized monochromatic light whose frequency corresponds to an absorption line of the atom at the ground state, a given excited level can be populated and the distribution between the Zeeman levels of the ground state changed. This electronic polarization can be transmitted by contact with nuclear spins of a monoatomic gas like xenon [8–10, 14–17]. One then speaks of "optical pumping of xenon", somewhat incorrectly, since it is the valence electron of the alkali metal which is optically pumped and not the nuclear spin of the xenon.

The optical pumping of rubidium is displayed schematically in Fig. 42. The electronic ground state $5^2S_{1/2}$ is split into two states, spin-down and spin-up. Only the spin-down state can absorb right circularly polarized light σ^+. Usually, the gas pressure applied in the experiments is so high that the population of the substrates of the excited state $5^2P_{1/2}$ is randomized by collisional mixing. Then, relaxation occurs to either of the ground states with the same probability. Since only the spin-down state, can be depleted upon irradiation, a net polarization builds up in the spin-up ground state. The radiative decay of the excited state generates photons without preferential spin orientation. To avoid depumping of the rubidium by reabsorption of these photons, a quenching gas such as nitrogen is often added.

In Fig. 43, the mechanism of spin exchange is illustrated schematically. A rubidium and a xenon atom collide in the presence of a buffer gas molecule (nitrogen) and form a van der Waals molecule, whose lifetime τ is limited by collision with another buffer gas molecule. During the lifetime of the van der

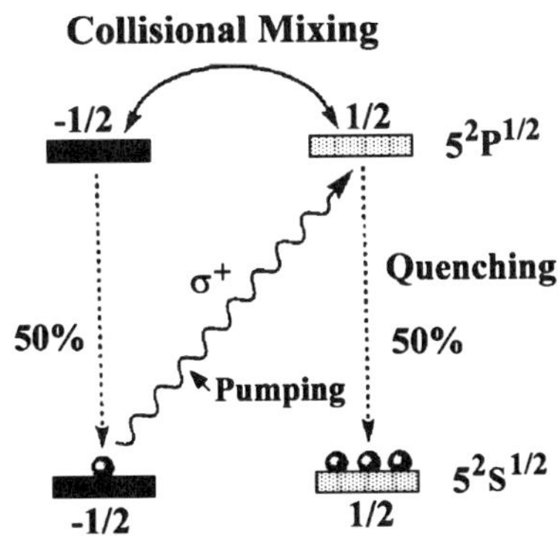

Fig. 42 Scheme of the optical pumping process, shown for σ^+ polarized light. Photons can only be absorbed by the – 1/2 ground state $5^2S_{1/2}$. Collisional mixing randomizes the population of the excited states $5^2P_{1/2}$ so that relaxation occurs to either ground state with the same probability. Due to depletion of only the – 1/2 state, a net polarization results in the 1/2 state

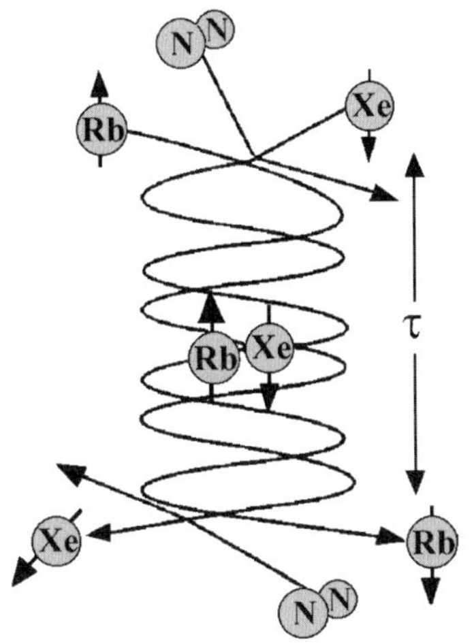

Fig. 43 Spin exchange between optically polarized Rb and Xe during the lifetime τ of the van der Waals pair

Waals pair, spin exchange occurs. The rubidium electronic spin $\vec{S}$ flips down, whereas the xenon nuclear spin $\vec{I}$ flips only partly up.

The Hamiltonian to be considered is:

$$\hat{H} = \hat{H}_{\text{Hyperfin}} + \gamma \hat{N}\hat{S} + \alpha \hat{J}\hat{S} + \hat{H}_{\text{Zeeman}} \tag{41}$$

$\hat{S} =$ spin of the valence electron of the rubidium atom

$\hat{N} =$ angular moment of rotation of the van der Waals complex

$\hat{J} =$ nuclear spin of the xenon atom.

In the right hand side the first term represents the hyperfine coupling between the electron spin and that of the nucleus of the Rb atom. The second term represents the spin-rotation interaction between the spin of the valence electron of Rb and the rotational angular moment of the complex. This interaction dominates the relaxation of rubidium and is, therefore, the main reason for the loss of polarization. The Fermi contact term, $\alpha\hat{I}\hat{S}$, is responsible for polarization transfer between Rb and Xe. By means of the optical pumping of rubidium the polarization of ^{129}Xe can reach values as high as 10^{-2} or 10^{-1}, that is, a polarization 4 orders of magnitude higher than that obtained at thermodynamic equilibrium.

The main applications of this technique to the study of solid surfaces are by A. Pines et al. In [11] there is a description of the apparatus used and, in the reviews of Raftery and Chmelka [7] and Pietras and Gaede [272], the presentation of much work.

References

1. Ito T, Fraissard J (1980) In: Rees LV (ed) Proc of the 5th Int Zeolite Conf. Heyden, London, p 510
2. Fraissard J, Ito T (1988) Zeolites 8:350 and references therein
3. Springuel-Huet MA, Bonardet JL, Fraissard J (1995) Appl Magn Reson 8:427
4. Bonardet JL, Fraissard J, Gedeon A, Springuel-Huet MA (1999) Catal Rev-Sci Eng 41(2):115–225
5. Barrie PJ, Klinowski J (1992) Progress in NMR spectroscopy 24:91 and references therein.
6. Dybowsky C, Bansal N, Duncan TM (1991) Ann Rev Phys Chem 42:433
7. Raftery D, Chmelka B (1993) In: Blunmich B, Kosfeld R (eds) NMR Basic Principles and Progress. Springer, Berlin Heidelberg New York
8. Brinkmann D (1963) Helv Phys Acta 36:413
9. Grover BC (1978) Phys Rev Lett 40:391
10. Happer W, Miron E, Schaefer S, Schreiber D, van Wingaarden WA, Zeng X (1984) Phys Rev A 29:3092
11. Raftery D, Long H, Meersmann T, Grandinetti PJ, Reven L, Pines A (1991) Phys Rev Lett 66:584
12. Raftery D, Long H, Reven L, Tang P, Pines A (1992) Chem Phys Lett 191:385
13. Raftery D, Reven L, Long H, Pines A, Tang P, Reimer JA (1993) J Phys Chem 97:1649
14. Driehuys B, Cates GD, Miron E, Sauer K, Walter DK, Happer W (1996) Appl Phys Lett 69:1668
15. Moudrakovski IL, Terskikh VV, Ratcliffe CI, Ripmeester JA, Wang LQ, Shin Y, Exarhos GJ (2002) J Phys Chem B 106(6):5938
16. Anala S, Pavlovskaya GE, Pichumani P, Dieken TJ, Olsen MD, Meersmann T (2003) JACS 125(43):13298–13302
17. Kaiser LG, Meersmann T, Logan JW, Pines A (2000) Proc Nat Acad Sci USA 97(6):2414
18. Bartlett N, Sladky FO (1976) The Chemistry of Krypton, Xenon and Radon. In: Bailar M (ed) Comprehensive Inorganic Chemistry, vol 1. Pergamon Press, Oxford, p 213 and references therein

19. Schrolbilgen GJ, Holloway JH, Granger P, Brévard C (1976) CR Acad Sci 519:282
20. Seppelt K (1979) Acc Chem Res 12:211
21. Schrobilgen GJ, Holloway JH, Granger P, Brévard C (1978) Inorganic Chem 17:980
22. Harris RK, Mann BE (eds) (1978) NMR and the Periodic Table. Academic Press, New York, p 439
23. Reisse J (1986) New Journal of Chemistry 10:665 and references therein
24. Ramsey NF (1950) Phys Rev 78:699; Ibid (1953) Nuclear moment. Wiley, New York
25. Diehl P, Ugolini R, Suryaprakash N, Jokisaari J (1991) Magn Res Chem 29:1163
26. Rummens FA (1975) Van der Waals and Shielding Effects. In: Diehl P, Fluck E, Kosfeld R (eds) NMR Basic Principles and Progress. Springer, Berlin Heidelberg New York, p 10
27. Dickinson WC (1951) Phys Rev 81:717
28. Bothner-By AA, Glick RE (1957) J Chem Phys 26:1651 and references therein
29. Bonardet JL, Fraissard J (1976) J Magn Res 22:1 and references therein
30. Proctor WG, Yu FE (1950) Phys Rev 78:471; bid (1951) 81:20
31. Streever RL, Carr HY (1961) Phys Rev 121:20
32. Hunt ER, Carr HY (1963) Phys Rev 130:2302
33. Kanegsberg E, Pass B, Carr HY (1969) Phys Rev Lett 23:572
34. Brinkmann D, Brun E, Staub HH (1962) Helv Phys Acta 35:431
35. Jameson AJ, Jameson CJ, Gutowsky HS (1970) J Chem Phys 53:2310
36. Adrian FJ (1964) Phys Rev A 136:980
37. Adrian FJ (1970) Chem Phys Lett 7:201
38. Jameson CJ, Jameson AK, Cohen SM (1973) J Chem Phys 59:4540
39. Cowgill DF, Norberg RE (1972) Phys Rev B 6:1636
40. Jameson CJ, Jameson AK (1971) Mol Phys 20:957
41. Buckingham AD, Koll PA (1972) Mol Phys 23:65
42. Jameson CJ, Jameson AK, Cohen SM (1975) Mol Phys 29:1919
43. Jameson CJ, Jameson AK, Cohen SM (1976) J Chem Phys 65:3397
44. Miller KW, Reo NV, Shoot Uiterkamp AJM, Stengle DP, Stengle TR, Williamson KL (1981) Proc Natl Acad Sci USA 78:4946
45. Stengle TR, Reo NV, Williamson KL (1981) J Phys Chem 85:3772
46. Muller N (1982) J Phys Chem 86:2109
47. Muller N (1976) J Magn Reson 22:479
48. Chen Q, Springuel-Huet MA, Fraissard J (1991) In: Öhlman E, Pfeifer H, Fricke R (eds) Proc ZEOCAT 1990, Leipzig, August 20–23, 1990. Elsevier, Amsterdam, p 219; Stud Surf Sci Catal 65:219
49. Springuel-Huet MA, Demarquay J, Ito T, Fraissard J (1988) In: Grobet PJ, Mortier WJ, Vansant EF, Schulz-Ekloff G (eds) Innovation in Zeolite Materials Science, Proc Int Symp, Niewport, September 13–17, 1987. Elsevier, Amsterdam, p 183; Stud Surf Sci Catal 37:183
50. Cheung TTP, Fu CM, Wharry S (1988) J Phys Chem 92:5170
51. Derouane EG, André JM, Lucas AA (1987) Chem Phys Lett 137:336
52. Derouane EG, B'Nagy J (1987) Chem Phys Lett 137:341
53. Ripmeester JA, Ratcliffe CI (1990) J Phys Chem 94:7652
54. Cheung TTP (1995) J Phys Chem 99:7089
55. Chen QJ, Fraissard J (1992) J Phys Chem 96:1815
56. Demarquay J, Fraissard J (1987) Chem Phys Lett 136:314
57. Johnson DW, Griffiths L (1987) Zeolites 7:484
58. de Menorval LC, Fraissard J, Ito T (1982) J Chem Soc Faraday Trans I 78:403; (1983) J Chim Phys 80:573

59. Derouane EG (1987) Chem Phys Lett 142:200
60. Chen QJ, Fraissard J (1992) Phys Chem 96:1809
61. Springuel-Huet MA, Fraissard J (1992) Zeolites 12:841
62. Vigné-Maeder F (1994) J Phys Chem 98:4666
63. Cheung TTP, Fu CM (1989) J Phys Chem 93:3740
64. Cheung TTP (1990) J Phys Chem 94:376
65. Heink W, Kärger J, Pfeifer H, Stallmach F (1990) J Am Chem Soc 112:2175
66. Pfeifer H, Freude D, Kärger J (1991) In: Öhlman E, Pfeifer H, Fricke R (eds) Catalysis and Adsorption by Zeolites. Proc ZEOCAT 90, Leipzig, August 20–23, 1990. Elsevier, Amsterdam, p 397; Stud Surf Sci Catal 65:397
67. Kärger J, Pfeifer H (1991) J Chem Soc Faraday Trans I 87:1989
68. Ripmeester JA, Ratcliffe CI (1993) Analytica Acta 283:1103
69. Shœmaker R, Apple T (1987) J Phys Chem 91:4024
70. Ryoo R, Pak C, Chmelka BF (1990) Zeolites 10:790
71. Chmelka BF, Pearson JG, Liu SB, Ryoo R, de Menorval LC, Pines A (1991) J Phys Chem 95:303
72. Tway C, Apple T (1992) J Catal 133:42
73. Chen QJ, Fraissard J (1990) Chem Phys Lett 169:595
74. Chmelka BF, Gillis JV, Petersen EE, Radke CJ (1990) AIChE J 36:1562
75. Chmelka BF, de Menorval C, Csencsits R, Ryoo R, Liu SB, Radke CJ, Petersen EE, Pines A (1989) In: Morterra C, Zecchina A, Costa G (eds) Structure and Reactivity of Surfaces. Proc Europ Conf, Trieste, September 13–16, 1988. Elsevier, Amsterdam, p 269; Stud Surf Sci Catal 48:269
76. Chmelka BF, Rosin RR, Went GT, Bell AT, Radke CJ, Petersen EE (1989) In: Jacobs PA, van Santen RA (eds) Zeolites: Facts, Figures, Future. Proc 8th Int Zeol Conf, Amsterdam, July 10–14, 1989, Part A and B. Elsevier, Amsterdam, p 995; Stud Surf Sci Catal 49:995
77. Ahn DH, Lee JS, Nomura M, Sachtler WMH, Moretti G, Woo SI, Ryoo R (1992) J Catal 133:191
78. Yang OB, Woo SI, Ryoo R (1992) J Catal 137:357
79. Jameson CJ, Jameson AK, Cohen SM (1975) J Chem Phys 62:4224
80. Ryoo R, Kwak JH, de Ménorval LC (1994) J Phys Chem 98:7101
81. Kim JG, de Ménorval LC, Ryoo R, Figueras F (1995) In: Beyer HK, Karge HG, Kiricsi I, B'Nagy J (eds) Catalysis by Microporous Materials. Proc ZEOCAT 95, Szombathely, July 9–13, 1995. Elsevier, Amsterdam, p 226; Stud Surf Sci Catal 94:226
82. Samant MG, de Menorval LC, Dalla Betta RA, Boudart M (1988) J Phys Chem 92:3937
83. Chmelka BF, Raftery D, Mc Comick AV, de Menorval LC, Levine RD, Pines A (1991) Phys Rev Lett 66:580; 67:931
84. Jameson CJ, Jameson AK, Gerald II R, de Dios AC (1992) J Chem Phys 96:1676
85. Ryoo R, de Menorval LC, Kwak JH, Figueras F (1993) J Phys Chem 97:4124
86. Larsen RG, Shore J, Schmidt-Rohr K, Emsley L, Long H, Pines A, Janicke M, Chmelka BF (1993) Chem Phys Lett 214:220
87. Jameson CJ, Jameson AK, Lim HM, Baello BI (1994) J Chem Phys 100:5965
88. Cheung TTP (1993) J Phys Chem 97:8993
89. Van Tassel PR, Davis HT, Mc Cormick AV (1992) Mol Phys 76:411; (1993) J Chem Phys 98:1
90. Chao K, Shy DS (1993) J Chem Soc Faraday Trans 89:3481
91. Seo G, Ryoo R (1990) J Catal 124:224
92. Alexander SM, Coddington JM, Howe RF (1991) Zeolites 11:368

93. Chen QJ, Springuel-Huet MA, Fraissard J, Smith ML, Corbin DR, Dybowski C (1992) J Phys Chem 96:10914
94. Annen MJ, Davis ME, Hanson BE (1990) Catal Lett 6:331
95. Chen QJ, Springuel-Huet MA, Fraissard J (1989) Chem Phys Lett 159:117
96. Fraissard J, Ito T, Springuel-Huet MA, Demarquay J (1986) In: Murakami Y, Iijima A, Ward JW (eds) New Developments in Zeolite Science and Technology. Proc 7th Int Zeol Conf, Tokyo, August 17–22, 1986. Elsevier, Amsterdam; Stud Surf Sc Catal 28:293
97. Ito T, Fraissard J (1982) J Chem Phys 76:5225
98. Ito T, Fraissard J (1986) J Chim Phys 83:441
99. Ito T, Fraissard J (1987) J Chem Soc Faraday Trans I 83:451
100. Pires J, Brotas de Carvalho M, Ribeiro FR, B'Nagy J, Derouane E (1993) Appl Catal A 95:75
101. Cotterman RL, Hickson DA, Cartlidge S, Dybowski C, Tsiao C, Venero AF (1991) Zeolites 11:27
102. Shertukde PV, Hall WK, Marcelin G (1992) Catal Today 15:491
103. Ito T, de Menorval LC, Guerrier E, Fraissard J (1984) Chem Phys Lett 111:271
104. Pellegrino C, Ito T, Gabelica Z, B'Nagy J, Derouane EG (1990) Appl Catal 61:L1
105. Chen QJ, Fraissard J, Cauffriez H, Guth JL (1991) Zeolites 11:534
106. Martens JA, Feijen E, Lievens JL, Grobet PJ, Jacobs PA (1991) J Phys Chem 95:10025
107. Chen QJ (1990) PhD Thesis, Université Paris 6, France
108. Dumont N, Ito T, Derouane EG (1989) Appl Catal 54:L1
109. Springuel-Huet MA, Fraissard J (1989) Chem Phys Lett 154:299
110. Davis ME, Saldarriaga C, Montes C, Hanson BE (1988) J Phys Chem 92:2557
111. Benslama R, Fraissard J, Albizane A, Fajula F, Figueras F (1988) Zeolites 8:196
112. Ito T, Springuel-Huet MA, Fraissard J (1989) Zeolites 9:68
113. Ito T, Fraissard J (1987) Zeolites 7:554
114. Tsiao CJ, Kauffman JS, Corbin DR, Abrams L, Carroll EE Jr, Dybowski C (1991) J Phys Chem 95:5586
115. Smith ML, Corbin DR, Abrams L, Dybowski C (1993) J Phys Chem 97:7793
116. Smith ML, Corbin DR, Dybowski C (1993) J Phys Chem 97:9045
117. Ripmeester JA (1984) J Magn Res 56:247
118. Moudrakovski IL, Ratcliffe CI, Ripmeester JA (1995) Appl Magn Reson 8:385
119. Bedard RL, Bowes CL, Coombs N, Holmes A, Jiang T, Kirkby SJ, Macdonald PM, Malek AM, Ozin GA, Petrov S, Plavac N, Ramik RA, Steele MR, Young D (1993) J Am Chem Soc 115:2300
120. Springuel-Huet MA, Ito T, Fraissard J (1984) In: Jacobs PA, Jaeger NI, Jiru P, Kazansky VB, Schulz-Elkoff G (eds) Structure and Reactivity of Modified Zeolites. Proc Int Conf, Prague, July 9–13, 1984. Elsevier, Amsterdam, p 13; Stud Surf Sc Catal 18:13
121. Chen QJ, Springuel-Huet MA, Fraissard J (unpublished data)
122. Campbell SM, Bibby DM, Coddington JM, Howe RF (1996) J Catal 161:350–358
123. Liu SB, Fung BM, Yang TC, Hong EC, Chang CT, Shih PC, Tong FH, Chen TL (1994) J Phys Chem 98:4393
124. Liu SB, Lee CS, Fung BM (1994) In: Hattori T, Yashima T (eds) Zeolites and Microporous Crystals. Proc Int Symp on Zeolites and Microporous Crystals, Nagoya, August 22–25, 1993. Elsevier, Amsterdam, p 233; Stud Surf Sci Catal 83:223
125. Gédéon A, Bonardet JL, Fraissard J (1989) In: Jansen JC, Moscou L, Post FA (eds) Zeolites for the Nineties, Recent Research Reports 8th Int Zeol Conf, July 10–14, Amsterdam, 1989. Moscou L, Akzo Chemicals Publi, Amsterdam, p 87
126. Scharpf EW, Crecely RW, Gates BC, Dybowski C (1986) J Phys Chem 90:9

127. Bansal N, Dybowski C (1988) J Phys Chem 92:2333
128. Gédéon A, Bonardet JL, Fraissard J (1988) J Chim Phys 85:871
129. Gédéon A, Bonardet JL, Ito T, Fraissard J (1989) J Phys Chem 93:2563
130. Bonardet JL, Gédéon A, Fraissard J (1995) In: Beyer HK, Karge HG, Kiricsi I, B'-Nagy J (eds) Catalysis by Microporous Materials. Proc ZEOCAT 95, Szombathely, July 9–13, 1995. Elsevier, Amsterdam, p 139; Stud Surf Sci Catal 94:139
131. Gallezot P, Imelik B (1973) J Phys Chem 77:2556
132. Chen QJ, Ito T, Fraissard J (1991) Zeolites 11:239
133. Shy DS, Chen SH, Lieves J, Liu SB, Chao KJ (1991) J Chem Soc Faraday Trans I 87:2855
134. Kim JG, Kompany T, Ryoo R, Ito T, Fraissard J (1994) Zeolites 14:427
135. Gédéon A, Burmeister R, Grosse R, Boddenberg B, Fraissard J (1991) Chem Phys Lett 179:191
136. Grosse R, Burmeister R, Boddenberg B, Gédéon A, Fraissard J (1991) J Phys Chem 95:2443
137. Grosse R, Gédéon A, Watermann J, Fraissard J, Boddenberg B (1992) Zeolites 12:909
138. Fraissard J, Gédéon A, Chen QJ, Ito T (1991) In: Jacobs PA, Jaeger NI, Kubelkova L, Wichterlova B (eds) Zeolite Chemistry and Catalysis, Proc Int Symp, Prague, September 8–13, 1991. Elsevier, Amsterdam, p 461; Stud Surf Sci Catal 69:461
139. Gédéon A, Bonardet JL, Fraissard J (1995) In: Beyer HK, Karge HG, Kiricsi I, B'-Nagy J (eds) Catalysis by Microporous Materials. Proc. ZEOCAT 95, Szombathely, July 9-13, 1995. Elsevier, Amsterdam, p 187; Stud Surf Sci Catal 94:187
140. Dieden R, Hevesi L (1989) Abstracts of Xth Meeting ISMAR, Morzine, France
141. Gédéon A, Bonardet JL, Fraissard J (1993) J Phys Chem 97:4254
142. Boddenberg B, Hartmann M (1993) Chem Phys Lett 203:531
143. Gédéon A, Fraissard J (1994) Chem Phys Lett 219:440
144. Hartmann M, Boddenberg B (1994) Microporous Materials 2:127
145. Gédéon A, Bonardet JL, Lepetit C, Fraissard J (1995) Solid State NMR 5:201
146. Liu SB, Lin TS, Yang TC, Chen TH, Hong EC, Ryoo R (1995) J Phys Chem 99:8277
147. Freitag A, Wüllen CV, Staemmler V (1995) Chem Phys 192:267
148. Liu SB, Lee CS, Shiu PF, Fung BM (1994) In: Hattori T, Yashima T (eds) Zeolites and Microporous Crystals. Proc Int Symp, Nagoya, August 22–25, 1993. Elsevier, Amsterdam, p 233; Stud Surf Sci Catal 83:233
149. Hartmann M, Boddenberg B (1994) In: Weitkamp J, Karger HG, Pfeifer H, Holderich W (eds) Zeolites and Related Microporous Materials: State of the Art. Proc 10th Int Zeol Conf, Garmisch-Parternkirchen, July 17–22, 1994. Elsevier, Amsterdam, p 509; Stud Surf Sci Catal 84
150. Seidel A, Rittner F, Boddenberg B (1996) J Chem Soc Faraday Trans 92:493
151. Seidel A, Boddenberg B (1996) Chem Phys Lett 249:117
152. Boddenberg B, Seidel A (1994) J Chem Soc Faraday Trans 90:1345
153. Boddenberg B, Sprang T (1995) J Chem Soc Faraday Trans 91:163
154. Sprang T, Boddenberg B (1995) J Chem Soc Faraday Trans 91:555
155. (a) Seidel A, Boddenberg B (1996) Chem Phys Lett 249:117; (b) Rittner F, Seidel A, Sprang T, Boddenberg B (1996) Appl Spect 50:1389
156. Seidel A, Kampf G, Schmidt A, Boddenberg B (1998) Catalysis Letters 51:213–218
157. Chen Q, Guth JL, Seive A, Caullet P, Fraissard J (1991) Zeolites 11:799
158. Chen Q, Guth JL, Fraissard J (1992) In: Guczi L, Solymosi F, Tetenyi P (eds) New Frontiers in Catalysis. Proc 10th Int Congress on Catal, Budapest, July 19–24, 1992. Elsevier, Amsterdam, p 2577, Stud Surf Sci Catal 75

159. Bonardet JL, Barrage MC, Fraissard J (1993) In: von Balmoos R, Higgins JB, Treacy MMJ (eds) Proc of the 9th Int Zeolite Conf, Montreal, July 5-10, 1992. Butterworth-Heinemann II:475
160. Bradley SM, Cambell SM, Hook J, Jackson K, Gorkom LV, Howe RF (1994) In: Hattori H, Misono M, Ono Y (eds) Acid-base catalysis II. Proc Int Symp on Acid-Base Catalysis, Sapporo, December 2–4, 1993. Elsevier, Amsterdam, p 129; Stud Surf Sci Catal 90
161. Ryoo R, Liu SB, de Menorval LC, Tahegoshi K, Chmelka B, Trecoske M, Pines A (1987) J Phys Chem 91:6575
162. Chmelka B, Ryoo R, Liu SB, de Menorval LC, Radke CJ, Petersen EE, Pines A (1988) J Am Chem Soc 110:4465
163. de Menorval LC, Raftery D, Liu SB, Tahegoshi K, Ryoo R, Pines A (1990) J Phys Chem 94:27
164. Lechert H, Basler WD, Wittern KP (1986) In: Murakami Y, Iijima A, Ward JW (eds.) New Developments in Zeolite Science and Technology. Proc 7th Int Zeol Conf Tokyo, August 17–22, 1986. Elsevier, Kodansha, p 595
165. de Mallmann A, Barthomeuf D. (1986) In: Murakami Y, Iijima A, Ward JW (eds) New Developments in Zeolite Science and Technology. Proc 7th Int Zeol Conf Tokyo, August 17–22, 1986. Elsevier, Kodansha, p 609
166. Wu JF, Chen TL, Ma LJ, Lin MW, Liu SB (1992) Zeolites 12:86
167. Liu SB, Ma LJ, Wu JF, Lin MW, Chen TL (1992) J Phys Chem 96:8120
168. Renouprez AJ, Jobic H, Oberthur RC (1985) Zeolites 5:222
169. Jobic H, Renouprez A, Fitch AN, Lauter HJ (1987) J Chem Soc Farad Trans 1 83:3199
170. Gédéon A, Ito T, Fraissard J (1988) Zeolites 8:376
171. Ito T, Bonardet JL, Fraissard J, Nagy JB, André C, Gabelica Z, Derouane EG (1988) Appl Catal 43:5
172. Barrage MC, Bonardet JL, Fraissard J (1990) Catal Lett 5:143
173. Barrage MC, Bauer F, Ernst H, Fraissard J, Freude D, Pfeifer H (1990) Catal Lett 6:201
174. Tsiao C, Dybowski C, Gaffney AM, Sofranco J (1991) J Catal 128:520
175. Liu SB, Prasad S, Wu JF, Ma LJ, Yang TC, Chiou JT, Chang JY, Tsai TC (1993) J Catal 142:664
176. Bonardet JL, Barrage MC, Fraissard J, Kubelkova L, Novakova J, Ernst H, Freude D (1992) Coll Czechos Chem Comm 57:733
177. Bonardet JL, Barrage MC, Fraissard J (1994) ACS Symposium Series 571:230
178. Miller JT, Meyers BL, Ray G (1991) J Catal 128:436
179. Ito T, Fraissard J, Kärger J, Pfeifer H (1991) Zeolites 11:103
180. Fraissard J, Ito T, de Ménorval LC, Springuel-Huet MA (1982) In: Jacobs PA, Jaeger NI, Jiru P, Schulz-Ekloff G (eds) Metal Microstructures in Zeolites. Proc Workshop, Bremen, September 22–24, 1982. Elsevier, Amsterdam, p 179; Stud Surf Sci Catal 12 and references therein
181. Fraissard J, Ito T, de Ménorval LC (1984) In: Proc of the 8th Int Congress on Catalysis Berlin July 2–6, 1984. Verlag Chemie, Dechema III:25
182. Ito T, de Ménorval LC, Fraissard J (1983) J Chim Phys 80:573
183. Boudart M, Samant MG, Ryoo R (1986) Ultramicroscopy 20:125
184. Boudart M, Ryoo R (private communication)
185. Rouabah D, Benslama R, Fraissard J (1991) Chem Phys Lett 179:218
186. Rouabah D, Fraissard J (unpublished data)
187. Pandya KJ, Heald SM, Hriljac JA, Petrakis L, Fraissard J (1996) J Phys Chem 100:5070
188. Chmelka BF, Ryoo R, Liv SB, de Ménorval LC, Radke CJ, Petersen EE, Pines A (1988) J Am Chem Soc 110:4465

189. Yang OB, Woo SI, Ryoo R (1990) J Catal 123:375
190. Boudart M, de Ménorval LC, Fraissard J, Valenca GP (1988) J Phys Chem 92:4033
191. de Ménorval LC, Fraissard J, Ito T, Primet M (1985) J Chem Soc Faraday Trans I 81:2855
192. Primet M, de Ménorval LC, Fraissard J, Ito T (1985) J Chem Soc Faraday Trans I 81:2867
193. Coddington JM, Howe RF, Yong YS, Asakura K, Iwasawa Y (1990) J Chem Soc Faraday Trans I 86:1015
194. Ichikawa M, Rao LF, Kimura T, Fukuoka A (1990) J Mol Catal 62:15
195. Peereboom PWE, Luigjes H, Prins KO (1989) Physica Acta 156:260
196. Stejskal ED, Tanner JE (1965) J Chem Phys 42:288
197. Kärger J, Pfeifer H, Heink W (1988) Adv Magn Reson 12:1
198. Kärger J, Pfeifer H, Stallmach F, Spindler H (1990) Zeolites 10:288
199. Kärger J, Pfeifer H (1987) Zeolites 7:90
200. Stallmach F, Kärger J, Datema KP, Bolt-Westerhoff JA, Nowack AK (1995) Zeitschrift für Physikalische Chemie 189:43
201. Dubinin MM, Gorlov VA, Voloshcuk AM (1980) In: Rees LV (ed) Proc of the 5th Int Zeolite Conf, Naples, July 2-6, 1980. Heyden, London, p 554
202. Ruthen DM, Derrah RJ (1975) J Chem Soc Faraday Trans I 71:2031
203. Bakaev VA, Smirnova LF (1978) Izv Akad Nauk SSSR Ser Khim 284
204. Kärger J, Pfeifer H, Stallmach F, Feoktistova NN, Zhdanov SP (1993) Zeolites 13:50
205. Hong U, Kärger J, Kramer R, Pfeifer H, Seiffert G, Muller U, Unger KK, Luck HB, Ito T (1991) Zeolites 11:816
206. Fenzke D, Kärger J (1993) Z Phys D25:345 and references therein
207. Ruthven DM (1993) Zeolites 13:594 and references therein
208. Fraissard J, Kärger J (1989) Zeolites 9:351
209. Kärger J, Richter R, Pfeifer H, Fürtig H, Roscher W, Seidel R (1988) AIChE J 34:1185
210. Kärger J, Pfeifer H, Caro J, Bülow M, Schlodder H, Mostowicz H, Völter J (1987) Appl Catal 29:21
211. Kärger J, Pfeifer H, Fraissard J (1987) Z Phys Chemie Leipzig 268:195
212. Bendoraitis JG, Chester AW, Dwyer FG, Garwood WE (1986) In: Murakami Y, Iijima A, Ward JW (eds) New Developments in Zeolite Science and Technology. Proc 7th Int Zeol Conf Tokyo, August 17–22, 1986. Elsevier, Kodansha, Tokyo p 669
213. Dyer A (1988) An Introduction to Zeolite Molecular Sieves. Wiley, Chichester
214. Nivarthi SS, McCormick AV (1994) Microporous Mater 3:47
215. Li FY, Berry R (1995) J Phys Chem 99:2459 and references therein
216. Jameson CJ, Jameson AK, Baello BJ, Lin HM (1994) J Chem Phys 100:5965
217. Jameson AK, Jameson CJ, Gerald II RE (1994) J Chem Phys 101:1775
218. Bansal N, Dybowski C (1990) J Magn Res 89:21
219. Kärger J (1991) J Magn Res 93:184 and references therein
220. Kärger J (1971) Ann Phys, Leipzig 27:107
221. Kärger J, Pfeifer H, Wutscherk T, Ernst S, Weitkamp J, Fraissard J (1992) J Phys Chem 96:5059
222. Förste C, Germanus A, Kärger J, Pfeifer H, Caro J, Pilz W, Zikánová A (1987) J Chem Soc Faraday Trans I 83:2301
223. Zikánová A, Bülow M, Schlodder H (1987) Zeolites 7:115
224. Beschmann K, Kokotailo G, Riekert L (1987) Chem Eng Process 22:223
225. Springuel-Huet MA, Nosov A, Kärger J, Fraissard J (1996) J Phys Chem 100:7200
226. Springuel-Huet MA, N'Gokoli-Kekele P, Mignon C, Bonardet JL, Fraissard J (2001) Stud Surf Sci Catal 135:3121–3128

227. Fetter G, Tichit D, de Menorval LC, Figueras F (1990) Appl Catal 65:L1
228. Barrie PJ, McCann GF, Gameson I, Rayment T, Klinowski J (1991) J Phys Chem 95:9416
229. Yamanaka S, Takahama KI, Kunii K, Shiotani M (1994) Clay Science 9:149; Fraissard J (1989) J Phys Chem 93:4138
230. Bonardet JL, Fraissard J, McGarvey GM, Moffat JB (1993) Colloid Surf, A 72:191; (1995) J Catal 151:147
231. Moffat JB (1989) J Mol Catal 52:169
232. Terskikh VV, Mastikhin VM, Timofeeva MN, Okkel LG, Fenelonov VB (1996) Catal Lett 42:99
233. Shibanuma T, Asada H, Ichi SI, Matsui T (1983) Jap J Appl Physics 22:1656
234. Venables JA, Kramer HM, Price GL (1976) Surf Sci 55:373
235. Neue G (1987) Z Physik Chem Neue Folge 152:271
236. Wernett PC, Larsen JW, Yamada O, Yue HJ (1990) Energy and Fuels 4:413
237. Suh DJ, Park TJ, Ihm SK, Ryoo R (1991) J Phys Chem 95:3767
238. Bansal N, Foley HC, Lafyatis DS, Dybowski C (1992) Catal Today 14:305
239. Horvath G, Kawazoe K (1983) J Chem Eng Japan 16:470
240. Romanenko K, d'Espinose J-B, Fraissard J, Reshetenko TV, Lapina O (2005) Microporous and Mesoporous Materials 81:41–48
241. Conner WC, Weist EL, Ito T, Fraissard J (1989) J Phys Chem 93:4138
242. Cheung TTP (1989) J Phys Chem 93:7549
243. Terskikh VV, Mudrakovski IL, Mastikhin VM (1993) J Chem Soc, Faraday Trans 89:4239
244. Terskikh VV, Moudrakovski IL, Mastikhin V M (1993) ^{129}Xe Nuclear Magnetic Resonance Studies of the Porous Structure of Silica Gels. J Chem Soc Faraday Trans 89(6):4239
245. Ripmeester JA, Ratcliffe CI (1993) ^{129}Xe NMR Spectroscopy in Microporous Solids: the Effect of Bulk Properties. Anal Chim Acta 283:1103
246. Terskikh VV, Moudrakovski IL, Breeze SR, Lang S, Ratcliffe CI, Ripmeester JA, Sayari A (2002) A General Correlation for the Xe-129 NMR Chemical Shift-Pore Size Relationship in Porous Silica-Based Materials. Langmuir 18(15):5653–5656
247. Kritzenberger J, Gaede HC, Shore JS, Pines A (1994) J Phys Chem 98:10173
248. Mansfeld M, Veeman WS (1994) Microporous Materials 3:85
249. Oepen SB, Gunther H (1996) Magn Res Imaging 14:993
250. Gregory DM, Gerald RE, Botto RE (1998) J Magn Res 131(2):327
251. Pavlovskaya G, Blue AK, Gibbs SJ, Haake M, Cros F, Malier L, Meersmann T (1999) J Magn Res 137(1):258
252. Kaiser LG, Meersmann T, Logan JW, Pines A (2000) Proceedings of the National Academy of Sciences of the United States of America 97(6):2414
253. Moudrakovski IL, Sanchez A, Ratcliffe CI, Ripmeester JA (2000) J Phys Chem B 104(31):7306
254. Filimonova SV, Knicker H, Hausler W, Kogel KI (2004) Geoderma 122(1):25
255. Wang LQ, Shin Y, Samuels WD, Exarhos GJ, Moudrakovski IL, Terskikh VV, Ripmeester JA (2003) J Phys Chem B 107(50):13793
256. Kresge CT, Leonowicz ME, Roth WJ, Vartuli JC, Beck JS (1992) Nature 359:710
257. Behrens P, Stucky GD (1993) Angew Chem, Int Ed Engl 32:696–699
258. Zhao D, Yang P, Huo Q, Chmelka BF, Stucky GD (1998) Curr Opin Solid Mater 3:111–121
259. Springuel-Huet MA, Fraissard J, Schmidt R, Stocker M, Conner WC (1997) Special Publications of the Royal Society of Chemistry 213:452

260. Springuel-Huet MA, Sun K, Fraissard J (1999) Microporous and Mesoporous Materials 33(1–3):89
261. Springuel-Huet MA, Bonardet JL, Gedeon A, Yue Y, Romannikov VN, Fraissard J (2001) Microporous and Mesoporous Materials, 44 Special Iss SI 77
262. Nossov A, Haddad E, Guenneau F, Gedeon A (2003) Physical Chemistry Chemical Physics 5(5):4479
263. Guo WP, Huang LM, Deng P, Xue ZY, Li QZ (2001) Microporous and Mesoporous Materials, 44 Special Iss SI 427
264. Davidson A, Berthault P, Desvaux H (2003) J Phys Chem B 107(51):14388
265. Chen F, Zhang MJ, Han Y, Xiao FS, Yue Y, Ye CH, Deng F (2004) Characterization of Microporosity in Ordered Mesoporous Material MAS-7 by ^{129}Xe NMR Spectroscopy. J Phys Chem B 108(12):3728
266. Nossov A, Haddad E, Guenneau F, Mignon C, Gedeon A, Grosso D, Babonneau F, Bonhomme C, Sanchez C (2002) Chem Comm 2476
267. Ryoo R, Ko CH, Kim JM, Howe R (1996) Catal Lett 37(1-2):29
268. Strange JH, Rahman M (1993) Phys Rev Lett 71:3589
269. Telkki V-V, Lounila J, Jokisaari J (2005) Behavior of Acetonitrile Confined to Mesoporous Silica Gels as Studied by ^{129}Xe NMR. J Phys Chem B 109:757
270. Saunavaara J, Jokisaari J to be published
271. Ripmeester JA, Ratcliffe CI (1990) J Phys Chem 94:8773
272. Pietrass T, Kneller JM (1999) J Phys Chem B 103(42):8837–8841

Mol Sieves (2007) 5: 249–364
DOI 10.1007/3829_005

Published online: 4 May 2006

Characterization of Coke on Zeolites

F. Bauer[1] (✉) · H. G. Karge[2] (✉)

[1]Leibniz Institute of Surface Modification, Permoserstr. 15, 04318 Leipzig, Germany
fbauer@rz.uni-leipzig.de

[2]Fritz Haber Institute of the Max Planck Society, Faradayweg 4–6, 14195 Berlin, Germany
karge@fhi-berlin.mpg.de

Abbreviations

AES	Auger electron spectroscopy
Al_F	framework aluminum
Al_{NF}	extra- or non-framework aluminum
BEA	beta, zeolite structure *
BET	Brunauer–Emmet–Teller (method)
CEELS	core electron energy loss spectroscopy
CMK-*n*	ordered mesoporous carbon, $n = 1, 2, 3, \ldots$
CP	cross polarization
Da	atomic mass unit
D_d, D_i	diffusion coefficients
DME	dimethyl ether
DRIFTS	diffuse reflectance infrared Fourier transform spectroscopy
δ	chemical shift [ppm]
EELS	electron energy loss spectroscopy
ERI	erionite, zeolite structure *
ESR	electron spin resonance (spectroscopy)
FAU	faujasite, zeolite structure *
FER	ferrierite, zeolite structure *
FTIR	Fourier transform infrared (spectroscopy)
GC	gas chromatography
IR	infrared (spectroscopy)
L	zeolite structure *
MALDI-TOF	matrix-assisted laser desorption/ionization time-of-flight
MAS	magic angle spinning
MCM-*n*	ordered mesoporous material, $n = 41, 48$
MFI	ZSM-5, zeolite structure *
MOR	mordenite, zeolite structure *
MRI	^1H magnetic resonance imaging
MS	mass spectrometry
MTG, MTO	methanol-to-gasoline, methanol-to-olefins
M41S	family of ordered mesoporous materials
NH_3-TPD	ammonia temperature-programmed desorption
NMR	nuclear magnetic resonance spectroscopy
OFF	offretite, zeolite structure *
PFG NMR	pulsed-field gradient ^1H NMR technique
RE	rare earth metal (cation)
SAPO-*n*	microporous silicoaluminophosphates, $n = 11, 34$
SCR	selective catalytic reduction
SEM	scanning electron microscopy
STEM	scanning transmission electron microscope
T	temperature [K]
TBPO	tributylphosphine oxide
TEM	transmission electron microscope
TEOM	tapered element oscillating microbalance
TG	thermogravimetry
TGA	thermogravimetric analysis
TMP, TMPO	trimethylphosphine, trimethylphosphine oxide
TON	Theta-1, ZSM-22, zeolite structure *
TOS	time on stream

TPD	temperature-programmed desorption
TPO	temperature programmed oxidation
USY	ultra-stable Y zeolite
UV-Vis	ultraviolet-visible spectroscopy
V, V_o	internal free volumes
W_c	coke content [wt %]
X	zeolite structure *
XPS	X-ray excited photoelectron spectroscopy
XRD	X-ray diffraction
Y	zeolite structure *
ZSM-n	zeolite structure, $n = 5, 8, 11, 22, 34$ *

* cf. Baerlocher C, Meier WM, Olson DH (2001) Atlas of zeolite framework types, 5th edn. Elsevier, Amsterdam, p 302

1
Introduction

Industrial requirements for a good catalyst are high activity, excellent selectivity and extended stability. Catalyst deactivation, which provokes loss of activity and change in selectivity, is usually inevitable, and the time scale of deactivation has profound consequences for process design. Catalyst deactivation is highly relevant for economical reasons, and scientifically it has many challenges [1–3]. In hydrocarbon processing, the undesired formation of carbonaceous materials produced by polymerization/dehydrogenation reactions and deposited on the catalyst surface plays a major role in deactivation. Therefore, a great number of studies have already been carried out on coke formation on the classic cracking catalysts, i.e. amorphous silica/alumina which was employed before the advent of zeolite catalysts [4]. In the particular case of acid-catalyzed hydrocarbon reactions on zeolite catalysts, the last few years has seen an increasing interest in this topic [5], in view of both industrial application and fundamental research on zeolite catalysis. Thus, in the context of catalytic cracking of hydrocarbon feed stocks, coke formation on the large pore-sized H-Y and the effects of the pore structure on the mode of coking has been widely studied [6]. The formation of coke on zeolites has been shown to be a "shape-selective reaction" and its rate an intrinsic property of the zeolite pore structure [7]. With an emphasis on the medium pore-size zeolite H-ZSM-5 and the methanol-to-gasoline (MTG) process, there are numerous reports concerning the effects of acid-site strength and concentration as well as the reaction conditions on the coking rate, the location and the nature of coke species [8]. Recently, the most frequently used techniques for the characterization of carbonaceous deposits on zeolite catalysts have been comprehensively illustrated by appropriate examples [9, 10]; another review article focuses on the contribution of NMR spectroscopy to the understanding of coke formation [11].

The economic point of view of coke formation in industrial hydrocarbon processing cannot be overestimated. Although the activity of a coked zeolite catalyst may be recovered by coke combustion, regeneration is always time consuming and, in any case, requires additional investment. Therefore, many efforts have been made to reduce the deactivation rate by variation of the reaction conditions and modification of the zeolite catalyst. The pathologic-like analysis of spent catalysts may give some hints for both the optimization of reaction conditions and the design of catalysts, because it reveals all of the deactivating effects that occurred during the "life time" of the catalyst.

In former deactivation studies, the characterization of coke was limited often to the measurement of its atomic H/C ratio. This information, although rather limited, may be used for a more general definition of coke: "*Coke consists of carbonaceous deposits which are mainly deficient in hydrogen compared with the coke-forming reactant molecule(s) and which are responsible for catalyst deactivation*". In this work the term "coke" will be used for all hydrocarbon compounds, irrespective of their chemical nature and molecular weight, which are located in and/or on the zeolite crystallites and reduce the catalytic activity. A more rigorous definition of "black" coke, generally designated to heavy polynuclear aromatics, does not include the deactivation of catalysts by paraffinic and olefinic species formed under relatively mild reaction conditions. Although the common use of the term coke may be misleading for these retained and deactivating by-products, the authors of this work do not regard it as necessary and helpful to introduce a special term, e.g., encapsulate [12] or retardate [13], for them. In particular, such low-molecular and hydrogen-rich coke molecules may participate in various reactions as catalytic entities and active sites [14, 15].

However, not only the nature of the carbonaceous residues is a matter of debate and the subject of many investigations. There are numerous problems related to coke formation on acidic zeolites: very important questions concern the effects of zeolite properties (such as structure, acidity and binder), the influence of the nature and composition of the reactants and intermediates, the effect of the operation conditions (temperature, pressure, carrier gas, time on stream, etc.), the location of carbonaceous deposits and the deactivation kinetics.

Although the formation and the nature of coke on zeolites have been widely studied during the last 20 years, many questions are still open and controversially discussed. Therefore, in this work particular attention will be paid to the various techniques which were frequently employed to tackle the problems related to coke on zeolite catalysts.

Thus, for the elucidation of the nature of coke a great variety of spectroscopic techniques is used which comprises, inter alia, infrared (IR), ultraviolet and visible spectroscopy (UV-Vis), electron spin resonance (ESR), nuclear magnetic resonance (NMR), and X-ray excited photoelectron spectroscopy (XPS). Adsorption measurements with various adsorbates on fresh and coked samples were employed, especially to determine the locus of coke deposition, i.e., internal pore structure vs. external surface of the zeolite crystallites. The deformations of the zeolitic lattice caused by the deposition of carbonaceous materials within the crystallites have been observed by X-ray diffraction (XRD). With a spatial resolution in the nm range, coke species can even be made visible by a transmission electron microscope (TEM). However, the number of methods that permit the investigation of coke formation under reaction conditions is still limited [16]. Besides all of these methods, which usually do not require the destruction of the coke/zeolite system, information on coke is also available via the combustion or the pyrolysis of carbonaceous residues, e.g., temperature-programmed oxidation (TPO), and via the liberation of coke-constituting species by dissolving the zeolitic matrix. The latter method allows us to obtain the distribution of the coke components which is essential to specify the route of coke formation and the deactivation effect of individual coke precursors.

This enumeration of tools for coke investigation is certainly not exhaustive, and new methods, such as neutron scattering and matrix-assisted laser desorption/ionization time-of-flight mass spectrometry (MALDI-TOF MS), have already been described or are likely to appear in the near future. In the following paragraphs the above-mentioned techniques will be briefly presented, and their contributions to the understanding of the phenomenon of coking will be illustrated by suitable examples and significant results. In conclusion, an attempt will be made to summarize some aspects of coke formation on acid zeolites which seem to be, at the present state of the art, of general validity.

2 Methods of Coke Investigation and Results

2.1 Determination of Coke Content and H/C Ratio

Elementary analysis is the basis for the elucidation of the structure of an unknown chemical substance. Likewise, the determination of the H/C ratio of carbonaceous deposits on catalyst surfaces should be the first step in characterizing coke residues, even though the composition of coke may be complex and it usually comprises constituents with different hydrogen content. Moreover, the elementary analysis of coke on zeolite should include the determination of nitrogen and sulfur content. Especially in industrial studies, small contents of organic bases and metal atoms in the feed (even in the sub-ppm range) may completely change the deactivation behavior found with pure substances.

The determination of the H/C ratio via oxidation and analysis of the formed H_2O and CO_2 by, e.g. gas chromatography or absorption into suitable materials is, in the case of coked zeolites, by no means trivial. The main source of error results from H_2O adsorption on the zeolite from the ambient atmosphere. But also desorption of low molecular or partially oxidized carbonaceous deposits and the possible formation of CO may corrupt the findings.

There are four unknown masses, viz. the masses of the right side of the equation

$$m_{\text{coked zeolite}} = m_{\text{zeol.}} + m_{\text{H}_2\text{O,zeol.}} + m_{\text{C}} + m_{\text{H}} \,, \tag{1}$$

where $m_{\text{zeol.}}$ is the weight of the zeolitic material and $m_{\text{H}_2\text{O,zeol.}}$ the weight of structural and adsorbed water, m_C and m_H the masses of carbon and hydrogen, respectively, of the coke on the zeolite sample (see also below). Thus, one needs four independent equations for a definite solution to determine the H/C ratio. After combustion, m_C can easily be obtained from the amount of CO_2 formed:

$$m_{\text{C}} = \frac{12}{44} m_{\text{CO}_2} \,. \tag{2}$$

The weight of released water is composed of the weight of structural and adsorbed water and increased by the amount of water which corresponds to the hydrogen contained in the coke:

$$m_{\text{H}_2\text{O}} = m_{\text{H}_2\text{O,zeol.}} + \frac{18}{2} m_{\text{H}} \,. \tag{3}$$

To determine m_H, Entermann and van Leuven [17] monitored the oxygen consumption in a recirculation apparatus during coke combustion.

Weitkamp et al. [18] developed another procedure for the determination of the H/C ratio. This method includes the following steps (Scheme 1):

(i) The coked zeolite is weighed, and the measured weight, M_1, contains the mass of the zeolite matrix, $m_{\text{zeol.}}$, the sum of the structural and adsorbed water, $m_{\text{H}_2\text{O,zeol.}}$ and the masses of hydrogen, m_H, and carbon, m_C (cf. Eq. 1).

(ii) The coked zeolite is subject to combustion, and the increase in weight of the absorption tubes (filled with sodium asbestos) for CO_2 and (filled with magnesium perchlorate) for H_2O is measured. This provides m_{CO_2} and therefrom m_C (cf. Eq. 2) as well as $m_{\text{H}_2\text{O}}$ (cf. Eq. 3).

(iii) Finally, the weight of the dry zeolite material, $m_{\text{zeol.}}$, is determined via "back weighing", generally after conversion into the amorphous state. This step is, in fact, the crucial one. However, if readsorption of water is carefully avoided, this step provides $m_{\text{zeol.}} = M_2$.

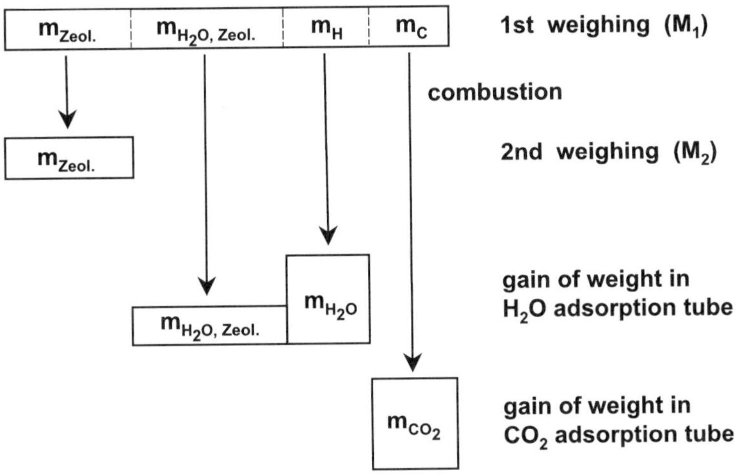

Scheme 1 Determination of the H/C ratio in coked zeolite catalysts (after [18])

Substitution of m_C and $m_{H_2O,zeol.}$ into Eq. 1 via Eqs. 2 and 3, respectively, gives

$$m_{\text{coked zeolite}} = m_{\text{zeol.}} + m_{H_2O} - 9m_H + \frac{12}{44}m_C + m_H \,, \tag{4}$$

$$m_H = \frac{1}{8}\left[m_{\text{zeol.}} - m_{\text{coked zeolite}} + m_{H_2O} + \frac{12}{44}m_{CO_2}\right], \tag{5}$$

and finally,

$$H/C = \frac{m_H}{\frac{1}{12}m_C} = \frac{12}{8}\left[m_{\text{zeol.}} - m_{\text{coked zeolite}} + m_{H_2O} + \frac{12}{44}m_{CO_2}\right] \Big/ \frac{12}{44}m_{CO_2} \,. \tag{6}$$

On the right side of Eq. 6 there are exclusively measurable data. The above procedure to obtain the H/C ratio of coke deposited on zeolites via "back weighing" of the burned sample was checked by the authors with several test hydrocarbons. It seems generally applicable for the determination of the H/C ratio of coke on zeolites and related materials.

More sophisticated experimental set-ups for the determination of the H/C ratio are combinations of thermogravimetric analysis (TGA) with an on-line detection of gas-phase products by GC, MS or IR. Darmstadt et al. [19] described a TGA/GC method with an intermediate trapping at 100 K. This procedure allows both the elucidation of coke combustion as well as the thermal treatment of spent samples under different carrier gases. When oxidation of carbonaceous deposits or desorption from the catalyst is completed, the trap is rapidly heated to 500 K and the trapped compounds purged into the GC for analysis. In this way, the H/C ratio of coke deposits was measured as a function of reaction temperature and time on stream. Table 1 demonstrates the weight gain of a H-ZSM-5 catalyst due to coke deposition during the reaction of ethylene at reaction temperatures increasing from 425 K to 675 K. Both the amount of coke deposited after a given time (e.g., 6 h) and the H/C ratio of carbonaceous residues continuously decreases with increasing reaction temperature.

Similar results concerning a decrease of the H/C ratio of coke as a function of reaction temperature were obtained during the propane transformation over ultrastable H-Y catalysts (H-USY) [20] and methanol conversion over H-ZSM-5, RE-Y, and H-MOR [21].

By passing a feed stream over a microbalance, the set-up can be used as a microreactor and the amount of deposited carbonaceous materials is measured continuously via the weight gain as a function of time [22–25]. Instead of conventional dynamic microbalances, modern equipment uses an oscillating tapered quartz element (TEOM: Tapered Element Oscillating Microbalance) for monitoring the change in mass via change in vibrational frequency [26–28]. In combination with on-line gas chromatography, the catalyst activity and selectivity can be determined. Gomm et al. [27] studied the conversion of isopropanol on H-Y, H-ZSM-5, H-ZSM-22, and H-MOR in the temperature range of 373–473 K and correlated catalyst deactivation by coking with mass changes of the catalyst. The catalytic experiments over zeolites with different pore systems revealed that deactivation does not necessarily occur with concomitant mass increase due to depositing carbonaceous materials. Among the zeolites, H-MOR having a unidimensional, 12-membered-ring channel system was the subject of pronounced deactivation, although the coke content was relatively constant at about 7 wt %. After raising the temperature by 50 K on H-Y, an initial decrease of about 10% of the mass of carbonaceous deposits is due to desorption and/or decomposition of coke species, whereas on H-ZSM-5, H-ZSM-22, and H-MOR such an effect of mass

Table 1 Amount and H/C ratio of carbonaceous deposits on H-ZSM-5 upon reaction of ethylene as a function of reaction temperature after 6 hours on stream (adapted from [19])

Reaction temperature [K]	Gain of weight by coke deposition [wt %]	H/C ratio
425	6.96	1.82
475	5.78	1.50
525	4.71	1.38
575	2.25	1.35
625	1.12	1.12
675	0.62	1.02

changes by temperature increase was observed to a minor degree. It is worth noting that such a rise of the reactor temperature by some tens of degrees over a short period of time in combination with the feed being cut off and flushing by recycle gases may partially restore the activity of spent catalysts without burning the carbonaceous deposits (vide infra, Sects. 3.4 and 3.5).

Table 1 also reveals an interesting effect of the reaction temperature on the amount of carbonaceous deposits. The yield of coke decreases with temperature in the range 400–600 K and reaches a relatively broad minimum between 600 and 700 K. As shown in [19], it increases at even higher temperatures. An explanation of this dependence is based on thermal conversion of hydrogen-rich coke species [29]. At lower temperatures these deposits are only accumulated within and/or on the outer surface of the zeolite crystallites. With increasing temperature isomerization and cracking of carbonaceous deposits occur, from which smaller coke constituents can leave the zeolitic pore system. The increase of coke content at temperatures higher than 700 K is caused by carbonaceous residues highly hydrogen-deficient in nature.

As shown in Fig. 1 the H/C ratio decreases not only with increasing reaction temperature but also with time on stream [9]. The zeolite material used was an H-ZSM-5 zeolite with an overall Si/Al ratio of 22.6 and an average diameter of the small crystallites of about 0.4 μm. The decline of the H/C ratio with time on stream, the so-called "aging" of coke, was seen by several authors [9, 21, 30, 31]. Whereas the total amount of coke may reach a certain plateau with respect to time (Fig. 2), its nature continuously undergoes changes. Here, an H-ZSM-5 catalyst was employed, which possessed an over-

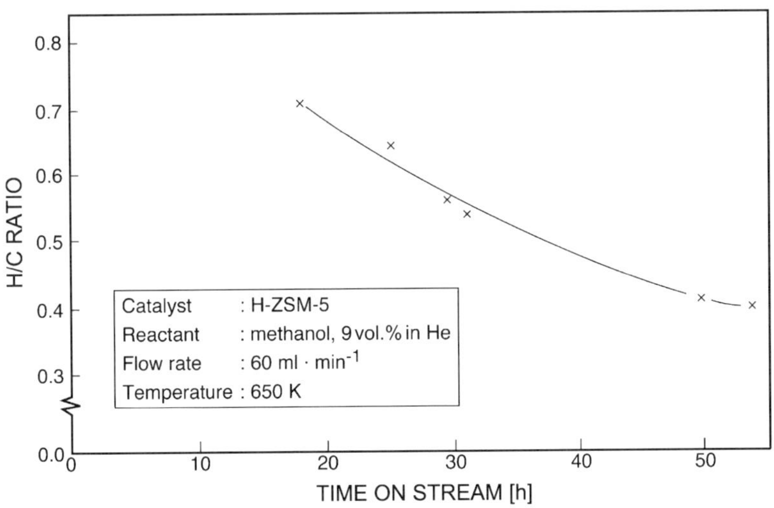

Fig. 1 H/C ratio of coke formed via reaction of methanol over H-ZSM-5 as a function of time on stream (after [9])

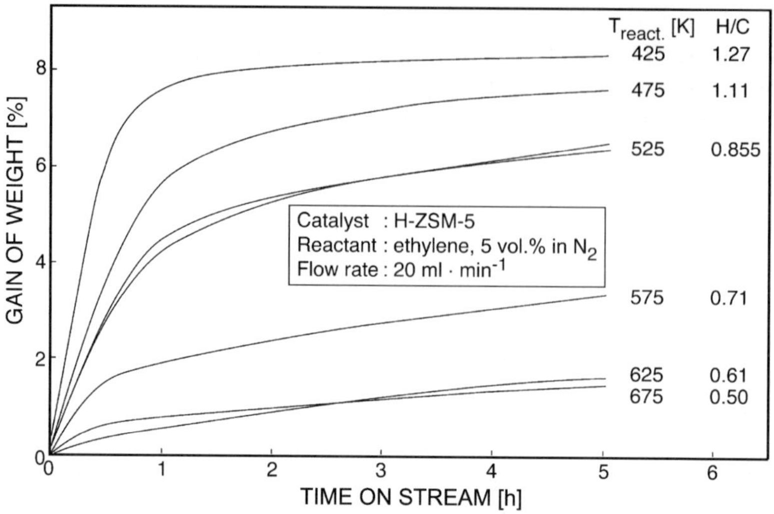

Fig. 2 TGA measurements of coke formation upon reaction of ethylene over H-ZSM-5 as a function of temperature (after [9])

all Si/Al ratio of 33.5 and consisted of well-shaped large crystallites with sizes of $8.8 \times 5.2 \times 3.2$ μm [9]. The decrease of the H/C ratio with time on stream supports the view that carbonaceous deposits are reaction partners within a complex reaction network having a low, but measurable conversion rate. Thus, the coke becomes more hydrogen deficient through hydrogen transfer reactions. Groten and Wojciechowski [31] studied the conversion of 1-hexene on H-USY zeolite at 582 K and found an initial H/C ratio of 1.85. Detailed stoichiometry of the initial reaction mechanism revealed that an average initial "coke molecule" can be represented by the formula $C_{5.1}H_{9.4}$ which is very close to that of the reactant [32, 33]. In the course of the reaction, the H/C ratio decreased to values of about 1.0. A similar decline of the H/C ratio from about 1.7 to 1.0 was observed during methanol conversion on H-ZSM-5 at 648 K, whereas on H-MOR and RE-Y the H/C ratio decreased from 1.0 to about 0.5 with increasing time on stream [21]. At much higher temperatures of about 953 K, Martin and Parlitz [30] observed a decrease from 0.34 to 0.24 during cracking of a methanol/*n*-butane mixture.

Carbonaceous materials formed on different zeolites under identical conditions are different in nature. Liu et al. [34] found an H/C ratio of 1.4 for H-ZSM-5, 1.1 for H-T, and 0.6 for H-Y during methanol conversion at 773 K. Because studies with different zeolites include a change in both acidity (number and strength) and pore structure, it is rather difficult to elucidate the pure influence of the number of acidic sites (H-ZSM-5 < H-Y < H-T) on the hydrogen content of coke (vide infra, Sect. 3.1).

The above-mentioned results indicate the wide range of H/C ratios found in coking studies and the need for further information on carbonaceous residues. Nevertheless, the temperature dependence of the content and the H/C ratio of coke denote different kinds of coke species:

(i) deposition of hydrogen-rich coke species or coke-type I (often called "low-temperature coke", "soft coke" or "white coke") with H/C ratio > 1.0 which are mainly formed at lower reaction temperatures (< 500 K) and
(ii) formation of hydrogen-deficient residues or coke-type II (so-called "high-temperature coke", "hard coke" or "black coke") with H/C ratio < 0.8 typically formed at higher temperatures (> 550 K).

Finally, Fig. 2 shows that the rate of coke deposition can be particularly high during the initial period of reactions followed by a period of much slower coking. The data suggest that the lower the reaction temperature the higher the initial coking rate and the shorter the time on stream after which the coke amount levels off. This phenomenon may be caused by sites of different catalytic properties present on the zeolite. It seems probable that very strong acidic sites or sites of enhanced activity, created by mild steaming, are more rapidly deactivated than normal acid sites [35]. At very low temperatures, particularly these sites catalyze oligomerization, cyclization and hydrogen transfer steps which are known to play an important role in coke formation on zeolites. Neuber et al. [32] found a rapid deactivation in a first stage of the reaction during cracking of 1-methylnaphthalene over H-Y and ZSM-20 at 573 K. However, Schulz et al. [21] observed a high initial deactivation upon methanol conversion over RE-Y only for a high reaction temperature (753 K).

2.2 Temperature-Programmed Oxidation

The knowledge of the coke content and H/C ratio is the starting point for the characterization of carbonaceous residues. Unfortunately, due to the total combustion of coke at temperatures > 1000 K both features are integral parameters, and the presence of two quite different coke species may yield the same overall H/C ratio as a uniform deposit. Therefore, the conditions for regeneration to eliminate the carbonaceous deposits should also require special attention. For example, Barbier [36], who studied the conversion of cyclopentane/cyclopentene mixtures on a bifunctional Pt/Al_2O_3 catalyst, observed two oxidation peaks, a low-temperature one at about 600 K and a high-temperature peak at about 800 K. Thus, temperature-programmed oxidation allowed discriminating between coke laid down on the metal and coke located on the support. Analogous investigations have been successfully applied to coke characterization on the H-forms of several zeolites.

It has been shown [37–39] that the oxidation of coke begins at temperatures of about 530 K with hydrogen removal from hydrogen-rich coke species, yielding water and oxygenated intermediates which subsequently decompose into CO and CO_2 (Fig. 3). Various oxygenated species such as ketonic, aldehydic, anhydride and phenolic compounds, have been identified in partially oxidized samples of coked H-Y [37] and H-ZSM-5 [40] zeolites. Because of the large variety of the components, the exact composition of partially oxidized coke is difficult to obtain. For the identification of gaseous oxidized products, infrared or mass spectroscopy is typically combined with thermogravimetric measurements, i.e., TG-MS [39] and TG-FTIR [41]. Pyrene oxidation in Y zeolites was suggested as a model reaction for coke oxidation [42]. To improve both sensitivity and resolution of TPO, all products generated during coke combustion were completely converted to methane on a nickel catalyst [43, 44]. The methane was then continuously analyzed by a flame ionization detector.

The combustion temperature depended obviously on the coke content: the higher the coke content, the more difficult the removal through oxidative treatment. This was related to a change in the location of coke (external vs. internal deposits) and/or in the coke composition (i.e., H/C ratio) with increasing coke content. In detail, hydrogen-poor coke components are burned at a higher temperature compared to hydrogen-rich residues. TPO profiles for spent H-FER coked during *n*-butene isomerization show two well-defined peaks (Fig. 4), one between 540 K and 600 K and the other at higher temperatures of about 900 K [44, 46]. The high-temperature peak mainly stems from

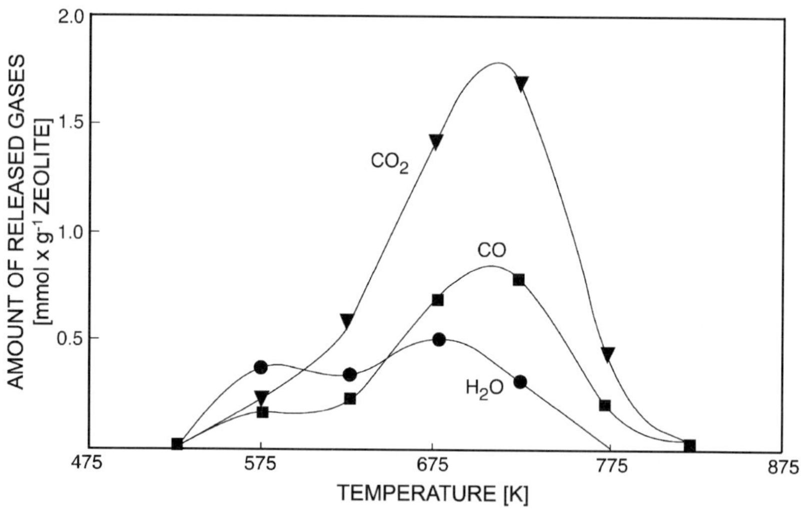

Fig. 3 Evolution of water, carbon monoxide and carbon dioxide as a function of the oxidation temperature applied on an H-Y zeolite sample with 8.5 wt % coke (after [45])

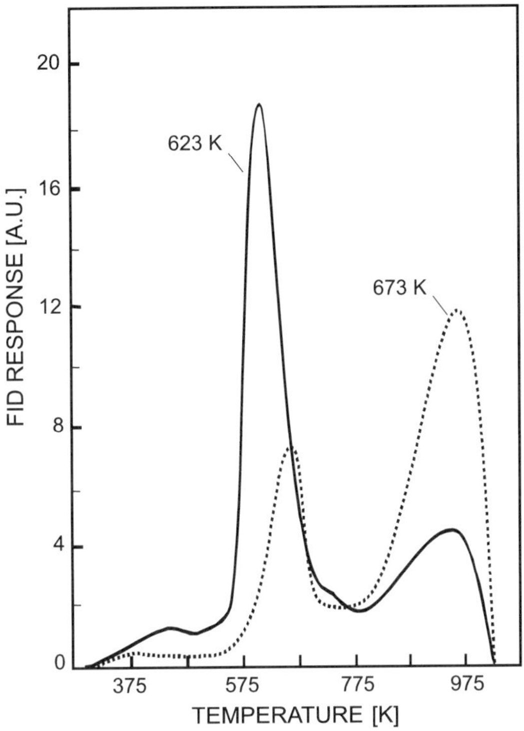

Fig. 4 TPO profiles of H-FER coked under a 1-butene/nitrogen flow during 180 min at 623 K and 673 K (after [44]); these reaction temperatures are indicated in the figure

aromatic coke formed during the reaction and/or during TPO analysis when heating is above the reaction temperature. As the reaction temperature increased from 623 K to 673 K, the size of the high-temperature peak became larger than the low-temperature one. The carbon content after 3 h on stream was 7.9 wt % and 8.4 wt % at temperatures of 623 K and 673 K, respectively. Despite a similar amount of carbonaceous deposits, the chemical nature of coke was significantly determined by the reaction temperature. This was also found on H-FER after *n*-butene isomerization by Xu et al. [47]. The H/C ratio of carbonaceous deposits, determined by measuring the total amount of water and CO_2 released during TPO, varied from as high as 4.1 via 1.4 to as low as 0.0 at 573 K, 673 K, and 873 K, respectively. Especially the finding at 573 K represents a maximum value of the H/C ratio (methane) and points, thereby, to uncertainties in the determination of H/C ratios of coke by TPO.

Figure 5 shows the thermogravimetric TPO profiles for H-ZSM-5 samples used in skeletal isomerization of *n*-butene at 548 K after 20 min and 100 h on stream (Bauer F, 2005, personal communication). The effect of the reaction temperature on the nature of coke is similar to that of the time of coking. Obviously, the rapid initial weight loss up to 450 K is characteristic of the release

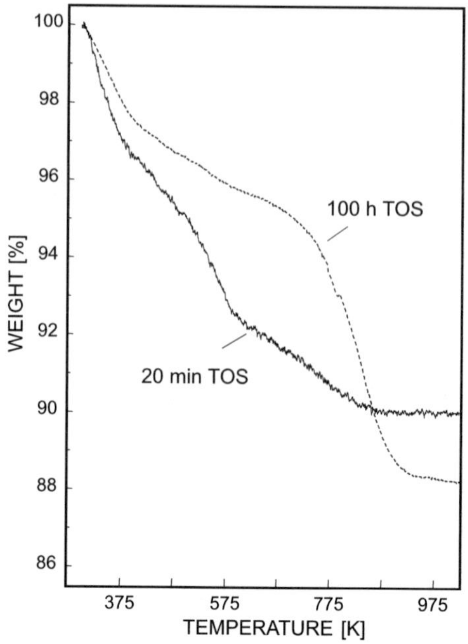

Fig. 5 TPO profiles of a silylated H-ZSM-5 coked under a 1-butene/nitrogen flow at 548 K after 20 min and 100 h TOS (after Bauer F, 2005, unpublished results)

of physisorbed water (vide supra, Sect. 2.1). The amount of low-temperature coke burning between 520 K and 570 K is significantly larger at shorter reaction times, i.e., with increasing time on stream hydrogen-rich coke species formed during the initial stage of a process are converted to more aromatic carbonaceous residues. This "aging" of coke indicated by a decrease of the H/C ratio with TOS was reported in several reactions [21, 46]. In addition, Fig. 5 visualizes a typical behavior of zeolite catalysts: the rate of coke formation is particularly high during the initial period of a process. In summary, temperature-programmed combustion of coke has been, inter alia, successfully applied to coke investigation on the H-forms of several zeolites and proved to be a suitable tool for discrimination between different types of coke:

(i) hydrogen-rich carbonaceous residues (coke-type I, low-temperature coke) which are burned at temperatures around 600 K, and;
(ii) highly condensed, polyaromatic residues (coke-type II, high-temperature coke) the oxidation of which starts at temperatures of about 700 K.

Similar to the effect of the aging of coke, the combustion of carbonaceous residues starts at higher temperatures after thermal treatment of spent catalysts under pyrolytic [39] or hydrogenolytic [48, 49] conditions. Both procedures result in catalyst reactivation due to splitting of some low-temperature coke components indicated by reduced coke content and a shift of the TPO profile to higher temperatures. An example is reported in [41]: after treatment of a spent H-ZSM-5 catalyst (deactivated during MTG reaction at 693 K) with propane at 693 K for 24 h, carbon dioxide evolution peaked at about 890 K compared with 830 K for the untreated sample (Fig. 6).

Furthermore, the pore structure of zeolites plays a significant role in the oxidation of coke. As oxidation of coke was more rapid for H-Y and H-MOR than for H-ZSM-5, it has been concluded that coke oxidation is, like coking, a shape-selective process [37]. The effect of the zeolite structure prevails over the role of the nature of the coke. In spite of their less condensed nature, carbonaceous deposits on H-ZSM-5 are oxidized at higher temperatures than those on H-Y. Higher burning temperatures may lead to different oxidation products. Thus, the zeolite structure is reflected not only in the oxidation temperature but also in the composition of the oxidation products, mainly in the CO/CO_2 ratio [39]. Even on the same zeolite type, regeneration requirements can significantly depend on coke location and deactivation conditions [50]. For example, Andy et al. [51] reported that carbonaceous deposits formed on H-FER during *n*-butene isomerization can completely be removed by burning at 773 K, while Xu et al. [47] needed 873 K for an oxidative treatment.

From the heterogeneous burning pattern of H-ZSM-5 coked upon alkylation of toluene by methanol and from sorption measurements of CO_2, benzene and *p*-diisopropylbenzene sorbed into partially regenerated samples, it was claimed that coke burning first occurred preferentially in the straight pores. Subsequently, coke was oxidized in the zig-zag pores of ZSM-5 [50]. Such subtle discrimination of the sites of coke formation from TPO data has

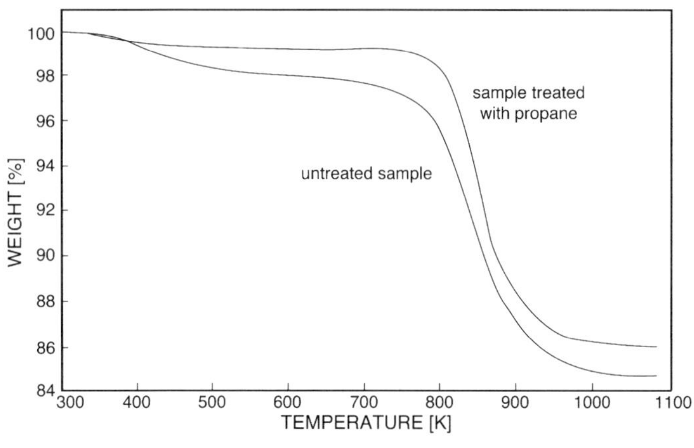

Fig. 6 TPO profile of a coked H-ZSM-5 sample before and after treatment with propane at 693 K (after [48])

to presume that the burning reaction occurs without any resistance to inter- and intracrystalline mass transfer.

While extra-framework aluminum species have no determining effect on the burning rate, coke is more easily oxidized with zeolites having a high density of acid sites [45]. According to isotopic studies with $^{16}O_2$ and $^{18}O_2$ on H-ZSM-5 [39], the activation of molecular oxygen on zeolites is apparently of minor importance at temperatures below 873 K. In order to explain the differences in coke oxidation over the different zeolites, a mechanism based on the activation of the hydrocarbon species must therefore be considered. A reaction scheme was proposed involving the formation of radical cations through reaction of molecular oxygen with protonated coke molecules [45].

For selective catalytic reduction (SCR) of NO_x by hydrocarbons on different zeolites at 673 K, TPO was applied to explore the SCR mechanism, especially the effect of carbon deposits on NO reduction [52]. When NO and O_2 were added to the hydrocarbon stream, the structure of the deposit did not appear to change but the deposition rate was strongly increased. This suggests that an SCR reaction may be involved in carbon deposition. Cu-ZSM-5 and H-MOR catalysts induced more coke formation than H-ZSM-5 and H-Y. Since N_2 was detected during TPO on Cu-ZSM-5 exposed to propene + NO + O_2, nitrogen compounds were either part of carbonaceous residues or they may adsorb on the deposit or zeolite. In general, the nitrogen peak (on H-ZSM-5 at 572 K for propene exposure and at about 673 K for toluene exposure) precedes the CO_2 peak (which appears at 723 K for propene and at 873 K for toluene). For the different zeolites, the CO_2 peaks appeared at different temperatures suggesting a difference in the nature of carbonaceous deposits. Cu-ZSM-5 and H-Y catalysts seem to be less active for combustion than H-ZSM-5 and H-MOR. On Cu-ZSM-5, coke entrapped in the channels was suggested to maintain the oxidation state of Cu at + 1 during SCR [53].

In addition to the conventional regeneration in oxygen, nitrous oxide and ozone [54] have been used to burn carbonaceous residues on zeolites. It is noteworthy that N_2O regenerates spent zeolites H-Y [55] and H-ZSM-5 [56] under milder conditions than oxygen. According to Ivanov et al. [56], the difference in activities between nitrous oxide and oxygen in the reaction with coke depends on the zeolite nature rather than on the chemical nature of coke. Moreover, in the case of nitrous oxide a complete restoration of catalytic activity for benzene hydroxylation requires the removal of only 30–35% coke compared to 60–65% using oxygen.

2.3
Infrared and Raman Spectroscopy

IR spectroscopy was one of the first techniques employed for the detailed elucidation of the mechanisms of catalyzed reactions. Information on the change of active centers, on the formation of adsorbed intermediates as well as on the building up of carbonaceous deposits during hydrocarbon reactions can be obtained. IR spectroscopy has not only the advantage of allowing the investigation of coke formation without destruction of the zeolite matrix containing the residues (vide infra, Sect. 2.9), but it can also be successfully employed for studying coke formation in situ, i.e. observing the deposition of carbonaceous materials on the working catalysts. This requires the use of a suitable IR flow reactor cell which may be connected to complementary equipment (MS, GC) for analysis of the product stream. Several types of cells and some of the experimental difficulties are described in the literature, see for instance [57–59]. Unfortunately, due to the difficulty to assign unambiguously an IR band to particular coke species the IR technique leads to limited information on the nature of carbonaceous deposits.

Compared with the traditional transmission technique, infrared spectroscopic studies using the diffuse reflectance Fourier transform infrared spectroscopy (DRIFTS) have the advantages of a simple preparation of samples and a flow-through reactor-like design of IR cells with temperature control within the catalyst bed. Nevertheless, there are only a few papers [60–62] concerning in-situ DRIFT investigation of zeolites mainly due to difficulties of quantitative analysis related with the conversion of the data by the Schuster–Kubelka–Munk relation [63, 64].

Carbonaceous deposits laid down on zeolite catalysts are investigated in the following spectra regions:

(i) 3000–3800 cm^{-1} (stretching modes of free acidic and nonacidic OH groups or those interacting with adsorbates);
(ii) 3000–3200 cm^{-1} (CH stretching modes of aromatics);
(iii) 2800–3000 cm^{-1} (CH stretching modes of paraffinic species) and;
(iv) 1300–1700 cm^{-1} (CC stretching modes of unsaturated hydrocarbons and CH bending modes of paraffinic species).

The contribution of each of the IR regions to the understanding of coking, especially as far as the dependence on the reaction temperature is regarded, will be given in the following sections.

2.3.1 Region between 3000 and 3800 cm^{-1}

For acidic zeolites, the decrease in the number of active acidic sites caused by the deposition of nonvolatile carbonaceous species is the main reason for catalyst deactivation in hydrocarbon processing. Although there are numerous reports concerning the effects of the pore structure of the zeolites, the aluminum content, crystallite size and the reaction conditions on the coking rate, the location and the nature of coke species, the influence of coke on the state and properties of zeolitic hydroxyls has not yet been sufficiently clarified.

The acidic bridging hydroxyls (Brønsted sites) of H-ZSM-5 zeolites absorb near 3610 cm^{-1} (Fig. 7). The band around 3725 cm^{-1} (usually observed at somewhat higher wave-numbers), indicates silanol groups on structural defects, amorphous impurities or the outer surface of zeolite crystallites, and a broad peak near 3500 cm^{-1} is attributed to hydrogen bonding between adjacent hydroxyl groups. Adsorption of benzene [65] and propene [66] on H-ZSM-5 at room temperature resulted in a reduction of the intensity of the band at 3610 cm^{-1} and in the formation of broad bands at 3300 cm^{-1} and 3460 cm^{-1}. Moreover, carbonaceous deposits derived from acetone decreased the intensity of the free bridging hydroxyls. Concomitantly, a weak broad band appeared between 3200–3300 cm^{-1}, i.e., in a region similar to that for benzene absorption. Analogously, upon contact with ethylene at low temperatures (300–450 K) the intensity of the band at 3605 cm^{-1} due to the acidic OH groups of H-MOR decreased [67]. Simultaneously a broad band around 3500 cm^{-1} appeared indicating a shift of a part of the 3605 cm^{-1} band to lower wavenumbers due to the formation of hydrogen-bonded adsorbate complexes. The situation changed significantly when the reaction temperature was increased above about 550 K: the bands arising from hydrogen-bonded complexes vanished, and the band indicative of acidic hydroxyls was essentially regenerated.

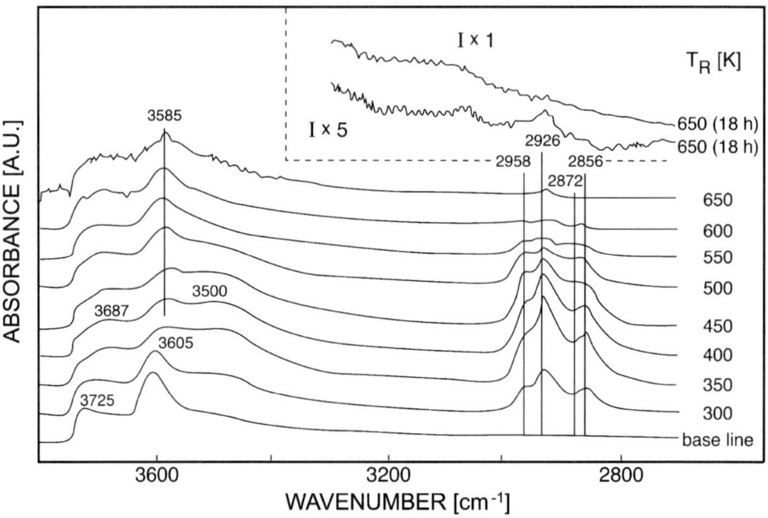

Fig. 7 IR spectra of carbonaceous deposits formed by reaction of ethylene over H-MOR at increasing temperatures (gas mixture: 5% ethylene in helium; flow rate: 16 ml min^{-1}; time on stream: 0.5 h; gas phase pumped off prior to scanning each spectrum; reactant flow renewed prior to each subsequent heating step. *Insert*: extended time on stream (18 h); *lower spectrum*: absorbance scale expanded by a factor of 5 (after [67])

Therefore, from the very beginning of the catalytic run a minor portion of acid sites is nondestructively covered by coke species via hydrogen bonding. Such reversible poisoning was claimed for coke derived from methanol conversion on H-ZSM-5 [68], where it was found that the acidic OH groups (band at 3610 cm^{-1}) were almost completely eliminated with increasing build-up of carbonaceous deposits up to about 5 wt % coverage. Subsequently, a reinforced intensity of the 3610 cm^{-1} band was observed at ca. 7.5 wt % followed by another fall at 10 wt %.

Karge and Boldingh [69] investigated the conversion of ethylbenzene on a working H-MOR catalyst. Inspection of the OH stretching region revealed that there was no substantial poisoning or consumption of active OH groups even though the activity decreased. Having confirmed this by studies with ethylene on spent catalysts the authors concluded that deactivation of the mordenite catalyst during the reaction of ethylbenzene was caused by deposition of coke blocking the active sites for the relatively large ethylbenzene molecules, whereas the smaller molecules still had access to the active centers.

Similar observations were made with H-ZSM-5 catalysts during acetone conversion at 623 K. Kubelková and Nováková [65] demonstrated that the initial steep decrease in the number of the acid OH groups is accompanied by a much slower diminution of the void volume. Below 9 wt % of carbonaceous deposits a direct poisoning of active sites can be assumed. Above this concentration the void volume becomes negligible, while an appreciable concentration of hydroxyl groups is indicated by the band at 3610 cm^{-1}. Even on the completely deactivated zeolite some acid OH groups (less than 20% of the original number) remain free but inaccessible for reactants due to pore blocking.

By contrast, the OH groups of H-Y and Pt/Ca-Y were irreversibly consumed by coke deposition upon olefin reaction [70] and during *n*-hexane conversion [71]. Okado et al. [62] tested ZSM-5 zeolites containing alkaline earth metal cations during methanol conversion at 873 K and observed for Ca-ZSM-5 much more coke resistance due to modification of the strong acid sites. The comparison of a regenerated catalyst with the fresh one revealed, however, that the band at 3605 cm^{-1} increased as a consequence of use and regeneration. Moreover, XPS data showed an increase in the Ca/Si ratio at the external crystallite surface of the regenerated sample. This implies that some of the calcium ions are released from their sites during methanol conversion.

2.3.2
Region between 2800 and 3000 cm^{-1}

Upon adsorption of olefins on acidic mordenite catalysts [72], in the region of the CH stretching modes the bands of paraffinic species developed ($\nu_{as}[CH_3] = 2958\ cm^{-1}$; $\nu_{as}[CH_2] = 2926\ cm^{-1}$; $\nu_s[CH_3] = 2872\ cm^{-1}$; $\nu_s[CH_2] = 2856\ cm^{-1}$). Therefore, the predominance of the IR bands of saturated hydrocarbons even at low reaction temperatures (about 350 K to 375 K) points to instantaneous polymerization of the olefin on contact with the zeolites (Fig. 7). However, no CH bands at wavenumbers higher than 3000 cm^{-1}, which would be indicative of olefins and/or aromatics, were observed. When the temperature was increased to 500 K, the intensities of the CH bands were generally weakened, whereby the proportion of the CH_3 bands increased. Obviously, at elevated temperatures cracking, isomerization and aromatization of the paraffinic carbonaceous residues occur to form more branched molecules and aromatics.

2.3.3
Region between 1300 and 1700 cm^{-1}

For the investigation of reactive adsorbates and, in particular, for the formation of coke, the IR range between 1300 and 1700 cm^{-1} is especially suited (Table 2). Using in-situ infrared spectroscopy, Karge et al. [67, 70] observed

Table 2 Assignment of IR bands due to formation of carbonaceous deposits (adapted from [38] and references therein)

IR band [cm^{-1}]	Assignments
1360–1390, 1420–1490	CH bending modes of paraffinic species
1440–1480, 1580–1620	carboxylate surface species
1490–1515	C = C of neutral and cationic polyene, non-condensed aromatics
1500–1530	alkyl, alkenyl and cycloolefinic carbocations
1510–1540	alkylnaphthalenes
1550–1580	allylic species
1570–1620	highly unsaturated polyenes and condensed aromatics
2850–2970	CH stretching modes of paraffinic species
3000–3200	aromatic CH stretching modes
3200–3500	hydrogen bonded surface species
3550–3680	SiOHAl Brønsted acid sites
3650	OH groups attached to non-framework Al species
3745	terminal Si – OH groups

an influence of the reaction temperature on the composition of carbonaceous deposits formed during the conversion of ethylene on H-MOR. Upon contact with the olefin at low temperatures (300–450 K), the region between 1300 and 1700 cm^{-1} exhibited CH deformation bands, viz. $\delta_{as}[CH_3] = 1485\ cm^{-1}$; $\delta_{as}[CH_2] = 1468\ cm^{-1}$; $\delta_s[CH_3] = 1382\ cm^{-1}$ and $\delta_s[CH_2] = 1359\ cm^{-1}$ (Fig. 8).

At around 450 K two bands developed near 1485 and 1510 cm^{-1}, these bands then disappeared at higher temperatures of about 550 K. As far as the assignment of bands in the 1505–1530 cm^{-1} range to surface species is concerned, a great variety of olefins and alcohols leads to the emergence of this band. The attribution to aromatic compounds may be questioned for several reasons [73]. Supported by ^{13}C MAS NMR [74] and UV-Vis measurements [75], a more accurate assignment to various carbocations (alkyl, allyl and cycloolefinic carbocations) can be given. The appearance of carbocations can occur even at lower reaction temperatures (in some cases as low as room temperature) as has been shown in the course of the isomerization of cyclopropane on H-MOR at 370 K [73]. The observed bands at 1505 cm^{-1} and 1535 cm^{-1} were ascribed to dienyl and monoenyl carbocations, respectively. Vedrine et al. [76] proposed a reaction scheme for the generation of di- and trienylic carbocations and their further transformation as precursors of aromatic compounds which are formed at higher temperatures. With respect to the IR spectroscopic detection of carbocations, deuterated acetonitrile CD_3CN has been reported to indicate secondary and tertiary carbenium ions by a band in the 2387–2377 cm^{-1} region besides its ability to probe Brøn-

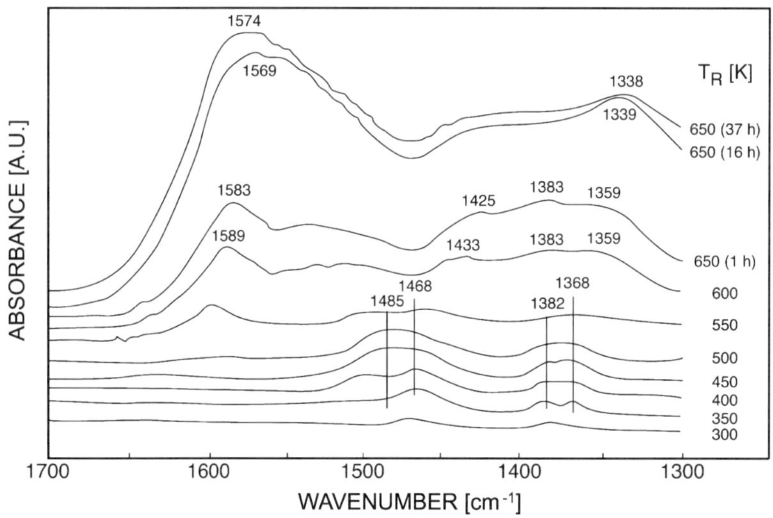

Fig. 8 IR difference spectra of carbonaceous deposits formed by reaction of ethylene over H-MOR (after [67]); conditions of the procedure as indicated in the legend of Fig. 7; time on stream 0.5 h, except for $T = 650$ K

sted and Lewis acid sites [77, 78]. This makes it a highly interesting probe molecule for studying the nature of active sites for selective isobutene formation on aged H-FER catalysts, i.e. to distinguish between Brønsted acidic sites and carbenium ions of catalytically active carbonaceous deposits (vide infra, Sect. 3.4). Van Donk et al. [79] demonstrated that on H-FER loaded with about 6 wt % coke no carbenium ions are present which excludes carbocations as active centers for butene skeletal isomerization over aged H-FER.

When the reaction temperature was raised above 550 K, the composition of the carbonaceous deposits changed significantly. The CH bands were generally weakened, indicating substantial cracking and branching of the carbonaceous residues on the surface. Finally, at reaction temperatures of about 550 K and/or extended times on stream, bands around 1600 cm^{-1} and 1540 cm^{-1} were observed. While the 1600 cm^{-1} band ("coke" band) can be ascribed to polyalkenes and/or aromatic species, the band at 1540 cm^{-1} has been assigned to alkylnaphthalenes. However, it may be also typical of CH stretching modes of polyphenylene structures [80]. Bands between 1500 cm^{-1} and 1540 cm^{-1} were also found by Fetting et al. [81], Rozwadowski et al. [82], and Schnabel et al. [83] under similar conditions. Aromatics represented by the band near 1500 cm^{-1} and formed during the conversion of methanol on H-ZSM-5 at 540 K can be removed from the zeolite surface by flushing the zeolite with nitrogen at 670 K [83].

At temperatures higher than 650 K, both bands at around 1600 and 1540 cm^{-1} collapse into one broad band with its center at about 1580 cm^{-1}. The assignment of the so-called "coke" band reflects the rather complex nature of coke. Eberly [84] ascribed it to highly unsaturated carbonaceous residues, most likely polyolefins. Eisenbach and Gallei [85] attributed the coke band to graphite-like deposits. In this context it is worth noting that the position of the band in all in-situ IR experiments of coke formation shifted from around 1600 cm^{-1} to lower wavenumbers when the time on stream was extended. This may indicate that the structure of the main components of the coke change with time on stream, corresponding to the aging of the coke (compare Sects. 2.1, 2.2, and 2.5). The literature is, however, contradictory concerning the formation of more or less unsaturated "graphitic" compounds at longer times on stream [84, 86, 87].

Remarkably, the bands in the region of 1310–1350 cm^{-1} and 1360–1390 cm^{-1} became prominent at higher reaction temperatures, the latter are usually attributed to $\delta_S[CH_3]$ of isopropyl or tertiary butyl groups. Such groups may be attached to the aromatic coke species and are probably removed during the thermal treatment of coke.

The characterization of the adsorbed species and deposits formed over H-FER during skeletal isomerization of *n*-butene brought about new aspects concerning the role of carbonaceous residues during catalyzed reactions. The proposal that the carbonaceous deposits could act as catalytic entities [14] required spectroscopic evidence. Adsorbed butenes, butene oligomers, polyenes, unsaturated carbocationic chains, and various aromatic species were reported [44, 51, 88–90]. The in-situ IR spectra [90] obtained over H-FER in the presence of *n*-butene at various temperatures and after 5 min on stream are shown in Fig. 9. The major bands observed at 423 K were at 1380, 1462, 1576, and 1641 cm^{-1}. The broad band around 1641 cm^{-1} can be assigned to C = C stretching modes of weakly adsorbed butenes, whereas the strong band at 1576 cm^{-1} is associated with C_4 allylic species [91]. The bands around 1462 and 1380 cm^{-1} can be assigned to typical CH_3 and CH_2 deformation modes of butenes and/or butene oligomers. At 523 K the band at 1576 cm^{-1} disappeared and a new band was formed at about 1500 cm^{-1} which is assigned to neutral and/or cationic polyenes resulting from butene oligomerization. At higher temperatures, the 1500 cm^{-1} band shifted to 1505–1510 cm^{-1} indicating further dehydrogenation. Therefore, Pazé et al. [91] and Wichterlova et al. [92] proposed that the carbonaceous species associated with the 1510 cm^{-1} band changed the nature to mono- and bicyclic aromatics. At temperatures *higher than 700 K and with longer times on stream* [90], new bands at 1588 cm^{-1} and 1540 cm^{-1} appeared (not yet visible in Fig. 9, i.e. after 5 min on stream), but the band at 1515 cm^{-1} was still

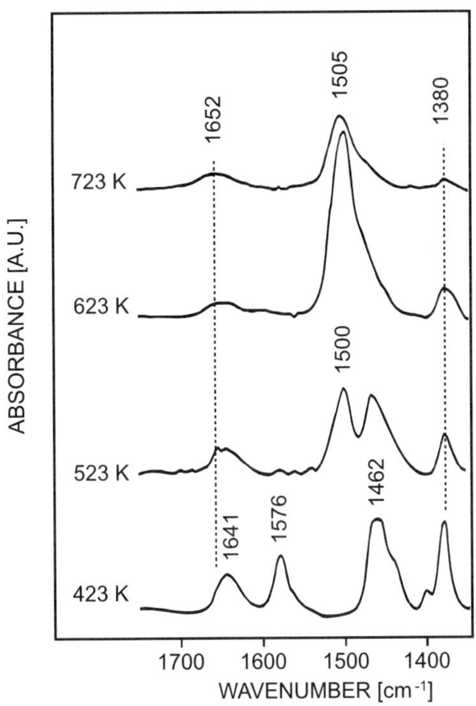

Fig. 9 In-situ IR spectra obtained from H-FER in the presence of *n*-butene at various temperatures and after 5 min on stream (after [90])

observed, i.e., all bands can be assigned to condensed aromatic coke species characteristic of zeolites deactivated under severe conditions [93].

In infrared spectroscopy, symmetrical vibrations tend to be weak. On the other hand, symmetric modes of weakly polar bonds such as C = C appear to be intense in the Raman spectrum. Therefore, Raman spectroscopy might be predestined for the characterization of polyolefinic coke. Two extra points of Raman spectroscopy for in-situ studies are that the spectra of adsorbed species interfere weakly with the signals from gas-phase molecules and that typical supports such as silica and alumina are weak Raman scatterers. However, surface fluorescence of the zeolite sample may seriously limit the detectability of the signals, in particular when the catalyst is contaminated with carbonaceous deposits. Recently, it was found that by using an ultraviolet wavelength below 260 nm the intensity of the Raman signals is enhanced and the background fluorescence is also avoided. Hence, UV Raman spectroscopy facilitates the identification of coke species formed during methanol conversion on H-ZSM-5, H-USY, and SAPO-34 [94–96].

Figure 10 shows the UV Raman spectra of the reaction of methanol on H-ZSM-5 at different temperatures. At 473 K, the 805 cm^{-1} band originating from the zeolite is only slightly attenuated, which is consistent with a small accumulation of carbonaceous residues at this temperature. To distinguish which of the observed bands are due to CH deformations and which are due to carbon-carbon stretching vibrations, experiments with deuterated methanol were also conducted. No significant isotope shift was observed for the bands between 1300 and 1700 cm^{-1}, indicating that all bands were associated with CC or ring stretching vibrations (Table 3). At 543 K, the 1550 cm^{-1} band intensity decreased and an additional band was observed at 1483 cm^{-1}, which was no longer detectable at the higher temperature of 733 K. While a plausible assignment for the band at 1545–1550 cm^{-1} refers to the C = C double bond stretching vibration of conjugated olefins or polyenes having six to eight double bonds [97], the behavior of the 1483 cm^{-1} band related to the reaction temperature is not consistent with an assignment to a polyene. Because the five-membered aromatic ring structure of cyclopentadiene has a strong Raman band in this spectral region (at about 1500 cm^{-1}) the 1483 cm^{-1} band, observed on H-ZSM-5 only in the temperature range between 450 K and about 600 K, is assigned to the in-phase stretch of the conjugated C = C bonds in an alkylated cyclopentadienyl species [97]. It is interesting that the cyclopentenyl cation, which is very prominent in the NMR measurements reported by several authors [98–100], was not detected in the Raman spectra during methanol conversion on H-ZSM-5 (Si/Al=90) at 543 K. Recently, the dimethylcyclopentenyl cation has been convincingly shown in the Raman spectrum of H-ZSM-5 (Si/Al = 14) after ethylene conversion at 623 K on the basis of prominent peaks at 1185, 1375, and 1440 cm^{-1} [101].

The Raman spectra of carbonaceous residues formed during methanol conversion on H-ZSM-5 exhibited an intensity pattern in which the intensity

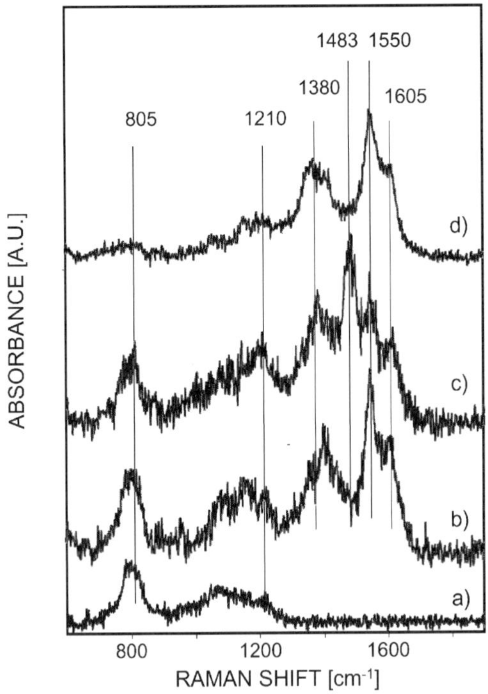

Fig. 10 UV Raman spectra of the reaction of methanol on H-ZSM-5. (a) Calcined H-ZSM-5 prior to reaction. (b) After dosing 720 molecules of CH_3OH per unit cell at 430 K. (c) After dosing 30 molecules of CH_3OH per unit cell at 543 K. (d) After dosing 1.5 molecules of CH_3OH per unit cell at 633 K (after [97])

Table 3 Assignment of Raman bands due to formation of carbonaceous residues (adapted from [96, 97])

Raman shift [cm^{-1}]	Raman band assignments
1200–1210	CC stretches of polyaromatic species
1360–1410	Ring stretches of polyaromatic species
1390–1470	CH deformation modes of paraffinic species
1483	In-phase C = C stretch of cyclopentadienyl species
1545–1550	C = C stretches of conjugated olefins
1600–1635	Polyolefinic and/or polyaromatic species

in the spectral region from 1300 to 1450 cm^{-1} is nearly equal to that in the region 1600 to 1650 cm^{-1} (Fig. 10). By contrast, the intensities are significantly different for the polyaromatic compounds with a two-dimensional, sheet-like topology such as pyrene and coronene. On H-ZSM-5, Chua and Stair [97] proposed chain-like topologies for the aromatic carbonaceous residues such as phenanthrene, anthracene, and pentacene. Such a chain-like structure would be expected for coke species deposited in the H-ZSM-5 pores, where the pore diameter is too small to accommodate the sheet-like topology. For H-USY, the polyolefinic and aromatic coke species gradually aggregate with increasing temperature into pregraphite species indicated by a band near the characteristic vibration of graphite at 1575 cm^{-1} [94].

The in-situ IR and Raman results described above suggest discrimination between at least two kinds of coke, viz.

(i) coke-type I constituted mainly of paraffinic (bands at 1380 cm^{-1} and 1460 cm^{-1}) and, to a lesser extent, aromatic, olefinic or polyolefinic species (bands around 1500 cm^{-1} and 1600 cm^{-1}) and
(ii) coke-type II which consists of polyalkenes, but predominantly of alkyl aromatics and polyaromatics (intense band below 1600 cm^{-1}).

The intensity of the coke band around 1600 cm^{-1} proved to be a suitable measure for the amount of coke deposited, even though the nature of the coke might be rather complex. A plot of the integrated absorbance of this band vs. the amount of coke deposited, measured for instance by a balance, usually provides straight lines [81, 90]. A slow change in the slope of such plots is observed at longer time on stream, which indicates a marked alteration in the nature of the coke due to aging.

The band near 1580 cm^{-1} (vide supra) is usually attributed to carbon-carbon stretching in hydrogen-deficient ring structures. Figure 11 shows bands at 1580 and 1460 cm^{-1} in the spectrum of coke produced by the exposure of alumina or Pt/alumina to acetylene at 523 K [102]. The bands grew at the same rate, as would be expected for a pair of bands from a single species. Both of the bands showed a shift of about 20 cm^{-1} when oxygen-16 in alumina was replaced by oxygen-18 [103]. This shift proved that these bands originate from vibrations where oxygen is involved, e.g., in the case of carboxylates. The production of the carboxylate bands reached a maximum at about 2 wt % coke, which suggests that these species are surface-bound. Above this value the total coke level increased, but the carboxylate band intensity did not. A calibration, based on comparison of bands with those obtained by addition of acetic acid, indicated that 15% of the total carbon was present as carboxylate.

The existence of carboxylate or acetate-like complexes was also assumed on dealuminated H-Y [86] and on Pt/Al_2O_3 [104]. Bands observed at 1450 and 1580 cm^{-1} on Pt/Al_2O_3 during cyclohexane dehydrogenation at 683 K and *n*-heptane dehydrocyclization at 748 K, remained after partial oxidation of the coked sample at 653 K [104]. This result was explained by the location of coke deposits on Al_2O_3 and not on Pt sites. Carbonaceous residues from acetone conversion on H-Y containing extra-framework aluminum exhibited a band at 1480 cm^{-1} which is absent if the zeolite does not contain non-framework Al species. Therefore, Novakova et al. [86] concluded that

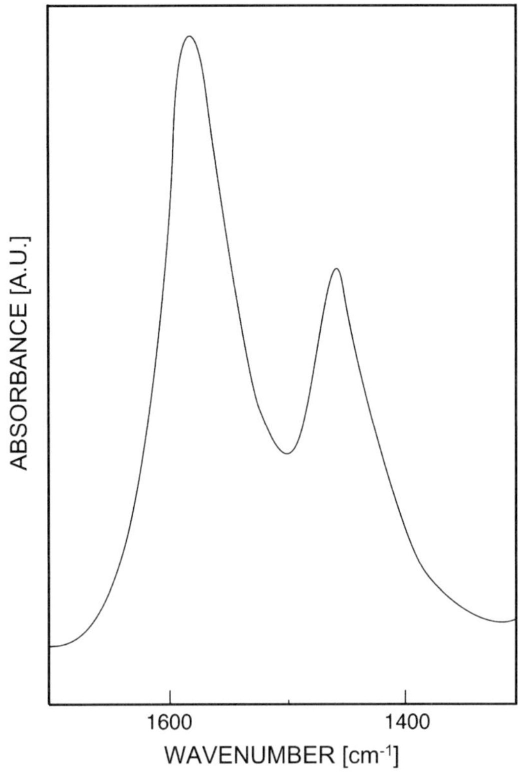

Fig. 11 Carboxylate IR bands on alumina (after [102])

dealuminated H-Y could contain a carboxylate-like complex bound to non-framework Al. In contrast to H-Y zeolites, H-ZSM-5 samples have a distinguishable band at 1645 cm^{-1} probably corresponding to residues of the coke precursors firmly bound to the zeolitic oxygen.

On the other hand, the observation of carboxylate species is not confirmed in infrared studies of coke on several zeolites [32, 70, 105]. For example, Bauer et al. [106] observed bands at 1585, 1455 and 1380 cm^{-1} of coke on H-ZSM-5 formed upon the methanol conversion at 693 K (Fig. 12). The intense band near 1585 cm^{-1} has been ascribed, in agreement with the literature, to the carbon-carbon stretching vibration in aromatic rings (vide supra). The coke bands with overlapping maxima at 1440–1455 cm^{-1} and 1380 cm^{-1} are of moderate intensity and attributed to carbon-hydrogen deformation vibrations. Blackmond and Goodwin [33] observed a small, hardly detectable band near 1460 cm^{-1} after exposure of H-Y zeolite to hexene at 573 K, contradictory to Rozwadowski et al. [82], who found a very intense band at 1455–1460 cm^{-1} during methanol conversion on H-Y at reaction temperatures between 523 K and 623 K. The quite different band intensity at

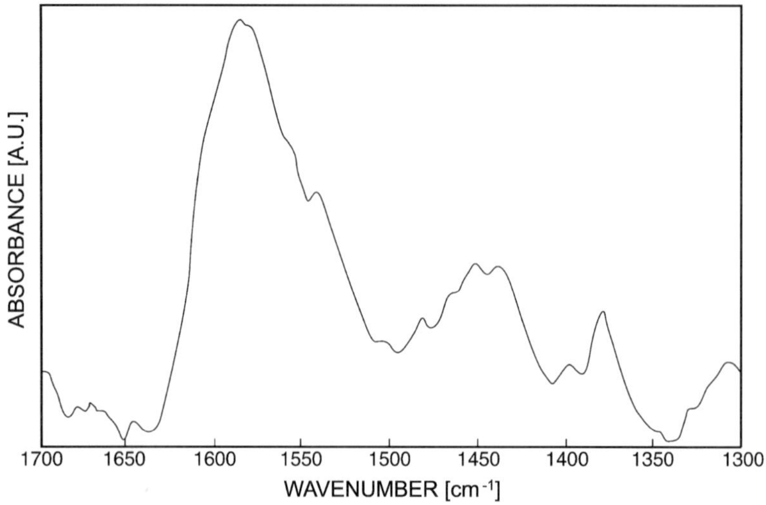

Fig. 12 IR spectra of carbonaceous deposits upon methanol conversion on H-ZSM-5 (after [106])

1460 cm^{-1} may be due to differences in the nature of coke, e.g., the H/C ratio. This conclusion is confirmed by the results of Mirth and Lercher [107], who observed, during the methylation of toluene with methanol on H-ZSM-5 at 573 K, in addition to bands of CH stretching vibrations at 2856–3086 cm^{-1} overlapping bands with absorption maxima at 1620 and 1485 cm^{-1}. By switching to pure helium, all these bands decreased in intensity, while the band of the acidic hydroxyl groups (3610 cm^{-1}) increased and the band of the silanol groups at 3745 cm^{-1} remained at reduced intensity. After this treatment the most intense bands were found at 2925 cm^{-1}, 1585 cm^{-1} and 1460 cm^{-1}, and the authors claimed that the mainly linear coke precursors were strongly adsorbed on terminal SiOH groups.

Regarding the discussion of the assignment of the coke bands, two important questions have to be answered. The first question concerns the difference between carboxylate bands particularly observed on alumina catalysts and bands of conventional coke, and the other has to do with the bonding of carbonaceous species on the surface of zeolites. According to Datka et al. [108], the observation of carboxylate is not attributable to differences in experimental procedures. Rather, it is due to differences in the surface properties between alumina and zeolites. The authors ascribed the absence of carboxylates during coke formation on zeolites under equivalent experimental conditions to the absence of basic hydroxyls on zeolites. However, it cannot be excluded that the differences are due to other reasons than differences in the Al – O – Al structure of alumina and the Si – O – Al structure of zeolites.

Concerning the interaction between the carbonaceous deposits and the zeolite surface, infrared studies on ^{18}O-labeled H-ZSM-5 were carried out [41, 109]. Chemical bonds involving an oxygen atom should yield a frequency shift of the corresponding bands to lower wavenumbers after ^{18}O-substitution. For example, the framework combination bands at 1980 cm^{-1} and 1880 cm^{-1} are shifted by more than 40 cm^{-1} on ^{18}O-exchanged H-ZSM-5 [41] indicating the substitution of ^{16}O by ^{18}O within the zeolite framework. After loading with 4 wt % carbonaceous deposits formed during *n*-hexene conversion at 693 K, typical coke bands at 1580 cm^{-1}, 1460–1420 cm^{-1} and 1390 cm^{-1} were observed (Fig. 13). Neither the positions nor the shape of these bands were influenced by ^{18}O isotope exchange which should occur if oxygen-containing bonds between the zeolite surface and carbonaceous residues existed. In addition, the IR spectrum of deuterated carbonaceous residues formed from deuterated *n*-hexene confirms the assignment of the bands between 1460–1420 cm^{-1} to CH vibration modes [41]. Hence, ^{2}H and ^{18}O isotopic studies on H-ZSM-5 give no indication of oxygen-containing bonds between the zeolite surface and coke. Moreover, the evidence of carbocations by IR, ^{13}C MAS NMR and UV-Vis spectroscopy clearly shows that hydrocarbon species are not strongly covalently bonded but rather weakly ionically bound on the zeolite surface.

Another interesting field of in-situ IR experiments concerning the characterization of coked zeolites is the investigation of the fragmentation and destruction of carbonaceous deposits by pyrolysis and oxidation. Demidov et al. [110] studied the effect of oxygen treatment on coke which had been formed on an (H,Na)-Y zeolite after methanol conversion at 573 K. The oxidation of coke begins already at 473 K, indicated by the reduced intensities of the methyl and methylene absorption bands. Bands at 1750 and 1780 cm^{-1}

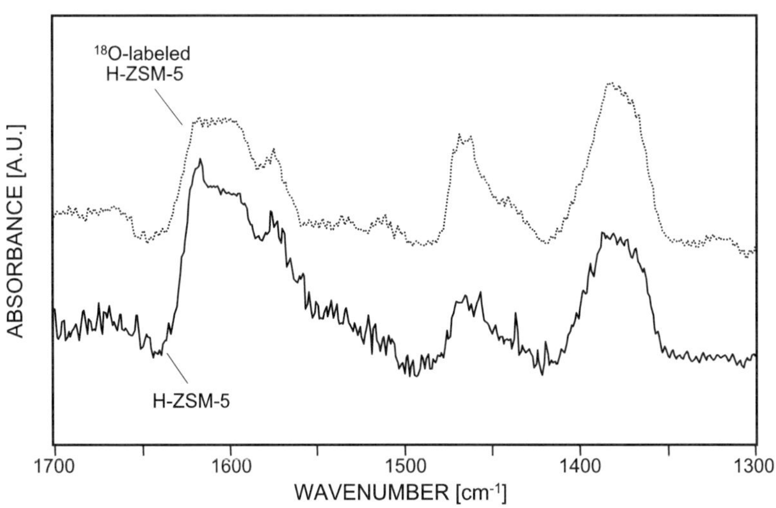

Fig. 13 DRIFT spectra of ^{18}O-exchanged H-ZSM-5 and parent H-ZSM-5 both coked during *n*-hexene conversion at 693 K (after [41])

typical of carbonyl groups appeared. The substantial reduction of the band at 1550 cm^{-1} indicated that the transformation of condensed aromatics was also occurring at these low temperatures. Increasing the temperature up to 673 K intensified the formation of carbonyl and carboxylate compounds (bands at 1440 and 1620 cm^{-1}) as well as ether linkages (band at 1280 cm^{-1}). Above 773 K the deep oxidation of coke began and only residual compounds with carboxylate or carbonate structure were observed.

Novakova et al. [86] investigated the effect of partial oxidation and pyrolysis on coke deposited on H-ZSM-5 and H-Y zeolites upon acetone conversion. The observed spectra reveal that the coke residues on H-ZSM-5 contain bands characteristic of lighter aromatic compounds, e.g., alkyl-substituted naphthalenes (1510–1540 cm^{-1}) together with more condensed aromatic rings and polyenes (1570–1620 cm^{-1}) that dominate in the H-Y zeolites. The bands at 1379 cm^{-1} and in the 2870–2970 cm^{-1} region corresponding to the CH_3 and CH_2 vibrations indicate the presence of aliphatic chains longer on H-Y than on H-ZSM-5 zeolites. Heating of coked zeolites in vacuum or in an inert atmosphere above 623 K leads to a release of volatile products on both zeolites connected with a disappearance of the bands at 2870, 2926 and 2970 cm^{-1} of aliphatic species, whereas the band at 1379 cm^{-1} ($\delta_s[CH_3]$) remained nearly unchanged. On heated H-ZSM-5, the bands at 1512 and 1541 cm^{-1} of the substituted aromatic compounds were missing. In the IR spectra of the partially oxidized coked zeolites new bands appeared at 1720–1760 cm^{-1} and were assigned to the presence of aldehydes or ketones. The same type of compounds were observed by Magnoux and Guisnet [37], who analyzed the methylene chloride extract of a partially oxidized coke after dissolving the H-Y zeolite in hydrofluoric acid.

For the 1300–1700 cm^{-1} region, strong bases like ammonia and pyridine (which are protonated by acid zeolites) can be used as titrants of the concentration of Brønsted and Lewis acid sites [111, 112] (cf. also Volume 4, Chapter 1 of the present series). For example, after pyridine adsorption the IR band at 1545 cm^{-1} assigned to pyridinium ions is often taken as a reference to characterize the evolution of Brønsted acidity, whereas the IR signal at 1450 cm^{-1} is attributed to Lewis acid sites. The relative strength of acid sites is compared with the maximum temperature to eliminate the 1545 cm^{-1} band on pumping off the pyridine. Note, that a substantial fraction of weakly basic carbonaceous residues can be replaced by more basic molecules such as ammonia and pyridine [113, 114].

2.4 Ultraviolet-Visible Spectroscopy

Compared with the number of IR studies, UV-Vis spectroscopy has been less frequently applied in zeolite research, although it is also able to provide valuable contributions to the detection and identification of surface intermediates formed during the reaction of hydrocarbons on catalysts (cf., Volume 4, Chapter 4 of the present series). UV spectroscopy has two advantages when dealing with double bond systems: (i) the $\pi - \pi^*$ transitions of different systems generally absorb in quite distinguishable ranges and (ii) the absorption coefficients of electronic transitions of unsaturated organic compounds in the visible and near UV are usually at least one order of magnitude higher than those associated with vibrational transitions. Therefore, UV-Vis spectroscopy can easily detect olefinic coke species. Moreover, it has turned out that UV-Vis spectroscopy is rather sensitive, particularly in the case of the positively charged compounds (Table 4). Hence, debates of some controversial assignments of bands to quite different surface species can only be settled by the combination of UV-Vis results with IR and ^{13}C NMR data [73, 91].

Even though the samples usually exhibit low transparency and, moreover, the transmittance is strongly reduced upon coke deposition, the formation of coke precursors on zeolites and the final coke structures have been studied by this technique in reflection mode [76] and in transmittance mode [115–119]. As an example, sets of difference spectra are depicted in Figs. 14 and 15. These spectra were obtained under static conditions with two different zeolite samples, H-ZSM-5(1) with Si/Al = 22.6 and H-ZSM-5(2) with Si/Al = 33.5, after loading with methanol and reacting at increasing temperatures [115]. The samples, characterized thoroughly by several techniques differed mainly in acidity and crystallite size. The density of acidic OH groups of H-ZSM-5(1), measured by the absorbance of acidic Brønsted sites at 3610 cm^{-1}, was higher than for H-ZSM-5(2). For the strength of Brønsted acidity temperature-programmed desorption of ammonia indicated the opposite order, i.e. H-ZSM-5(1) had more, but weaker acidic sites than H-ZSM-5(2).

Immediately upon adsorption of methanol (not shown in the figures) a small band at 308 nm appeared. Absorbance in this range (around 310 nm) was usually observed after interaction of olefins with acidic zeolites at am-

Table 4 Assignment of UV-Vis bands due to formation of carbonaceous deposits (adapted from [38, 91] and references therein)

Band [nm]	Assignment
200–1000	aromatics
250–290	neutral di- and trienes
290–330	monoenyl cations, cyclopentenyl and cyclohexenyl cations
350–390	dienyl and aromatic cations
400–430	diphenyl and aromatic cations, polyalkylaromatics
440–470	tri- or polyphenyl cations, alkylaromatic cations
500–575	polyphenyl and polyaromatic cations, condensed aromatics

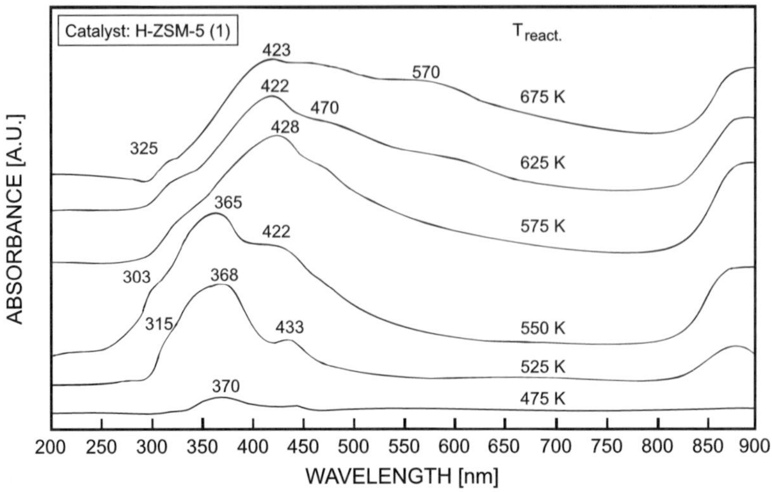

Fig. 14 UV-Vis spectra of carbonaceous deposits after reaction of methanol on H-ZSM-5(1) with Si/Al ratio of 22.6 and crystallite size of $8.8 \times 5.2 \times 3.2$ μm (after [115, 130])

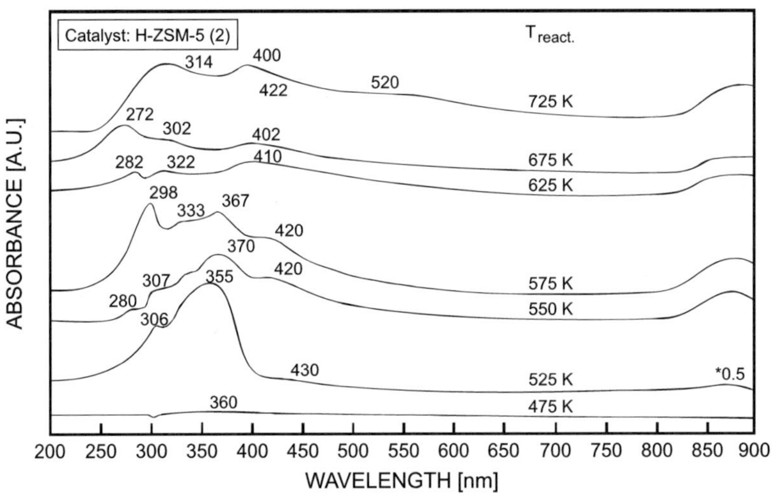

Fig. 15 UV-Vis spectra of carbonaceous deposits after reaction of methanol on H-ZSM-5(2) with Si/Al ratio of 33.5 and crystallite size of about 0.4 μm (after [115, 130])

bient temperature and proved the creation of unsaturated carbocations [118, 120–122]. The bands were assigned to the $\pi - \pi^*$ transitions of alkenyl carbocations. Kiricsi et al. [119] obtained a good correlation between the appearance of the UV-Vis absorption near 310 nm (observed in both superacids and zeolites and assigned to monoenyl carbocations), and the IR band around 1530 cm^{-1} (cf. Table 2).

Subsequent stepwise heating developed bands typical of positively charged intermediates. At low temperature, e.g., 525 K, the most prominent band is that around 360 nm. UV-Vis bands in the range 350–390 nm are ascribable to di- or polyenylic cations [120, 123–126]. While ^{13}C MAS NMR is unable to detect any polyenylic cations, only a small fraction of the polyalkenes needs to be present as carbocations in order to be identified by UV-Vis spectroscopy. The weak band around 310 nm may still contain some contributions of allylic cations, but at reaction temperatures of about 550 K mainly arises from cyclohexenyl cations (315 nm), since such cyclization products of olefin interaction on acidic catalysts are indeed indicated by UV-Vis bands in the range of 275–330 nm [76, 122].

The alkenyl carbocations with chain and cyclic structures absorb in almost the same UV-Vis region. Therefore, it is difficult to assign the observed bands, which are usually rather broad and overlapping, to well-defined structures. The shoulder at 420–430 nm is most probably due to bulkier species such as diphenylcarbenium ion [127] or polyalkylaromatics and condensed aromatic compounds [76, 120] which, to a small extent, may form even at low temperatures.

Similar results were observed during the reaction of 1-hexene on H-Y zeolites with different framework Si/Al ratios [128]. After olefin adsorption at room temperature, allyl carbocations were immediately formed in the zeolite cavities. At 373 K polyenic carbocations and low-condensed aromatics could be detected. Further elevation of the temperature up to 423 K resulted in bands at 225 nm, 265 nm and 425 nm which may be ascribed to dienes, cylohexadiene and/or benzene and cations of low-condensed aromatics as important precursors of further coking. The absorption bands related to coke were weaker for increased framework Si/Al ratios.

The UV-Vis spectra of coke products formed during conversion of propylene on H-Y and dealuminated mordenite at 420 K showed intense bands at 300 nm and 370 nm. Kustov et al. [129] assigned these bands to the presence of conjugated carbocations formed from polyenes. The assignment of bands at 420–460 nm and 540–590 nm to acenes or polyphenylenes is more questionable. The comparison with amorphous aluminosilicate catalysts showed that the Lewis acid sites were responsible for dehydrogenation of olefinic oligomers which resulted in the formation of aromatic hydrocarbons.

Signals of such species, however, become predominant at higher reaction temperatures at the expense of the polyenic species which are weakened and, finally, vanish upon the transformation of "low-temperature coke" into "high-temperature coke". Pertinent assignments of the bands, which appear at higher reaction temperatures, are as follows: a pair of bands (325 and, more intense, 410–420 nm) is indicative of bulky aromatic species (diphenyl or polyphenyl carbocations, polyalkylaromatics, and condensed aromatic ring systems). Evidence of formation of such species at high temperatures (600–700 K) is also provided by bands at 520 and 570 nm [76, 127]. Bands observed at 270 and 470 nm may be ascribed to cyclohexadiene and/or benzene and cations of substituted benzenes, respectively.

In agreement with the preceding considerations of the IR results, the UV-Vis investigations illustrate again that two kinds of coke should be distinguished:

(i) coke-type I with bands of olefinic/polyolefinic constituents, viz. monoenyl and dienyl carbocations around 300 nm and 360–370 nm, respectively, were mainly observed in a low-temperature range (< 550 K) and
(ii) spectra of coke-type II with signals of predominantly small aromatic carbocations and bulky polyaromatics, viz. around 420 nm and above 500 nm, respectively, were obtained at higher reaction temperature (> 550 K).

The nature of the coke residues, as identified by UV-Vis spectroscopy did not depend significantly on whether methanol or olefins were used as the feed (Karge HG, 1991, personal communication). This is particularly true for relatively high reaction temperatures. Here, however a significant difference between the two H-ZSM-5 samples was observed. From Fig. 15 it can be seen that on H-ZSM-5(2) the intense bands of monoenylic and dienylic carbocations (around 300 nm and 370 nm) persisted at temperatures as high as 675 K. This suggests that the formation of coke-type I is favored on an H-ZSM-5 sample with a lower density but higher strength of Brønsted sites even at a relatively high reaction temperature.

In this context it is worth noting that the preferential occurrence of one coke type or the other depends not only on the reaction temperature but on properties of the catalysts as well. As mentioned above, on H-ZSM-5(2) with fewer but more strongly acidic sites than those on H-ZSM-5(1) the dealkylation of alkyl aromatics and the formation of volatile species via cracking of high-molecular coke are facilitated at higher temperatures. Thus, the effect of strongly acidic sites during coking may be twofold: (i) easier formation of coke precursors at lower temperatures and (ii) cracking of these at higher temperatures. This feature may explain why H-ZSM-5(2) had a much lower deactivation rate than H-ZSM-5(1) in the reaction of methanol at 650 K [130]. However, the samples had quite different crystallite sizes, viz. H-ZSM-5(1) showed large crystallites of $8.8 \times 5.2 \times 3.2$ μm and H-ZSM-5(2) small crystallites of an average diameter of about 0.4 μm. Therefore, it cannot be excluded that the large crystallite size in the case of H-ZSM-5(1) additionally leads to a higher coking rate.

The debate on the role of coke during isomerization of *n*-butene to isobutene on H-FER (vide infra, Sect. 3.4) favored the identification of coke species using UV-Vis spectroscopy [91]. The UV spectra obtained after interaction of butene at increasing temperatures are shown in Fig. 16. At 423 K, the band centered at 310 nm can be assigned to monoenic allyl carbocation species [131]. In the temperature range 473–573 K several clearly distinct

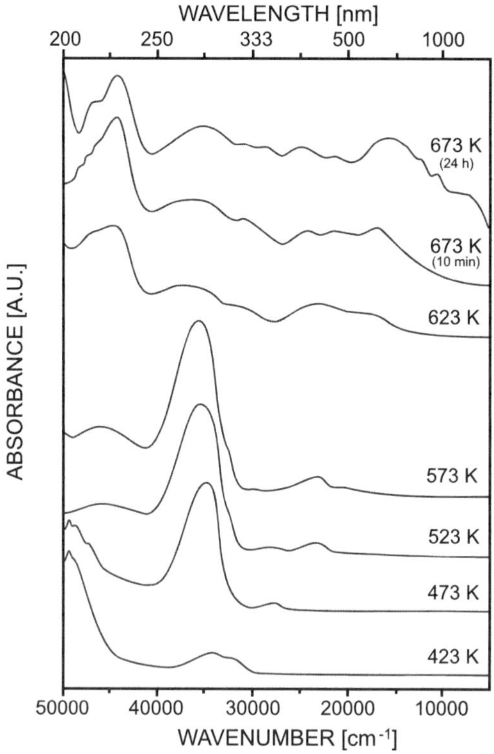

Fig. 16 UV-Vis spectra of interaction of 1-butene with H-FER at increasing temperatures (after [91])

components are evident. Bands present in the 250–330 nm and 330–500 nm range can be summarized in two main groups. Their assignment can be confidently made to neutral and carbocationic alkenes [132]. Absorptions in the range 250–330 nm (which consist of a band centered at about 285 nm) are assigned to neutral conjugated double bond oligomers having a different number of conjugated double bonds [133]. Bands present in the 330–500 nm range are assigned to absorption by carbocationic conjugated double-bond oligomers having a different number of conjugated double bonds. In particular, the band around 370 nm (present at 473 and 523 K) is assigned to dienic ions, while that at about 435 nm (present at 523 and 573 K) is assigned to trienic ions. The carbocationic nature is confirmed by the dosage of NH_3 on the coked zeolites: all bands are then subjected to a partial decrease of intensity due to a displacement by a more basic molecule (vide infra, Sect. 2.10).

UV spectra obtained at 623 and 673 K (Fig. 16) are mainly characterized by three groups of absorptions, at 200–240, 240–370, and 370–1000 nm. It is noteworthy, that the well-defined strong component observable in the

250–330 nm range at lower temperatures is no longer evident. The complexity of the spectra reveals, as in the case of the IR spectra, a considerable number of hydrocarbon species, both neutral and carbocationic, having aromatic character. Absorptions at highest energies (200–240 nm) can be easily assigned to neutral molecules having a single aromatic ring (alkylated benzenes), absorptions in the region between 250 and 500 nm can be assigned to carbocationic and neutral species (aromatics but also linear) having an extensive conjugated double bond system. Prolonged heating at 673 K (24 h) leads to an increase of the intensities in the region between 285 nm and 1000 nm, which indicates the formation of highly polycyclic, graphitic-like carbonaceous deposits [91].

Finally, the visual inspection of spent catalysts within a fixed-bed microreactor made of Pyrex glass may be helpful, e.g., with respect to the coking kinetics. For example, depending on the content and the nature of coke the color of spent zeolites may change from gray-white or yellow-brown at the reactor inlet to black at the outlet or vice versa. Interesting to note, coke with a green and green-brown color has been observed on H-Y [134] and on SAPO-34 [135].

2.5 Nuclear Magnetic Resonance Spectroscopy

Among the many techniques that have been employed to study coking, NMR occupies a particularly important place because of the many nuclei that can be involved and the development of high-resolution solid-state NMR. Catalyst deactivation may be investigated by NMR via the detection of coke-forming nuclei (^{1}H, ^{13}C, ^{15}N), via zeolite modification due to coke deposition (^{1}H, ^{17}O, ^{27}Al, ^{29}Si) and via probe molecules (^{1}H, ^{13}C, ^{31}P, ^{129}Xe). Recently, in a comprehensive review article by Bonardet et al. [11] a special effort has been made to demonstrate the contribution of each of the nuclei to the understanding of the phenomenon of coking.

The development of sophisticated techniques such as magic angle spinning (MAS) and cross polarization (CP) as well as in some cases the use of strong magnetic fields compensates for the low natural abundance of some nuclei, e.g., ^{13}C with 1.1%, and makes NMR studies possible even with zeolite catalysts used in the chemical industry. In-situ MAS NMR experiments have been performed by cyclic heating of samples sealed in glass ampoules between ambient and higher temperatures (outside of the spectrometer), by measurements with high-temperature spinning modules up to 673 K [98], by short-time heating up to 800 K using a laser beam [136] and, more recently, under continuous-flow conditions within an MAS rotor up to 423 K [137].

2.5.1 ^{13}C MAS NMR

The nucleus that is best suited for the elucidation of the nature of the carbonaceous deposits is of course carbon-13 (spin 1/2). The natural abundance of only 1.108% possibly demands the usage of ^{13}C-enriched substances for mechanistic studies. For decades, ^{13}C MAS NMR has turned out to be a powerful tool for the investigation of coke and coke precursors [138]. This method was elegantly applied, for instance by Neuber et al. [32], Derouane et al. [139], van den Berg et al. [140], and Maixner et al. [141] when investigating coke formation during methanol conversion. It was similarly used in studies of coke formation upon reaction of alkenes over H-mordenite or H-ZSM-5 and conversion of polynuclear aromatics.

The nature of the coke residues characterized by chemical shifts, δ, between 0 and 300 ppm (Table 5) depends on several parameters, such as the type of zeolite, the reactants (including intermediates and products) and the operation conditions. Thus, a wide distribution of aliphatic compounds (δ = 10–40 ppm), olefinic/aromatic compounds (δ = 125–145 ppm), polyalkylated aromatics (δ = 140–150 ppm), straight-chain and branched olefins (δ about 150 ppm) and traces of oxygen-containing hydrocarbon species ($\delta \leq$ 50 ppm) and carbocations (δ = 150–180, > 200 ppm) can be observed on different samples. Unfortunately, ^{13}C NMR spectroscopy cannot distinguish between olefinic and aromatic carbons since their signals appear in the same region of the spectra. This discrimination can be made by UV-Vis spectroscopy (vide supra, Sect. 2.4). Moreover, polyaromatic coke species having very long

Table 5 Assignment of ^{13}C MAS NMR signals due to formation of carbonaceous deposits (adapted from [115] and references therein)

Signal * [ppm]	Assignment
12–15	methyl groups
25–30	methylene groups (short chains)
30–40	methylene groups (long chains)
35–50	tertiary CH groups and quaternary aliphatic carbon
45–50	carbocations (C atoms close to positive C atoms)
55–90	alkoxy groups
130–135	neutral C atoms in alkenes and/or aromatics
140–150	polyalkylated aromatics
140–250	enylic, cyclic and aromatic carbocations
160	carbons near O atoms
280–340	alkyl carbocations

* referenced to TMS

relaxation times and graphite-like carbonaceous residues interacting with conducting electrons are not always completely detected by ^{13}C NMR spectroscopy.

The ^{13}C MAS NMR results obtained by Karge et al. [67, 115] with ^{13}C-enriched hydrocarbons demonstrate that the reaction temperature has a dominant effect on the nature of coke. In Fig. 17 sets of spectra are displayed which show the formation of coke during the conversion of methanol and ethylene on H-ZSM-5 catalysts. The series of experiments started with samples contacted with methanol or ethylene at 300 K. The most interesting difference between the interaction of methanol and ethylene with H-ZSM-5 is related to the formation of alkoxy groups. There were in the spectra of ethylene weak signals at about 51 and 63–69 ppm which might be due to alkoxy groups, but no sharp signal at 81–83 ppm appeared as was the case upon methanol interaction. Instead, prominent signals were observed around 55 ppm originating probably from rapid interchange between two adsorption structures of ethoxy groups.

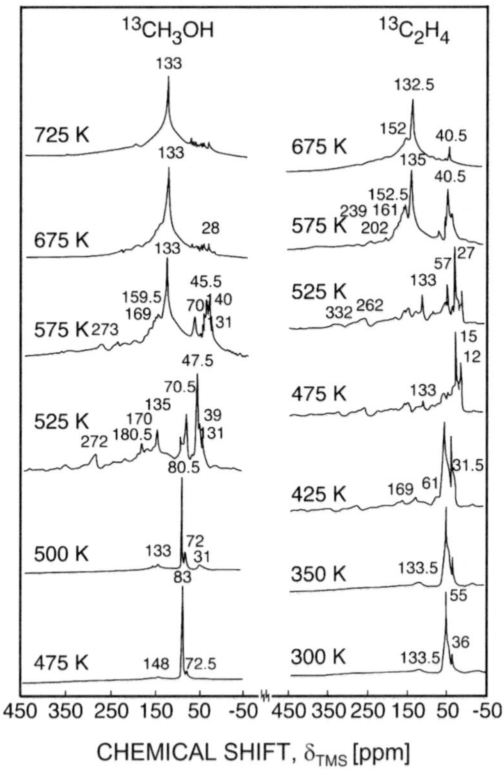

Fig. 17 ^{13}C-MAS NMR spectra of H-ZSM-5 after coking through reaction of $^{13}CH_3OH$ and $^{13}C_2H_4$ (after [115])

Upon increasing the temperature, significant changes in the ^{13}C MAS NMR spectra occurred. At low temperatures, i.e. up to 500 K, signals were observed which gave evidence for the formation of branched and possibly also linear alkanes. The carbons of paraffinic chains may contribute to the NMR lines observed at 12–15, 25–27, and 30 to 40 ppm. A large fraction of branched chains was indeed indicated by the signal around 40 ppm. Most importantly, signals at 31 and 39 ppm started to appear, resulting from alkyl chains which must have been rather long, since the signal of terminating methyl groups (15–25 ppm) was not observed. The signal at 45–47 ppm is most likely due to carbon atoms which are adjacent to positively charged carbon rather than to tertiary or quaternary carbon atoms. The oligomeric deposit remained nearly unchanged upon heating to 450 K.

Raising the temperature from 500 to 575 K resulted in a decrease and, finally, in the disappearance of the alkoxy groups. The carbonaceous species isomerized and subsequently cracked whereby smaller molecules were formed. This is derived from the change in the spectra between 0 and 40 ppm, where the lines significantly decreased. At higher reaction temperatures the amount of aliphatic structures in the coke is much lower than observed under less severe conditions.

At 525–575 K, the presence of various carbocations is indicated in Fig. 17 by signals at about 152, 161, 169, 180 and > 200 ppm, in agreement with UV-visible spectroscopic results (vide supra, Sect. 2.4). Oliver et al. [98] observed signals at 250, 148, and 48 ppm in the spectra at 523 and 573 K during cracking of ethylene-oligomers on H-ZSM-5. These signals were assigned to methyl-substituted cyclopentenyl carbocations which have been proposed as intermediates in the formation of aromatic hydrocarbons. Similar NMR results were obtained for the conversion of *n*-butene on H-FER [100]. Characteristic signals at 152 and 252 ppm point to the appearance of alkyl-substituted cyclopentenyl cations in the temperature range between 500 K and 600 K (Fig. 18). After 1 h reaction at 623 K, the signals from the cyclopentyl cation as well as from fluorenic and/or biphenylic compounds (signal at 144 ppm) disappeared, whereas the signals of butene oligomers (signals at 13, 25, and 33 ppm) and condensed aromatics (130 ppm) can be still observed at this temperature.

At temperatures above 600 K, enylic and aromatic carbocations (lines between 150 and 180 ppm) and alkyl carbocations (250–280 ppm) disappear. The existence of polyenic carbocations is not clearly supported by the results presented in Fig. 17, since this would require not only the appearance of the line at about 133 ppm but also the presence of a line in the range between 80 and 100 ppm [142]. The occurrence of uncharged polyenes, however, is not excluded.

At temperatures of about 650 K the signal of neutral, aromatic species (130 ppm) largely predominated. For instance, Anderson and Klinowski [143] observed during methanol conversion on H-ZSM-5 at 643 K penta- and even

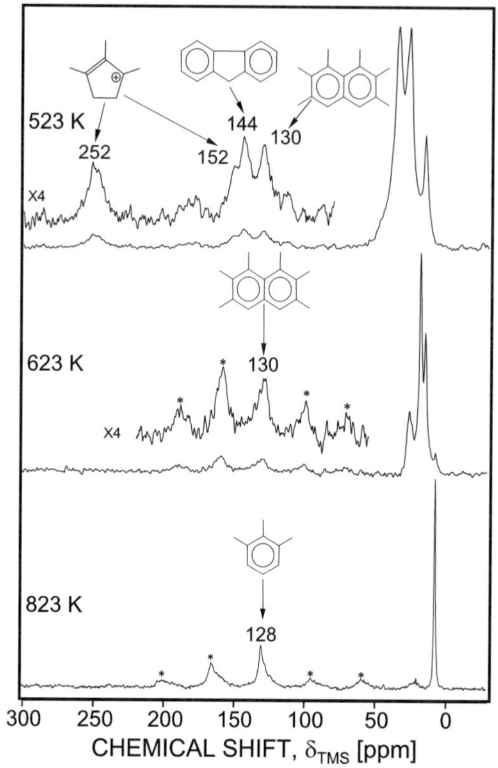

Fig. 18 ^{13}C CP/MAS NMR spectra of *n*-butene(1-^{13}C) on H-FER after successive heating of the zeolite sample for 1 h at 523 K, 623 K, and 823 K (after [100]); asterisks denote spinning side bands

hexamethylbenzenes adsorbed within the zeolite crystallite, whereas only very low concentrations of pentamethylbenzene and no hexamethylbenzene were found in the gas phase. Stepanov et al. [100] reported that polycyclic aromatic coke species formed during *n*-butene isomerization on H-FER were converted at higher temperatures to simple methyl-substituted aromatics, exhibiting a signal centered at 128 ppm (Fig. 18), with methane and ethane evolution. All these NMR results confirm the discrimination of carbonaceous deposits into two types

(i) hydrogen-rich coke ($10\,\text{ppm} < \delta < 40\,\text{ppm}$) is formed at low reaction temperatures ($T \leq 550\,\text{K}$) and;
(ii) polyaromatic coke ($\delta \approx 130\,\text{ppm}$) is built up at higher temperatures ($T \geq 550\,\text{K}$).

While the NMR results clearly indicate that beyond a certain reaction temperature the carbonaceous residues are mainly (poly)aromatic regardless of the type of zeolite and the reactant, the effect of the time on stream or the amount of coke on the composition of the coke, i.e. the aging of coke, is controversially discussed. For the conversion of acetone on H-ZSM-5, Novakova et al. [86] found that the fraction of condensed aromatics is higher at low coking levels. On the other hand, spectra of coke taken at various times on stream on an ultrastabilized H-Y (H-USY) indicate that the structure of the carbonaceous residues reaches a steady state within a very short time and does not change subsequently [144].

The decrease of the H/C ratio with time on stream (vide supra, Sect. 2.1) is mainly due to hydrogen transfer reactions from coke species to unsaturated intermediates. Bauer et al. [48] and Aguayo et al. [49] showed that hydrogen transfer can also take place from hydrogen or light alkanes (used as carrier gases) to carbonaceous deposits. For example, H-ZSM-5 samples deactivated during the methanol conversion at 693 K were treated with propane or *n*-hexane at the same reaction temperature. The ^{13}C CP MAS NMR spectra revealed that the nature of coke has changed into a more aromatic one by this treatment (Fig. 19). Moreover, the alkane treatment resulted, via dealkylation of bulky coke species, in a partial reactivation of spent samples [48]. Similar results of catalyst rejuvenating were obtained by a hydrogen sweeping treatment at 823 K on H-ZSM-5 deactivated by 6 wt % coke during acetone transformation at 673 K [49]. The thermal treatment in a hydrogenating medium avoided further dehydrogenation reactions of carbonaceous residues indicated by a nearly constant H/C ratio throughout the sweeping. Helium

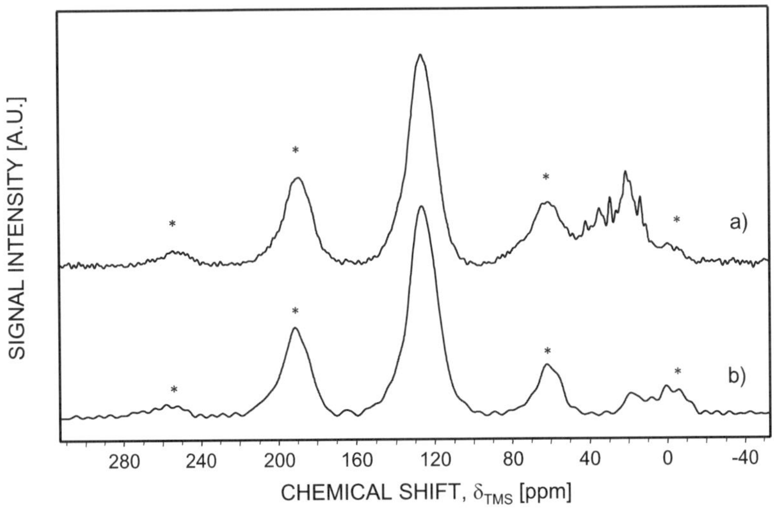

Fig. 19 ^{13}C CP MAS NMR spectra of coke formed upon methanol conversion at 693 K on H-ZSM-5 (a) and after treatment of the spent zeolite with propane at 693 K (b) (after [48]); asterisks denote spinning side bands

was also efficient in decreasing the coke content but the remaining deposits with an H/C ratio of about 0.7 yielded rapid deactivation after the sweeping.

During acetone conversion at 623 K on H-ZSM-5 and H-Y, the poorly resolved band with a maximum at 130 ppm is present on both zeolites and is even broader on H-Y than on H-ZSM-5, which indicates the presence of more condensed coke components [86]. A peak at about 154 ppm is clearly separated from that at 130 ppm on H-ZSM-5 and could correspond to olefins. Because of the resistance of this peak to bromine vapor, the assignment to olefins is less probable, and the signal at 154 ppm was considered to be characteristic of phenolic compounds or aromatic ethers. The coke on H-Y exhibits a poorly resolved signal at 140 ppm. This signal also appears in the case of coke on H-ZSM-5 and could be assigned to polyacetylenes or oxygen-containing aromatic compounds. Moreover, in this case a signal at 180 ppm was assigned to acetate-like complexes on non-skeletal Al species (vide supra, Sect. 2.3).

The use of the cross polarization technique enhances the sensitivity of ^{13}C MAS NMR. It must be noted, however, that it makes quantitative analysis more difficult. Meinhold and Bibby [87] showed that the amount of carbon observed by NMR increased much more slowly than expected from the buildup of coke. Whereas all the coke has been detected below a coke content of 0.85 wt %, only one third of the coke was "NMR visible" for the highest amount of coke (23.6 wt %). The loss of carbon signal intensity could be due to probe detuning by the presence of organic free radicals, low values of the transversal relaxation time of the proton and/or the formation of hydrogen-deficient regions. Measurements of samples with different coke contents showed that as coking proceeded, the proportion of aromatic methyl groups increased initially, followed by condensation to form polycyclic aromatics with more ring junction carbons and fewer methyl groups. The authors concluded that the formation of highly condensed, hydrogen-deficient polycyclic aromatic hydrocarbons on the surface is the reason for low ^{13}C signal areas. At very high coke levels, caused mainly by the buildup of "graphitic" coke on the external surface of zeolite crystallites, the samples became partially conducting, resulting in an additional loss of signal intensity.

This result has been confirmed by Richardson and Haw [145], whose experiments with butadiene on H-Y revealed that the ^{13}C CP MAS NMR spectra of coke residues can be analyzed quantitatively only at very low reaction temperatures. At higher temperatures the aromatic carbon fraction is underestimated. Already at 423 K only 78% of the carbon is detected by NMR.

2.5.2
^{1}H MAS NMR and ^{1}H NMR Imaging

Despite the low range of chemical shift, δ, between 0 and 10 ppm, different hydroxyl groups in zeolites can be separated by ^{1}H MAS NMR lines: (a) lines between ca. 1.8 and 2.3 ppm are caused by non-acidic silanol groups at the outer surface of zeolite crystallites and at framework defects; (b) lines at 2.5 to 3.6 ppm represent OH groups associated with extra-framework aluminum species; (c) peaks at 3.8 to 4.4 ppm are ascribed to Brønsted acid sites, known to be of bridging SiOHAl type; (d) lines at 4.8 to 5.6 ppm are from acidic OH groups under the influence of an additional electrostatic interaction, e.g., in the case of formation of hydrogen bonds to neighboring oxygen atoms [146].

In addition to the number of Brønsted acid sites still active after coking, ^{1}H MAS NMR spectra of spent samples provide information about the nature of coke and the location of the carbonaceous deposits on coked zeolites (vide infra, Sect. 3.3). Typically, proton magnetic resonance spectroscopy discriminates among aliphatic, alkylaromatic, and aromatic protons (Table 6). Also, high-resolution liquid ^{1}H NMR has been successfully employed for chemical analysis of hydrocarbon extracts from coked zeolites (vide infra, Sect. 2.9).

Compared to ^{13}C MAS NMR there are only few studies using ^{1}H NMR analysis for the characterization of fresh and coked zeolites, maybe due to the problems of remaining water as well as the overlap of signals from zeolitic and carbonaceous protons [147–149]. In Fig. 20, peak (a) at 2 ppm corresponds to non-acidic silanol groups, whereas peak (b) at 4.2 ppm is due to the protons of Brønsted sites on dealuminated H-ZSM-5. Ernst et al. [147] showed that the intensity of the peak at 4.2 ppm decreased after coking with *n*-hexane. The broad signal (represented by a broken line) in the spectrum of the coked sample is not affected by MAS and must be caused by the protons of the coke or by the protons of acidic hydroxyl groups having a strong homonuclear dipolar interaction due to the weak mobility of carbonaceous residues. Moreover, the authors deduced from the decreased intensity of the ^{27}Al MAS NMR peak that the carbonaceous deposits have poisoned the Brønsted acid sites inside the zeolite crystallites.

Table 6 Assignment of ^{1}H MAS NMR signals on zeolites and due to formation of carbonaceous deposits

Signal * [ppm]	Assignment
0.8–2.2	aliphatic protons
1.3–2.2	non-acidic Si – OH groups
2.2–2.8	alkylaromatic protons
2.6–3.6	OH groups attached to non-framework Al species
3.3–5.2	protons attached to C near O atoms
3.8–4.3	SiOHAl Brønsted acid sites
4.6–7.0	acidic OH groups with electrostatic interaction
6.5–8.5	aromatic protons

* referenced to TMS

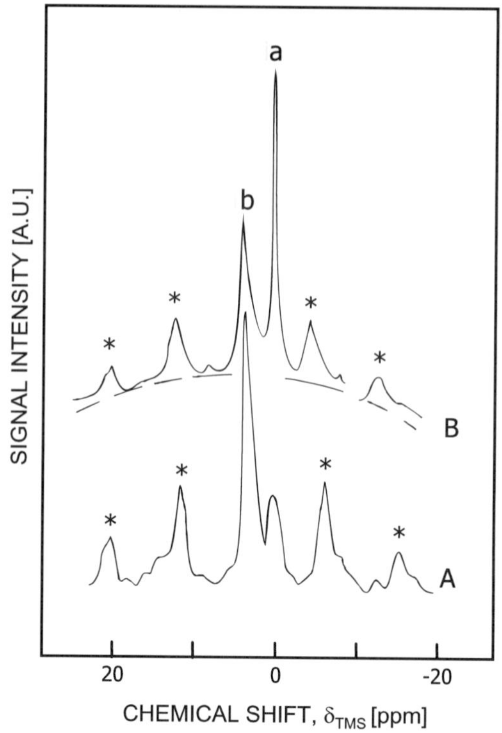

Fig. 20 ^{1}H MAS NMR spectra of a non-steamed H-ZSM-5 zeolite: (A) before coking; (B) after coking (after [147]); asterisks denote spinning side bands

An excellent method to study molecular migration in zeolites is the pulsed field gradient ^{1}H NMR (PFG NMR) technique [150]. The signal intensity is expected to follow the equation

$$\ln(I/I_0) = -\gamma^2 G^2 \delta^2 D/(\Delta - \delta/3)\,, \tag{7}$$

where I is the observed intensity, I_0 is the intensity in the absence of gradient pulses, γ is the gyromagnetic ratio of hydrogen, G is the applied field gradient amplitude, δ is the length of the gradient pulses, Δ is the interval between the gradient pulses, and D is the diffusivity. Due to the presence of both inter- and intracrystalline diffusion, it is necessary to analyze the experimental data assuming two distinct diffusion components. Kärger and co-workers [151–154] measured the intracrystalline self-diffusion coefficient D_i and the effective diffusion coefficient D_d of methane in H-ZSM-5 before and after coking with n-hexane (at 773 K) or mesitylene (at 803 K). Whereas in the initial reaction period D_i and D_d decreased in parallel, at longer reaction times the intracrystalline self-diffusion coefficient ceased to decrease and remained nearly constant. In this case, the molecular passage through the crystallites was reduced in addition to intracrystalline mass transport by a surface diffusion barrier which has been taken as proof of external coke deposits. For coking with mesitylene, such deposits on the outer crystal surface were present from the very beginning of the reaction.

Self-diffusion measurement using PFG NMR has the advantage that the measurement is performed under equilibrium conditions and that the intracrystalline diffusivity can normally be obtained. However, PFG NMR results for methane on deactivated H-USY [155] showed that interparticle effects had the major influence on the measured diffusivity as the mean displacement during the measurement was larger than the crystallite size of the sample (0.5–1 μm), which meant that an average of inter- and intraparticle effects was observed. For highly spent samples with more than 10 wt % coke, most of the methane is in the interparticle region, thus a fast diffusion component was observed. The intracrystalline diffusivity of methane was found to be in the range of $0.9–2.0 \times 10^{-7}$ m^2/s for the coked samples compared with 8.5×10^{-7} m^2/s for fresh H-USY. This is still a rapid diffusion of methane through the zeolite loaded with carbonaceous residues. Hence, *n*-butane as a larger adsorbate was used as a probe molecule (Fig. 21). Even on the fresh zeolite, both inter- and intraparticle effects are apparent. The curves could in principle be affected by coke deposition within the pore channels, carbonaceous residues at the crystallite surface as well as coke formation in the interparticle space. The fast interparticle diffusion component was found to be 1×10^{-8} m^2/s for the fresh and spent catalysts. The slower intracrystalline

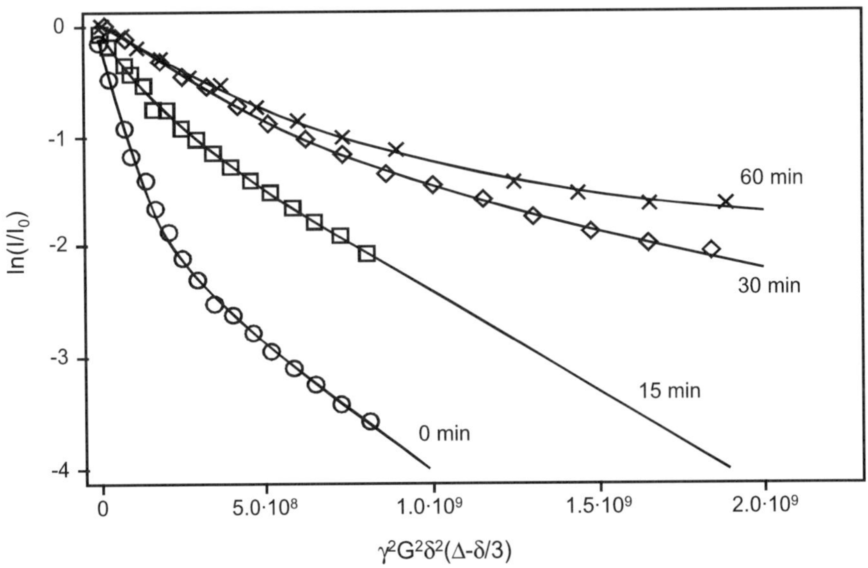

Fig. 21 PFG NMR intensity results for adsorbed *n*-butane after different time on stream in ethylene conversion at 773 K on H-USY (after [155])

diffusivity decreased by one order of magnitude (from 23.0×10^{-10} m^2/s to 2.0×10^{-10} m^2/s at a coke content of about 13 wt % after 60 min on stream). Nevertheless, diffusion of methane and *n*-butane is rapid in this spent H-USY sample, and fairly homogeneous coking throughout the internal pore structure can be assumed. At about 18 wt % coke, the diffusivity of *n*-butane became very slow ($< 8 \times 10^{-12}$ m^2/s) indicating that pore blockage was now significant.

^1H magnetic resonance imaging (MRI) has been successfully applied to visualize internal structures and to monitor dynamic processes in a wide range of medical systems. Coke deposits can be made visible by MRI in two different ways. In the direct way, the small amount of hydrogen present in the coke species might be detected in a dehydrated sample. However, short spin-spin relaxation times and low spin density of hydrogen within the carbonaceous residues are major obstacles to the direct visualization of coke species [156]. In the indirect way, the dehydrated zeolite is preloaded with a hydrogen-containing adsorbate, e.g., hydrocarbons or water. By taking advantage of the difference in the relaxation times of the adsorbed molecules as well as the different adsorption affinities of fresh and coked zeolite surfaces, the distribution of coke within a zeolite pellet (Fig. 22) and catalyst bed can been obtained [157–159]. A spatial resolution less than 1 mm is an important argument for the assessment of MRI as a non-destructive method for axial and radial concentration profiling at both the milli- and micrometer scale. Furthermore, due to the spectrometric background it is even possible to trace different signals, e.g., aliphatic or aromatic hydrogen, and to visualize different coke species. From the industrial point of view, the potential of MRI can be applied to optimize the effectiveness factor of different shapes of catalyst pellets.

For reason of completeness, positron emission tomography (PET) using ^{11}C-labeled compounds has been applied to adsorption studies and axial

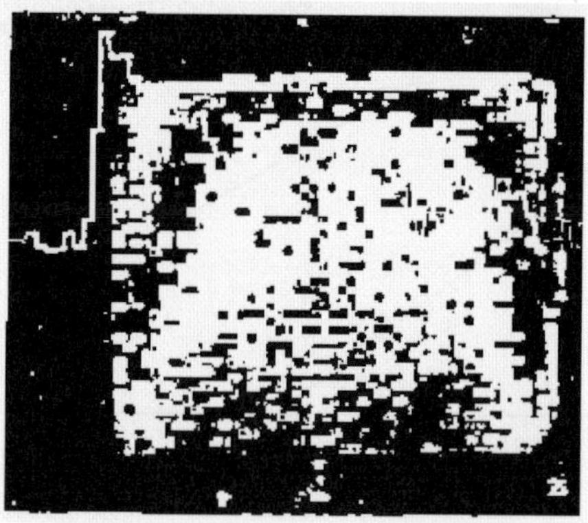

Fig. 22 2D image of a 2.5 wt % coked H-Y pellet (12 × 6 mm; *darker zones* correspond to the more coked regions) (after [158])

concentration profiling within chemical reactors [160]. In principal, PET is capable of generating in-situ 3D images of carbonaceous deposits with a resolution of about 3 mm. However this method is relatively expensive, because a cyclotron is required to generate the short-lived isotope ^{11}C ($t_{1/2}$ = 20.4 min) (cf. also Volume 7, Chapter 7 of the present series). The use of such short-lived isotopes limits applications to rapid processes, whereas deactivation due to coke deposition is generally slow.

2.5.3 ^{27}Al MAS NMR and ^{29}Si MAS NMR

^{27}Al (spin 5/2, natural abundance 100%) and ^{29}Si (spin 1/2, natural abundance 4.7%) belong to the zeolite framework-building atoms. Both nuclei can be used to determine the effect of coking on the zeolite lattice. ^{17}O (spin 5/2), which is intimately involved in adsorption and catalysis, has the drawbacks of low natural abundance of 0.037% and strong anisotropic line broadening due to quadrupole interactions. Therefore, relatively few ^{17}O NMR studies have been performed on zeolites [109, 161–163].

Framework Al species (Al_F) in tetrahedral environments are characterized by a signal at about 60 ppm (referenced to $Al(NO_3)_3$ solution). Filling the channels with carbonaceous deposits shifted the ^{27}Al signal up-field, increased the width of the signal and decreased the signal intensity [164, 165]. These modifications are probably due to the effect of an increase in the electric field gradient, which induces quadrupolar shifts and an increase in the transversal relaxation time. For example, H-ZSM-5 loaded with 23.6 wt %

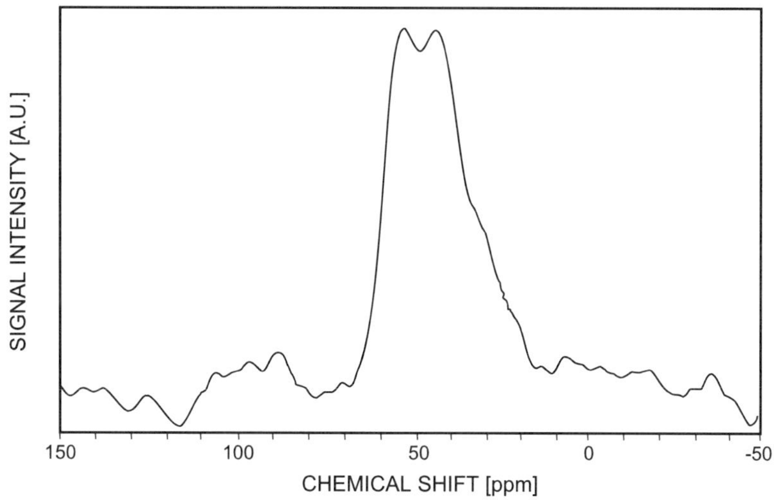

Fig. 23 ^{27}Al MAS NMR spectrum of H-ZSM-5 with 23.6 wt % coke (after [165])

coke formed during the MTG process revealed a splitting of the peak into components at 48 ppm and 38 ppm (Fig. 23). Rehydration of the sample yielded a symmetric peak at 50 ppm.

^{27}Al MAS NMR is particularly suitable for the elucidation of the role of extra- or non-framework Al species (Al_{NF}) in the formation of coke. Echevskii et al. [166] demonstrated that most of the Al_{NF} species in an octahedral environment, indicated by a signal at 0 ppm, are not involved in the formation of coke. The authors found no direct relationship between the Al_{NF}/Al_F ratio and the deactivation rate in the methanol conversion on H-ZSM-5. Contrary to this result, Bonardet et al. [167] observed changes of intensity of the signal at 0 ppm depending on the degree of coking or partial regeneration of a dealuminated H-ZSM-5 coked during acetone conversion. Hence, they concluded that the extra-framework species are involved in the formation of coke.

Besides the ^{27}Al MAS NMR signals of Al in an octahedral and a tetrahedral environment at 0 and 60 ppm, respectively, the ^{27}Al NMR spectrum of a dealuminated H-Y sample coked by *n*-hexane cracking showed a third signal at 30 ppm [167]. Its relative intensity increased with the coke content. Because the 30 ppm signal has been attributed to Al species of a strongly deformed lat-

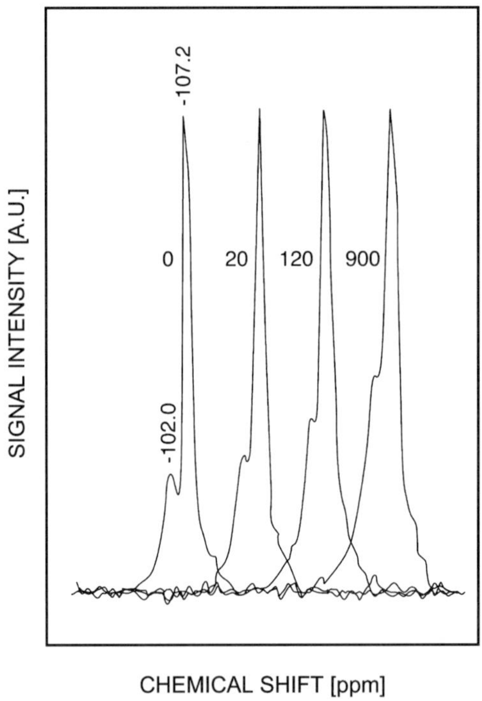

Fig. 24 Broadening of ^{29}Si MAS NMR signals with increasing time on stream (in s) during conversion of *n*-hexene on H-USY zeolite. Spectra are not magnified (after [144])

tice it can be assumed that coke formation causes the displacement of some lattice Al atoms [168].

The amount and structure of an adsorbate have been found to affect the ^{29}Si MAS NMR spectra of zeolites [169]. Analogously, carbonaceous deposits influence the electron distribution around framework silicon atoms in some fashion. The peak broadening observed in the ^{29}Si MAS NMR spectra (see Fig. 24) is interpreted as being due to the presence of free radicals in the coke, a redistribution of initially highly localized coke [144] and, at coke contents higher than 16 wt %, to a slight distortion of the zeolite channels [165] as a result of the formation of more rigid coke molecules.

Meinhold and Bibby [165] demonstrated that ^{29}Si CP MAS NMR allows the in-situ measurement of the hydrogen content of coke and the H/C ratio on H-ZSM-5. For this purpose an empirical relationship between the intensity of the ^{29}Si NMR signal and the hydrogen concentration of adsorbed benzene was set up [164]. Under the assumption that this relationship holds for coke and the cross-polarization signal of silanol groups is weak, the number of protons per unit cell provided by the coke is equivalent to the number of ^{29}Si/u.c. obtained by ^{29}Si CP MAS NMR and increases linearly with the coke content up to at least 9.6 wt %. From the level of coke, which corresponds to carbon atoms/u.c., and from the ^{29}Si signal, which implies the protons/u.c., the H/C ratio of the sample can be calculated. H-ZSM-5 samples of 0.85 wt % and 9.6 wt % coke gave H/C ratios of 0.9 and 0.5, respectively, consistent with ratios found by chemical analysis. For a highly coked sample (23.6 wt %) the total number of cross-polarizing protons is considerably reduced, giving an H/C ratio of 0.1, indicating a large "graphitic" component on the zeolite surface. In fact, it should be taken into account that the reliability of quantitative measurements by cross-polarization techniques is questionable.

Ernst and Pfeifer [170] have shown for slightly dealuminated H-ZSM-5 that *n*-hexane cracking does not modify the ^{29}Si MAS NMR spectra, whereas the intensity of the ^{27}Al signal of the lattice aluminum ($\delta_{Al_F} = 60$ ppm) decreases when the amount of coke increases. The carbonaceous deposits reduce the number of NMR-visible framework Al atoms, which is probably due to the effect of an increase in the electric field gradient and/or due to some perturbation of the tetrahedral symmetry of the Al atoms.

2.5.4
^{31}P MAS NMR and ^{129}Xe NMR

The sensitivity of ^{31}P (spin 1/2, natural abundance 100%) is much greater than that of ^{13}C, and the ^{31}P nucleus has a large chemical shift range being suitable for the detection of subtle varieties of the chemical environment. For ^{129}Xe NMR studies, an enrichment of ^{129}Xe (spin 1/2, natural abundance 26.44%) is required.

Methods for the characterization and quantification of acid sites in zeolites involve the loading of a basic probe molecule, such as ammonia and pyridine, followed by spectroscopic detection and/or by temperature-programmed desorption (cf. Volumes 4 and 6 of this series). Similar investigations using solid-state ^{13}C and ^{15}N NMR methods have been applied to the study of solid acids [171]. However, these studies are hampered by low sensitivity and/or limited chemical shift ranges. The use of phosphorus-containing bases in conjunction with solid-state ^{31}P MAS NMR [172–175] overcomes the experimental limitations associated with ^{13}C and ^{15}N studies.

Comparable in basicity to ammonia, trimethylphosphine (TMP) having an effective diameter of 0.55 nm is sufficiently small to fit into the pores or channel apertures of several zeolites, including ZSM-5. Being an effective probe molecule for unambiguously identifying both Brønsted and Lewis acid sites [172], TMP is a highly flammable air-sensitive liquid at room temperature and requires great care in handling. The use of solid trialkylphosphine oxides [173, 174] offers the advantage of being not susceptible to oxidation, while their basicity is of the same order of magnitude as that of TMP. Similar to TMP, adsorbed trialkylphosphine oxides are capable of identifying different strengths of acid sites on zeolites [175, 176].

Recently, Zhao et al. [177, 178] proposed a method for qualitative and quantitative characterization of internal and external acid sites in solid acid catalysts using ^{31}P MAS NMR of adsorbed trimethylphosphine oxide (TMPO) and tributylphosphine oxide (TBPO) probe molecules. The size of TMPO (kinetic diameter of about 0.55 nm) is in the range of the pore aperture of zeolite ZSM-5; therefore it can diffuse into the intracrystalline channels, thus allowing the simultaneous detection of both the internal and external acid sites. On the other hand, TBPO (ca. 0.82 nm) is too large to penetrate into the channels and hence can merely detect acid sites located on the external surface of the zeolite crystallites. Thus, by combining the ^{31}P NMR results obtained from these two different molecules together with elemental analysis data, detailed information of acid sites (namely nature, location, strength and concentration) present in the sample can be obtained.

To improve the selectivity in hydrocarbon processing over zeolite catalysts normally requires the inactivation of nonselective external active sites [179]. Surface modifications by chemical vapor/liquid deposition of organosilicon compounds or the pre-coking treatment are the most common and effective ways to passivate acid sites present on the external surface of zeolite crystallites (vide infra, Sect. 3.5). ^{31}P MAS NMR spectra obtained from the parent H-ZSM-5 and modified samples after loading with TBPO are shown in Fig. 25. The dashed curves represent results of simulation by Gaussian deconvolution method. Three characteristic resonance peaks at 90, 72 and 55 ppm were observed which strictly represent adsorption sites located on the external surface. Whereas the peak at 55 ppm can be unambiguously assigned to physisorbed TBPO, the other peaks observed at lower field are ascribed to

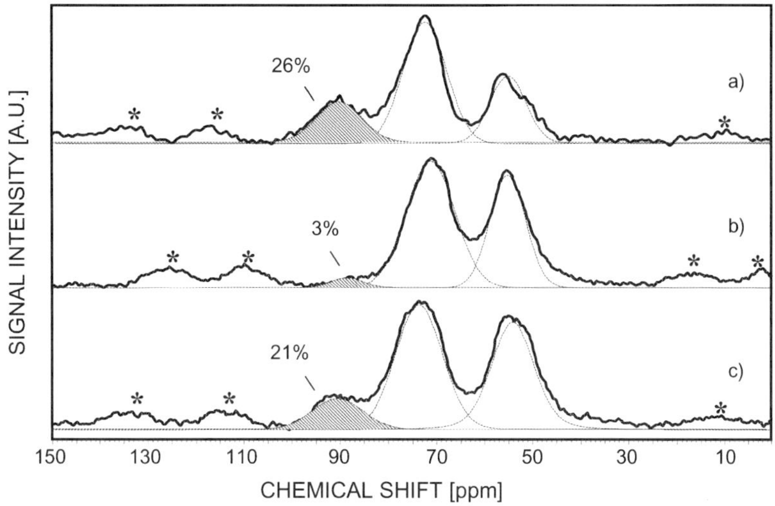

Fig. 25 ^{31}P MAS NMR spectra of TBPO adsorbed on (a) parent H-ZSM-5, (b) pre-coked and (c) silylated samples (after [181]); *asterisks* denote spinning sidebands

complexes between TBPO and Brønsted acid sites. A higher observed chemical shift would represent acid sites with the higher acidic strength, i.e., only two types of acid sites (90 and 72 ppm) can be identified on the external surface of H-ZSM-5 by ^{31}P MAS NMR spectroscopy. In comparison, at least four different acid sites (85, 77, 66 and 56 ppm) have been identified by adsorbed TMPO [175, 180].

The spectra of TBPO adsorbed on parent H-ZSM-5 and modified samples reveal a significant decrease in the amount of strongest acid sites, i.e. of the resonance peak at 90 ppm, after pre-coking treatment. The integrated area (which reflects acid site concentration) of the strong acid sites decreases from 26% for the parent H-ZSM-5 to 3% for the pre-coked sample, whereas only marginal reduction to 21% was found for the sample treated with organosilanes. This implies that the majority of carbonaceous residues are selectively deposited at the strongest acid sites located on the external surface of the zeolite, while chemical liquid deposition of hydrolyzed TEOS is less effective in inactivating the external acidity [181].

The technique of ^{129}Xe NMR of xenon adsorbed in zeolites, developed by Fraissard and Ito [182, 183] (cf. J.-L. Bonardet, A. Gedeon, M.-A. Springuel-Huet, 2005, in this Volume, Chapter 4), has been successfully employed to gain insight into the free intra-crystalline volume of coked zeolites. Unlike other techniques, ^{129}Xe NMR can be used to probe the local environments inside porous materials and thereby define the location of coke [31, 184–189]. This indispensable information is obtained by the variation of the ^{129}Xe chemical shift, δ, as a function of the xenon atoms adsorbed. The observed ^{129}Xe NMR chemical shift, $\delta(\rho)$, of adsorbed xenon can be expressed by [175, 190]:

$$\delta(\rho) = \delta_o + \delta_s(\rho = 0) + \sigma_{Xe}\rho\,, \tag{8}$$

where $\delta_o = 0$ is the chemical shift reference and ρ represents the xenon loading density. The term $\delta_s(\rho = 0)$ is the chemical shift at zero Xe loading and can be divided into the sum of two contributions: namely Xe-wall and Xe-guest interactions. The value of $\delta_s(\rho = 0)$ for each sample can readily be obtained by extrapolating the chemical shift curve to $\rho = 0$ (i.e. the vertical axis). The last term, which is proportional to ρ, arises from the chemical shift contribution due to Xe – Xe interactions. The slope of the $\delta_s(\rho)$ vs. ρ therefore yields σ_{Xe}, which is inversely proportional to the effective free volume of the sample, that is:

$$V/V_o = (\sigma_{Xe})_{parent}/(\sigma_{Xe})_{coked}\,, \tag{9}$$

where V and V_o are the internal free volumes of the modified and the parent samples, respectively. Hence, the inspection of the $\delta(\rho)$ curve gives two parameters: (i) δ_s, the chemical shift extrapolated to zero coverage, which can be interpreted in terms of a mean free path of xenon molecules characterizing the size of the zeolitic cavities and/or channels; and (ii) the slope, $d\delta/d\rho$, which is inversely proportional to the free volume of intra-crystalline cavities accessible to the xenon atoms. Furthermore, the mean-free-path (Δ, in nm) of Xe adsorbed in intra-crystalline channels of zeolites can be estimated by the

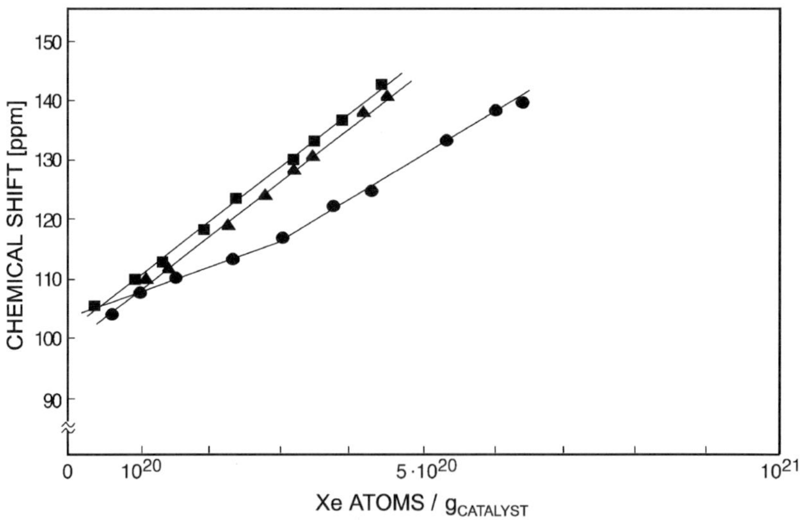

Fig. 26 ^{129}Xe NMR chemical shift as a function of sorbed xenon atoms (after [185]) •, fresh H-ZSM-5 catalyst; ▲, coked in nitrogen (7 wt% coke); ■, coked in hydrogen (14 wt% coke)

empirical equation [190]:

$$\delta(\rho) = 243 \times [2.054/(2.054 + \Delta)] . \tag{10}$$

For example, Barrage et al. [185] showed for H-ZSM-5 samples coked in methanol conversion at 670 K under the carrier gases nitrogen (7 wt % coke) or hydrogen (14 wt % coke) that the $\delta(\rho)$ curves are rectilinear and parallel, whereas for the initial sample this plot presents a break (Fig. 26), generally attributed to the possibility of multiple collisions between xenon atoms at the channel intersections. From the observation that the break is absent after coking, the authors concluded that the carbonaceous deposits are probably located near the channel intersections. A minimum of the $\delta(\rho)$ plot at low xenon loading, which was found for an initial H-Y sample, has been explained by strong adsorption sites, e.g., extra-framework aluminum species [184].

On the other hand, the nearly identical values of the slopes $d\delta/d\rho$ for samples with quite different amounts of coke (Fig. 26) imply that the internal microporous volume is the same. This was confirmed by the evolution of the adsorption isotherms which were identical at higher xenon pressure. However, the slopes were greater than that of the initial sample by a factor of 1.35 which corresponds to a 27% loss of internal volume. This decrease of void volume was independent of the coke content of the samples. The authors deduced from this that, when there is more than 7 wt % coke, the coke is mainly deposited on the outside of the ZSM-5 crystallites.

The ^{129}Xe data were used for a very precise estimation of coke locations. Tsiao et al. [186] claimed for H-ZSM-5 that the deposition of coke occurs mainly in the channels or at intersections blocking a large part of the internal volume even at low coking levels, whereas surface coking still permits access to most of the internal volume. Liu et al. [188] and Pradhan et al. [191] investigated the formation of coke during disproportionation of ethylbenzene and *n*-propylbenzene on H-ZSM-5 and H-USY and concluded that the coke on H-ZSM-5 tends to be deposited at the midchannel position instead of channel intersections. For zeolite USY, the broadening of the ^{129}Xe line and the increase in the chemical shift of the coked samples vs. uncoked parent USY indicate the presence of coke within the USY supercage. Moreover, the broadening of the ^{129}Xe line with increasing temperature was attributed to a rearrangement of coke. Such an assumption of redistribution of coke on the catalyst surface was supported by ^{29}Si NMR results which also show an effect of reaction time on line shape [144].

2.6 Electron Spin Resonance Spectroscopy

Upon coke formation, radicals are generated on the surface of zeolites. The transformation of intermediates into radicals occurs through an electron transfer to (or from) the zeolite or a hydrogen abstraction at suitable acceptor sites such as aluminum atoms at framework distortions or extra-framework aluminum. As a consequence of this radical formation, electron spin resonance spectroscopy allows us to draw conclusions on the structure of adsorbed hydrocarbon molecules and carbonaceous residues. Experimental setups are described for static ESR cells of fused silica [192] and for in-situ measurements [193, 194].

The adsorption of ethylene or propylene on H-mordenite at ambient temperature gives rise to the appearance of an ESR spectrum with a well-resolved 15-line signal with a hyperfine splitting constant of $a = 0.8$ mT and a g-value of 2.0028 [195–197]. The multiplicity of lines proved that the olefins must have oligomerized even at 195 K [198]. In similar experiments by Lange et al. [192] the multi-line signal of oligomeric radicals was well reproduced (Fig. 27). The obtained spectrum can be attributed to a mixture of variously substituted olefinic or allylic intermediates, e.g., $[>C \cdots C<]^{\bullet +}$ or $[>C \cdots C \cdots C<]^{\bullet +}$, resulting in an overlap of several spectra characterized by $a = 1.6$ mT. Upon heating a decrease in hyperfine splitting was observed. Under static conditions but with renewed ethylene atmosphere prior to each heating step, a spectrum developed with seven hyperfine lines ($a = 1.6$ mT) and a g-value of 2.0025 at reaction temperatures between 350 K and 450 K.

When temperatures higher than 500 K were reached, the hyperfine structure vanished completely and the signals collapsed into a single line with a g-value close to that of free electrons, indicating species with a highly delocalized π-electron system. The spin concentration of these coke residues

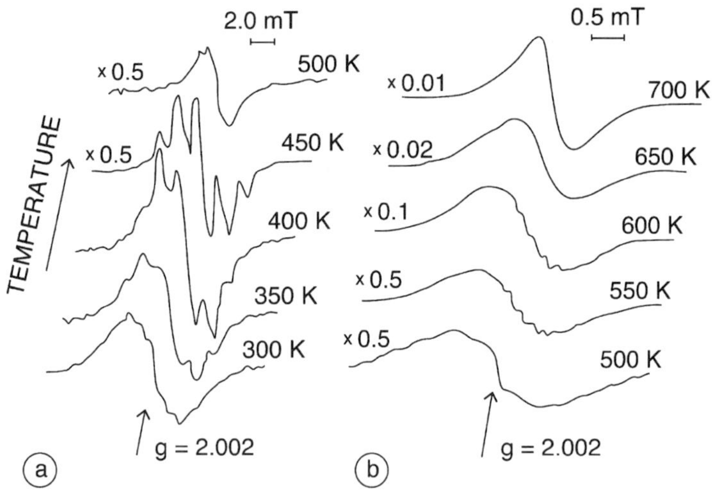

Fig. 27 ESR spectra recorded after coke formation via reaction of ethylene over H-MOR under an atmosphere of 20 kPa ethylene, renewed after each heating step; **a** low-temperature coke; **b** high-temperature coke; note: different settings for the sensitivity of detection and the scale of the magnetic field (after [192])

steeply increased with temperature combined with a sharpening of the structureless signal.

For carbonaceous deposits on zeolites it is reasonable to assume that

(i) multiple line ESR spectra and broad signals with linewidth $\Delta H \geq 0.8$ mT which appear from oligomeric olefinic/allylic and hydrogen-rich coke-type I species, formed at reaction temperatures below 550 K, whereas
(ii) a narrow single line spectrum ($\Delta H \leq 0.5$ mT, $g = 2.002$) is due to highly condensed aromatics (coke-type II) formed at temperatures above 550 K.

The suggested discrimination between types of coke by the ESR linewidth is supported by studies with coals and chars, where a linear correlation between linewidth and hydrogen content due to unresolved electron-proton hyperfine interactions has been found [199]. For highly coked samples it cannot be excluded that the main part of the high-temperature coke has a multilayer graphite-like structure which would be indicated by very broad lines due to interaction with conduction electrons.

Even though only a small fraction of the carbon atoms of the coke deposits correspond to radicals (about 0.1 to 1%; cf. Table 7) and the composition of the deposits is likely to be heterogeneous, the ESR results can be regarded as representative of the formation of carbonaceous residues on the zeolite catalyst. A close correlation was found between the concentration of radicals of high-temperature coke (determined via ESR) and the total amount of coke formed on hydrogen mordenite at 573 K (determined via TGA) [192].

As demonstrated by Karge et al. [194], the number of radicals typical of coke-type I decreased with a decreasing number of acid sites. The effect of the zeolite acidity (Lewis and Brønsted type, number and strength of the acidic sites) on the formation of coke-type II is much more complex. Thus, for instance, Bauer [200] studied the conversion of methanol

Table 7 Characterization of carbonaceous deposits on several H-ZSM-5 samples coked during methanol conversion at 693 K and 1.5 $g_{methanol}/(g_{cat}\ h)$ (adapted from [200])

Zeolite samples	H-ZSM-5(A)	H-ZSM-5(B)	H-ZSM-5(C)	H-ZSM-5(D)
Si/Al ratio	56	56	80	120
Time on stream [h]	267	433	60	55
Acid site density [$\mu g\ NH_3/m^2$]	2.0	2.7	3.6	3.3
Strong acid sites [% TPD area]	68	74	54	59
Coke content [wt%]	5.0	16.3	4.6	4.5
H/C ratio of coke	0.50	0.59	0.91	1.4
Spin/$g_{catalyst} \times 10^{18}$	1.8	8.1	1.0	1.1
C atoms/spin	1400	1000	2500	1700
Linewidth [mT]	0.41	0.29	0.74	0.53

at 693 K on several H-ZSM-5 samples prepared with γ-Al_2O_3 as a binder (Table 7). At reaction temperatures of about 650 K the preferential occurrence of coke-type II was expected. The H/C ratios of the carbonaceous deposits on the samples H-ZSM-5(C) and H-ZSM-5(D) indicated, however, substantial amounts of non-aromatic coke species. Moreover, the samples H-ZSM-5(A) and H-ZSM-5(B), both having according to NH_3-TPD a lower total amount of acid centers, but more acid sites with higher strength than H-ZSM-5(C) and H-ZSM-5(D), showed higher spin densities and sharper ESR signals. This means that at temperatures of about 600 K first the strength of acid sites, and secondly the Si/Al ratio, determine whether the reaction temperature is high enough to generate coke-type II. The author concluded that the higher number of strong acid sites of the former samples would facilitate dealkylation and cracking of coke precursors resulting in a weaker propensity to deactivation. This feature also explains that highly condensed aromatic coke remains on H-ZSM-5(A) and H-ZSM-5(B).

Similar studies of coke formation on H-ZSM-5 zeolite catalysts for methanol conversion at 653–693 K were done by Echevskii et al. [201]. The lifetime of catalysts until the appearance of 3% methanol in the products strongly depended on the Si/Al ratio. Samples with low Si/Al ratios (15–25) survived only short times on stream (24–100 h). The carbonaceous deposits were characterized by broad signals ($\Delta H = 1.0$ mT). For zeolites with Si/Al = 30–100, which were on stream for 130–260 h, linewidths from 1.05 to 0.85 mT were observed. Narrow signals ($\Delta H = 0.5$ mT) and lifetimes up to 460 h were found for samples with Si/Al ratios greater than 100. These results

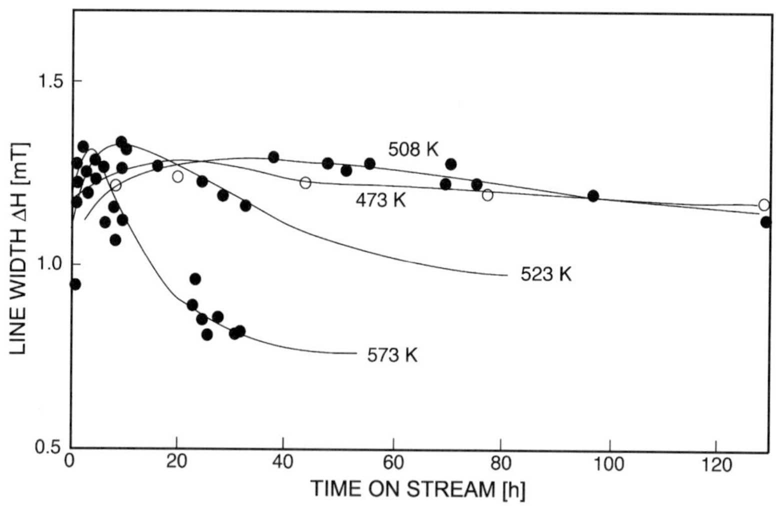

Fig. 28 Width of the high-temperature coke signal in ESR spectra as a function of time on stream, during reaction of ethylene over H-MOR at various temperatures (after [194])

suggest that, when the density of active sites is high, a fast condensation of coke precursors takes place. At lower site densities coke species cannot condense into polyaromatic structures, and the observed narrow ESR signals on high-silica zeolites can be ascribed largely to polycondensed deposits on the outer surface of the crystallites. The authors conclude that zeolite deactivation by coke deposits can be decelerated either by an increase in the external surface area or by a decrease in the number of framework Al atoms on the external surface of the crystals.

During the reaction of ethylene on H-MOR at higher reaction temperatures, Karge et al. [194] observed that the linewidth first increased to a maximum of about 1.3 mT with time on stream, but was reduced at longer times on stream (Fig. 28). During the entire process the *g* value of the line remained at 2.0024. Similar results were observed by Meinhold and Bibby [87] for the conversion of methanol on H-ZSM-5. The linewidth reached a maximum of about 0.95 mT at 3 wt % coke on the zeolite and decreased to a value of about 0.1 mT in a sample with 23.6 wt % coke. Moreover, a correlation between signal linewidth, ΔH, and percentage hydrogen in carbonaceous residues on zeolites has been established, i.e., the smaller the line the more deficient in hydrogen is the coke [199]. Related ESR studies of coke formation on H-ZSM-5 during methanol conversion at 693 K are shown in Fig. 29 [202]. Structureless signals with small linewidths of about 0.22 mT and 0.37 mT were obtained for the heavily (21.5 wt %) and weakly (0.5 wt %) coked sample, respectively. Although both spectra show the presence of polyaromatic carbonaceous residues, i.e., $\Delta H \leq 0.5$ mT, the spectral linewidth of the weakly

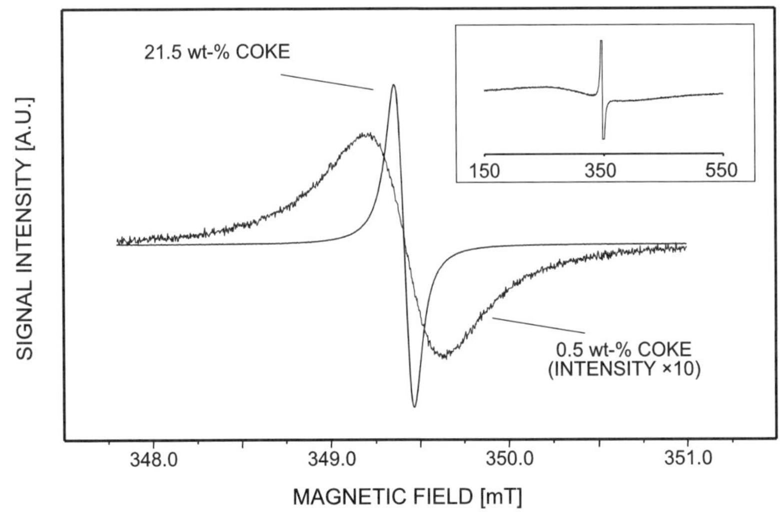

Fig. 29 ESR spectra of weakly and heavily coked H-ZSM-5 after methanol conversion at 693 K. *Insert*: scale down of spectrum showing the overall lineshape (after [203])

coked sample is broader indicating a higher hydrogen content and/or small aromatic areas. For a coke content of 21.5 wt %, it cannot be excluded that the main part of the coke has a multilayer graphite-like structure which is indicated by a very broad line of about 150 mT due to interaction with conduction electrons. The g-values of both ESR signals are the same (2.0026). The significant narrowing of the signal with time on stream or coke content may indicate that the so-called aging of coke during the reaction period corresponds to a change in the composition of carbonaceous deposits in the direction to more aromatic compounds.

Similar to the problem of "NMR visible" carbon atoms in cross-polarization MAS NMR experiments, ESR studies may suffer from the fact that only a small fraction of the carbon atoms of the coke residues yields radicals. The radical density of coke formed at higher temperatures corresponds approximately to one radical per 500–10 000 carbon atoms. Karge et al. [194] observed, however, a close (linear) correlation between the concentration of radicals of high-temperature coke and the total amount of coke formed on H-MOR at 573 K (determined via TGA). Therefore, the ESR results were regarded as representative of the formation of carbonaceous residues on zeolite catalysts.

Furthermore, ESR investigations allow some insight into the nature of paramagnetic centers of zeolites formed by impurities of Fe, Ti, V and Zr. Because commercial zeolite samples contain iron impurities up to the 100 ppm range their role in coke formation needs elucidation. Lange et al. [192] observed two lines of iron cations on H-MOR, a sharp one with a g-value of 4.34 and a broad one with $g = 2.05$. They were ascribed to tetrahedrally and octahedrally coordinated Fe^{3+} ions, respectively. While at lower temperature both lines remained unchanged after adsorption of ethylene, at reaction temperatures above 500 K both iron lines vanished. Simultaneously a broad line with approximately $g = 3.0$ appeared which could be attributed to amorphous iron oxide. The authors concluded that the paramagnetic centers of H-MOR do not react with olefins and are not responsible for the formation of coke radicals.

2.7 Electron Microscopy and Electron Spectroscopy

Advanced microscopic techniques such as transmission and scanning transmission electron microscopy (TEM, STEM), electron and core electron energy loss spectroscopy (EELS, CEELS), X-ray photoelectron spectroscopy (XPS) and Auger electron spectroscopy (AES) have been applied in catalyst characterization. The spatial resolution of the analysis can be as small as 1 nm^2, i.e., it is possible to detect the location of coke. The benefits of these sophisticated techniques for surface science are obvious, but some complications may question their suitability for the characterization of carbonaceous deposits. The use of an ultra-high vacuum may result in a migration of some coke species from the interior catalyst pore system to the external surface. Electron beam damage may appear and the assignments of signals might be erroneous because of severe conductivity problems or the presence of carbon contamination in the spectrometer [204]. For example, the disturbance of XPS spectra of coked zeolite caused by charging effects depends on the degree and topology of the surface coverage with coke. Nevertheless, photoelectron spectroscopy is a promising and attractive method for coking studies. Particularly modern spectrometer devices are equipped with several of these techniques and permit essentially in-situ measurements by the combination with a reaction chamber [205].

Applying TEM, Gallezot et al. [206] were able to show that external coke deposits formed upon *n*-hexane cracking over H-OFF (13 wt% coke) and H-ZSM-5 (7 wt% coke) are like a shell or cocoon which is sufficiently rigid to stand even if the zeolite crystallite has been dissolved by hydrofluoric acid treatment [207]. Carbonaceous residues on H-USY (16 wt% coke) exhibit a more heterogeneous morphology. Even filaments of coke species, about 1–2 nm thick, were observed suggesting that some coke is produced inside the zeolite micropores and then emerges from the zeolite like an extrudate. Further examinations with TEM by Behrsing et al. [208] revealed that coke on the external surface of H-ZSM-5 crystals became noticeable at about 6 wt% coke. At contents of more than 10 wt%, coke was present as a continuous layer thicker than 10 nm on the zeolite crystallites. One of the samples with 21 wt% coke was still active in the methanol conversion showing that the carbonaceous deposits on the external surface might be transmittant for the passage of reactants to and from the channels. TEM images of H-FER obtained after 1-butene skeletal isomerization [15] showed for a sample containing about 9 wt% coke that the surface of the FER crystallites was covered by a layer of amorphous material. Due to XPS data was this probably amorphous coke [209].

Surface analysis by XPS is limited to the first few atomic layers and predisposed both to elucidate surface framework dealumination and to distinguish between carbonaceous deposition in the interior pore system and on the external surface of the zeolite crystallites (vide infra, Sect. 3.3). Peaks at 285 eV (C 1*s*), 530 eV (O 1*s*), 153 or 103 eV (Si 2*s*/2*p*), and 120 or 75 eV (Al 2*s*/2*p*) can be taken to represent the coke and the surface of zeolitic crystallites. Sexton et al. [210] determined the C/Si ratio as a function of the coke content of H-ZSM-5. This ratio increases linearly at a low rate with the coke content up to 8 wt% coke, and above about 15 wt% coke exponentially. Since methanol conversion into hydrocarbons ceased at the latter coke content, the authors suggested that the external coke was probably formed by the thermal cracking of methanol. The C 1*s* binding energy obtained from samples with high external coke content was 284.3 eV, whereas that from low-coke samples with predominantly internal coke was 284.5 to 284.7 eV. Although there is little variation in the C 1*s* binding energy levels of paraffinic and aromatic hydrocarbons, the authors concluded that this systematic difference provides some support for an overall chemical difference between internal and external coke. Mooiweer et al. [211] investigated *n*-butene isomerization on the medium-pore zeolite FER. XPS and TEM studies indicated that up to a carbon level of about 7.5 wt% the majority of the carbonaceous residue is located in the pores. At higher carbon levels surface coking becomes dominant.

XPS spectra of spent H-ZSM-5 samples showed reduced peak areas compared to the corresponding ones of fresh catalysts but still exhibited Si and Al signals (Schlögl R, 1991, personal communication). While the Si intensity of H-Y in CEELS measurements was only a little affected by coking, the decrease in the Al intensity has been attributed to coke deposits preferentially on Al sites and to electronic interactions of these mainly aromatic carbon species with the Al ions [212]. This assignment of coke to Al sites due to CEELS results in a contradiction to XPS studies of partially coked H-ZSM-5 catalysts in which a decrease of the Si(2*s*) and O(1*s*) binding energies by 1.0 eV was found to be due to electron charge transfer from the coke to the framework atoms [213]. The latter results suggested that the coke was deposited on SiO_4 tetrahedra. Since the depth of observation of both techniques is, however, restricted to a few monolayers, the coverage of the external surface of the zeolite grains with a thick homogeneous layer of coke would result in a complete suppression of the Si, Al, and O signals. Therefore, one has to conclude that the external zeolite surface was not completely covered by a homogeneous coke layer; rather a topologically heterogeneous coverage is suggested.

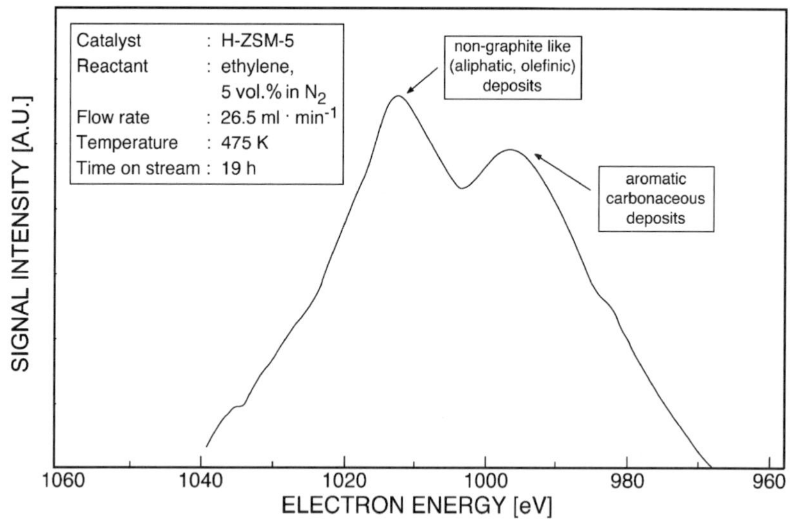

Fig. 30 Carbon Auger spectrum of an H-ZSM-5 catalyst after coking in an ethylene stream at 475 K (after Schlögl, personal communication)

The carbon AES spectrum of a coked H-ZSM-5 sample is shown in Fig. 30. The spectrum exhibits two broad peaks, viz. at 995 eV and around 1010 eV. The former peak is ascribable to highly aromatic carbonaceous deposits. The high energy peak around 1010 eV is indicative of deposits richer in hydrogen and aliphatic or olefinic in nature.

In qualitative agreement with the spectroscopic results described above, surface analysis of coked zeolites by CEELS, XPS, and AES shows the appearance of both

(i) paraffinic or olefinic, hydrogen-rich coke-type I (with lower C binding energies) and
(ii) carbon residues with dominating aromatic species and higher C binding energies but still non-graphite species (corresponding to coke-type II).

Electron microscopy and spectroscopy will certainly allow for further valuable and very detailed in-situ investigations of the nature of coke and the mechanism of its formation.

2.8 X-Ray Diffraction

The molecular and crystalline structure of zeolites has been elucidated with single-crystal and powder diffractometry (cf. Volume 1, Chapter 1 of this series). For example, zeolite ZSM-5 is classified as orthorhombic with several variations in structural parameters according to the Si/Al ratio. A change in framework symmetry between orthorhombic and monoclinic depends on the temperature and is reversible. It has further been shown that the adsorption of bulky molecules results in changes in the diffraction pattern [214]. Therefore, XRD should be capable of detecting coke deposited inside the zeolitic crystallite.

The presence of internal coke has been demonstrated by Bibby et al. [215] from diffraction patterns of H-ZSM-5 coked during the conversion of methanol. Changes in the relative positions and intensities of the peaks in

Table 8 Displacement of XRD lines of H-ZSM-5 in the region of $2\Theta = 23°–24°$ due to coke deposition (from [200])

H-ZSM-5	Observed 2Θ [°]			
Fresh sample	23.10	23.76	23.94	24.40
Sample with 7.4 wt% coke (H/C ratio = 0.80)	23.06	23.59	23.84	24.29
Sample with 13.4 wt% coke (H/C ratio = 0.67)	23.10	23.63	23.88	24.31

the XRD spectra of the coked samples were found and attributed to a distortion of the ZSM-5 structure. These findings agree with the observations of Bauer et al. [106, 200]. The diffraction patterns of H-ZSM-5 after coking by methanol showed peak shifts in the region of $2\Theta = 23°$ and $2\Theta = 24°$. The poorly resolved doublet at $2\Theta = 23.1°$ and 23.3° of the fresh sample vanished with coking. Table 8 reveals the considerable displacement of the line $2\Theta = 23.76°$ to smaller angles and shows how coke causes lattice distortions in the zeolite.

Supported by structural calculations, Alvarez et al. [216] showed that a phase change was caused by the deposition of carbonaceous materials within the zeolitic pores, and the crystalline structure of coked ZSM-5 became tetragonal. In Fig. 31 obvious changes in the diffraction pattern can be seen in the region between $2\Theta = 23°$ and $2\Theta = 24°$. The doublet corresponding to the lines $2\Theta = 23.08°$ and 23.32° appears after coking as a single line at $2\Theta = 23.207°$. The displacement of the line $2\Theta = 23.720°$ to $2\Theta = 23.635°$ is evident. The final cell constants were calculated as $a = 1.9965$ nm and $c = 1.3380$ nm, which are close to those reported by Pollack et al. [217] with $a = 2.003$ nm, $b = 1.999$ nm, and $c = 1.341$ nm. Regeneration processes reverse this structural change.

Similar results were obtained by Uguina et al. [218] for the conversion of toluene and *i*-butene on H-ZSM-5. During the conversion of mesitylene, however, both the doublet at $2\Theta = 23.1°$ and 23.3° and the position of the peak at $2\Theta = 23.7°$ remained nearly unchanged. Coke built from mesitylene did not seem to change the cell dimensions of ZSM-5, which also suggests its loca-

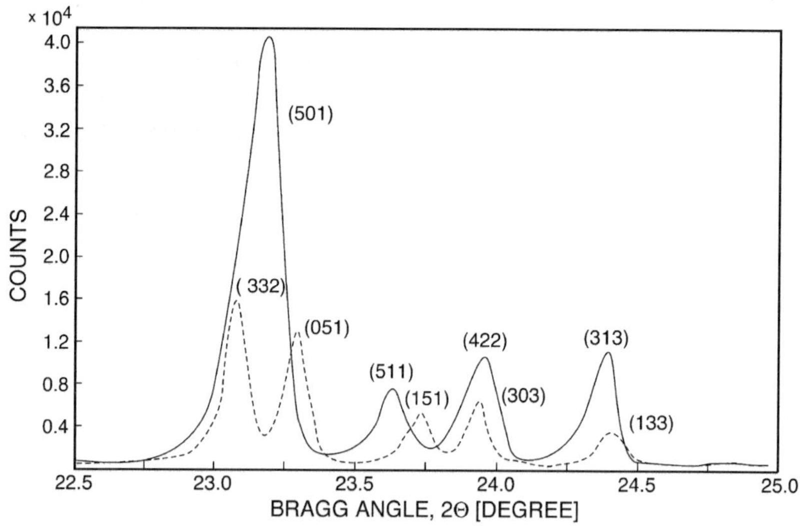

Fig. 31 Comparison of the diffraction pattern of ZSM-5 in fresh, orthorhombic (*dashed line*) and coked, tetragonal (*solid line*) forms (after [216])

tion on the outer surface of the crystallites (compare [152]). Okkel et al. [219] observed this doublet in the X-ray pattern of ZSM-5 during conversion of ethylene up to coke contents of about 5 wt %, i.e. at low coke coverage no distortion of the zeolite lattice was detected.

The deformation of the ZSM-5 framework caused by the deposition of carbonaceous material in channels and intersections may depend not only on the amount but also on the nature of coke (indicated by the H/C ratio in Table 6). Rozwadowski et al. [82] investigated the methanol conversion on H-Y and observed for temperatures up to 473 K that the unit cell parameter remains close to that of the non-treated zeolite, viz. $a = 2.460$ nm. For the temperature range of 523–593 K the unit cell parameter is higher by about 0.01 nm and, above 593 K, it increases with the reaction temperature. The authors concluded that coke formed at temperatures below 593 K influences the zeolite lattice in a different way than that built above 593 K, i.e. different kinds of carbonaceous residues are present in the two temperature ranges. For the detection of such subtle effects further XRD investigations have to be carried out using the Rietveld procedure.

After coking H-Y and H-MOR during *n*-hexane cracking and dissolution of the zeolite matrix with hydrofluoric acid, Mori et al. [220] used XRD for the characterization of the liberated coke. The diffraction peak at $2\Theta = 26.37°$ assigned to ordered graphite was found to be higher in H-Y than in H-MOR and related to the difference in the pore structure of both zeolites.

2.9 Extraction of Coked Zeolite Catalysts

Spectroscopic techniques as described in the preceding sections have the advantage of not modifying the spent catalyst samples and of allowing an in-situ characterization of carbonaceous deposits. These methods give valuable information on the chemical identity of pure substances; unfortunately, none of them is able to give the detailed composition of a mixture of coke species. However, the knowledge of the composition of carbonaceous residues is essential to specify the route of coke formation and the deactivation effect of individual coke precursors. Unfortunately, only a small part of the carbonaceous residues can be recovered by direct soxhlet treatment of coked zeolites with organic solvents.

Holmes et al. [207] used CCl_4, a bulky molecule, too large to enter the pores of H-ZSM-5, for removing external carbonaceous residues formed during hexadecane conversion at 623 K. A second extraction of the coked zeolite with CH_2Cl_2, which can in principle enter the channels, was expected to remove tiny coke species formed within the pore system.

Comparable with solvent extraction, a headspace and high-temperature GC-MS method for analyzing the volatile components of carbonaceous residues was described [221]. The coked zeolite was heated to 773 K under

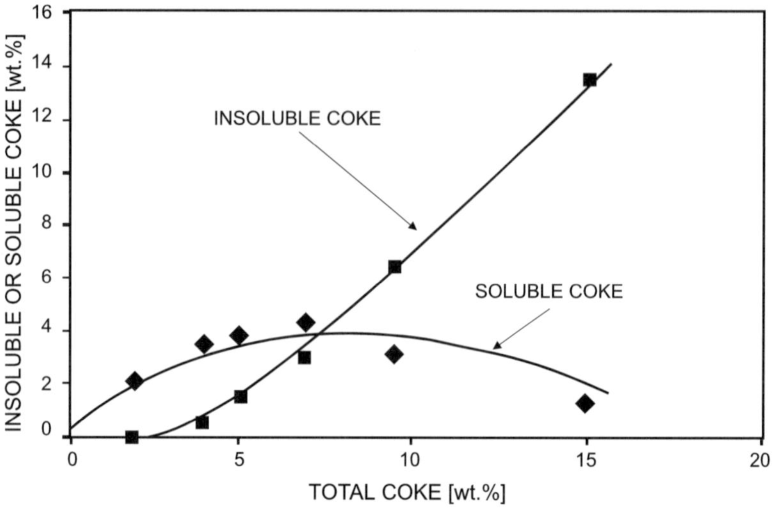

Fig. 32 Percentage of coke soluble and insoluble in CH_2Cl_2 as a function of the percentage of coke formed during *n*-heptane cracking at 723 K on H-USY (after [227])

a flow of helium, and an open-tubular trap with a thick liquid film was employed as a preconcentration trap in on-line dynamic headspace sampling. The on-line purge and trap system with thermal desorption allowed coke components of low volatility to be transferred to the GC column, and the high temperature capillary GC-MS permitted these high-molecular weight components to be detected. The proposed method can identify polynuclear aromatics with up to nine rings. However, changes in the nature of coke during thermal evaporation at about 773 K cannot be excluded (vide supra, Sect. 2.1).

To determine the genuine distribution of bulky coke species, the carbonaceous deposits must be released from the zeolite framework, e.g., by dissolving the zeolite in acid solutions or by destroying the zeolitic structure with ball-milling. Soluble coke components can then be extracted by different solvents, separated by chromatographic methods (GC, HPLC), and identified through appropriate techniques such as IR, UV-Vis, NMR and MS. It should be kept in mind, however, that the identification of coke species for a given reaction temperature and time on stream is not necessarily related to the deactivation process, which in many cases is initiated by feed impurities.

Venuto and Hamilton [222] were the first to dissolve coked zeolite catalysts and analyze the extracts of the solution by GC. Since they used RE-X catalysts, a relatively mild acid treatment was sufficient to destroy the zeolite matrix. The method of acid treatment and extraction of coked zeolite catalysts was further developed and extensively used by Guisnet and coworkers in a series of systematic studies on coke formation [6, 51, 223–230]. The carbonaceous deposits were liberated by dissolution of the aluminosilicate matrix in hydrofluoric acid (40%) at room temperature. Subsequently, the soluble components were extracted by CH_2Cl_2 as a solvent; in some cases "insoluble coke" remained in the form of black particles. Figure 32 shows the percentage of soluble and insoluble coke as a function of the percentage of coke formed during *n*-heptane cracking at 723 K on H-USY [227].

2.9.1
Composition of Soluble Coke

A serious problem related to the dissolution/extraction method, however, is whether or not changes in the original chemical nature of the coke occur upon the rigorous treatment with mineral acids. Magnoux et al. [20] have carried out a series of tests which seem to confirm that the procedure employed does not modify the coke or generate any artifact. It is, however, questionable whether this is true in every situation. Ball-milling of spent zeolite catalyst has shown to be insufficient for a total release of retained carbonaceous materials [200].

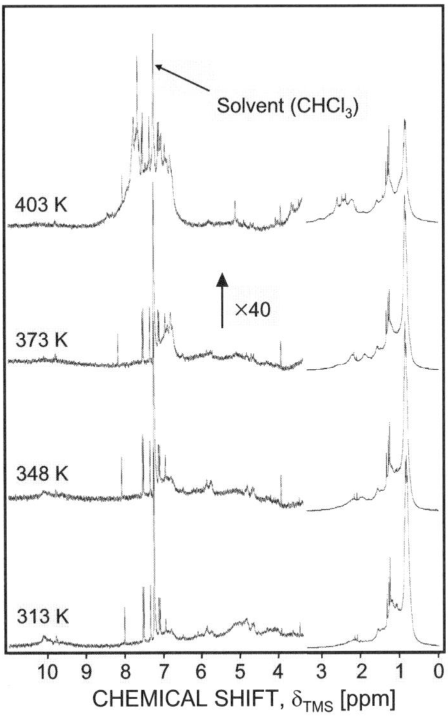

Fig. 33 ^{1}H NMR spectra of released deposits after La,H-X zeolite dissolution recovered after isobutene/butene alkylation at different temperatures (after [114])

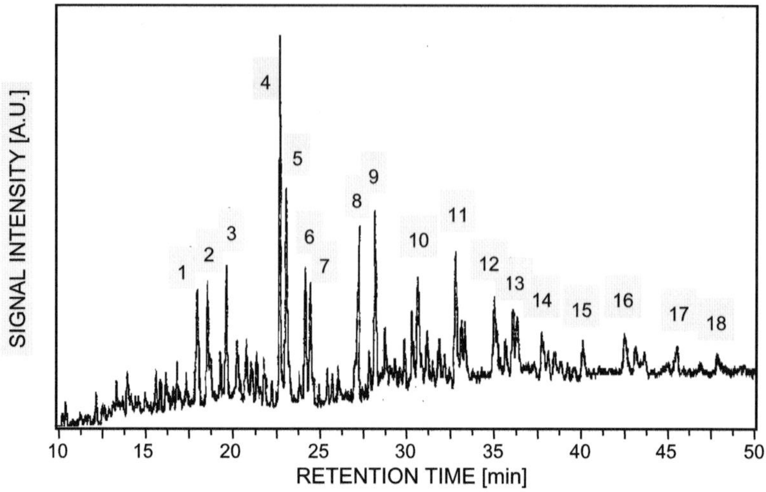

Fig. 34 Total ion chromatogram of volatile components of coke after aromatics transformation on zeolite catalyst (after [221]) 1, acenaphthylene; 2, acenaphthene; 3, fluorene; 4, phenanthrene; 5, anthracene; 6, 2-/3-methylphenanthrene; 7, 1-/2-methylanthracene; 8, fluoranthene; 9, pyrene; 10, 2.3-/3.4-benzofluorene; 11, 3.4-benzophenanthrene; 12, benzo(b)fluoranthene; 13, benzo(a)pyrene; 14, indeno(1.2.3-cd)pyrene; 15, benzo(ghi) perylene; 16, coronene; 17, benzo(a)coronene and 18, ovalcene

The composition of coke depends, among other parameters, on the pore structure of the zeolites used. This has been demonstrated by many studies with large-pore size zeolites, e.g., H-Y and H-MOR, with intermediate-pore size pentasil zeolites, e.g., H-ZSM-5, H-ZSM-8 and H-ZSM-11, and with zeolites of small pore apertures, e.g., H-ERI or H-ZSM-34. Naturally, these zeolites have not the same acidity. Certain differences in coking could be due to differences in acidity. Therefore, Guisnet's group studied *n*-heptane cracking on zeolites having similar initial activity. This was achieved by different percentages of protonic exchange of the zeolites.

As shown in Fig. 33, the composition of the soluble coke was found to be rather complex in view of ^{1}H NMR spectroscopic characterization [37, 114, 231]. However, GC-MS provides useful information on both the mass and the structure [221]. Depending on the coke loading, however, a number of families of polynuclear aromatic compounds constituting the carbonaceous deposits were deduced from the GC-MS experiments (Fig. 34). The main components found at low and at high coke contents as well as their size and the approximate value of their boiling point under normal pressure are indicated in Table 9.

For all zeolites, the H/C ratio of the carbonaceous deposits decreases with increasing time on stream. The general trend of aging of coke towards higher aromaticity is also indicated by the composition of soluble coke. On the other

Table 9 Main components of the soluble coke for low and high coke contents; size and boiling point (adapted from [6])

Zeolite	Coke content	
H-USY	2 wt.%	9 wt.%
	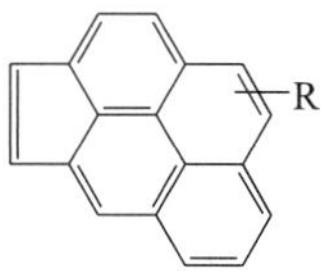	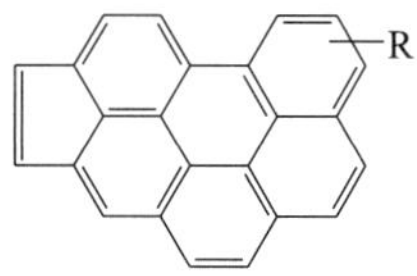
	(0.85 × 0.95 nm) (673-723K)	(0.85 × 1.2 nm) (793 K)
H-MOR	2 wt.%	4.5 wt.%
	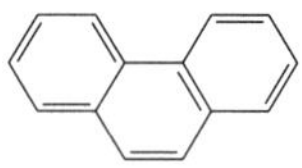	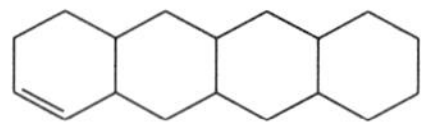
	(0.65 × 0.85 nm) (613 K)	(0.6 × 1.2 nm) (573 - 623 K)
H-ZSM-5	1 wt.%	7 wt.%
	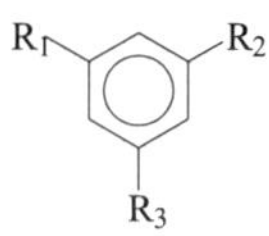	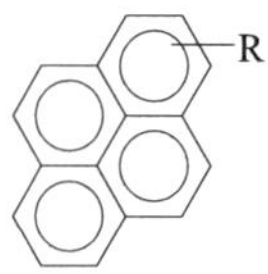
	(1.0 nm) (473 - 523 K)	(0.85 × 0.85 nm) (673 K)
H-ERI	0.5 wt.%	6 wt.%
	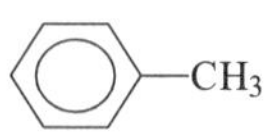	
	(0.65 nm) (384 K)	(0.65 × 1.25 nm) (723 K)

hand, $C_5 - C_{16}$ alkanes have been found even at reaction temperatures above 500 K [200, 232, 233]. This finding confirms the existence of paraffinic coke species.

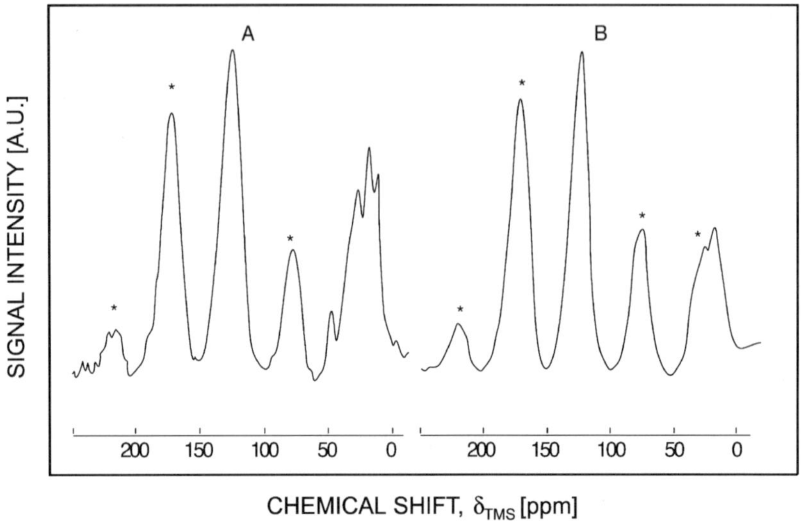

Fig. 35 ^{13}C CP MAS NMR spectra of coke on H-ZSM-5 formed during methanol conversion (A) and extracted coke after dissolution of the zeolite catalyst (B) (after [185]); *asterisks* denote spinning sidebands

2.9.2
Composition of Nonsoluble Coke

Only about 30 wt% of the carbonaceous residues formed during methanol conversion on H-ZSM-5 at 723 K were soluble in methylene chloride indicating, thereby, the polyaromatic nature of the carbonaceous residues [185]. The characterization of this insoluble coke can be carried out by the aforementioned techniques. For example, ^{13}C MAS NMR confirms the hydrogen-deficient nature of the insoluble coke particles (Fig. 35). Compared with the coke on the spent zeolite before zeolite dissolution, the shape of the signal in the aromatic region is not changed, whereas a significant reduction of aliphatic carbon atoms was observed for the insoluble coke [185].

To estimate the number of coke molecules, Cerqueira et al. [234] assumed a value of 600 g mol^{-1} for the molecular weight of insoluble coke formed during *m*-xylene transformation on H-USY. In this context, matrix-assisted laser desorption/ionization time-of-flight mass spectrometry (MALDI-TOF MS) proved to be an efficient tool. This technique has provided valuable information in the field of polymer analysis [235]. It utilizes the energy of a laser beam for desorbing and ionizing sample molecules, which are subsequently analyzed in a time-of-flight mass spectrometer. In MALDI-TOF MS, a matrix absorbs the laser power and minimizes thereby fragmentation reactions. Thus, the obtained mass spectrum directly points to the molecular weight of individual species as well as the molecular weight distribution of the analyzed mixture. Moreover, MALDI-TOF MS yields qualitative information about repeat units and end groups of polymeric species [236].

In the case of carbonaceous deposits formed during isobutane/butene alkylation on La,H-X between 303 and 403 K, Feller et al. [114] applied MALDI-TOF MS to analyze the deactivated catalysts and carbonaceous residues after zeolite dissolution. Peaks in the mass range from 150 Da to 450 Da (carbon number distribution in the range of $C_{12} - C_{35}$) and a repetitive pattern of 14 Da (CH_2 group) were obtained. The main peaks were related to series such as C_nH_{2n-14} and C_nH_{2n-18}, which was in good agreement with the distribution measured by GC-MS. After zeolite dissolution, mass spectra of the free carbonaceous deposits resembled the spectrum of the deactivated catalyst with masses of the formula C_nH_{2n-22} and C_nH_{2n-24}. However, the assignment of individual masses to certain types of compounds was ambiguous, and the interpretation of MALDI TOF mass spectra of catalysts with purely aromatic coke was assumed to be of lesser complexity.

The MALDI-TOF mass spectrum of such hydrogen-deficient, polyaromatic carbonaceous residues formed at 723 K on H-ZSM-5, which were insoluble in CH_2Cl_2 after HF treatment of the coked zeolite, revealed a broad molecular weight distribution (Fig. 36). When the laser beam (without any matrix) was used for desorption and ionization of the released deposits, MS signals appeared which may have originated from radical cations (Bauer F, 2005, personal communication). Starting with a signal at 202 Da assigned to pyrene

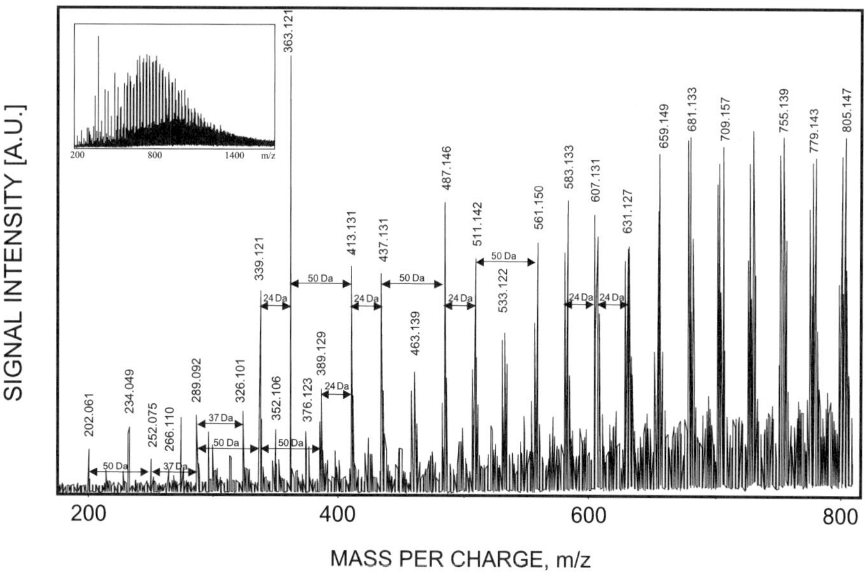

Fig. 36 MALDI-TOF mass spectrum of nonsoluble coke obtained from H-ZSM-5 after precoking treatment at 723 K (after Bauer F, unpublished results)

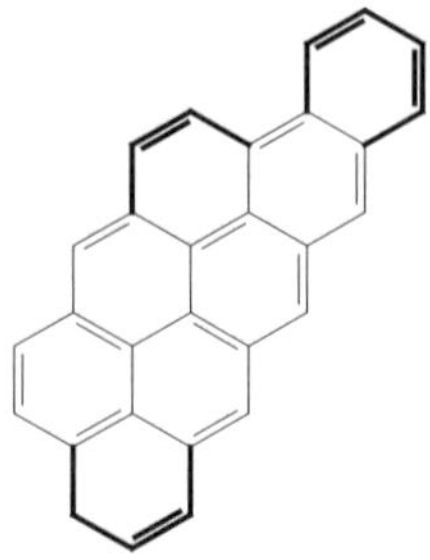

Fig. 37 $C_{29}H_{16}$ (MS signal at 363 Da) as an example for schematic build-up of polyaromatic coke on H-ZSM-5 via C_2 (24 Da), C_3H_2 (37 + 1 Da), and C_4H_2 (50 Da) building units

($C_{16}H_{10}$), the mass distribution had a sharp maximum at 363 Da and a second broad maximum at about 750 Da. Molecular ion peaks up to 1300 Da corresponding to species with $n_C < 100$ were detected. The complex pattern of the MS signals with repeating mass increments of 24 Da, 37(+1) Da, and 50 Da is quite different from that observed for carbonaceous residues formed at temperatures $T < 400$ K with CH_2 increments [114]. These building units can be assigned to increments of two, three, and four carbon atom entities (marked in bold lines in Fig. 37) attached at different positions to a parent polyaromatic structure. The appearance of odd mass signals from condensed aromatics can be explained by splitting off of one hydrogen (similar to what can be observed with phenalene, $C_{13}H_{10}$). Hence, the prevailing MS signal at 363 Da can be assigned to $C_{29}H_{16}$ with the possible structure shown in Fig. 37. This two-dimensional, belt-like topology of polyaromatic deposits can be formed inside the channels of ZSM-5, and yet a migration of slightly polyaromatic molecules towards the crystallite surface may be assumed. Obviously, large carbon entities of $n_C > 100$ are preferentially deposited on the external surface and/or in the pore-mouth region (Bauer F, 2005, personal communication).

The method of coke extraction confirms the discrimination of carbonaceous species into

(i) low-molecular aliphatic residues (coke-type I) which are totally soluble in organic solvents and the aromaticity of which increases with the coke content and the reaction temperature, and
(ii) highly polyaromatic deposits (coke-type II) which are partly insoluble and the amount of which increases with reaction temperature and time on stream.

Upon *n*-heptane cracking on H-ZSM-5 at 723 K, insoluble coke can form an external envelope around the zeolite crystal which remains after zeolite dissolution [206]. EELS showed that its structure is similar to that of coronene (pre-graphitic). For the non-soluble part of coke on H-USY, which appeared in filaments protruding from the zeolite pores, a more pentacene-like structure (linear polyaromatic) was found under identical reaction conditions. At even higher reaction temperatures (1073–1173 K), carbonaceous materials formed on zeolite Y during thermal decomposition of different organic substrates and released by shaking in concentrated hydrofluoric acid, exhibited remarkably high specific surface areas of about 1300–2000 m^2/g [237–239]. Although the shape of particles was often retained after dissolution of the zeolite framework, X-ray diffraction patterns showed that the long-range order of the zeolite was not preserved by the carbon polymers. However, on mesoporous MCM-48, Ryoo et al. [240] obtained a three-dimensional regular carbon structure (denoted CMK-1) having uniform pores 3 nm in diameter after conversion of sucrose at 1073–1373 K under vacuum and dissolution of the silica framework by an aqueous/ethanol solution of NaOH. Elementary analysis gave the empirical formula of $C_{16}H_1O_1$ for CMK-1 materials obtained after heating at 1373 K. Using the mesoporous structure of SBA-15 as a template, the formed mesoporous carbon material CMK-3 consisted of uniformly sized carbon rods arranged in a hexagonal pattern and maintained by interconnecting micropores [241]. Obviously, carbonization reactions at temperatures above 1000 K yield a structure replica of the templating zeolitic material. After template removal, the resulting carbon materials with high surface areas and uniform pores promise to be suitable as adsorbents and catalyst supports [241, 242]. It is interesting to point to the possibility of using the mesoporous carbons themselves as templates for mesoporous inorganic materials such as alumina, titania, zirconia, etc. because the carbons have well-ordered mesostructures and rigidity [243].

Only a few papers describe the formation of coke during hydrocarbon reactions on microporous molecular sieves other than aluminosilicates, e.g., on silicoaluminophoshates (SAPOs). Similar to chabazite, the pore structure of H-SAPO-34 consists of elongated cages (1.0 nm × 0.66 nm) linked via relatively small eight-ring intercage windows (free diameter of 0.43 nm). Anderson et al. [232] treated H-SAPO-34 with 1-hexene at 513 K and applied the method of extraction in their study on retained carbonaceous products. The cage size is large enough for rather bulky coke molecules. In fact, a striking result after dissolution of the zeolite framework was a substantial proportion of adamantane and alkyladamantane in the soluble coke. At 673 K only trace amounts of adamantanes could be detected, whereas alkylnaphthalenes and anthracenes/phenanthrenes dominated. At a lower reaction temperature of 453 K, the main product classes were (cyclo)alkanes/olefins, alkylbenzenes, and alkylcyclopentadienes. Since alkylcyclopentadienes have been identified as a substantial product class during the reaction of 1-hexene on H-USY and H-ZSM-5 [244], in agreement with earlier interpretations [76] the authors proposed a generalized pathway of coke formation with cyclopentadienes as important intermediates. Furthermore, no adamantanes were observed in the coke on H-USY and H-ZSM-5. As demonstrated by Schnabel et al. [135], the

IR spectra of carbonaceous residues from methanol conversion on H-SAPO-34 and H-SAPO-11 differed significantly from that of adamantane.

Finally, the in-situ regeneration of deactivated zeolites by dissolving coke or coke precursors with supercritical fluids has been proposed. Niu et al. [245] studied the disproportionation of ethylbenzene on H-Y under supercritical conditions (573–673 K, 60–80 bar) with butane as an inert component and observed longer times on stream compared with normal pressure. However, a complete coke removal from zeolite catalysts by supercritical fluid extraction with benzene or ethylbenzene did not occur [246].

2.10 Adsorption Measurements

Notwithstanding all of the sophisticated spectroscopic and hyphenated techniques (e.g., such as TG-FTIR and GC-MS), simple sorption measurements belong to the most important methods for studying fresh and spent zeolites. On the basis of theoretical models, sorption isotherms can yield information about the surface area and the pore radius distribution which is of great importance for the development of catalysts as well as for the elucidation of catalyst deactivation. The applicability of the theories of Brunauer–Emmet–Teller (BET) and Polanyi–Dubinin to sorption studies on microporous materials has been critically examined [247, 248] (cf. also volume 7 of this series).

Adsorption measurements are suitable to clarify whether the coke is predominantly laid down in the zeolite pores or on the outer surface. Using small adsorbate species such as nitrogen, water, ammonia, trimethylamine, pyridine, and small paraffins, the inter- and intra-crystalline surface of zeolites can be investigated. Bulkier probe molecules such as methyldiisopropylamine, substituted benzenes, and methylene blue, which have no access to the particular internal volume, were used to estimate the external surface of the zeolite crystallites.

As an example, Bibby and Pope [249] used trimethylamine for monitoring the change in the total number of acid sites on H-ZSM-5 caused by coke deposition. Ethyldiisopropylamine, which is too large to penetrate into the zeolite channels, gave independent information about the exterior surface sites. In the initial stages of coking, some empty regions of the crystals were sealed off, but thereafter most of the decrease in sorption capacity resulted from simple pore filling. Only a little coke was deposited externally. The authors' proposal for catalyst design with respect to long times on stream in the methanol conversion is an H-ZSM-5 catalyst with a relatively low density of active sites and small crystallite size (vide infra, Sect. 3.2).

Strong basic adsorbates such as ammonia and pyridine, however, cannot only interact with acid sites but also replace some adsorbed coke species [113, 114]. This conclusion has to be drawn from IR studies aimed at the estimation of acid site concentration. The concentration of both Brønsted and

Lewis sites is commonly determined from the interaction of pyridine with OH groups, i.e., the formation of pyridinium ions (band at 1545 cm^{-1}) and pyridine bonded to Lewis acid sites (band at 1450 cm^{-1}, cf. also volume 4 of this series). Whereas carbonaceous deposits on H-USY affected neither the concentration nor the strength of Lewis sites, coke formation is preferentially associated with the strongest Brønsted sites. However, weakly basic coke molecules can be displaced by more basic molecules such as pyridine. Since a substantial fraction of carbonaceous residues could be replaced by pyridine [113], some of them obviously were not too bulky and therefore easily desorbed from the zeolite pores. As demonstrated by Feller et al. [114], pyridine with its high proton affinity of 930 kJ/mol can replace almost all carbenium ions (vide supra, Sect. 2.5.1).

Nitrogen, argon, krypton, and xenon as nonpolar and non-basic molecules fulfill the requirements of non-specific physical adsorption. For most purposes, nitrogen is probably the best choice for surface area measurements. For example, Bauer et al. [106] compared H-ZSM-5 samples which had spent quite different times on stream in the methanol conversion. The void volumes and the BET analogous surface areas of the fresh and coked catalysts were probed by nitrogen adsorption. Results are presented in Table 10. From these data the authors concluded that coke deposition proceeded preferentially inside the crystals, and in the case of sample C and D (both having higher acid site density, see Table 7) blocking of channel entrances occurred which sealed off some intra-crystalline regions.

Post and van Hooff [250] compared different zeolite structures (H-Y, H-MOR, H-ZSM-5) with respect to pore filling by *n*-butane before and after cracking of *n*-hexane at 573 K. In the case of H-Y and H-MOR with 2.59 and 4.75 wt % coke, respectively, carbonaceous residues were located inside the pore structure, whereas coke deposition occurred only to a minor ex-

Table 10 Time on stream, coke deposition and decrease of surface area and pore volume after methanol conversion at 693 K for different H-ZSM-5 samples (cf. Table 6), adapted from [200]

	Sample A	Sample B	Sample C	Sample D
Time on stream [h]	267	433	60	57
Coke deposition [wt%]	5.0	16.3	4.6	4.5
Initial surface area [m^2/g]	260	270	190	180
Final surface area [m^2/g]	230	160	140	100
Initial pore volume [ml/g]	0.296	0.269	0.207	0.178
Final pore volume [ml/g]	0.235	0.123	0.176	0.155
Loss of pore volume [%] per Coke deposition [wt%]	1.8	1.1	3.3	2.9

tent in the channels of H-ZSM-5 (about 2 wt % coke). Okkel et al. [251] and Soni et al. [252] confirmed the occurrence of these external and internal coke residues on H-ZSM-5 by sorption measurements with argon and nitrogen. In contrast, Bibby and Pope [249] and Bauer et al. [106] claimed preferentially intra-crystalline coke deposition up to 5–10 wt % coke content.

The *n*-butane adsorption measurements of Uguina et al. [218] clearly showed that the location of the carbonaceous residues strongly depended on the hydrocarbon used as the coke-forming agent. Whereas coking with isobutene led to deposits located within the internal pore structure, coke from toluene is deposited both on the external and the internal zeolite surface, and the reaction of mesitylene on H-ZSM-5 yielded coke on the external zeolite surface partially closing the entrance to the channel system.

H-Y and La,H-Y catalysts used in the alkylation of benzene with ethanol and propylene, respectively, were characterized by low-pressure N_2 and di-isopropylbenzene adsorption. While on coked H-Y blocking of the pores was revealed [252], the deactivation of La,H-Y under nearly the same conditions was suggested to be due to coverage of the active sites by carbonaceous deposits [253]. Lechert et al. [254] compared the sorption capacities and the nuclear relaxation times of *n*-butane and benzene in fresh and coked H-Y samples. The almost unchanged sorption capacity for *n*-butane and the small effect on the mobility of benzene supported the conclusion that the coke deposits on H-Y were large entities of relatively high density accommodated in adjacent cavities, leaving an appreciable amount of zeolitic voids free.

Mignard et al. [255] used molecules with different effective molecular diameters (*n*-hexane, 3-methylpentane) in order to determine not only the change in adsorption capacity as a function of coke deposition but also to discriminate elegantly between the various locations of the deposited coke components in, for instance, the different cavities of hydrogen offretite. Mori et al. [220] compared the adsorption capacities of nitrogen, ammonia and *n*-hexane on fresh and coked H-MOR, H-Y and H-ZSM-5. The adsorption of *n*-hexane abruptly declined on the spent H-MOR after *n*-hexane cracking at temperatures above 773 K, whereas only a low decrease of the adsorbed amount of nitrogen and ammonia was observed. Therefore, the carbonaceous deposits on H-MOR prevent *n*-hexane molecules from entering the channels, but most of the acid sites remain intact and accessible for smaller molecules such as ammonia. This meets exactly the observations made by Karge and Boldingh [69] during the conversion of ethylbenzene on H-MOR. Whereas carbonaceous deposits blocked the active sites for the relatively bulky ethylbenzene molecules, the smaller ethylene molecules still had access to the active centers on the spent catalyst.

Xenon adsorption isotherms and ^{129}Xe NMR of adsorbed xenon (vide supra, Sect. 2.5.4) are a striking combination of techniques to probe the local environments inside porous materials. Room temperature xenon adsorption isotherms of fresh H-ZSM-5, deactivated by 21.5 wt % coke formed

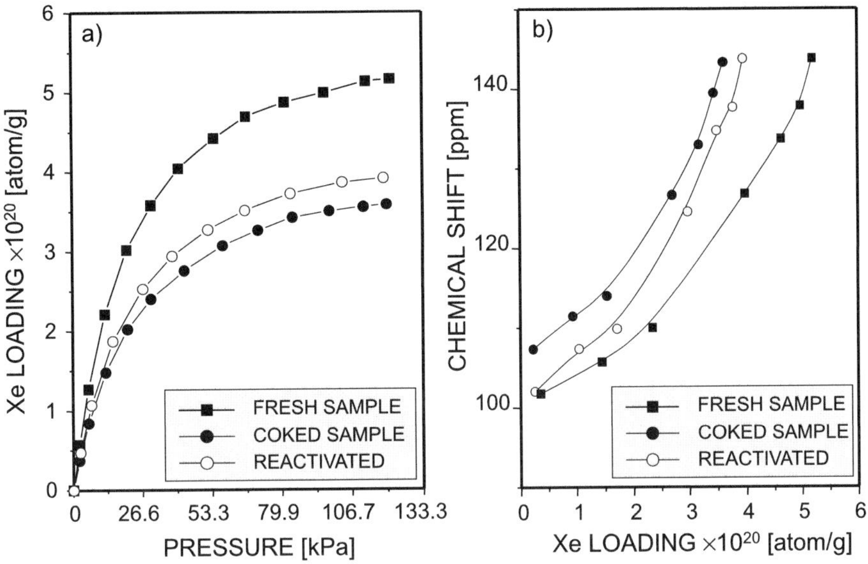

Fig. 38 Room temperature xenon adsorption isotherms (**a**) and variations of ^{129}Xe NMR chemical shifts (**b**) with xenon loading of fresh, spent, and reactivated H-ZSM-5 catalysts (after [203])

Table 11 NMR parameters of H-ZSM-5 samples used in methanol conversion, before and after the additional propane rejuvenation treatment [a] (adapted from [203])

H-ZSM-5 sample	Fresh	Spent	Reactivated
Coke content [wt %]	–	21.5	19.2
δ_S [ppm]	83.1	91.8	86.6
σ_{xe}	11.2	13.6	13.2
V/V_o [%]	100	82.4	84.8

[a] done at 693 K for 24 h

during methanol conversion at 693 K as well as the spent sample after reactivation by propane at 693 K are displayed in Fig. 38a [203]. In general, all isotherms show the presence of Langmuir-type adsorption within the pressure range covered. Notable decreases in both adsorption capacity and adsorption strength were found for the heavily coked sample and the reactivated one compared to the parent catalyst [256]. Moreover, the propane treated sample exhibited a slight increase in adsorption capacity and adsorption strength compared to the spent sample. Therefore, the post-treatment by propane was effective in removing carbonaceous deposits from the channels of the zeolite (vide infra, Sects. 3.4 and 3.5).

Figure 38b displays the variations of ^{129}Xe NMR chemical shifts with xenon loading for the various samples. The NMR parameters obtained from various samples are listed in Table 11. The variations in δ_S for the coked samples are consistent with the corresponding changes in coke content; an increase in δ_S indicates the presence of a larger amount of coke in the intra-crystalline channels of the catalyst and vice versa. After methanol conversion, V/V_0 decreased from 100 to 82.4 indicating that substantial amounts of coke are located within the intra-crystalline channels of the zeolite. Furthermore, a slight increase in V/V_0 reveals the removal of intra-crystalline carbonaceous deposits by the reactivation treatment with propane.

3 Particular Phenomena Related to Coke Formation

In the following sections some aspects of coke on zeolites will be discussed with particular emphasis on the various relations between them. For example, the effect of pore structure on the deactivation of zeolites ought to be investigated on zeolite samples having the same acidity (number and strength of acidic sites). Otherwise, a clear distinction between both effects is not possible. Moreover, coke formation depends not only on the feed composition but also on the nature of reaction intermediates which may vary due to different acid sites, pore systems and reaction conditions. Therefore, any conclusion on coking in zeolite catalysis requires a complex consideration of this phenomenon and needs more information than supplied by only one of the above-mentioned techniques.

3.1 Selectivity of Coke Formation

The idea that in zeolite catalysis both desired reactions and undesired coke formation are influenced by the steric constraints imposed by the zeolite framework was first advanced by Rollmann [7] and Rollmann and Walsh [257]. In general the formation of bulky coke species will be minimized in the case of zeolites with small apertures of a tridimensional pore system in which there are no large cavities.

Reactant selectivity in coke formation involves the influence of both the pore geometry of zeolites and the molecular size of substances. Thus, in the conversion of complex feedstocks, small pore-sized zeolites (ferrierite, erionite), which exhibit high shape selectivity, were also characterized by low deactivation rates [7]. By contrast, large-pore zeolites (mordenite, faujasite) did not selectively convert the feed mixture and, correspondingly, suffered from severe coking. On the contrary, narrow-pore zeolites (e.g., such as chabazite and ZSM-34) are effective for the conversion of methanol to olefins but have short catalyst life due to rapid coke deposition, whereas a pentasil-type H-Fe-silicate yielded almost exclusively ethylene and propylene [258]. Uguina et al. [218] investigated coking on H-ZSM-5 with mesitylene, toluene and *i*-butene as coke-forming reactants. XRD results clearly indicated that *i*-butene and toluene lead to carbonaceous residues located within the pore structure, whereas coke formed from mesitylene, which cannot enter the channel system, is deposited on the external zeolite surface.

The concept of restricted transition-state shape selectivity in zeolite catalysis claims that the steric constraints around the active sites inhibit the formation of spacious transition states [259]. The remarkably low propensity of H-ZSM-5 to deactivate is often mentioned to confirm this view of shape selectivity in coke formation [257]. H-ZSM-5 with a medium size of pore openings and channel intersections, possessing no large cavities, is thought to exclude bimolecular coking reactions. Magnoux et al. [224] investigated the influence of the nature of reactants on the formation of coke on H-ZSM-5 at low and high reaction temperatures. At 393 K coking occurred very rapidly from propene and a propene/toluene mixture, slowly from toluene transformation. The coke, in all cases located inside the pores and released by HF treatment, consisted mainly of: (i) $C_{12}-C_{35}$ aliphatic hydrocarbons from propene conversion; (ii) diisopropyltoluenes from the transformation of the mixture, and; (iii) methyldiphenyl- and triphenylmethane from toluene conversion. The appearance of the diphenyl and triphenyl compounds in the channels of H-ZSM-5 showed the formation of bulky intermediates and confirms that the coking tendency of reactants depends on both the steric constraints and the activation energy of coking which may be low for condensation and oligomerization reactions. At higher temperatures cyclization and hydrogen transfer reactions were involved and, consequently, at 723 K the rate of coking did not significantly depend on the nature of the reactants. Methylpyrenes were the main coke components.

To estimate the coking tendency of a single compound is not an easy task because under reaction conditions products and intermediates will be present and influence the results. This problem may be solved by using ^{14}C-labeled compounds. Walsh and Rollmann [260, 261] investigated the conversion of ^{14}C-*n*-hexane/benzene mixtures on H-MOR, H-Y and H-ZSM-5 at 633 K. The amount of ^{14}C atoms deposited on the zeolites was determined via combustion of coke and monitoring the radioactivity of CO_2. For the mordenite structure an increase in the aromatic contribution to coke was observed with increasing aluminum content (Table 12), whereas coke on H-ZSM-5 derived from aromatics decreased with increasing aluminum content. Although the coke contents on the zeolites differed significantly, the nearly equal participation of both paraffins and aromatics in coking led the authors to the conclusion that the formation of alkylaromatics is an intermediate step in the pathway to coke. Similar radiotracer studies on the methylation of toluene with ^{14}C-labeled methanol on H-ZSM-5 confirmed that, at reaction tempera-

Table 12 Yield and origin of coke during conversion of ^{14}C-labeled *n*-hexane/benzene mixtures at 633 K (adapted from [260, 261])

Zeolite	Si/Al	Coke yield [g/100 g conversion]	Coke from aromatics [%]
H-ZSM-5	12.5	0.22	30
	36	0.23	49
H-Y	2.65	37	78
	6.6	33	76
H-MOR	9	16.9	63
	54	6.6	47

tures of about 700 K, the origin of carbon atoms forming internal carbonaceous deposits was almost equally distributed among the reactants [200].

Coke produces variations in the shape selectivity and might consequently itself modify the mechanism of coking [69]. For example, the selective formation of para-isomers exhibited by medium pore-sized zeolites changes with time on stream. The para-selectivity was increased by coking with mesitylene due to the deposition of the coke species on the external surface of H-ZSM-5, whereas internal coke from *i*-butene caused a decrease of conversion almost without changing the para-selectivity [218]. Sotelo et al. [262] found that changes in the para-selectivity and in the nature and composition of the coke simultaneously occurred. Pore filling arising at longer times on stream was suggested to lead to a decrease of the diffusional pathway length, which can explain the observed decline in para-selectivity. In fact, a reduced length of diffusional pathways would require that coke deposition preferentially starts to occur in the center or at least in some depth of the pores.

Coke formation, however, reduces also the acid strength of the catalyst and suppresses the isomerization of the primarily produced para-isomers. Bhat et al. [263] observed that special coking procedures were allowed to attack only the very strongly acidic sites and are able to stabilize the para-selectivity and the catalytic activity over long times on stream. Additionally, the deactivation of the non-shape-selective sites on the external surface by modifier agents or carbonaceous deposits may enhance the para-selectivity.

The effect of shape selectivity with respect to coke formation seems to be primarily connected with the phenomenon of reactant selectivity. If the reactants are able to enter the zeolitic pore system, then the deactivation rate and composition of coke are characteristic of the reactants in the case of coke-type I, formed mainly at low temperatures. In the case of coke-type II, formed under severe reaction conditions, the coking tendency of the substances does not vary significantly.

On mesoporous materials (e.g., of the M41S family) containing a narrow pore size distribution, but within the dimensions of about 1.5–10 nm, a minor effect of shape selectivity on deactivation and the formation of voluminous polyaromatic coke species may be expected. Thus, the group of Rozwadowski and co-workers published a series of interesting papers on coke formation on aluminum-containing mesoporous molecular sieves, in particular on Al-MCM-41 [264–271]. Their methods of investigation were TGA, DRIFT, UV-Vis, EPR, ^{13}C solid state NMR and a sorption technique. As a coke-forming reaction, the authors used the conversion of cyclohexene. Rozwadowski et al. showed that the nature of coke formed on mesoporous molecular sieves of the MCM-41 type upon cyclohexene conversion at various temperatures is comparable with that of coke on zeolites. The aluminum content did not significantly affect the chemical composition of the carbonaceous deposits. Rather, the composition depended on the reaction temperature and the time on stream [266] (cf. also Sect. 2.1). Thus, the authors discriminated low-temperature coke ($T < 500$ K) and high-temperature coke ($T > 500$ K), similar to the results which were obtained with microporous aluminosilicates (cf. Sect. 2.1). At lower reaction temperatures of about 483 K, a large quantity of coke (ca. 50 wt %), containing both aliphatic and aromatic species, was formed within the pores. Part of the deposits was weakly bound to the surface and could be removed by purging with helium. However, the higher the reaction temperatures, the higher the content of aromatic species that could not be removed upon heating. At higher reaction temperatures of about 633 K, only about 14 wt % carbonaceous residues were obtained and a fraction of the coke migrated out of the pores. Both the coke remaining at higher temperatures in the pores and the coke detected on the external surface formed multilayered polyaromatic structures, which were strongly bound to the Al-MCM-41 materials [266]. The coke formation decreased with increasing reaction temperature [264, 269]. However, EPR showed that the spin density increased when the temperature was raised [270]. The amount of carbonaceous deposits increased with increasing aluminum content [264, 268, 269]. Simultaneously, the aromaticity of the coke deposits decreased [271]. Coke was preferentially formed in the Al-rich pores and, in fact, coke formation started in the vicinity of the acid centers [264, 268]. Deposition of carbonaceous materials lowered the density of Brønsted and Lewis acid sites [268, 269], reduced both the surface area and pore volume [270] and lowered the sorption capacity for benzene, nitrogen and water [265, 267]. But even substantial coking, i.e. to about 50 wt %, did not lead to the development of microporosity.

Furthermore, during isopropylation of pyrene inside the mesopores of Al-MCM-41 and Al-MCM-48 coke formation caused catalyst deactivation and changes of the product distribution indicating that shape selectivity occurs inside the regular mesopores of these M41S types [272].

Finally, the presence of extra-framework aluminum, tortuosity of the pore structure and possible differences in crystal size can also contribute to some diffusional hindrance on zeolites, resulting in changes in selectivity and retention of coke-forming molecules.

3.2
Effect of Acidity

The formation of coke on acid zeolites is considered by most of the authors as an acid-catalyzed reaction. A simple relationship, however, between the nature, the number and the strength of acidic sites and the deactivation rate of zeolites is far from being evident, although the activity of many other reactions correlates directly with the initial acid site concentration. For instance, in *n*-hexane cracking on H-ZSM-5 a linear correlation over four orders of magnitude between the aluminum content of the framework and the first-order rate constant has been established [273]. To characterize the acid properties of spent zeolite samples, temperature-programmed desorption (TPD) of basic molecules and calorimetric measurements have been used. One of the inherent problems in TPD studies is of course the thermal decomposition of carbonaceous deposits. Moreover, the remaining sodium level after preparing the H-forms of zeolites may particularly decrease the number of strongly acidic sites and influence clearly the catalytic and deactivation behavior [274, 275].

The low concentration of acid sites has been considered to explain the high resistance of H-ZSM-5 to deactivation [276]. For example, Bibby et al. [215] reported that the deposition of coke during methanol conversion over H-ZSM-5 was the higher the lower the Si/Al ratio was. Moreover, TPD of ammonia from fresh and coked H-ZSM-5 samples showed a reduced intensity of the TPD peak due to coking but no measurable shift in the peak position or variation of the peak width [277]. Similar results of Bauer et al. [106] and Topsøe et al. [278] suggest that carbonaceous residues deposited during methanol conversion deactivate the acid sites of H-ZSM-5 in a non-selective way, i.e. with no change in the strength and/or distribution of the remaining acid sites. However, not only the total number of acid sites and the distribution of acid strength remaining on the coked zeolite samples are essential features for the elucidation of the effect of acidity on coke formation, but also changes in the nature of acid sites (Brønsted or Lewis centers, hydroxyl groups associated with extra-framework aluminum species) and their distribution within the zeolite crystallite produced during catalyst pretreatment or on stream. Thus, Bauer [200] observed minimal coking during methanol conversion on H-ZSM-5 samples with Si/Al ratios between 50 and 80, associated with the dealumination of the framework during time on stream. Lukyanov [279] observed nearly the same deactivation rate on two quite different H-ZSM-5 samples (Si/Al = 17 and 120). Enhanced activity sites, which may be generated in H-ZSM-5 by mild steaming and attributed to an association of Brønsted acid sites with non-framework aluminum species [280], were shown to be the reason for a remarkably high aromatization and coking activity. Moreover, non-uniformity of aluminum distribution within ZSM-5 crystallites may disguise the original effect of acidity on the coking rate [210]. These findings with H-ZSM-5 under nearly identical conditions may reveal the difficulties encountered in the comparison of deactivation properties of different zeolite types.

Choudhary and Akolekar [281] compared low-silica zeolites (viz., H-Y, Ce,Na-Y, Ce,Na-X, and H-L) and high-silica zeolites (viz., H-MOR, H-ZSM-5, H-ZSM-8, and H-ZSM-11) with respect to their catalytic activity/selectivity and deactivation in the cracking of cumene at 600 K. These zeolites differ widely in the number of acid sites and in the distribution of acid strength. The involvement of strong acid sites in cracking has been investigated by selective poisoning of acid sites with pyridine at 623 K. While the conversion decreased in the order H-Y > Ce,Na-X > Ce,Na-Y > H-MOR > H-ZSM-8 > H-L > H-ZSM-5 > H-ZSM-11, the deactivation of the zeolites was found to occur in the order H-MOR > H-L > H-Y > Ce,Na-Y > Ce,Na-X > H-ZSM-8 > H-ZSM-11 > H-ZSM-5. The authors showed that the catalytic activities per framework Al were not the same and depended strongly on both the structure and the topology of the zeolite and the chemical environment of the framework Al. Moreover, the location of carbonaceous deposits on the zeolites was quite different. The deactivation of H-Y, Ce,Na-Y, and Ce,Na-X was due to both coke deposition in the zeolite channels and pore blockage, and that of H-MOR and H-L was mostly due to pore-mouth blockage. In the case of the pentasil zeolites, their deactivation was attributed to the blockage of some of the zeolite channels and to the deposition of coke on the external surface of the zeolite crystallites. Similar observations were made by Karge and Boldingh [69], who found via in-situ IR measurements that the number of acidic OH groups in H-MOR and H-ZSM-5 remained essentially unaffected in spite of severe coking. This was explained by deactivation through blocking of pores rather than poisoning of sites.

The aluminosilicate framework of zeolites is more or less unstable in the presence of steam. Non-framework aluminum species (Al_{NF}), produced during synthesis or catalyst treatment to generate a catalytically active form, may account for many of the reported effects in zeolite reactivity behavior. Addison et al. [282] and Corma et al. [283] tested H-USY samples in gas oil cracking. The amounts of Al_{NF} in the zeolite catalyst were varied by acid extraction. The catalyst activity and the selective conversion of the gas oil feed were the higher the higher the Al_{NF} content. The removal of Al_{NF} produced an increase in the gas and coke formation. By contrast, according to Machado et al. [284] Al_{NF} species increased the initial rate of coking during propene conversion on dealuminated H-Y. The authors claimed that this is due to both the generation of superacid sites and the reduction of pore openings. Similar studies on the formation of coke from propene at 823 K on a series of H-Y samples differing by the number of acid sites, i.e. framework aluminum (Al_F) and Al_{NF} atoms, were reported by Moljord et al. [45]. While the deactivation rate and the maximum amount of carbonaceous deposits increased with the number of acid sites, a very limited participation of the Al_{NF} in the formation of coke has been observed. This agrees with the results of Echevskii et al. [166] on the methanol conversion on H-ZSM-5, who found no direct relationship between the Al_{NF}/Al_F ratio and the coking rate. In addition, upon formation of non-framework aluminum a mesopore system is simultaneously formed which may have some effect on the selectivity [282], and positively charged Al_{NF} may act, like Fe cations, as electron acceptor sites and promote coke formation [285, 286].

As shown above, it is difficult to separate clearly the effect of acidity from other factors of deactivation. However, it is generally accepted that

(i) In the case of coke-type I, the yield of coke increases with the number of acid sites and the stronger the acid site the easier the formation of carbonaceous residues.

In the intermediate region between the two coke-types, i.e. at moderate temperatures, cracking of oligomeric coke species into smaller fragments may occur which reduces consequently the coke content.

(ii) At higher reaction temperatures, even on weakly acidic sites cyclization and dehydrogenation results in carbonaceous deposits of coke-type II.

Strongly acidic sites seem to be able to crack alkylaromatic coke species and cause hydrogen transfer from reactants to less condensed aromatic coke at higher temperatures which may lead to lower deactivation rates [115, 287].

At this point it is, however, appropriate to emphasize that there still exist serious contradictions between the results with respect to the effect of the Si/Al ratio and the role of Al_{NF} in coke formation. In the case of zeolites used in the MTG process some discrepancies in findings may be explained by an extensive dealumination caused by the product water during extended reaction periods [48, 185, 288]. The non-framework aluminum species thereby formed are claimed to differ from those produced by calcination or mild hydrothermal treatment [288]. Changes in the structure and properties of zeolites have been observed during prolonged MTG cycling, i.e. the methanol conversion itself may be visualized as an alternative modification step in the catalyst preparation.

3.3 Localization of Coke Deposition

The localization of the coke deposits is not only an interesting scientific problem but has also some consequences for the design of zeolite catalysts. If, for example, the carbonaceous residues were preferentially formed on the external surface of a zeolite it would appear possible to improve the catalyst stability. For example, the crystallite size may be changed and the acid sites on the external crystallite surface may be removed by dealuminating agents or covered by inert silica.

Early observations of coke formation on the medium pore-sized pentasil zeolites were interpreted with respect to the small coke loadings of about 2 wt % on deactivated samples in terms of coke deposition entirely on the external surface [289]. Meanwhile, working H-ZSM-5 samples with about 15–20 wt % coke have been reported for the methanol-to-gasoline process [48, 130, 215]. In such an event it is obvious that not all of the coke on H-ZSM-5 can be deposited in the internal channel system. Rather, a great part of the coke must be located on the external surface and fill the intercrystalline void volume. On the basis of the information of the various techniques used in the characterization of carbonaceous deposits on zeolites, a description more detailed than the two extreme assumptions on the location of coke formation is possible.

One effect of the presence of internal coke is to modify the framework of zeolites. Changes in the XRD pattern of H-ZSM-5 after coking were found by several authors [106, 215, 218, 219] and attributed to a distortion of the ZSM-5 structure. More detailed structural calculations showed that the crystalline structure of coked ZSM-5 became tetragonal [216] and that both the circular and the oval pores significantly changed their shape. Carbonaceous residues from the conversion of bulky mesitylene molecules on H-ZSM-5 did not change the cell dimensions of ZSM-5 suggesting their location to be on the outer surface of the crystallites [218].

Surface analysis of spent samples by XPS showed that coking on the external crystallite surface occurred predominantly at higher coke loadings which was indicated by an increase of the C/Si ratio (Fig. 39). For *n*-butene coking on the medium pore-sized H-FER, Mooiweer et al. [211] derived from the C/Si ratios that up to a carbon level of about 7.5 wt % the majority of the carbonaceous residues is located in the pores. This agrees with the result of Sexton et al. [210] for methanol conversion on H-ZSM-5 at 643 K which indicated that up to 8 wt % coke the filling of the pores predominated. The C/Si ratio showed, however, some external deposits even at low coke contents. By application of XPS, AES, and CEELS, the nature of the external carbonaceous materials on zeolites was found to be highly polyaromatic, but not graphite-like [206, 210].

Applying the PFG NMR technique with methane and propane as probe molecules, Kärger et al. [151, 152, 290] measured the intracrystalline self-diffusion coefficient D_i and the effective diffusion coefficient D_d in H-ZSM-5 before and after coking with *n*-hexane or mesitylene. While D_i is affected only by carbonaceous deposits inside the pore system and not by external coke layers, D_d is sensitive to both types of coke location. Carbonaceous deposits from mesitylene had very little effect on intracrystalline molecular

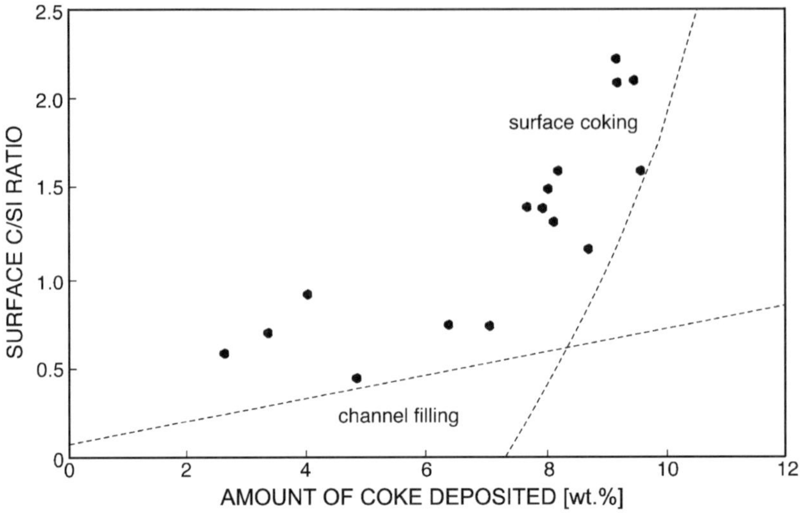

Fig. 39 C/Si ratio of coke formed from butene at 623 K on H-FER and determined via XPS (after [211]); boundaries for channel filling by coke and surface coking are indicated by *dashed lines*, in agreement with [210]

self-diffusion, showing that the coke was formed only on the external crystallite surface. During cracking of *n*-hexane, which can penetrate the pore structure of zeolite ZSM-5, initial coke formation occurred within the channel of H-ZSM-5, indicated by a continuous decrease in the intracrystalline self-diffusion coefficient D_i. At longer coking times, the build-up of external coke deposits exclusively led to a decrease of the effective diffusion coefficient. The curve D_i/D_d vs. coke content indicated that for *n*-hexane cracking at coke contents above 1.5 wt % further coke residues are preferentially deposited on the outer surface. Moreover, the molecular mobility of propane in polycrystalline grains, which was higher than in polyhedral twins and explained by additional diffusion paths in secondary pores, decreased with increasing observation time. The authors concluded from this, that coke, which forms outside the crystals, is deposited preferentially in the intercrystalline void volume of H-ZSM-5 catalysts [291].

The ^{129}Xe NMR technique has also provided evidence for coke formation occurring within the pore system of H-ZSM-5. Barrage et al. [185] observed that samples with quite different amounts of coke formed during methanol conversion showed nearly identical slopes $d\delta_s/dn$ implying that their internal microporous volume was the same (vide supra, Sect. 2.5.4). The authors deduced from this that for samples with more than 7 wt % coke, the coke is mainly deposited on the outside of the ZSM-5 crystallites. Tsiao et al. [186] showed that H-ZSM-5 loses after coking with *n*-butene up to 50% of its internal pore volume. A Na,H-ZSM-5 sample, prepared by heating Na-ZSM-5 with the sterically hindered salt (*n*-butyl)$_3$N·HCl and having acid sites only on the external crystallite surface, showed even at coke levels of 12 wt % only negligible loss of internal pore volume on coking—consistent with external coke formation in this case. In addition, the channels of such a strongly coked Na,H-ZSM-5 sample remained effectively accessible to xenon, i.e. only a minor amount of coke on the external surface led to pore mouth blocking.

The technique of zeolite dissolution and solvent extraction of coke has been extensively used by Guisnet and co-workers [6, 24, 45, 51, 223–230] (vide supra, Sect. 2.9). While all the coke components formed at low temperatures, i.e., coke-type I species, were soluble in methylene chloride after dissolution of the zeolitic framework in hydrofluoric solution, only a small part of this coke could be removed by direct Soxhlet treatment of the undissolved spent sample. This showed that carbonaceous residues were blocked within the pore system. At longer TOS, however, some coke species formed on or very close to the outer surface of the crystallites and on the surface of mesopores were extractable by Soxhlet treatment. Pater et al. [292] found that there was practically no difference in composition between the extractable and the non-extractable part of coke.

After dissolution of the zeolite framework, the major components of soluble coke observed on H-ZSM-5, H-ERI and H-MOR have boiling points below the reaction temperature. A comparison between the size of the soluble coke species and the pore structure of the various zeolites suggests that these coke molecules are located in the cavities (H-USY and H-ERI), at the channel intersections (H-ZSM-5) or in the channels (H-MOR). Insoluble coke-type II molecules may result from coupling reactions of soluble coke species trapped at adjacent locations [45]. Therefore, at least some part of the insoluble coke is deposited in the micropores of the zeolites.

Anderson et al. [244] also applied the method of coke extraction and showed that on H-ZSM-5, which had been $SiCl_4$-treated to deacidify the external surface, the carbonaceous species formed upon reaction of 1-hexene at 513 K were the same as without $SiCl_4$ treatment. At higher reaction temperatures of about 593 K, surface deacidification resulted in a large decrease in the proportion of higher aromatics (naphthalenes, phenanthrenes, anthracenes) in the carbonaceous residues and in an increase in the proportion of alkylnaphthalenes. Hence, in the case of the non-modified H-ZSM-5 those bulky aromatic species seem to form preferentially outside the pore system. Moreover, alkylnaphthalenes formed and accumulated within the pore system may migrate at higher temperatures to the external zeolite surface to undergo further condensation reactions to polyaromatics.

Finally, sorption studies with quite different probe molecules indicate that, where the size of the reactants made it possible, internal coking took place [7, 68, 218, 251, 252]. For H-ZSM-5, the amount of carbonaceous residues up to which coke formation occurred preferentially inside the channel system differ between about 2 wt % (Karge HG, 1991, personal communication) and 5–10 wt % coke [69, 106, 249].

Thanks to the sophisticated techniques it seems to be well established that, when there are no steric constraints of shape selectivity, coke formation occurs both on the outer surface of crystallites and within the zeolitic pore system. While coke-type I is mainly located inside the crystals, coke-type II is initially deposited within the cavities and channels. Under more severe conditions coke formation occurs preferentially on the external surface. Studying the location of coke involves, however, not only the question concerning internal or external coke deposition. When various types of cavities are possible loci of carbonaceous residues adsorption experiments, employing adsorbates with suitably selected sizes, shapes and adsorptive properties, promise to provide more detailed information [255, 293].

Martin and Guisnet [294] developed a molecular modeling method to estimate and visualize the location of coke molecules and to follow their growth inside zeolite pores. For toluene transformation on H-ZSM-5 at low temperature, coke molecules, such as methyldiphenylmethane or methyltriphenylmethane, were preferentially located at the intersections between the straight channels and the zigzag channels. At higher temperatures of about 723 K, pyrene molecules, which were the main components of the coke composition [224], were shown to be only located at the intersection between a straight channel and a zigzag channel. Furthermore, it was shown that due to steric hindrance, pyrene molecules can bear three methyl groups at the maximum, two being located in the straight channel direction and the third one being located in one of the zigzag channels.

3.4
Mechanism and Kinetics

Thanks to the knowledge of coke composition obtained by the zeolite dissolution/solvent extraction technique it is possible to discuss the mode of coke formation in a more detailed way [24, 45, 223, 245]. Although the very complex network of coking reactions depends on the specific situation (zeolite, feed, temperature etc.), some general conclusions may be drawn. For coke-type I generally formed at lower temperatures, the main coke constituents result from condensation, oligomerization and alkylation reactions of reactants and/or intermediates. The mode and the rate of deactivation are more or less specific for the zeolite and the reaction under study. At higher temperatures leading to coke-type II, reaction steps having higher activation energies such as cyclization, cracking and hydrogen transfer are involved in coke formation. Under these more severe conditions similar coke species occur, regardless of whether the feed contains paraffins, olefins or aromatics. The current view of the pathways of coke formation is presented in the scheme of Fig. 40.

Besides the fact that the hydrogen content of coke decreases with time on stream (aging of coke), carbonaceous species have been more or less consid-

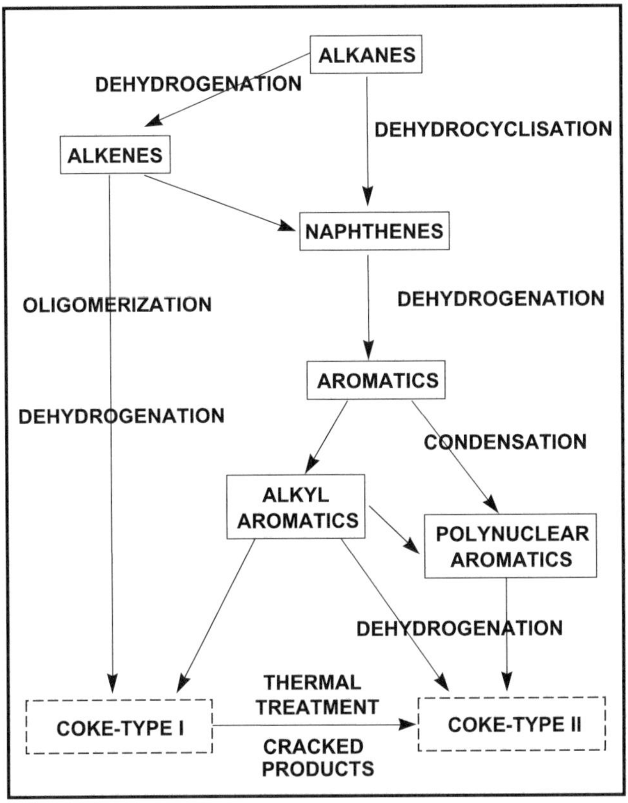

Fig. 40 Current view about coke formation on zeolites

ered as inert final products of a complex reaction scheme. Less attention has been paid to the reactivity of carbonaceous deposits. Novakova et al. [295] showed that all (or most) of the hydrogen atoms of coke on H-ZSM-5 can be exchanged for deuterium of deuterated olefins and aromatics at the relatively mild temperature of 573 K. The authors proposed that the isotope exchange proceeds on hydroxyl groups located on the surface and in near-surface layers of the zeolite crystallites. The limited migration of hydrogen through the coke is assumed to explain the observed deviations from first-order law kinetics. A very fast isotope exchange was found between the hydroxyl groups of the fresh H-ZSM-5 and deuterated compounds, even when their dimensions did not allow entering the pores of H-ZSM-5. This was explained by a rapid jumping of the hydrogen of hydroxyl groups throughout the zeolite framework.

Rhodes and Rudham [296] observed after some hours on stream a clear maximum in activity for toluene disproportionation on H-Y at 673 K (Fig. 41). The sigmoid curve of coke deposition shows that total poisoning occurred at more than 10 wt % coke, whilst the maximum activity appeared at

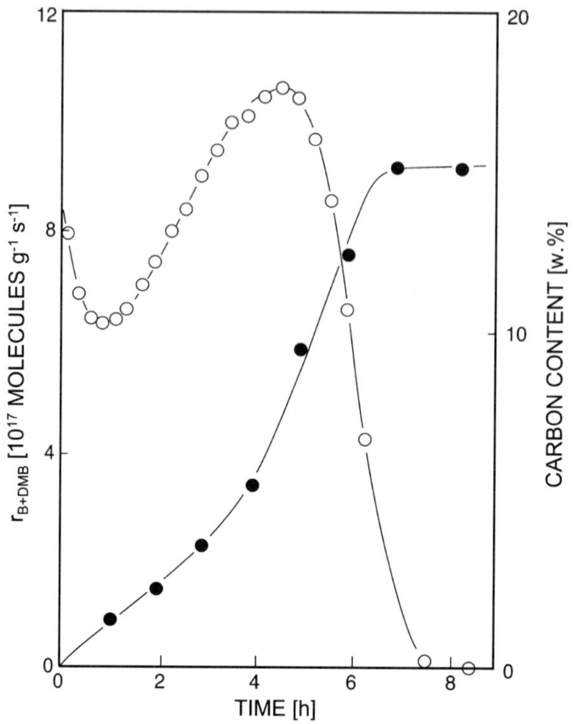

Fig. 41 Catalytic activity of toluene disproportionation and carbonaceous residue formation on H-Y as a function of time on stream at 673 K; o, formation rate of benzene and xylenes; •, coke content. r_{B+DMB} is the rate of formation of benzene (B) and dimethylbenzenes (DMB) (after [296])

about half of that carbon content. Therefore, the authors claimed that the reaction is associated with "catalytically active coke". The time to achieve this maximum rate depended on the extent of extra-framework aluminum and the carrier gas (He or H_2) and occurred after considerable coke formation. The use of hydrogen almost doubled the time required to attain the maximum activity that was observed with the He carrier.

Such an effect of hydrogen-containing stream gases on the coking rate of acid zeolites, that are free of any hydrogenating metal, is confirmed by the results of Bauer et al. [297] obtained during methanol conversion on H-ZSM-5. Using hydrogen-containing recycle gases resulted in longer times on stream compared with the use of nitrogen (see Table 13).

Moreover, spent zeolite catalysts were partially reactivated by a treatment with hydrogen or alkanes at the same reaction temperature as used during the MTG process [48, 287] (vide supra). Typical of MTG, deactivation takes place in a small reaction zone moving through the catalyst bed [8]. Corresponding to the movement of the reaction zone, the olefin content of the MTG

Table 13 Effect of different recycle gases on time on stream during the methanol conversion on H-ZSM-5 at 690 K and 1.5 $g_{methanol}/(g_{cat}\,h)$; adapted from [282]

	Recycle gas Nitrogen	Hydrogen	Propane/*i*-butane
Time on stream *[h]	127	267	290
Coke content [wt %]	7.4	14.3	13.4
Loss of BET surface area [%]	30	45	25
H/C ratio	0.80	0.82	0.67

* end of run when the gas phase concentration of dimethyl ether exceeded 5%

products increases with TOS [48]; both findings can be taken as a measure of deactivation. After stopping the methanol feed at the end of the MTG cycle, the reactor was held at the reaction temperature of 693 K and the recycle gas containing a mixture of hydrogen, propane, and *i*-butane was circulated over the catalyst bed for about 12 hours. After restarting the methanol feed, the position of the maximum temperature in the catalyst bed was shifted towards the reactor inlet and the olefin content, determined by the amount of bromine (in g) which can be added to 100 g hydrocarbons, was reduced indicating an increase of catalyst activity (Fig. 42). Although the removal of coke was far

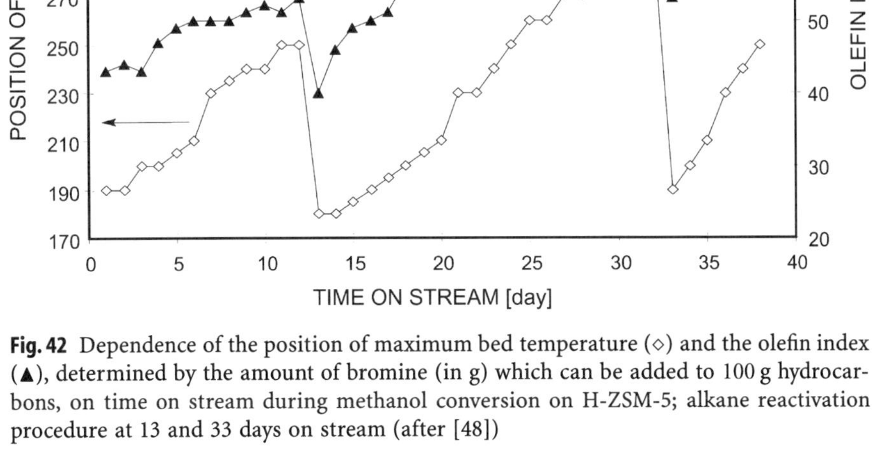

Fig. 42 Dependence of the position of maximum bed temperature (◇) and the olefin index (▲), determined by the amount of bromine (in g) which can be added to 100 g hydrocarbons, on time on stream during methanol conversion on H-ZSM-5; alkane reactivation procedure at 13 and 33 days on stream (after [48])

from being complete, the reactivation procedure could be repeated several times [256]. After two reactivation steps with alkane-containing recycle gas, more than 1500 $g_{methanol}/g_{cat}$ had been converted on H-ZSM-5 without an oxidative regeneration step. The finally spent zeolite was covered with 21.5 wt % coke. A similar rejuvenation process of deactivated zeolite catalysts by hydrogen treatment was observed by Aguayo et al. [49], Haag et al. [298], and Fang et al. [299]. Also, it was carried out in the commercial New Zealand MTG plant, where the temperature of the partially coked H-ZSM-5 bed was raised and a hydrogen-containing light-gas re-circulated [8]. To explain the rejuvenation effect, dissociative adsorption of hydrogen and alkanes was assumed to take place on a few very strong acid sites, even present on spent H-ZSM-5, according to:

$$\mathrm{H_2} + [\mathrm{zeolite}] \longrightarrow (2\mathrm{H}^+)[\mathrm{zeolite}]^{2-}$$
$$\text{R-coke} + (2\mathrm{H}^+)[\mathrm{zeolite}]^{2-} \longrightarrow \text{R-H} + \text{H-coke} + \text{zeolite}\,,$$

where R = aliphatic groups.

These species should be responsible for the retardation of coke formation by hydrocracking of carbonaceous residues and desorption of volatile cracked products [48]. A migration of positively charged or radical-like hydrogen-rich species through the voids of the crystallite may explain such reactivation steps in coked areas located far away from the sites of dissociative adsorption. However, a direct interaction between the carbonaceous deposits and hydrogen molecules might also explain the results as indicated by IR studies of H/D exchange on coked zeolites (vide supra, Sect. 2.3.3).

As opposed to deactivation, beneficial effects of coke include an improved selectivity towards desired products. During skeletal isomerization of *n*-butenes into isobutene on H-FER a partial deactivation by carbonaceous deposits has been reported to achieve higher isobutene selectivity [51, 300]. There are, however, different opinions regarding the mechanism by which coke deposits might promote the selective formation of isobutene. Coke may reduce the number of non-selective acid sites of higher strength or those located on the external surface resulting in a lower extent of undesired reactions [301], such as olefin dimerization and cracking, which lead to by-product formation. Consequently, lowering the strength and density of acid sites has been reported to give better initial selectivities and lower deactivation rates. This can be achieved by replacing framework Al^{3+} for another trivalent element (such as B, Ga, and Fe), by partial ion exchange with alkaline and alkaline earth cations and by increasing the Si/Al ratio [302–304].

However, in all these cases the formation of carbonaceous deposits has been claimed to be a necessary prerequisite for obtaining the highly selective and stable behavior characteristic of H-FER. On the other hand, carbonaceous deposits increase the diffusion barriers at the pore openings and within the channels and/or reduce the free space around the acid sites limiting the competitive bimolecular reaction via bulky C_8 intermediates [305–307]. To handle the crucial problem of the monomolecular isomerization reaction of C_4 olefins, i.e. avoiding the formation of primary carbenium ions, Guisnet et al. [51, 226, 228] proposed a pseudo-monomolecular isomerization mechanism involving, as catalytically active sites, carbonaceous carbenium ions formed from coke molecules located in the zeolite pores. Thus, isomerization of the formed C_{5+} intermediates (Fig. 43) could occur via transformation of secondary and tertiary carbenium ions which have already been detected by ^{13}C MAS NMR during *n*-butene conversion [100, 308]. On the contrary, van Donk et al. [79] used CD_3CN as an IR spectroscopic probe molecule for secondary and tertiary carbenium ions and did not observe any carbenium ions present on H-FER loaded with about 6 wt % coke.

To explain an increase of isobutene yield observed on H-FER at short times on stream, Guisnet et al. [229] have proposed a reaction scheme involving a tert-butyl cation (formed from isobutene) and cracking of the resulting trimethylpentyl cation. On the other hand, Meunier et al. [90] detected various unsaturated coke species by IR spectroscopy which were readily formed upon contacting fresh H-FER with *n*-butene, while a poor selectivity to isobutene was observed. It is noteworthy that a change in reaction mechanism with increasing TOS was confirmed by experiments with ^{13}C-labeled butenes [209, 306, 309, 310]. Whereas on fresh H-FER intense ^{13}C scrambling was observed, on aged samples the original labeling position of *n*-butene remained in the reaction product isobutene, which is characteristic

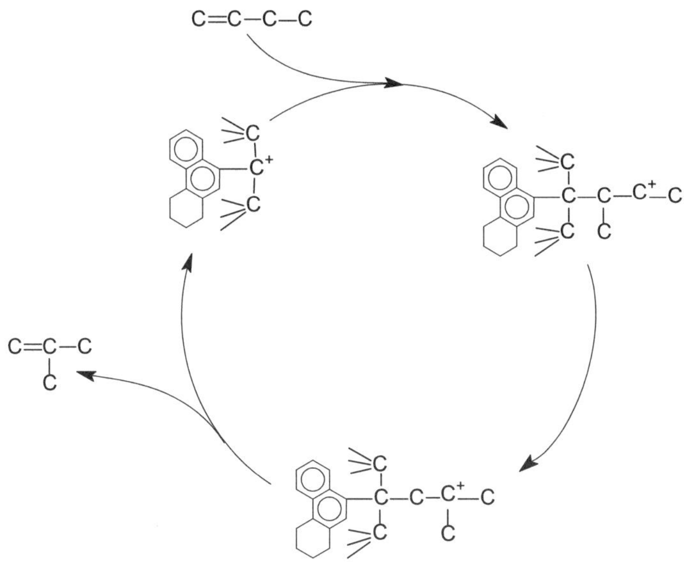

Fig. 43 Pseudo-monomolecular mechanism of *n*-butene isomerization into isobutene on coked H-FER (after [228])

of a monomolecular mechanism. All these data suggest that initially formed hydrogen-rich polyenic coke species and/or unsaturated aliphatic carbenium ions are involved in unselective isobutene formation and are gradually converted into more inert aromatic deposits at longer TOS [90]. Guisnet [14] and van Donk et al. [15] have assessed the pros and cons of active coke species during *n*-butene isomerization. The remarkable discussion about the nature of the active sites responsible for the selective isobutene formation on H-FER [311, 312] has initiated further studies on the implication of coke in reaction mechanisms [313].

With respect to the technique of pre-coking (vide infra, Sect. 3.5), Finelli et al. [44] observed that the presence of residual carbonaceous deposits (about 0.9 wt %) after coke combustion at 933 K improved catalytic activity, isobutene yield and catalyst stability after some minutes on stream. Moreover, van Donk et al. [28, 314] revealed that short TOS reactive carbonaceous residues [47] largely contributed to the non-selective, overall isobutene production on H-FER. With prolonged TOS, the deposits became non-reactive and on the aged catalyst no carbenium ions, but some remaining accessible Brønsted acid sites were observed. Hence, the occurrence of a pseudo-monomolecular pathway was excluded and those few remaining acid sites were held responsible for the selective formation of isobutene [28]. Nevertheless, all these findings about the "truly" active sites on H-FER during *n*-butene isomerization illustrate the above-mentioned close relationship between selectivity and reactant diffusivity [307].

The incorporation of deactivation kinetics into reaction simulation is an essential step for global comprehension of a catalytic process. The aim of the study of deactivation kinetics is to gain a relationship for the activity decay which includes fundamental interrelations of physical, chemical and transport processes leading to catalyst deactivation. Since coke on catalysts is the main source of activity loss, activity decay is frequently described by the kinetics of coke deposition, e.g., the empirical equation

$$W_c = k_c \cdot t^{n_c} \tag{11}$$

of Voorhies [315], who observed that the weight percent of coke, W_c, laid down on the catalyst can be satisfactorily correlated with the time on stream (t) by means of a simple power-law function where k_c and n_c are appropriately fitted constants. This simple deactivation behavior has also been reported for zeolite catalysts [69, 194]. Some deviations from Voorhies' correlation may be attributed to the dependence of coke formation on space velocities through the reactor and to a multi-step deactivation process. For example, Benito et al. [316, 317] observed that the coke content during the MTG reaction decreased with increasing contact time using an isothermal fixed-bed reactor. Therefore, the deposition of coke was assumed to be higher at the entrance of the reactor and should decrease with longitudinal position, i.e., deactivation occurs in parallel to the main reaction. The evolution of coke content with time on stream has been well fitted with time on stream (t) and exponentially with reaction temperature (T) to

$$W_C = k_C \cdot t^{n_C} \cdot \exp(-k'_C/T) \,. \qquad (12)$$

Often it is necessary to introduce different deactivation functions, for the reaction and the coke formation [23] or for reversible and permanent deactivation of the zeolite [318]. However, the activity decay should be expressed in terms of true deactivation variables, which are the concentrations of the reactants, the coke amount, the structure of the zeolite, and the operation conditions and not only the time on stream. Semi-empirical decay functions, which allow simulating such phenomena as site coverage, pore blockage and internal/external coking, have been used to describe the deactivation of zeolite catalysts [275, 319–322]. For example, Lopes et al. [275] compared the cracking activity, selectivity and coking rate of H-ZSM-20 and H-USY during the transformation of n-heptane at 723 K. Although there was no significant difference in the coke composition on both zeolites, the deactivating effect of carbonaceous residues on the cracking process was lower for H-ZSM-20. As can be seen in Fig. 44, the time-on-stream dependence of the cracking activity was well described by an empirical two-stage decay. A simplified model of the cracking kinetics with parameters for diffusion limitations and pore-blocking effects was proposed which fits well both experimental results for the coke content vs. time on stream (Fig. 45) and for cracking activity vs. coke content. The obtained model parameters express the different cracking and coking behavior of both zeolites.

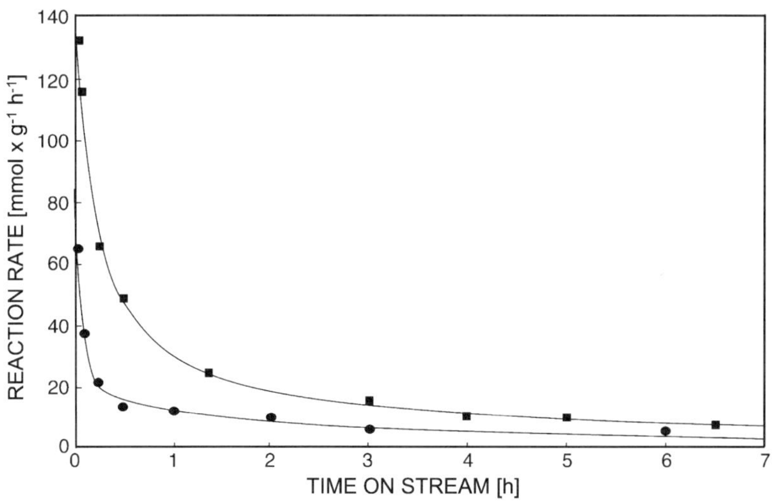

Fig. 44 Cracking activity for n-heptane conversion (in mmol h^{-1} g^{-1}) on H-ZSM-20 (■) and H-USY (•) at 723 K; experimental values and model fit (after [275])

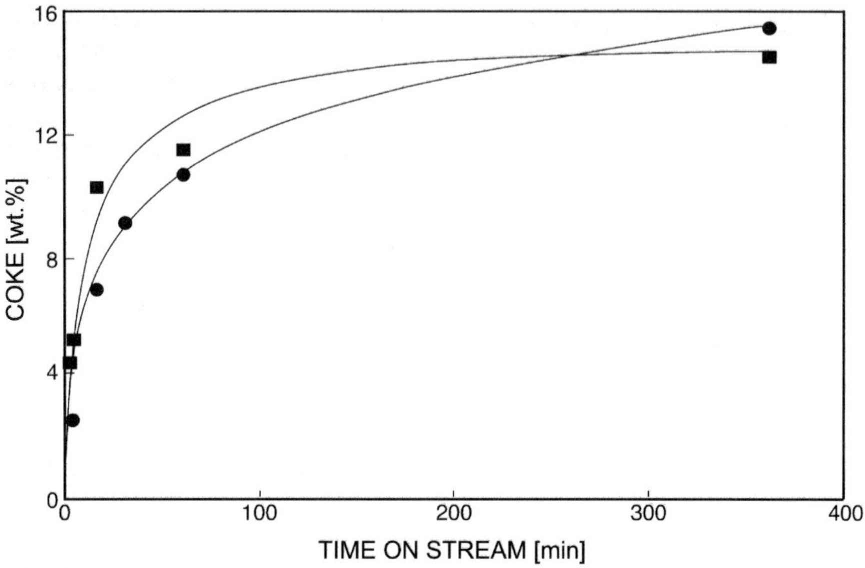

Fig. 45 Level of coke deposition (wt %) during *n*-heptane transformation at 723 K on H-ZSM-20 (■) and H-USY (•) as a function of time on stream; experimental values and kinetic model fit (after [275])

Benito et al. [317] and Pinheiro et al. [323] investigated simplified three-parameter models for the interpretation of the kinetic data of coke formation during transformation of light olefins on H-USY and methanol on H-ZSM-5.

Catalyst activity (a) is proposed to depend on the concentration of three lumps (oxygenates (Ox), light olefins (Ol), and aromatics (A)):

$$-\frac{\mathrm{d}a}{\mathrm{d}t} = (k_{\mathrm{Ox}}X_{\mathrm{Ox}} + k_{\mathrm{Ol}}X_{\mathrm{Ol}} + k_{\mathrm{A}}\,X_{\mathrm{A}})a\,. \tag{13}$$

In the temperature range between 573 K and 673 K, the order $k_{\mathrm{Ox}} > k_{\mathrm{A}} > k_{\mathrm{Ol}}$ revealed that olefins are the least active components in coke formation during the MTG process [317]. Moreover, Pinheiro et al. [323] proposed an autocatalytic growth of coke for the conversion of ethene, propene, and *n*-butene on H-USY:

$$\frac{\mathrm{d}W_{\mathrm{c}}}{\mathrm{d}t} = (k_{\mathrm{Ol}}X_{\mathrm{Ol}} + k_{\mathrm{C}}X_{\mathrm{Ol}}W_{\mathrm{c}}) \cdot \left(1 - \frac{W_{\mathrm{c}}}{W_{\mathrm{max}}}\right)\,. \tag{14}$$

The rate of coke formation was a maximum at the beginning of reaction and decreased with time on stream going to zero as the zeolite surface reached coke saturation. The maximum amount of coke W_{max} increased with the number of carbon atoms of the olefins. This fact may be related to a different composition of carbonaceous residues.

Finally, the present knowledge of coke formation on zeolites is characterized by the precision of the mathematical description of the deactivation process. Computer modeling and the application of percolation theory in zeolite coking promise to be a fruitful "experimental" technique with respect to a better understanding of zeolite catalyst deactivation.

3.5 Coke as Catalytically Active Species in Hydrocarbon Conversion

In homogeneous catalysis, it is generally accepted that the working catalyst is not the as-synthesized one. Similar statements are seldom true in heterogeneous catalysis, and the formation of high molecular hydrocarbons is regularly seen with respect to catalyst deactivation; but as just mentioned, carbonaceous deposits do not always have a detrimental effect on the catalytic behavior of zeolite catalysts. A well-known example is the pre-coking technique applied in toluene disproportionation on H-ZSM-5 which yields a very high para-xylene selectivity [324, 325]. The enhancement in selectivity is based on poisoning the non-shape selective acid sites present on the outer surface and improving the sieving properties of the zeolite due to changes in product diffusivity. However, carbonaceous residues in zeolites are not inert deposits, i.e., coke molecules maintain reactivity and can participate in reactions with sometimes a positive effect on selectivity. A comprehensive review by Guisnet [14] disclosed as examples of active coke species:

(i) isopropylation of naphthalene and alkylation of toluene with long chain *n*-alkenes over H-FAU and H-BEA;
(ii) selective skeletal isomerization of *n*-butene over H-FER;
(iii) selective hydroisomerization of long chain *n*-alkanes over Pt/H-TON;
(iv) selective methanol conversion into light alkenes over SAPO-34 and H-ZSM-5.

Two liquid-phase alkylation reactions (naphthalene by isopropanol [326, 327] and toluene with 1-heptene [328, 329]) carried out at 623 K over large pore H-FAU and H-BEA were considered to include carbonaceous species in transalkylation reactions. For isopropylation of naphthalene, the activity of both zeolites increased with TOS although the micropore volume accessible to nitrogen was practically zero after 1 h of reaction. Therefore, it could be concluded that isopropylnaphthalenes were not formed in the micropores but on active sites located on the outer surface of the zeolite crystallites or at the entrance of the pores (pore mouth catalysis). Moreover, only a small amount of coke was recovered by direct soxhlet treatment of the spent samples, whereas a large part was soluble in methylene chloride after dissolution of the zeolites in hydrofluoric acid. This soluble coke was mainly constituted of polyisopropyl naphthalenes (two to four isopropyl groups) as well as polyisopropyl pyrenic and indenopyrenic compounds. Hence, transalkylation reactions between polyalkylaromatic carbonaceous residues adsorbed on Brønsted acid sites near the entrance of the micropores and aromatic molecules may be assumed (Fig. 46). On the other hand, the initial increase in activity could not be explained if the active sites were pure acid sites on the external surface. Furthermore, the loss of the isopropyl group could be rapidly compensated by realkylation of the entrapped carbonaceous deposits with isopropanol. Similar transalkylation reactions with participation of polyalkylated carbonaceous species may play a significant role in alkylation of toluene by *n*-dodecenes [328], ethylbenzene disproportionation on La,Na-Y, H-L and H-BEA [330, 331] as well as in the reactivation of spent alkylation catalysts to remove the side-chains from alkylaromatic coke species resulting in dealkylated residues which might desorb easily from the zeolite pores [329]. Curiously, for the disproportionation of toluene and xylene on zeolite catalysts (both reactions have received much interest, not only from an industrial point of view but also scientifically) methyl group transfer by carbonaceous deposits has up to now not been described.

For hydroisomerization of *n*-decane, the maximum yield on Pt/H-ZSM-5 was about 20%, whereas over Pt/H-TON an exceptionally high yield of isodecanes (75%) could be obtained [332]. To explain the high selectivity of Pt/H-TON towards isodecanes, pore mouth catalysis was proposed. This assumption was based on the GC-MS analysis of carbonaceous compounds released from the zeolite pores by HF dissolution, which showed that the carbonaceous compounds were mainly constituted of linear and monobranched C_{12}–C_{20} alkanes. Traces of alkenes were also detected. Furthermore, the pore volume occupied by the carbonaceous deposits was much lower than the pore

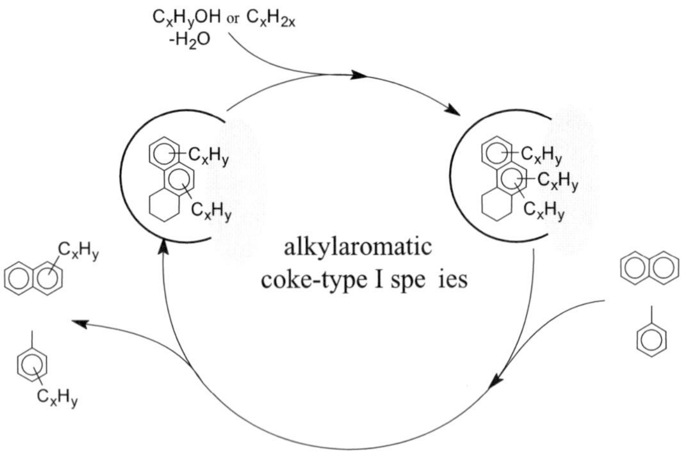

Fig. 46 Scheme of transalkylation reactions involving coke species during isopropylation of naphthalene and alkylation of toluene with long chain *n*-alcohols or *n*-alkenes over H-FAU and H-BEA

volume made inaccessible to nitrogen, which suggested a preferential location of coke species near the pore mouth yielding pore blocking. A simple isomerization mechanism involving as active sites tertiary carbenium ions formed at the pore mouth by adsorption of methyl alkenes and trapped in the zeolite pores during the first minutes of reaction [332, 333] was proposed. This would be similar to the mechanism in the skeletal isomerization of *n*-butene over H-FER (Fig. 43). However, recent investigations dispute this mechanistic proposal [334, 335].

Similar to skeletal isomerization of *n*-butene on H-FER, an induction period was observed during methanol conversion to gasoline or to olefins on H-ZSM-5 and SAPO-34 catalysts indicating a modification of the fresh into the working catalyst [13, 336–338]. Moreover, there was evidence for a reaction scheme based on a pool of olefinic intermediates which react with methanol (Fig. 47). The resulting products split after rearrangement into various olefins (carbon pool mechanism [200, 339–341]). Hydrogen-rich carbonaceous residues (coke-type I) with low temperature combustion (vide supra, Sect. 2.2.) were proposed as pool components [342]. Paraffinic/olefinic coke species formed by olefin oligomerization and acting as co-catalysts via an oligomerization/cracking mechanism for the transformation of methanol into light alkenes did not require extensive rearrangement reactions. By contrast, methylarenes or polymethylbenzenes trapped in the zeolite channels could only have a significant effect on the rate of methanol conversion via side-chain methylation of alkylbenzenes [338] and/or carbon scrambling reactions of aromatics and naphthenic compounds [13, 338, 341, 343, 344] followed by olefin elimination. For example, the transformation of hexamethylbenzene into ethyltetramethylbenzene, i.e., 6,5-ring contraction followed by 5,6-ring expansion, is most likely the limiting reaction step [345].

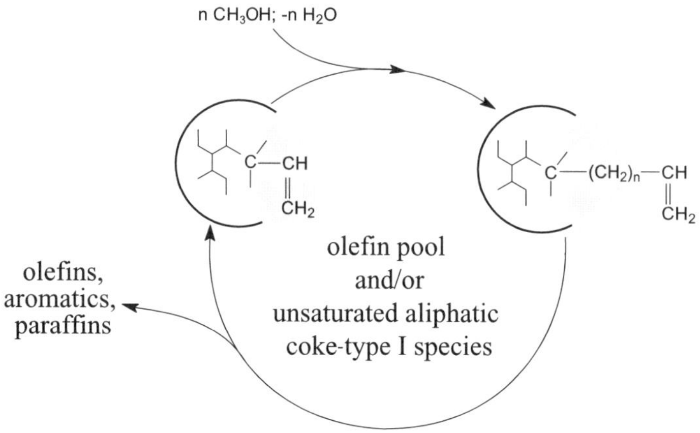

Fig. 47 Scheme of reaction mechanism involving coke species during the MTG/MTO process over SAPO-34 and H-ZSM-5

3.6 Coke-Induced Selectivation

The change in selectivity due to coke formation can be explained by changes in the diffusivity of the reactants or in the zeolite pore volume [346–348]. Unfortunately, changes in selectivity are normally coupled with a decrease in conversion and the various shape selective effects cannot be distinguished by kinetic analysis only. Froment and Bischoff [349] discussed the selectivity problem for deactivation using rate constants which were dependent on the amount of coke on the catalyst. For zeolite catalysts, a selectivity change caused by a change in conversion, acid site density, or acid site strength distribution can be classified as an intrinsic selectivity change, while a selectivity change caused by a change in zeolite pore structure can be classified as a (transition-state) shape selectivity change [350]. Three possible cases have to be considered as reasons for selective deactivation: (i) role of reaction steps requiring a different number or different strength of active sites; (ii) domination of product selectivity with different product diffusivities; (iii) domination of transition-state shape selectivity with different sizes of the intermediates. The transition-state shape selectivity is normally considered to be important for the medium pore size zeolite ZSM-5 [351]. Moreover, it is reasonable to assume that coke is mostly deposited at strong acid sites, whereas the deactivation rate for a reaction on weak acid sites might be lower, and a higher selectivity will result from coke formation. One example is the higher selectivity to dimethyl ether (DME) at high coke content during the MTG/MTO process on SAPO-34 [352]. Another example is the decrease of the yield of DME on SAPO-34 with coke content as a result of the stronger effect of carbonaceous deposits on DME diffusion than on methanol diffusion due to the larger diameter of the DME molecule [353].

However, the effect of the external surface on selectivity has also to be considered. The intrinsic selectivity of a zeolite catalyst might be totally changed by undesired secondary reactions on the surface of its crystallites. By pretreating the external surface these acid sites can be eliminated; but it was found that the modified samples frequently reacted in a way very similar to the unmodified sample, and only a slightly longer lifetime was found for the modified samples [354]. These observations show that reactions at the external surface are often not significant.

In situations where reactants are subject to substantial intracrystalline diffusion limitations, the discovery of synthesis methods that produced very small crystals in the range of 100 nm brought about an increase of overall activity. While reducing the effective diffusion path length, small crystallites increase the impact of the external surface on catalytic activity. Even more dramatic are the effects on catalytic selectivity. Reactant and product shape selectivity can markedly change as a result of a small number of non-selective active sites on the external surface. Hence, various surface modification techniques have been developed to fine-tune acidity and selectivity of zeolite catalysts [179]. Among them, chemical vapor deposition of organosilicon compounds [355, 356], the pre-coking treatment [357, 358], and the modification by (non)metal oxides [359, 360] are particularly relevant. Recent studies by Chen et al. [175], Bauer et al. [181], Jones and Davis [361], Niwa et al. [362], and Zheng et al. [363] have attempted to address the impact of non-shape-selective reactions occurring on the crystal surface on xylene isomerization and toluene disproportionation.

By means of economic and technical feasibility studies [364], the pre-coking technique has been demonstrated as a viable zeolite selectivation procedure and has been applied in Mobil's Selective Toluene Disproportionation Process (MSTDP^{SM}) producing para-xylene far in excess of xylene equilibrium distribution. Pre-coking can be done preferentially on the external surface by using voluminous coke-forming agents the molecular sizes of which are greater than the pore aperture of the zeolite. From the industrial point of view, pre-coking is typically performed by contacting the catalyst with aromatic feedstocks under high severity conditions during the initial time on stream [357, 365]. Besides covering any surface acidity, which would rapidly isomerize the desired product para-xylene back to an equilibrium mixture of xylenes, carbonaceous residues are also deposited close to the pores, thereby increasing both the length of the crystal diffusion path and its tortuosity. Because the diffusivity of para-xylene in H-ZSM-5 is significantly larger than that of ortho- or meta-xylene, the result is a reduction of the effectiveness factor for ortho- and meta-xylene formation. However, the predominantly uncontrolled coke deposition during pre-coking treatment has the drawback of affecting pore entrance and internal active sites that, in turn, reduces the activity of the catalyst. Nevertheless, MSTDP is an excellent example of the application of shape selectivity to preferred products by playing off diffusion rates against reaction rates.

Although pre-coking is a rather straightforward procedure, the detailed features of the carbonaceous deposits, e.g., nature, location and the mechanism of their formation, are rather complicated. Haag and Olson [357] disclosed that controlled pre-coking might be achieved by exposing the catalysts to organic compounds at elevated temperature (813 K). For H-ZSM-5 catalysts, coke is assumed to be uniformly deposited by decomposition of monocyclic aromatic compounds. Moreover, it was concluded that a coke content exceeding ca. 2 wt % is sufficient for enhancement of para-xylene selectivity during toluene disproportionation. Low et al. [365] revealed a pre-treatment method for the isomerization of C_8 aromatics, in which the zeolite catalyst was during the initial time on stream contacted with the aromatic feedstock under high severity conditions, e.g., at 677 K compared with the normal mode of an operation temperature of 655 K. In addition to aromatics, other feedstocks such as olefins, paraffins and alcohols have also been used for the selectivity enhancement through pre-coking [357, 366]. However, the amount of carbonaceous deposits claimed for desirable selectivity enhancement on various catalysts spanned over a wide range from 0.1 to 60 wt % [357, 365].

The use of smaller molecules as coke-forming agents has the drawback of forming an excessive amount of internal coke that, in turn, reduces the activity of the catalyst. Moreover, the coke lay-down may also involve rearrangement and/or migration of the carbonaceous compounds. Haag et al. [298, 367] and Bauer et al. [48, 256] showed that the catalytic activity of the spent H-ZSM-5 catalysts can be partially restored by additional thermal treatment with hydrogen or alkanes (vide supra, Sect. 3.4). In terms of efficiency in coke removal [41, 368], different gases follow the order: argon < nitrogen < hydrogen < propane (Fig. 48). It was found that both the activation of molecular hydrogen and alkanes by the strong acid sites and partial hydrocracking of carbonaceous deposits were responsible for the preferential reactivation of internal acidity. Liu and co-workers [299, 369, 370] further demonstrated that selective coking could be manipulated either by regeneration treatment under a hydrogen environment or by incorporating hydrogen into the feedstock as a carrier gas. Such hydrogen treatment of coked catalysts tends to provoke partial hydrocracking, rearrangement and/or migration of carbonaceous deposits. Consequently, the intracrystalline aliphatic coke (coke-type I) was effectively removed, while the bulkier aromatic or polyaromatic coke (coke-type II) remained on the external surface, as a result of which the catalyst activity was improved while maintaining a desirable selectivity (Fig. 49). Hence, a passivation of surface acid sites is achieved by even

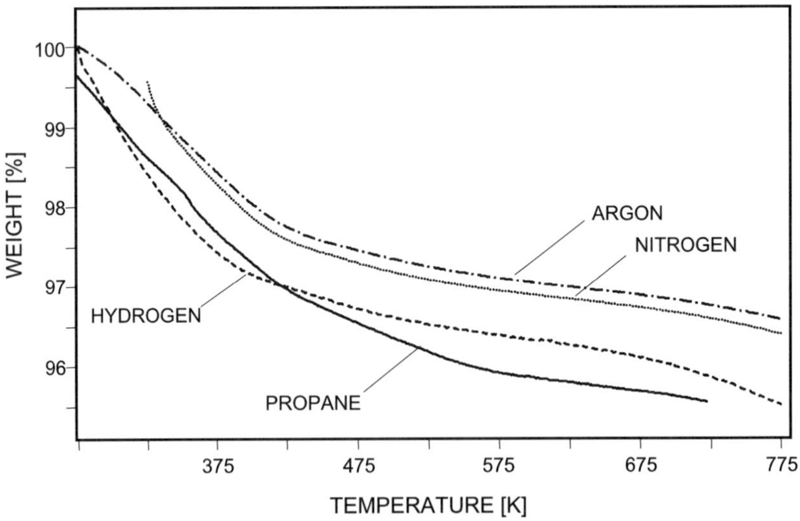

Fig. 48 Rejuvenation effect of different gases for spent H-ZSM-5 deactivated during MTG process at 693 K (after [203])

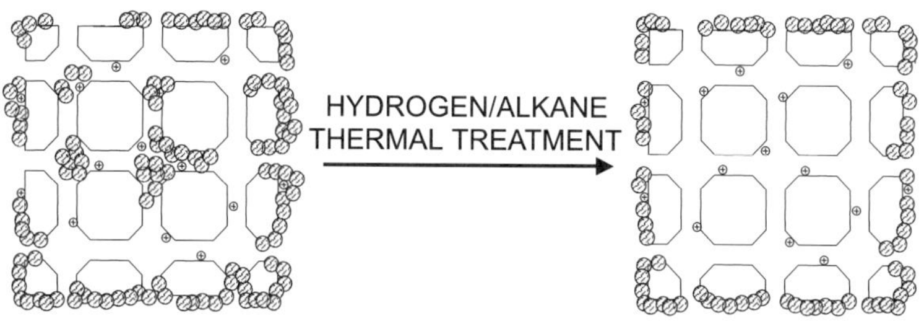

Fig. 49 Schematic depiction of location of coke ⊘ in spent H-ZSM-5 zeolite before and after modification treatments with hydrogen or alkanes. Acid sites are denoted by the ⊕ symbol (after [203])

low coke contents of about 0.5 wt % while sustaining a reasonable catalyst activity [203].

In terms of xylene isomerization on H-ZSM-5, it was anticipated that inactivation of external acid sites serves the purpose of suppressing undesirable bimolecular disproportionation reactions which would result in loss of xylenes, i.e., in the formation of toluene and trimethylbenzenes (TMBs). However, a few strong acid sites are desirable for dealkylation reactions to remove during xylene isomerization as much ethylbenzene (about 10 wt % of the industrial C_8 aromatics feed) as possible. This should avoid an ethylbenzene accumulation during the recycle loop. For example, modern xylene isomerization technologies normally require ethylbenzene conversion in the range of 50–70% [371]. Figure 50 shows a comparison between the conventional pre-coking procedure of H-ZSM-5 catalysts (performed at 723 K using industrial xylene feed) and the combined pre-coking/post-modification propane treatment [202, 203]. Obviously, conventional in-situ pre-coking during initial time on stream was characterized by both high ethylbenzene conversion and xylene loss, whereas the combined pre-coking/post-modification treatment of the sample was performed ex-situ resulting in a coke content of about 0.5 wt %. While sustaining a desirable activity for ethylbenzene conversion (> 55%), the pre-coked H-ZSM-5 sample modified by propane treatment yielded a xylene loss of only about 1.1% even after more than 120 h on stream as compared to 2.5% using the sample pre-coked by the conventional procedure. An excellent criterion for comparing the performance of xylene isomerization catalysts is the ratio (ε) of ethylbenzene conversion vs. xylene loss. At 55% ethylbenzene conversion, the combined pre-coking/hydrogen treatment resulted in an ε value of about 50/1, whereas samples pre-coked by the conventional (high severity) method led to typical ε values of about 25/1 [367]. For comparison, the current xylene isomerization technology yields xylene losses of about 2% and a ratio ε of about 35/1 [371]. Thus, modifying the nature and the location of carbonaceous deposits under hydrocracking conditions improves common pre-

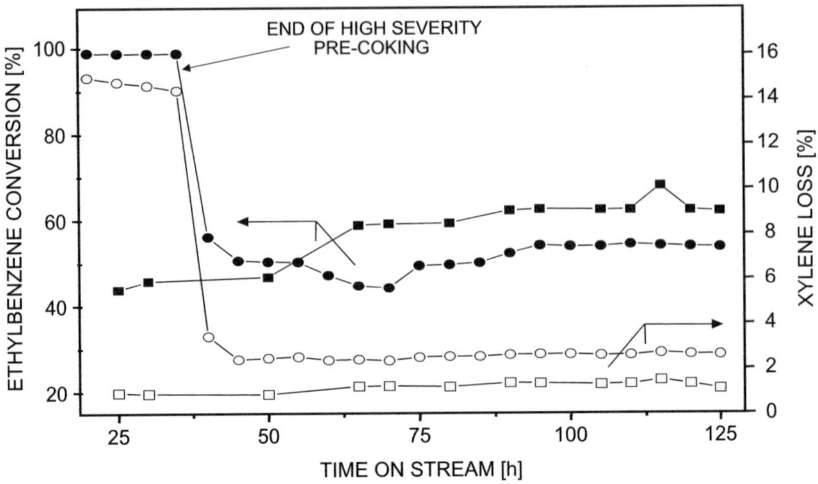

Fig. 50 Effects of conventional pre-coking (●,○) and combined pre-coking/propane treatment (■,□) methods on ethylbenzene conversion (*closed symbols*) and xylene loss (*open symbols*) over H-ZSM-5 (Si/Al = 20) zeolites during xylene isomerization (after [202])

coking procedures and facilitates selective passivation of the external active sites responsible for the undesired transalkylation and secondary isomerization reactions.

4 Summary and Concluding Remarks

To develop zeolite catalysts with higher tolerance in coking, the understanding of different phenomena related to coke, such as nature, location, and formation mechanism, is needed. Various analytical techniques and their manifold combinations are the prerequisite to gain an insight into these topics. The principal information that each technique can provide and some relevant references are summarized in Table 14.

Thanks to these methods and several examples of a multitechnique approach used in deactivation studies on zeolites the following features seem to be well established:

(i) At least two classes of carbonaceous residues may be distinguished, viz. hydrogen-rich coke-type I (“low-temperature”, “white” or “soft” coke) and hydrogen-poor coke-type II (“high-temperature”, “black” or “hard” coke), even though the borderline between the two types of coke may not always be very clear.

(ii) Coke-type I consists of a variety of less bulky paraffinic, olefinic, naphthenic and alkylaromatic species preferentially formed under mild reaction conditions, e.g., low temperature.
(iii) Coke-type II occurs at severe reaction conditions and is constituted mainly of bulky polyaromatics, frequently of condensed ring systems.
(iv) Coke formation is a shape-selective process and is significantly influenced by the architecture of the particular zeolite catalyst.
(v) Carbonaceous deposits, especially coke-type I species, are not inert molecules and completely grafted within the zeolite pores and on the external surface of its crystallites, but exhibit some reactivity and mobility.

Table 14 Characterization of coke on zeolites by different techniques

Technique	Information	Refs.
Elementary analysis	Coke content, H/C ratio N,S-containing deposits	[17, 18, 21]
TGA (+ IR or MS)	Volatile coke species, Coke decomposition	[19, 34]
TPO	Coke nature, location	[37, 44]
Microbalance (TEOM)	Deactivation kinetics	[26–28]
IR, DRIFTS, Raman	Coke nature (aliphatic, aromatic) Acid site deactivation	[70, 96, 107]
UV-Vis	Coke nature (olefinic, carbenium ions)	[91, 115]
^{1}H NMR	Coke nature (aliphatic, aromatic)	[37, 147]
^{1}H PFG NMR	Coke location (intracrystalline diffusion)	[152–154]
^{1}H NMR Imaging	Coke location (macroscopic)	[157, 158]
^{13}C (CP) MAS NMR	Coke nature (aliphatic, aromatic)	[11, 32]
^{27}Al, ^{29}Si MAS NMR	Coke nature, location	[144, 165, 166]
^{129}Xe NMR	Coke location (intracrystalline)	[188, 191]
ESR	Coke nature, radical formation	[192, 194]
TEM	Coke location, distribution	[15, 206]
XPS	Coke location (surface), nature	[209–211]
AES, CEELS	Coke nature	[205]
XRD	Coke location (intracrystalline)	[215, 216, 218]
Solvent extraction + GC/MS	Composition of soluble coke fraction	[37, 207]
Zeolite dissolution + GC/MS, MALDI-TOF MS	Coke composition and structure	[37, 227, 230]
Sorption measurements + Pore volume analysis	Coke location (pore mouth or site blocking)	[220, 250, 255]

Nevertheless, the present knowledge of deactivation of acid zeolites is far from being satisfactory. The most important phenomena, where controversial findings are still encountered, are the effect of Si/Al, effect of acidity, effect of Lewis sites, and the location of coke. The coke formation upon a process under consideration depends on a large variety of specific factors. Therefore, investigation under nearly identical conditions may lead to contradictory results. Such aspects of catalyst deactivation, as the effects of acidity and selectivity, the location of coke deposition, kinetics and mechanism of coke formation, are highly correlated and small differences between zeolite samples and/or reaction conditions may result in quite different findings. Therefore, studies on deactivation require the comprehensive characterization of fresh and spent samples to obtain a consistent and satisfying description of these phenomena. Due to the relevant contribution of catalyst deactivation to an improvement of the overall industrial process, the formation of carbonaceous deposits will further receive considerable attention in zeolite catalysis.

References

1. Bartholomew CH, Fuentes GA (eds) (1997) Catalyst deactivation 1997, Proc 7th Int Symp, Oct 5–8, 1997, Cancun, Mexiko. Elsevier, Amsterdam, Stud Surf Sci Catal 111
2. Delmon B, Froment GF (eds) (1999) Catalyst deactivation 1999 Proc 8th Int Symp, Oct 10–13, 1999, Brugge, Belgium. Elsevier, Amsterdam, Stud Surf Sci Catal 126
3. Moulijn JA, van Diepen AE, Kapteijn F (2001) Appl Catal A 212:3
4. Wolf EE, Alfani F (1982) Catal Rev Sci Eng 24:329
5. Bhatia S, Beltramini J, Do DD (1990) Catal Rev Sci Eng 31:431
6. Guisnet M, Magnoux P (1989) Appl Catal 54:1
7. Rollmann LD (1977) J Catal 47:113
8. Bibby DM, Howe RF, McLellan GD (1992) Appl Catal A 93:1
9. Karge HG (2001) In: van Bekkum H, Flanigen EM, Jansen JC (eds) Introduction to zeolite science and practice, 2nd edn. Elsevier, Amsterdam, 2001, p 707; Stud Surf Sci Catal 137:707
10. Guisnet M (1997) In: Ertl G, Knözinger H, Weitkamp J (eds) Handbook of heterogeneous catalysis, vol 2. Wiley, Weinheim, p 626
11. Bonardet JL, Barrage MC, Fraissard J (1995) J Mol Catal A 96:123
12. Schulz H, Wei M, Kraut M (1996) Proc DGMK-Conf, Catalysis on solid acids and bases, Berlin, 187 pp
13. Schulz H, Wei M (1999) Micropor Mesopor Mater 29:205
14. Guisnet M (2002) J Mol Catal A 182–183:367
15. van Donk S, Bitter JH, de Jong KP (2001) Appl Catal A 212:97
16. Hunger M, Weitkamp J (2001) Angew Chem 113:3040
17. Entermann W, van Leuven HCE (1972) Anal Chem 44:589
18. Weitkamp J (1988) In: Grobet PJ, Mortier WJ, Vansant EF, Schulz-Ekloff G (eds) Innovation in zeolite materials Science. Proc Int Symp, Nieuwpoort, Belgium, September 13–17, 1987. Elsevier, Amsterdam, 1988, p 515; Stud Surf Sci Catal 37:515
19. Darmstadt H, Karge HG, Łaniecki M, Amberg R (1991) Chem Ing Tech 63:742

20. Magnoux P, Roger P, Canaff C, Fouche V, Gnep NS, Guisnet M (1987) In: Delmon B, Froment GF (eds) Catalyst deactivation 1987. Proc 4th Int Symp, Antwerp, Sept 29–Oct 1, 1987. Elsevier, Amsterdam, 1987, p 317; Stud Surf Sci Catal 34:317
21. Schulz H, Böhringer W, Baumgartner W, Siwei Z (1987) In: Delmon B, Froment GF (eds) Catalyst deactivation 1987. Proc 4th Int Symp, Antwerp, Sept 29–Oct 1, 1987. Elsevier, Amsterdam, 1987, p 479; Stud Surf Sci Catal 34:479
22. Holmen A, Schanke D, Sundmark G (1989) Appl Catal 50:211
23. Froment GF, De Meyer J, Derouane EG (1990) J Catal 124:391
24. Dimon B, Cartraud P, Magnoux P, Guisnet M (1993) Appl Catal A 101:351
25. Garcia-Ochoa F, Santos A (1995) AIChE J 41:286
26. Chen D, Grønvold A, Rebo HP, Moljord K, Holmen A (1996) Appl Catal A 137:L1
27. Gomm S, Gläser R, Weitkamp J (2002) Chem Eng Technol 25:962
28. van Donk S, Bus E, Broersma A, Bitter JH, de Jong KP (2002) Appl Catal A 237:149
29. Schulz H, Barth D, Siwei Z (1991) In: Bartholomew CH, Butt JB (eds) Catalyst deactivation 1991. Proc 5th Int Symp, Evanston, IL, June 24–26, 1991. Elsevier, Amsterdam, 1991, p 783; Stud Surf Sci Catal 68:783
30. Martin A, Parlitz B (1993) React Kinet Catal Lett 51:245
31. Groten WA, Wojciechowski BW (1990) J Catal 122:362
32. Neuber M, Ernst S, Geerts H, Grobet PJ, Jacobs PA, Kokotailo GT, Weitkamp J (1987) In: Delmon B, Froment GF (eds) Catalyst deactivation 1987. Proc 4th Int Symp, Antwerp, Sept 29–Oct 1, 1987. Elsevier, Amsterdam, 1987, p 567; Stud Surf Sci Catal 34:567
33. Blackmond DG, Goodwin JG (1982) J Catal 78:34
34. Liu Z, Chen G, Liang J, Wang Q, Cai G (1991) In: Bartholomew CH, Butt JB (eds) Catalyst deactivation 1991. Proc 5th Int Symp, Evanston, IL, June 24–26, 1991. Elsevier, Amsterdam, 1991, p 815; Stud Surf Sci Catal 68:815
35. Lukyanov DB (1991) Zeolites 11:325
36. Barbier J (1987) In: Delmon B, Froment GF (eds) Catalyst deactivation 1987. Proc 4th Int Symp, Antwerp, September 29–October 1, 1987. Elsevier, Amsterdam, 1987, p 1; Stud Surf Sci Catal 34:1
37. Magnoux P, Guisnet M (1988) Appl Catal 38:341
38. Darmstadt H (1992) PhD Thesis, Technische Hochschule Darmstadt
39. Novakova J, Dolejsek Z (1990) Zeolites 10:189
40. Carlton L, Copperthwaite RG, Hutchings GJ, Reynhardt EC (1986) J Chem Soc Chem Commun:1008
41. Bauer F, Geidel E, Geyer W, Peuker C (1999) Micropor Mesopor Mater 29:109
42. Moljord K, Magnoux P, Guisnet M (1994) Catal Lett 28:53
43. Fung SC, Querini CA (1992) J Catal 138:240
44. Finelli ZR, Querini CA, Figoli NS, Comelli RA (1999) Appl Catal A 187:115
45. Moljord K, Magnoux P, Guisnet M (1995) Appl Catal A 121:245
46. Finelli ZR, Querini CA, Figoli NS, Comelli RA (2001) Appl Catal A 216:91
47. Xu WQ, Yin YG, Suib SL, O'Young CL (1995) J Phys Chem 99:758
48. Bauer F, Ernst H, Geidel E, Schödel R (1996) J Catal 164:146
49. Aguayo AT, Gayubo AG, Erena J, Atutxa A, Bilbao J (2003) Ind Eng Chem Res 42:3914
50. Hashimoto K, Masuda T, Mori T (1988) Chem Eng Sci 43:2275
51. Andy P, Gnep NS, Guisnet M, Benazzi E, Travers C (1998) J Catal 173:322
52. Ayuthaya SIN, Mongkolsiri N, Praserthdam N, Silveston PL (2003) Appl Catal B 43:1
53. Dossi C, Fusi A, Moretti G, Recchra S, Psaro R (1999) Appl Catal 188:107
54. Mariey L, Lamotte J, Chevreau T, Lavalley JC (1996) React Kinet Catal Lett 59:241

55. Copperthwaite RG, Foulds G, Themistocleous T, Hutchings GJ (1987) J Chem Soc Chem Commun 9:748
56. Ivanov DP, Sobolev VI, Panov GI (2003) Appl Catal A 241:113
57. Karge HG, Klose K (1973) Z Physik Chem [N.F.] 83:92
58. Hecker WC, Bell A, Hicks RF, Kellner CS, Savatsky BJ (1981) J Catal 71:216
59. Mirth G, Eder F, Lercher JA (1994) Appl Spectrosc 48:194
60. Moser WR, Chiang CC, Thompson RW (1989) J Catal 115:532
61. Hoser H, Innocenti A, Riva A, Trifiro F (1987) Appl Catal 30:11
62. Okado H, Shoji H, Sano T, Ikai S, Hagiwara H, Takaya H (1988) Appl Catal 41:121
63. Boroumand F, Moser JE, van den Bergh H (1992) Appl Spectrosc 46:1874
64. Horr TJ, Ralston J, Smart RSC (1992) Coll Surf 64:67
65. Kubelková L, Nováková J (1992) J Mol Catal 75:53
66. Gosh AK, Kydd RA (1987) J Catal 100:185
67. Lange JP, Gutsze A, Allgeier J, Karge HG (1988) Appl Catal 45:345
68. Bibby DM, McLellan GD, Howe RF (1987) In: Delmon B, Froment GF (eds) Catalyst deactivation 1987. Proc 4th Int Symp, Antwerp, Sept 29–Oct 1, 1987. Elsevier, Amsterdam, 1987, p 651; Stud Surf Sci Catal 34:651
69. Karge HG, Boldingh EP (1988) Catal Today 3:53
70. Karge HG, Boldingh EP (1988) Catal Today 3:379
71. Chen C (1983) Diploma Thesis, University of Karlsruhe
72. Karge HG (1977) In: Katzer J (ed) Molecular Sieves II. Proc 4th Int Zeolite Conf, Chicago, Ill, USA, April 18–22, 1977; ACS Symp Series 40:584
73. Kiricsi I, Förster H, Tasi G (1991) J Mol Catal 65:L29
74. Lombardo EA, Dereppe JM, Marcelin G, Hall WK (1988) J Catal 14:167
75. Medin AS, Borovkov VY, Kazanskii VB (1989) Kinet Katal 30:177
76. Vedrine JC, Dejaifve P, Garbowski ED (1980) In: Imelik B, Naccache C, Ben Taarit Y, Vedrine JC, Coudurier G, Praliaud H (eds) Catalysis by zeolites. Proc Int Symp, Ecully (Lyon), Sept 9–11, 1980. Elsevier, Amsterdam, 1980, p 29; Stud Surf Sci Catal 5:29
77. Bystrov DS (1992) Zeolites 12:328
78. Jolly S, Saussey J, Lavalley JC (1994) Catal Lett 24:141
79. van Donk S, Bus E, Broersma A, Bitter JH, de Jong KP (2002) J Catal 212:86
80. DMS Arbeitsatlas der Infrarot-Spektroskopie (1972) Verlag Chemie, Weinheim
81. Fetting F, Gallei E, Kredel P (1984) Ger Chem Eng 7:32
82. Rozwadowski M, Włoch J, Erdmann K, Kornatowski J (1992) Coll Czech Chem Commun 57:959
83. Schnabel KH, Schreier E, Peuker C (1988) Catal Today 3(5):513
84. Eberly PEJ (1967) J Phys Chem 71:1717
85. Eisenbach D, Gallei E (1979) J Catal 56:377
86. Nováková J, Kubelková L, Bosacek V, Mach K (1991) Zeolites 11:135
87. Meinhold RH, Bibby DM (1990) Zeolites 10:121
88. Seo G, Kim MY, Kim JH (2000) Catal Lett 67:207
89. Ivanov P, Papp H (2001) Appl Surf Sci 179:235
90. Meunier FC, Domokos L, Seshan K, Lercher JA (2002) J Catal 211:366
91. Pazé C, Sazak B, Zecchina A, Dwyer J (1999) J Phys Chem B 103:9978
92. Wichterlová B, Zilková N, Uvarová E, Cejka J, Sarv P, Paganini C, Lercher JA (1999) Appl Catal A 182:297
93. Karge HG, Niessen W, Bludau H (1996) Appl Catal A 146:339
94. Li C, Stair PC (1997) In: Chon H, Ihm SK, Uh YS (eds) Progress in zeolite and microporous materials. Proc 11th Int Zeolite Conf, Seoul, Korea, Aug 12–17, 1996. Elsevier, Amsterdam, 1997, p 599; Stud Surf Sci Catal 105:599

95. Li C, Stair PC (1997) Catal Today 33:353
96. Li J, Xiong G, Feng Z, Liu Z, Xin Q, Li C (2000) Micropor Mesopor Mater 39:275
97. Chua YT, Stair PC (2003) J Catal 213:39
98. Oliver FG, Munson EJ, Haw JF (1992) J Phys Chem 92:8106
99. Haw JF, Nicholas JB, Song W, Deng F, Wang Z, Heneghan CS (2000) J Am Chem Soc 122:4763
100. Stepanov AG, Luzgin MV, Arzumanov SS, Ernst H, Freude D (2002) J Catal 211:165
101. Chua YT, Stair PC, Nicholas JB, Song W, Haw JF (2003) J Am Chem Soc 125:866
102. Eischens RP (1989) In: Jacobs PA, van Santen RA (eds) Zeolites: Facts, Figures, Future. Proc 8th Int Zeolite Conf, Amsterdam, The Netherlands, July 10–14, 1989. Elsevier, Amsterdam, 1989, p 51; Stud Surf Sci Catal 49:51
103. Najbar J, Eischens RP (1988) In: Philips MJ, Ternan M (eds) Proc 9th Int Congress on Catalysis, Calgary, Canada, 1988. The Chemical Institute of Canada, Ottawa, 1988, p 1434
104. Zarkov BB, Medschinskii WL, Butocnikova LF, Oranskaja OM, Marischev WB (1988) Chim Technol Topliv Masel 18
105. Kubelková L, Nováková J, Tupá M, Tvaruzková Z (1985) In: Fejes P, Kalló D (eds) Proc Int Symp Zeolite Catalysis, Siófok, Hungary, May 13–16. Petöfi Nyomda, Kecskemét, p 649
106. Bauer F, Kanazirev V, Vlaev L, Hanisch R, Weis W (1989) Chem Techn 41:297
107. Mirth G, Lercher JA (1991) J Catal 132:244
108. Datka J, Sarbak Z, Eischens RP (1994) J Catal 145:544
109. Bauer F, Ernst H, Geidel E, Peuker C, Pilz W (1997) In: Chon H, Ihm SK, Uh YS (eds) Progress in zeolite and microporous materials. Proc 11th Int Zeolite Conf, Seoul, Korea, Aug 12–17, 1996. Elsevier, Amsterdam, 1997, p 487, Stud Surf Sci Catal 105:487
110. Demidov AV, Davydov AA, Kurina LN (1985) Zh Prikl Spektrosk 43:845
111. Farneth WE, Gorte RJ (1995) Chem Rev 95
112. Corma A (1995) Chem Rev 95:559
113. Cerqueira HS, Ayrault P, Datka J, Guisnet M (2000) Micropor Mesopor Mater 38:197
114. Feller A, Barth JO, Guzman A, Zuazo I, Lercher JA (2003) J Catal 220(1):192
115. Karge HG, Darmstadt H, Gutsze A, Vieth HM, Buntkowsky G (1994) In: Weitkamp J, Karge HG, Pfeifer H, Hölderich W (eds) Zeolites and related microporous materials: State of the art 1994. Proc 10th Int Zeolite Conf, Garmisch-Partenkirchen, Germany, July 17–22, 1994. Elsevier, Amsterdam, 1994, p 1465; Stud Surf Sci Catal 84:1465
116. Karge HG, Trevizan de Suarez S, Dalla Lana IG (1984) J Phys Chem 88:1782
117. Karge HG, Ziołek M, Łaniecki M (1988) J Catal 109:252
118. Förster H, Seebode J, Fejes P, Kiricsi I (1987) J Chem Soc Farady Trans I 83:1109
119. Kiricsi I, Förster H, Tasi G (1989) In: Karge HG, Weitkamp J (eds) Zeolite as catalysts, sorbents and detergent builders— applications and innovations, Proc Int Symp, Würzburg, Germany, Sept 4–8, 1988. Elsevier, Amsterdam, 1989, p 355; Stud Surf Sci Catal 46:355
120. Garbowski ED, Praliaud H (1979) J Chim Phys Phys Chim-Biol 76:687
121. Garbowaki FU, Candy JP, Primet M (1983) J Chem Soc Farad Trans I 79:835
122. Asmolov GN, Krylov OV, Bragin OV, Furmen DB (1977) Kinet Katal 18:730
123. Förster H, Franke S, Seebode J (1983) J Chem Soc Farad Trans I 79:373
124. Łaniecki M, Karge HG (1987) In: Shopov D (ed) Proc 6th Int Symp Heterogeneous Catalysis, Sofia, Bulgaria, July 13–17, 1987. Publishing House Bulgarian Acad. Sci., 1987, p 129
125. Sorensen TS (1965) J Am Chem Soc 87:5075

126. Förster H, Kiricsi I (1988) Catal Today 3:65
127. Olah GA, Pittmann JCU, Waack R, Doran M (1966) J Am Chem Soc 88:1488
128. Luan Z, Zhang Y, Ni Y, Zheng L (1992) Cuihua Xuebao 13(6):479
129. Kustov LM, Zholobenko VL, Kazanskii VB (1988) Kinet Katal 29:928
130. Karge HG, Łaniecki M, Ziołek M, Onyestyák G, Kiss A, Kleinschmit P, Siray M (1989) In: Jacobs PA, van Santen RA (eds) Zeolites: Facts, Figures, Future. Proc 8th Int Zeolite Conf, Amsterdam, The Netherlands, July 10–14, 1989. Elsevier, Amsterdam, 1989, p 1327; Stud Surf Sci Catal 49:1327
131. Förster H, Seebode J, Fejes P, Kiricsi I (1986) J Chem Soc Farad Trans 83:1109
132. Zecchina A, Geobaldo F, Spoto G, Bordiga S, Ricchiardi G, Buzzoni R, Petrini G (1996) J Phys Chem 100:16584
133. Bordiga A, Ricchiardi G, Spoto G, Scarano D, Carnelli L, Zecchina A (1993) J Chem Soc Farad Trans 89:1843
134. Manos G, Hofmann H (1990) Chemiker-Zeitung 114:183
135. Schnabel KH, Fricke R, Girnus I, Jahn E, Löffler E, Parlitz B, Peuker C (1991) J Chem Soc Farad Trans 87:3569
136. Ernst H, Freude D, Mildner T, Pfeifer H (1994) In: Weitkamp J, Karge HG, Pfeifer H, Hölderich W (eds) Zeolites and related microporous materials: State of the art 1994. Proc 10th Int Zeolite Conf, Garmisch-Partenkirchen, Germany, July 17–22, 1994. Elsevier, Amsterdam, 1994, p 1717; Stud Surf Sci Catal 84:1717
137. Hunger M, Horvath T (1995) J Chem Soc Chem Commun:1423
138. Breitmeier F, Völter W (1972) In: 13C-NMR spectroscopy, methods and application in organic chemistry. Wiley Interscience, New York, p 221
139. Derouane EG, Gilson JP, Nagy JB (1982) Zeolites 2:42
140. van den Berg JP, Wolthuizen JP, Claque ADH, Hays GR, Huis R, van Hooff JHC (1983) J Catal 80:130
141. Maixner S, Chen CY, Grobet PJ, Jacobs PA, Weitkamp J (1986) In: Murakami Y, Iijima A, Ward JW (eds) New developments in zeolite science and technology. Proc 7th Int Zeolite Conf, Tokyo, Japan, August 17–22, 1986. Elsevier, Amsterdam, 1986, p 693; Stud Surf Sci Catal 28:693
142. Young RN (1979) Progr Nucl Magn Reson Spectrosc 12:261
143. Anderson MW, Klinowski J (1990) J Am Chem Soc 112:10
144. Groten WA, Wojciechowski BW, Hunter BK (1990) J Catal 125:311
145. Richardson BR, Haw JF (1989) Anal Chem 61:1821
146. Pfeifer H (1994) NMR of solid surfacs. In: NMR Basic principles and progress, vol 31. Springer, Berlin Heidelberg New York
147. Ernst H, Freude D, Hunger M, Pfeifer H (1991) In: Oehlmann G, Pfeifer H, Fricke R (eds) Catalysis and adsorption by zeolites, Proc Int Conf ZEOCAT '90, Leipzig, August 20–23, 1990. Elsevier, Amsterdam, 1990, p 397; Stud Surf Sci Catal 65:397
148. Ernst H, Freude D, Mildner T, Wolf I (1995) In: Beyer HK, Karge HG, Kiricsi I, Nagy JB (eds) Catalysis by microporous materials. Proc Int Symp ZEOCAT '95, Szombathely, Hungary, July 9–13, 1995. Elsevier, Amsterdam, 1995, p 413; Stud Surf Sci Catal 94:413
149. Ma D, Shu YY, Han XW, Liu XM, Xu YD, Bao XH (2001) J Phys Chem B 105:1786
150. Kärger J, Ruthven DM (1992) Diffusion in zeolites and other microporous solids. Wiley, New York; especially Chapters 5 and 6
151. Kärger J, Hunger M, Freude D, Pfeifer H, Caro J, Bülow M, Spindler H (1988) Catal Today 3:493
152. Völter J, Caro J, Bülow M, Fahlke B, Kärger J, Hunger M (1988) Appl Catal 42:15
153. Caro J, Jobic H, Bülow M, Kärger J, Zibrowius B (1993) Adv Catal 39:351

154. Snurr RQ, Hagen A, Ernst H, Schwarz HB, Ernst S, Weitkamp J, Kärger J (1996) J Catal 163:130
155. Paweewan B, Barrie PJ, Gladden LF (1998) Appl Catal A 167:353
156. Bär NK, Bauer F, Ruthven DM, Balcom BJ (2002) J Catal 208:224
157. Domeniconi T, N'Gokoli-Kekele P, Bonardet J-L, Springuel-Huet M-A, Fraissard J (1999) In: Tracy MMJ, Marcus BK, Bisher ME, Higgins JB (eds) Proc 12th Int Zeolite Conf, Baltimore, Maryland, July 5–10, 1998. Materials Research Society, Warrendale, PA, p 2991
158. Bonardet J-L, Domeniconi T, N'Gokoli-Kekele P, Springuel-Huet M-A, Fraissard J (1999) Langmuir 15:5836
159. Cheach KY, Chiaranussati N, Holleward MP, Gladden LF (1994) Appl Catal A 115:147
160. Anderson BG, van Santen RA, van Ijzendoorn LJ (1997) Appl Catal A 160:125
161. Bull LM, Cheetham AK (1997) In: Chon H, Ihm SK, Uh YS (eds) Progress in zeolite and microporous materials. Proc 11th Int Zeolite Conf, Seoul, Korea, Aug 12–17, 1996. Elsevier, Amsterdam, 1997, p 471; Stud Surf Sci Catal 105:471
162. Bull LM, Cheetham AK, Anupold T, Reinhold A, Samoson A, Sauer J, Bussemer B, Lee Y, Gann S, Shore J, Pines A, Dupree R (1998) Chem Phys Lett 285:10
163. Pingel UT, Amoureux JP, Anupold T, Bauer F, Ernst H, Fernandez C, Freude D, Samoson A (1998) Chem Phys Lett 294:345
164. Meinhold RH, Bibby DM (1990) Zeolites 10:74
165. Meinhold RH, Bibby DM (1990) Zeolites 10:146
166. Echevskii GV, Nekipelov VM, Ione KI, Zamaraev KI (1987) React Kinet Catal Lett 33(1):233
167. Bonardet JL, Barrage MC, Fraissard JF (1992) In: von Ballmoos R, Higgins JB, Treacy MMJ (eds) Proc 9th Int Zeolite Conf., Montreal, Canada, July 5–10, 1992, Butterworth-Heinemann, Boston, 1993, Vol II, p 475
168. Man PP, Klinowsky J (1988) Chem Phys Lett 147:581
169. West GW (1984) Aust J Chem 37:455
170. Ernst H, Pfeifer H (1990) Z Phys Chem (Leipzig) 271:1145
171. Dawson WH, Kaiser SW, Ellis PD, Inners RR (1981) J Am Chem Soc 103:6780
172. Lunsford JH (1997) Top Catal 4:91
173. Rakiewicz EF, Peters AW, Wormsbecher RF, Sutovich KJ, Mueller KT (1998) J Phys Chem B 102:2890
174. Osegovic JP, Drago RS (2000) J Phys Chem B 104:147
175. Chen WH, Tsai TC, Jong SJ, Zhao Q, Tsai CT, Wang I, Lee HK, Liu SB (2002) J Mol Catal A 181:41
176. Sutovich KJ, Peters AW, Rakiewicz EF, Wormsbecher RF, Mattingly SM, Mueller K (1999) J Catal 183:155
177. Zhao Q, Chen WH, Huang SJ, Wu YC, Lee HK, Liu SB (2002) J Phys Chem B 106:4462
178. Zhao Q, Chen WH, Huang SJ, Wu YC, Lee HK, Liu SB (2002) In: Sayari A, Jaroniec M (eds) Nanoporous materials III. Proc 3rd Int Symp Nanoporous Materials, Ottawa, Ontario, Canada, June 12–15, 2002. Elsevier, Amsterdam, 2002, p 453; Stud Surf Sci Catal 141:453
179. Tsai TC, Liu SB, Wang I (1999) Appl Catal A 181:355
180. Chen WH, Bauer F, Bilz E, Freyer A, Huang SJ, Lai CS, Liu SB (2004) In: van Steen E, Callanan LH, Claeys M (eds) Recent advances in the science and technology of zeolites and related materials, Proc. 14th Int Zeolite Conf, Cape Town, South Africa, 25–30 April 2004. Elsevier, Amsterdam, 2004, p 2269; Stud Surf Sci Catal 154:2269
181. Bauer F, Chen WH, Ernst H, Huang SJ, Freyer A, Liu SB (2004) Micropor Mesopor Mater 72:81

182. Ito T, Fraissard J (1982) J Chem Phys Lett 76:5225
183. Fraissard JP, Ito T (1988) Zeolites 8:350
184. Barrage MC, Bonardt JL, Fraissard J (1990) Catal Lett 5:143
185. Barrage MC, Bauer F, Ernst H, Fraissard J, Freude D, Pfeifer H (1990) Catal Lett 6:201
186. Tsiao C, Dybowsski C, Gaffney AM, Sofranko J (1991) J Catal 128:520
187. Bonardet JL, Barrage MC, Fraissard JP, Kubelkova L, Novakova J, Ernst H, Freude D (1992) Coll Czech Chem Commun 57:733
188. Liu SB, Prasad S, Wu JF, Ma LJ, Yang TC, Chiou JT, Chang JY, Tsat TC (1993) J Catal 142:664
189. Raffery MD, Chmelka BF (1993) In: Bluemich B, Kosfeld R (eds) NMR Basic principles and progresss, Vol 30. Springer, Berlin Heidelberg New York, p 111
190. Demarquay J, Fraissard J (1987) J Chem Phys Lett 136:314
191. Pradhan AR, Wu JF, Jong SJ, Chen WH, Tsai TC, Liu SB (1997) Appl Catal A 159:187
192. Lange JP, Gutsze A, Karge HG (1988) J Catal 114:136
193. Karge HG, Boldingh EP, Lange JP, Gutsze A (1985) In: Fejes P, Kalló D (eds) Proc Int Symp Zeolite Catalysis, Siófok, Hungary, May 13–16, 1985. Petőfi Nyomda, Kecskemét, p 639
194. Karge HG, Lange JP, Gutsze A, Łaniecki M (1988) J Catal 114:144
195. Leith IR (1972) J Chem Soc Chem Commun:1282
196. Kucherov AV, Slinkin AA (1982) Kinet Katal 23:1172
197. Kucherov AV, Slinkin AA (1983) Kinet Katal 24:947
198. Slinkin AA, Kucherov AV, Kondrat'ev DA, Bondarenko TN, Rubinshtein AM, Minachev KM (1986) In: Murakami Y, Iijima A, Ward JW (eds) New developments in zeolite science and technology. Proc 7th Int Zeolite Conf, Tokyo, Japan, August 17–22, 1986. Elsevier, Amsterdam, 1986, p 819; Stud Surf Sci Catal 28:819
199. Retcofsky HL, Thompson GP, Raymond R, Friedel RA (1975) Fuel 54:126
200. Bauer F (1990) ZfI Mitteilungen Leipzig 156:1
201. Echevskii GV, Kalinina NG, Anufrienko VF, Poluboyarov VA (1987) React Kinet Catal Lett 33(2):305
202. Bauer F, Freyer A (2001) In: Galarneau A, Di Renzo F, Fajula F (eds) Zeolites and mesoporous materials at the dawn of the 21st century. Proc 13th Int Zeolite Conf, Montpellier, France, July 8–13, 2001. Elsevier, Amsterdam, 2001, p 307; Stud Surf Sci Catal 135:307
203. Bauer F, Chen WH, Zhao Q, Freyer A, Liu SB (2001) Micropor Mesopor Mater 47:67
204. Choudhary VR, Mayadevi S, Akolekar DB (1993) J Catal 144:16
205. Muhler M, Schlögl R, Ertl G (1987) Surface Sci 189:69
206. Gallezot P, Leclerq C, Guisnet M, Magnoux P (1988) J Catal 114:100
207. Holmes SM, Garforth A, Maunders B, Dwyer J (1997) Appl Catal A 151:355
208. Behrsing T, Jaeger H, Sanders JV (1989) Appl Catal 54:289
209. de Jong KP, Mooiweer HH, Buglass JG, Maarsen PK (1997) In: Bartholomew CH, Fuentes GA (eds) Catalyst deactivation 1997. Proc 7th Int Symp, October 5–8, 1997, Cancun (Mexico). Elsevier, Amsterdam, 1997, p 127; Stud Surf Sci Catal 111:127
210. Sexton BA, Hughes AE, Bibby DM (1988) J Catal 109:126
211. Mooiweer HH, DeJong KP, Kraushaar-Czarnetzki B, Stork WH, Krutzen BCH (1994) In: Weitkamp J, Karge HG, Pfeifer H, Hölderich W (eds) Zeolites and related microporous materials: State of the art 1994. Proc 10th Int Zeolite Conf, Garmisch-Partenkirchen, Germany, July 17–22, 1994. Elsevier, Amsterdam, 1994, p 2327; Stud Surf Sci Catal 84:2327
212. Zajac GW, Fleisch TH, Meyers BL, Ray GJ, Miller JT (1989) J Catal 115:254

213. Kulkarni SJ, Hattori H, Toyoshima I (1987) Zeolites 7:178
214. Fyfe CA, Kennedy GJ, DeSchutter CT, Kokotailo GT (1984) J Chem Soc Chem Commun:541
215. Bibby DM, Milestone NB, Patterson JE, Aldridge LP (1986) J Catal 97:493
216. Alvarez AG, Viturro H, Bonetto RD (1992) Mater Chem Phys 32:135
217. Pollack SS, Gormeley RJ, Wetzel EL, Newbury D (1984) Zeolites 4:181
218. Uguina MA, Serrano DP, van Grieken R, Venes S (1993) Appl Catal A 99:97
219. Okkel LG, Fenelonov VB, Romannikov VN, Soloveva LP (1992) Kinet Katal 33:176
220. Mori N, Nishiyama S, Tsuruya S, Masai M (1991) Appl Catal 74:37
221. Zhu ZR (2001) Analyst 126:1669
222. Venuto PB, Hamilton LA (1967) Ind Eng Chem Prod Res Develop 6:190
223. Magnoux P, Canaff C, Machado F, Guisnet M (1992) J Catal 134:286
224. Magnoux P, Machado F, Guisnet M (1993) In: Guczi K, Solymosi F, Tetenyi P (eds) New frontiers in catalysis, Part A–C. Proc 10th Int Congress on Catalysis, Budapest, Hungary, July 19–24, 1992. Elsevier, Amsterdam, Akademiai Kiado, Budapest, 1993, p 435; Stud Surf Sci Catal 75:435
225. Moljord K, Magnoux P, Guisnet M (1995) Appl Catal A 122:21
226. Guisnet M, Andy P, Gnep NS, Travers C, Benazzi E (1995) J Chem Soc Chem Commun 16:1685
227. Guisnet M, Magnoux P, Martin D (1997) In: Bartholomew CH, Fuentes GA (eds) Catalyst deactivation 1997. Proc 7th Int Symp, Oct 5–8, Cancun. Elsevier, Amsterdam, 1997, p 1; Stud Surf Sci Catal 111:1
228. Guisnet M, Andy P, Gnep NS, Travers C, Benazzi E (1997) In: Chon H, Ihm SK, Uh YS (eds) Progress in zeolite and microporous materials. Proc 11th Int Zeolite Conf, Seoul, Korea, August 12–17, 1996. Elsevier, Amsterdam, 1997, p 1365; Stud Surf Sci Catal 105:1365
229. Guisnet M, Andy P, Boucheffa Y, Gnep NS, Travers C, Benazzi E (1998) Catal Lett 50:159
230. Guisnet M, Magnoux P (2001) Appl Catal A 212:83
231. Magnoux P, Guisnet M, Mignard S, Cartraud P (1989) J Catal 117:495
232. Anderson JR, Chang YF, Western RJ (1990) J Catal 124:259
233. Anderson JR, Dong QN, Chang YF, Western RJ (1991) J Catal 127:113
234. Cerqueira HS, Ayrault P, Datka J, Magnoux P, Guisnet M (2000) J Catal 196:149
235. Nielen MWF (1999) Mass Spec Rev 18:309
236. Hanton SD (2001) Chem Rev 101:527
237. Kyotani T, Nagai T, Inoue S, Tomita A (1997) Chem Mater 9:609
238. Johnson SA, Brigham ES, Ollivier PJ, Mallouk TE (1997) Chem Mater 9:2448
239. Rodriguez-Mirasol J, Cordero T, Radovic LR, Rodriguez JJ (1998) Chem Mater 10:550
240. Ryoo R, Joo SH, Jun S (1999) J Phys Chem B 103:7743
241. Ryoo R, Joo SH, Kruk M, Jaroniec M (2001) Adv Mater 13:677
242. Kruk M, Jaroniec M, Kim TW, Ryoo R (2003) Chem Mater 15:2815
243. Kang M, Yi SH, Lee HI, Yie JE, Kim JM (2002) Chem Commun:1944
244. Anderson JR, Chang YF, Western RJ (1989) J Catal 118:466
245. Niu F, Kolb G, Hofmann H (1995) Chem Techn 47:68
246. Manos G, Hofmann H (1991) Chem Eng Technol 14:73
247. Seifert J, Emig G (1987) Chem Ing Techn 59:475
248. Rozwadowski M, Wojsz R, Wisniewski KE, Kornatowski J (1989) Zeolites 9:503
249. Bibby DM, Pope CG (1988) J Catal 116:407

250. Post JG, van Hooff JHC (1986) In: Japan Association of Zeolite (ed) New developments in zeolite science and technology. Preprints of Poster Papers. 7th Int Zeolite Conf, Tokyo, Japan, August 17–22, 1986, p 249
251. Okkel LG, Garribov VY, Fenelov VB, Echivskii GV (1986) React Kinet Catal Lett 30:269
252. Soni HS, Samuel V, Ramaswamy AV (1994) Indian J Chem Sect A 33:844
253. Pradhan AR, Rao BS (1991) J Catal 132:79
254. Lechert H, Basler WD, Jia M (1988) Catal Today 3:23
255. Mignard S, Cartraud P, Magnoux P, Guisnet M (1989) J Catal 117:503
256. Bauer F, Ernst H, Geidel E, Hanisch R, Onderka E, Schödel R (1996) Erdöl Erdgas Kohle 112:516
257. Rollmann LD, Walsh DE (1979) J Catal 56:139
258. Inui T (1991) In: Graselli RK, Sleight AW (eds) Structure-activity and selectivity relationships in heterogeneous catalysis. Proc ACS Symp on Structure-Activity Relationships in Heterogeneous Catalysis, Boston, MA, April 22–27, 1990. Elsevier, Amsterdam, 1990, p 233; Stud Surf Sci Catal 67:233
259. Csicsery S (1986) Pure Appl Chem 58:841
260. Walsh DE, Rollmann LD (1977) J Catal 49:369
261. Walsh DE, Rollmann LD (1979) J Catal 56:195
262. Sotelo JL, Uguina MA, Valverde JL, Serrano DP (1994) Appl Catal A 114:273
263. Bhat YS, Das J, Halgeri AB (1995) J Catal 155:154
264. Rozwadowski M, Lezanska M, Włoch J, Erdmann K, Golembiewski R, Kornatowski J (2000) Phys Chem Chem Phys 2:5510
265. Rozwadowski M, Lezanska M, Włoch J, Erdmann K, Kornatowski J (2000) In: Kowalak S, Łaniecki M, Stankiewicz K, Szymkowiak E, Malecka-Grycz M (eds) Proc 7th Forum Zeolitowe, University im A Mickiewicza, Oct. 1–4, 2000, Kolobrzeg (Poland); ISBN 83-907325-4-7, p 35
266. Rozwadowski M, Lezanska M, Włoch J, Erdmann K, Golembiewski R, Kornatowski J (2001) Chem Mater 13:1609
267. Rozwadowski M, Lezanska M, Włoch J, Erdmann K, Golembiewski R, Kornatowski J (2001) Langmuir 17:2112
268. Rozwadowski M, Datka J, Lezanska M, Włoch J, Erdmann K, Kornatowski J (2001) Phys Chem Chem Phys 3:5082
269. Lezanska M, Rozwadowski M, Włoch J, Erdmann K (2001) In: Krysciak-Czerwenka J, Derewinski M (eds) Proc 8th Forum Zeolitowego, Sept 23–27, 2001, Wysowa-Zdroj (Poland); ISBN 83-907325-2-1, p 111
270. Nowak I, Lezanska M, Rozwadowski M, Rozploch F, Marciniak W (2002) Appl Surf Sci 201:182
271. Lezanska M, Rozwadowski M, Włoch J, Erdmann K (2003) In: Łaniecki M, Wojtowski M, Zalas M (eds) Proc 10th Forum Zeolitowego, Sept 21–26, 2003, Tuczno (Poland); ISBN 83-918771-3-2, p 317
272. Pu SB, Kim JB, Seno M, Inui T (1997) Micropor Mater 10:25
273. Haag WO, Lago RM, Nagy JB, Weisz PB (1984) Nature 309:589
274. Lopes JM, Lemos F, Ribeiro FR, Dewaele N, Derouane EG, Guisnet M (1988) J Mol Catal 48:373
275. Lopes JM, Lemos F, Ribeiro FR, Derouane EG, Magnoux P, Guisnet M (1994) Appl Catal A 114:161
276. Magnoux P, Cartraud P, Mignard S, Guisnet M (1987) J Catal 106:242
277. McLellan GD, Howe RF, Parker LM, Bibby DM (1986) J Catal 99:486

278. Topsøe NY, Pedersen K, Derouane EG (1981) J Catal 70:41
279. Lukyanov DB (1992) Zeolites 12:287
280. Lago RM, Haag WO, Mikovsky RJ, Olson DH, Hellring SD, Schmitt KD, Kerr GT (1986) In: Murakami Y, Iijima A, Ward JW (eds) New developments in zeolite science and technology. Proc 7th Int Zeolite Conf, Tokyo, Japan, August 17–22, 1986. Elsevier, Amsterdam, 1986, p 677; Stud Surf Sci Catal 28:677
281. Choudhary VR, Akolekar DB (1990) J Catal 125:143
282. Addison SW, Cartlidge S, Harding DA, McElhiney G (1988) Appl Catal 45:307
283. Corma A, Fornes V, Mocholi FA, Monton JB, Rey F (1991) In: Occelli ML (ed) Concepts in catalyst design: Fluid Cracking Symp, Washington, DC, USA, August, 26–31, 1990; ACS Symp Ser 452:12
284. Machado F, Magnoux P, Guisnet M (1991) In: Bartholomew CH, Butt JB (eds) Catalyst deactivation 1991. Proc 5th Int Symp, Evanston, IL, June 24–26, 1991. Elsevier, Amsterdam, 1991, p 807; Stud Surf Sci Catal 68:807
285. Nedomová K, Wichterlová B, Beran S, Bednarová S (1988) Catal Today 3:373
286. Chen Q, Guth JL, Fraissard J (1993) In: Guczi K, Solymosi F, Tetenyi P (eds) New Frontiers in catalysis, Part A–C. Proc 10th Int Congress on Catalysis, Budapest, Hungary, July 19–24, 1992. Elsevier, Amsterdam, 1993, Akademiai Kiado, Budapest, 1993, p 2577; Stud Surf Sci Catal 75:2577
287. Bauer F, Geidel E, Petzold E (1994) In: Weitkamp J, Karge HG, Pfeifer H, Hölderich W (eds) Zeolites and related microporous materials: State of the art 1994. Proc 10th Int Zeolite Conf, Garmisch-Partenkirchen, Germany, July 17–22, 1994. Elsevier, Amsterdam, 1994, p 1749; Stud Surf Sci Catal 84:1749
288. Campbell SM, Bibby DM, Coddington JM, Howe RF (1996) J Catal 161:350
289. Auroux A, Vedrine JC, Gravelle PC (1981) In: Rouquerol J, Sing KSW (eds) Proc Int Symp, Aix-en-Provence, France, September 21–23, 1981. Elsevier, Amsterdam, 1982, p 305; Stud Surf Sci Catal 10:305
290. Bülow M, Caro J, Völter J, Kärger J (1987) In: Delmon B, Froment GF (eds) Catalyst deactivation 1987. Proc 4th Int Symp, Antwerp, Belgium, September 29–October 1, 1987. Elsevier, Amsterdam, 1987, p 343; Stud Surf Sci Catal 34:343
291. Völter J, Lietz G, Kürschner U, Löffler E, Caro J (1988) Catal Today 3:407
292. Pater J, Cardona F, Canaff C, Gnep NS, Szabo G, Guisnet M (1999) Ind Eng Chem Res 38:3822
293. Magnoux P, Guisnet M (1989) Zeolites 9:329
294. Martin D, Guisnet M (1999) J Catal 188:346
295. Novaková J, Kubelková L (1992) J Catal 136:258
296. Rhodes NP, Rudham R (1994) J Chem Soc Farad Trans 90:809
297. Bauer F, Ernst H, Geidel E, Hanisch R, Onderka E (1996) Chem Techn 48:185
298. Haag WO, Olson DH, Rodewald PG (1985) US Patent No. 4,508,836 assigned to Mobil Oil Corporation
299. Fang LY, Liu SB, Wang I (1999) J Catal 185:33
300. Pellet RJ, Casey DG, Huang HM, Kessler RV, Kuhlman EJ, O'Young CL, Sawicki RA, Ugolini JR (1995) J Catal 157:423
301. Houzvicka J, Hansildaar S, Nienhuis JG, Ponec V (1999) Appl Catal A 176:83
302. Houzvicka J, Nienhuis JG, Hansildaar S, Ponec V (1997) Appl Catal A 165:443
303. Asensi MA, Corma A, Martinez A, Derewinski M, Krysciak J, Tamhankar SS (1998) Appl Catal A 174:163
304. Xu WQ, Yin YG, Suib SL, Edwards JC, O'Young CL (1996) J Catal 163:232
305. Xu WQ, Yin YG, Suib SL, Edwards JC, O'Young CL (1995) J Phys Chem 99:9443

306. Mériaudeau P, Naccache C, Le HN, Vu TA (1997) J Mol Catal A 123:L1
307. Onyestyák G, Valyon J, Pál-Borbély G, Rees LV (2002) Appl Surf Sci 196:401
308. Philippou A, Dwyer J, Ghanbari-Siahkali A, Paze C, Anderson MW (2001) J Mol Catal A 174:223
309. Mériaudeau P, Bacaud R, Hung LN, Vu AT (1996) J Mol Catal A 110:L177
310. Cejka J, Wichterlová B, Sarv P (1999) Appl Catal A 179:217
311. Houzvicka J, Ponec V (1998) Ind Eng Chem Res 37:303
312. Guisnet M, Andy P, Gnep NS, Travers C, Benazzi E (1998) Ind Eng Chem Res 37:300
313. Asensi MA, Martinez A (1999) Appl Catal A 183:155
314. van Donk S, Bus E, Broersma A, Bitter JH, de Jong KP (2002) In: Aiello R, Testa F, Giordano G (eds) Impact of zeolites and other porous materials on the new technologies at the beginning of the new millennium. Proc 2nd Int FEZA Conf, Taormina, Italy, September 1–5, 2002. Elsevier, Amsterdam, 2002, p 573; Stud Surf Sci Catal 142:573
315. Voorhies A (1945) Ind Eng Chem 37:318
316. Benito PL, Gayubo AG, Aguayo AT, Castilla M, Bilbao J (1996) Ind Eng Chem Res 35:81
317. Benito PL, Gayubo AG, Aguayo AT, Olazar M, Bilbao J (1996) Ind Eng Chem Res 35:3991
318. Schipper PH, Krambeck FJ (1986) Chem Eng Sci 41:1013
319. Beyne AOE, Froment GF (1990) Chem Eng Sci 45:2089
320. Forissier M, Bernard JR (1991) In: Bartholomew CH, Butt JB (eds) Catalyst deactivation 1991. Proc 5th Int Symp, Evanston, IL, June 24–26, 1991. Elsevier, Amsterdam, 1991, p 359; Stud Surf Sci Catal 68:359
321. Nelson PH, Bibby DM, Kaisar AB (1991) Zeolites 11:337
322. Reyes SC, Scriven LE (1991) Ind Eng Chem Res 30:71
323. Pinheiro CI, Lemos F, Ribeiro FR (2000) Reac Kin Catal Lett 69:39
324. Olson DH, Haag WO (1984) In: Whyte TE (ed) Catalytic materials: Relationship between structure and reactivity, State-of-the-art Symp, San Francisco, USA, June 13–16, 1983; ACS Symp Ser 248:275
325. Chen NY, Garwood WE, Dwyer FG (1989) Chemical industries. Marcel Dekker, New York, p 36
326. Colon G, Ferino I, Rombi E, Selli E, Forni L, Magnoux P, Guisnet M (1998) Appl Catal A 168:81
327. Ferino I, Monaci R, Rombi E, Solinas V, Magnoux P, Guisnet M (1999) Appl Catal A 183:303
328. Da Z, Magnoux P, Guisnet M (1999) Appl Catal A 182:407
329. Da Z, Magnoux P, Guisnet M (1999) Catal Lett 61:203
330. Weiss U, Weihe M, Hunger M, Karge HG, Weitkamp J (1997) In: Chon H, Ihm SK, Uh YS (eds) Progress in zeolite and microporous materials. Proc 11th Int Zeolite Conf, Seoul, Korea, Aug 12–17, 1996. Elsevier, Amsterdam, 1997, p 973; Stud Surf Sci Catal 105:973
331. Karge HG, Ernst S, Weihe M, Weiss U, Weitkamp J (1994) In: Weitkamp J, Karge HG, Pfeifer H, Hölderich W (eds) Zeolites and related microporous materials: State of the art 1994. Proc 10th Int Zeolite Conf, Garmisch-Partenkirchen, Germany, July 17–22, 1994. Elsevier, Amsterdam, 1994, p 1805; Stud Surf Sci Catal 84:1805
332. Martens JA, Parton R, Uytterhoeven L, Jacobs PA (1991) Appl Catal 76:95
333. Martens JA, Souverijns W, Verrelst W, Parton R, Froment GF, Jacobs P (1995) Angew Chem Int Ed Engl 34:2528

334. Shenk M, Smit B, Vlugt TJH, Maesen TLM (2001) Angew Chem Int Ed 40:736
335. Sastre G, Chica A, Corma A (2000) J Catal 195:227
336. Kolboe S (1987) In: Methane conversion. Proc Symp Production of Fuels and Chemicals from Natural Gas, Auckland, New Zealand, April 27–30, 1987. Elsevier, Amsterdam, 1987, p 189; Stud Surf Sci Catal 36:189
337. Prinz D, Riekert L (1988) Appl Catal 37:139
338. Song W, Haw JF, Nicholas JB, Heneghan CS (2000) J Am Chem Soc 122:10726
339. Dessau RM (1986) J Catal 99:111
340. Dahl IM, Kolboe S (1993) Catal Lett 20:329
341. Arstad B, Kolboe S (2001) Catal Lett 71:209
342. Mikkelsen Ø, Ronning PO, Kolboe S (2000) Micropor Mesopor Mater 40:95
343. Mole T, Bett G (1983) J Catal 84:435
344. Arstad B, Kolboe S (2001) J Am Chem Soc 123:8137
345. Pines H (1985) J Catal 93:205
346. Dadyburjor DB (1992) Catalysis 9:229
347. Rene' Bos AN, Tromp PJJ, Akse HN (1995) Ind Eng Chem Res 34:3808
348. Xu WQ, Yin YG, Suib SL, O'Young CL (1995) J Phys Chem 99:758
349. Froment GF, Bischoff KB (1962) Chem Eng Sci 17:105
350. Chen D, Rebo HP, Moljord K, Holmen A (1997) Ind Eng Chem Res 36:3473
351. Degnan TF (2003) J Catal 216:32
352. Nawaz S, Kolboe S, Kvisle S, Lillerud KP, Stöcker M, Øren HM (1991) In: Holmen A, Jens KJ, Kolboe S (eds) Natural gas conversion. Proc Symp Natural Gas Conversion, Oslo, Norway, August 12–17, 1990. Elsevier, Amsterdam, 1990, p 421; Stud Surf Sci Catal 61:421
353. Chen D, Rebo HP, Moljord K, Holmen A (1998) In: Parmaliana A, Sanfilippo D, Frusteri F, Vaccari A, Arena F (eds) Natural gas conversion V. Proc 5th Int Gas Conversion Symp, Giardini-Naxos, Taormina, Italy, September 20–25, 1998. Elsevier, Amsterdam, 1998, p 521; Stud Surf Sci Catal 119:521
354. van Niekerk MJ, Fletcher JCQ, O'Connor CT (1996) Appl Catal A 138:135
355. Niwa M, Ito H, Kato S, Hattori T, Murakami Y (1982) J Chem Soc Chem Commun:819
356. Begum HA, Katada N, Niwa M (2001) Micropor Mesopor Mater 46:13
357. Haag WO, Olson DH (1978) US Patent No. 4,097,543 assigned to Mobil Oil Corporation
358. Gorra F, Breckenridge LL, Guy WM, Sailor RA (1992) Oil Gas J 90(41):60
359. Kaeding WW, Chu CC, Young LB, Weinstein B (1981) J Catal 67:159
360. Kaeding WW, Young LB, Chu CC (1984) J Catal 89:267
361. Jones CW, Davis ME (1998) Nature 393:52
362. Niwa M, Kunieda T, Kim JH (1999) In: Song C, Garcés JM, Sui Y (eds) Proc Symp on Shape Selective Catalysis in Hydrocarbon Processing and Chemicals Synthesis, Dallas, USA, March 29–April 02, 1998; ACS Symp Ser 738:181
363. Zheng S, Jentys A, Lercher JA (2003) J Catal 219:310
364. Venkat C (1994) 19th Dewitt Petrochemical Review, 22–24 March:A 1
365. Low CD, Lawson RJ, Kuchar PJ, Gray GL (1994) US Patent No. 5,321,184 assigned to UOP
366. Beck JS, Crane RA, Mathias MF, Kowalski JA, Lissy DN, Stern DL (1999) WO Patent No. 99 52842 assigned to Mobil Oil Corporation
367. Haag WO, Olson DH, Rodewald PG (1982) US Patent No. 4,358,395 assigned to Mobil Oil Corporation

368. Chen WH, Pradhan A, Jong SJ, Lee TY, Wang I, Tsai TC, Liu SB (1996) J Catal 163:436
369. Jong SJ, Pradhan AR, Wu JF, Tsai TC, Liu SB (1998) J Catal 174:210
370. Pradhan AR, Lin TS, Chen WH, Jong SJ, Wu JF, Chao KJ, Liu SB (1999) J Catal 184:29
371. Nacamuli GJ, Vogel RF (1998) WO Patent No. 98 43932 assigned to Chevron Chemical Company LLC

Mol Sieves (2007) 5: 365–478
DOI 10.1007/3829_006

Published online: 17 February 2006

Isomorphous Substitution in Zeolites

J. B.Nagy[1] (✉) · R. Aiello[2] · G. Giordano[2] · A. Katovic[2] · F. Testa[2] · Z. Kónya[3] · I. Kiricsi[3]

[1]Laboratoire de RMN, Facultes Universitaires Notre-Dame de la Paix, 61 rue de Bruxelles, 5000 Namur, Belgium
janos.bnagy@fundp.ac.be

[2]Department of Applied Chemistry, University of Calabria, Via Pietro Bucci, 87030 (CS) Arcavacata di Rende, Italy

[3]Department of Applied and Environmental Chemistry, University of Szeged, Rerrich Bela ter 1., 6720 Szeged, Hungary

Abbreviations[1]

1D NMR	One-dimensional nuclear magnetic resonance (spectroscopy)
2D 3QMAS NMR	Two-dimensional three quantum magic angle spinning nuclear magnetic resonance (spectroscopy)
3QMAS NMR	Three quantum magic angle spinning nuclear magnetic resonance (spectroscopy)
$A_{\|\|}$	Electron-nucleus coupling constant for the component parallel to the symmetry axis
AAS	Atomic absorption spectroscopy
^{27}Al MAS NMR	Aluminum magic angle spinning nuclear magnetic resonance (spectroscopy)
Al_O	Octahedrally coordinated framework aluminum atom
Al_T	Tetrahedrally coordinated framework aluminum atom
$AlPO_4$-11	Microporous aluminophosphate zeolite-like structure (cf. [70])
[Al]-ZSM-5	Zeolite structure (MFI, cf. [70]) containing aluminum in the framework[2]
Amp	Peak-to-peak amplitude
AS	As synthesized
AST	Page 86 zeolite structure (cf. [70])
AV-1	Sodium yttrium silicate structure (cf. [156])
^{9}Be NMR	Beryllium nuclear magnetic resonance (spectroscopy)
BET	Brunauer-Emmett-Teller specific surface measurement
BEA	Zeolite structure, acronym for zeolite Beta (cf. [70])
[B]-BEA	Zeolite structure (BEA, cf. [70]) containing boron in the framework
[B]-EUO	Zeolite structure (EUO, cf. [70]) containing boron in the framework
[B]-FER	Zeolite structure (FER, cf. [70]) containing boron in the framework
[B]-LEV	Zeolite structure (LEV, cf. [70]) containing boron in the framework
[B]-MEL	Zeolite structure (MEL, cf. [70]) containing boron in the framework
[B]-MFI	Zeolite structure (MFI, cf. [70]) containing boron in the framework (cf. [181–183]; Testa F, Chiappetta R, Crea F, Aiello R, Fonseca A, Bertrand JC, Demortier G, Guth JL, Delmotte L, B.Nagy J, submitted for publication)
[B]-SSZ24	Zeolite structure (SSZ24, cf. [70]) containing boron in the framework

[1] Unfortunately, many of the above-indicated abbreviations have various meanings (vide supra); in view of the current conventions in the literature, this is hardly avoidable. However, the correct meaning of the abbreviations should follow from the respective context.

[2] Presenting an element symbol in square brackets should indicate that the respective element is supposed to be incorporated into the framework of the material designated by the subsequent acronym or abbreviation. For instance, "[B]-ZSM-5" is indicating that boron is incorporated into the framework of ZSM-5.

[B]-ZSM-5	Zeolite structure (MFI, cf. [70]) containing boron in the framework
BU	Chemical identity of the Fe(III) species
^{13}C MAS NMR	Carbon magic angle spinning nuclear magnetic resonance (spectroscopy)
Cs-[Fe]-silicalite-1	Zeolite structure (cf. [70]) containing iron in the framework and charge-compensating cesium ion in extra-framework position
Cs-[Fe]-ZSM-5	Zeolite structure (cf. [70]) containing iron in the framework and charge-compensating cesium ion in extra-framework position
CIT-6	Zeolite structure (BEA structure, cf. [70])
[Co]-MFI	Zeolite structure (MFI, cf. [70]) containing Co in the framework (cf. [196, 197])
Cu-TON	Zeolite structure (TON, cf. [70]) containing Cu in charge-compensating extra-framework position (cf. [203])
CVD	Chemical vapor deposition
CCVD	Catalytic chemical vapor deposition
CQ	Quadrupole coupling constant
DAS	Dynamic angle spinning (spectroscopy)
deferrization	Removal of iron
DR	Diffuse reflectance (spectroscopy)
DSC	Differential scanning calorimetry
DTA	Differential thermal analysis
DTG	Differential thermogravimetry
EFW	Extra framework
EG	Ethylene glycol
EMT	Zeolite structure; hexagonal faujasite (cf. [70])
EPR	Electron paramagnetic resonance (spectroscopy) (acronym for ESR)
EPMA	Electron probe micro-analysis
ESEM	Environmental scanning electron microscopy
ESCA	Electron spectroscopy for chemical analysis (acronym for XPS)
ESR	Electron spin resonance (spectroscopy) (acronym for EPR)
ETS-10	Zeolite structure (cf. [70])
EUO	Zeolite structure (cf. [70])
EXAFS	Extended X-ray absorption fine structure
FAAS	Flame atomic absorption spectroscopy
FAU	Zeolite structure; acronym for faujasite (cf. [70])
[Fe]-BEA	Zeolite structure (BEA, cf. [70]) containing iron in the framework (cf. [194])
[Fe,Al]-BEA	Zeolite structure (BEA, cf. [70]) containing iron and aluminum in the framework
[Fe,Al]-MOR	Zeolite structure (MOR, cf. [70]) containing iron and aluminum in the framework
[Fe]-MCM-22	Zeolite structure (acronym or IZA structure code is MWW; cf. [70]) containing iron in the pore walls
[Fe,Al]-MCM-22	Zeolite structure (acronym or IZA structure code is MWW; cf. [70]) containing iron and aluminum in the pore walls
[Fe]-MCM-41	Mesoporous MCM-41 material containing iron in the pore walls
[Fe]-MFI	Zeolite structure (MFI, cf. [70]) containing iron in the framework (cf. [185, 186])

[Fe]-MTW	Zeolite structure (MTW, cf. [70]) containing iron in the framework (cf. [189])
[Fe]-TON	Zeolite structure (TON, cf. [70]) containing iron in the framework (cf. [189])
FER	Zeolite structure; acronym for ferrierite (cf. [70])
FID	Flame ionization detector (GC)
FTIR	Fourier transform infrared (spectroscopy)
FW	Framework
FWHM	Full line width at half-maximum (of a band)
g	g factor
$g_{\|\|}$	g factor for the component parallel to the symmetry axis
$g_{\perp}$	g factor for the component perpendicular to the symmetry axis
^{71}Ga NMR	Ga nuclear magnetic resonance (spectroscopy)
[Ga]-BEA	Zeolite with Beta (BEA) structure containing gallium in the framework, (cf. [146])
[Ga]-MCM-22	Zeolite structure (acronym or IZA structure code is MWW; cf. [70]) containing boron in the pore walls
[Ga]-MFI	Zeolite with MFI structure containing gallium in the framework, (cf. [183, 184])
[Ga]-ZSM-5	Zeolite with MFI structure containing gallium in the framework, (cf. [183, 184])
GC	Gas chromatography
H	Magnetic field (in Tesla)
^{1}H MAS NMR	Proton magic angle spinning nuclear magnetic resonance (spectroscopy)
HMI	Hexamethylene imine
HT	High temperature
I	Intensity
I_{rel}	Relative intensity
ICP-AES	Inductively coupled plasma atomic emission spectroscopy
IR	Infrared (spectroscopy)
IS	Isomer shift (Mössbauer spectroscopy)
K-[Fe]-silicalite-1	Zeolite structure (cf. [70]) containing iron in the framework and charge-compensating potassium ion in extra-framework position
L	Length
L	Liter
L/W	Aspect ratio
LEV	Zeolite structure (acronym of levyne; cf. [70])
LT	Low temperature
LTL	Linde-type L zeolite (cf. [70])
M	Metal or metal cation
MAS NMR	Magic angle spinning nuclear magnetic resonance (spectroscopy)
MFI	Zeolite structure (of, e.g., ZSM-5 or silicalite, cf. [70])
MCM-22	Zeolite structure (acronym or IZA structure code is MWW; cf. [70])
MCM-41	Mesoporous material with hexagonal arrangement of the uniform mesopores (cf. Volume 1, Chapter 4 of this series)
MCM-48	Mesoporous material with cubic arrangement of the uniform mesopores (cf. Volume 1, Chapter 4 of this series)
MCM-58	Zeolite structure (acronym or IZA structure code is IFR, cf. [70])
MEL	Zeolite structure (cf. [70])

MeQ^+	Methyl quinuclidinium cation
MOR	Zeolite structure; acronym for mordenite (cf. [70])
MQMAS	Multiquantum magic angle spinning (NMR)
MTT	Zeolite structure (cf. [70])
MTW	Zeolite structure (cf. [70])
Na-[Fe]-silicalite-1	Zeolite structure (cf. [70]) containing iron in the framework and charge-compensating sodium ion in extra-framework position
NCL-1	High-silica (n_{Si}/n_{Al} = 20 to infinity) zeolite (cf. [70])
NH_4-[Fe]-silicalite-1	Zeolite structure (cf. [70]) containing iron in the framework and charge-compensating ammonium ion in extra-framework position
NMR	Nuclear magnetic resonance
O_h	Octahedrally coordinated species
OFF	Zeolite structure, acronym for offretite (cf. [70])
PIGE	Proton induced γ-ray emission
PIXE	Proton induced X-ray emission
PQ	Quadrupole-quadrupole interaction
PTFE	Polytetrafluorethylene
PULSAR	NMR simulation program (cf. [284])
Q_{cc}	Quadrupole coupling constant
QS	Quadrupole shift (Mössbauer spectroscopy)
R	Crystallization rate
REDOR	Rotational-echo double-resonance NMR experiments (cf. [87])
RI	Spectral contribution (Mössbauer spectroscopy)
SAM	Scanning Auger microscopy
SEM	Scanning electron microscopy
^{29}Si MAS NMR	Silicon magic angle spinning nuclear magnetic resonance (spectroscopy)
Si(1Ga)	Si with 1 Ga in the neighborhood
Sil-1	Zeolite structure (acronym of SIL-1, cf. [70])
Silicalite-1	Zeolite structure (cf. [70])
^{119}Sn NMR	Tin nuclear magnetic resonance (spectroscopy)
SiOM	Defect group (M = NH_4, Na, K, Cs)
SiOTPA	Defect group
SiOX	Defect group (X = NH_4, Na, K, Cs, H, TPA, ...)
SOD	Zeolite structure, acronym for sodalite (cf. [70])
SSIMS	Static secondary ion mass spectroscopy
SSR	Solid-state reaction
SSZ-*n*	Series of zeolite structures; aluminosilicates, e.g., SSZ-24 and SSZ-13, isostructural with corresponding aluminophosphates, $AlPO_4$ (AFI) and $AlPO_4$-34 (CHA structure) (cf. [70])
T	Tetrahedrally coordinated framework atom (cation) such as Si, Al, Ti, Fe, V, B
T	Absolute temperature, in Kelvin (K)
T^{III}	Tetrahedrally coordinated trivalent framework atom (cation) such as Al, B, Ga
T_h	Tetrahedrally coordinated species
t_1	Longitudinal relaxation time
t_{ind}	Reaction induction time
t_{pulse}	Pulse length
TA	Thermal analysis

TCD	Thermal conductivity detector (GC)
TEAOH	Tetraethylammonium hydroxide
TEOS	Tetraethyl orthosilicate
TEM	Transmission electron microscopy
TG	Thermogravimetry
TMA	Tetramethyl ammonium
TON	Zeolite structure; acronym for theta-1 (cf. [70])
TPA	Tetrapropyl ammonium
TPABr	Tetrapropyl ammonium bromide
TPD	Temperature-programmed desorption
TPR	Temperature-programmed reduction
TS-1	ZSM-5 (MFI) structure containing small amounts of titanium besides silicon in the framework
TsG-1	Zeolite structure (BEA, cf. [70])
VS-1	Zeolite structure (MFI, cf. [70]) containing vanadium besides silicon in the framework
W	Width
UV	Ultraviolet (spectroscopy)
UV Res Raman	Ultraviolet resonance Raman (spectroscopy)
UV-Vis	Ultraviolet-visible (spectroscopy)
X	Zeolite structure (faujasite type structure with $n_{Si}/n_{Al} \leq 2.5$, cf. [70])
XANES	X-ray absorption near edge spectroscopy
XRD	X-ray diffraction
XRF	X-ray fluorescence spectroscopy
XPS	X-ray photoelectron spectroscopy
Y	Zeolite structure (faujasite-type structure with $n_{Si}/n_{Al} \geq 2.5$, cf. [70])
YAG	Yttrium aluminum garnet (laser)
[Zn]-MFI	Zeolite structure (MFI, cf. [70]) containing Zn in the framework (cf. [198–200])
ZSM-5	Zeolite structure (MFI, cf. [70])
ZSM-12	Zeolite structure (cf. [70])
α	Indicates the large cage in the structure of zeolite A (cf. [70])
α	The main channel of ZSM-5 zeolite
β	Indicates the sodalite cage in, e.g., A-type or faujasite-type structure (cf. [70])
β	Mid positions in the six-membered rings of ZSM-5 zeolite
γ	Mid positions in the five-membered rings of ZSM-5 zeolite
δ	Chemical shift (NMR)
δ_{CS}	Chemical shift (NMR)
2Θ	Degree
Θ	Pulse angle
λ	Wavelength (in μm)
ν	Resonance frequency
ΔH	Full line width at half-maximum (of a band)
ν_L	Larmor frequency
ν_Q	Quadrupole frequency
ν_{rot}	Rotation frequency
ν_{RF}	Radio frequency

1 Introduction

The isomorphous substitution of Si by other tetrahedrally coordinated heteroatoms such as B^{III} [1, 2], Al^{III} (ZSM-5) [3], Ti^{IV}(TS-1) [4–9], Ga^{III} [10–14] and Fe^{III} [15–18] in small amounts (up to 2–3 wt %) provides with new materials showing specific catalytic properties in oxidation and hydroxylation reactions related to the coordination state of the heteroatom [19]. Moreover, MFI-type materials with trivalent metal present in tetrahedral (T) sites have had tremendous impact as new shape-selective industrial catalysts having tunable acidic strength. In fact, the acidic strength of the protons in the bridged $Si(OH)T^{III}$ (T = B, Al, Fe, Ga) groups depends on the nature of the trivalent heteroatom. Indeed, the choice of T^{III} critically affects this property according to the sequence of Al > Fe = Ga ≫ B [20–23]. The recent discovery of an Al-containing natural zeolite (mutinaite) with the MFI topology [24] also makes this structure relevant to the mineralogy.

[Ga]-ZSM-5 zeolites are interesting materials as selective catalysts in the transformation of low molecular weight alkanes to aromatics [25–27]. These catalysts were mostly synthesized in alkaline media, however, several fluorine-containing media (adding either HF or NH_4F to the initial gel) have already been used [28, 29]. Note that the incorporation of gallium into the ZSM-5 structure is less effective than the incorporation of aluminum in the same reaction media [30]. The fluorine-containing reaction medium is generally made using either HF or NH_4F as a source of F^- ions [28, 29, 31]. Guth et al. have published a series of very interesting papers in which T^{III} elements (T = B, Al, Fe, Ga) were partially substituted for silicon in the MFI framework [32]. We have previously initiated a series of studies where the role of alkali cations was systematically explored. These studies include the synthesis of silicalite-1 [33–35], silicalite-2 [36], borosilicalite-1 [37, 38], ferrisilicalite-1 [39], ZSM-5 [40] and zeolite Beta [40, 41]. The differences in the catalytic activity of iron-containing and iron-supported zeolites are also very interesting, and several methods of preparation have been developed [42–44]. [Fe]-silicates with MFI [45, 46], MOR [47], BEA [48], MTT [49], TON [50] and MWW [51] structures have been synthesized in alkaline media. However, despite the fact that isomorphous substitution seems to be easier in fluoride-containing media [52], only [Fe]-ZSM-5 has been synthesized so far in the presence of NH_4F as a mineralizing agent [53]. Although the introduction of boron, gallium, or iron is relatively easy and well documented [19], few studies are devoted to the introduction of Co(II) into the framework of zeolites [54]. As both the framework and the extra-framework Co-species seem to be active in catalysis [55], it is of paramount importance to synthesize and thoroughly characterize Co-containing zeolites [56]. Zinc has been reported as a component of various molecular sieves such as zincophosphates, zincoarsenates [57–60], zincoalumino-silicates [61–63], zincosilicates [64–68], and zincoaluminonophosphates [19]. In some cases crystalline analogs of zeolite structures have been obtained under unusually mild conditions and crystallization occurred almost spontaneously on mixing the substrate solutions [57] or even on grinding the substrates [69]. The resulting zincophosphates and zincoarsenates, however, were unstable and usually decomposed above 200 °C. The reported zincosilicates were more stable, although most novel structures showed a narrow pore system [54, 64–68], not suitable for catalysis and adsorption. The MFI structure (zeolites ZSM-5) [70] has been very often used as a catalyst. Besides the efficiency of active sites (mainly strong acid sites), the medium-sized channels provide shape selective effects for the reactions of commercial importance. Therefore, the preparation of the zincosilicalite structure is also of interest. Due to the double negative charge of the tetrahedral lattice zinc, it could be modified with various cations including protons and might be considered as catalysts for various reactions. Moreover, some redox activity could result from the presence of zinc in the lattice. The zinc-modified MFI zeolites have been applied as active catalysts in the Cyclar process [62, 63, 71], which consists in the formation of aromatics from light paraffins. The catalysts used in methanol synthesis contain mostly zinc and copper oxides [72]; it is conceivable that MFI zincosilicate modified with copper cations could be efficient for this reaction. The well-ordered crystalline structure as well as the uniform pore system could be advantageous for the catalyst performance. Attempts to synthesize MFI aluminosilicate with some admixture of zinc [62, 73–75] as well as zincosilicate [68, 76] have been reported.

Due to environmental problems in the last years great attention has been devoted to air pollution. The automotive air pollutants (NO_X, CO and hydrocarbons) give large contribution to the total air pollutants. In order to reduce emission of pollutants, the trend in the automotive industry is to substitute traditional engines with engines operating under lean burn conditions. However, under these conditions the traditional three-way catalysts are not effective. With this new kind of engines, interesting results were obtained by using Cu- or Co-zeolite catalysts at the engine exhaust [77–79]. Unfortunately, one of the most active and selective catalysts (i.e., [Cu]-MFI-type), exhibits very rapid deactivation in the presence of water that is, of course, present in the automotive exhaust [80]. In a large number of papers on Cu zeolites, the introduction of Cu is carried out by ionic exchange from the Na form to obtain the Cu form. On the other hand, literature indicates that the solid-state reaction is a very good method for metal incorporation into the zeolites [81–83]. It is also indicated that during the zeolite synthesis with alcohols, the presence of sodium can occlude the zeolitic channels [84] and that the ionic exchange to the ammonium form followed by calcination opens the zeolitic channels. As an example the Na^+-TON presents a micropore volume equal to 0.05 ml g^{-1}, on the contrary the H^+-TON shows a value equal to 0.91 ml g^{-1}.

Isomorphous substitution was essentially performed with the MFI structure. Table 1 gives an overview of additional references to be used for entering into the subject. It can be seen that boron, gallium, vanadium and iron are the most commonly used elements. It is worthwhile to mention that the introduction of other elements such as Ti, In, Be, Mn, Sn, Cr, Mo, Ge and Zn, was also successful.

The second most studied zeolite for isomorphous substitution is the zeolite BEA [70] (Table 2). However, the number of publications remains far smaller than that dealing with ZSM-5. The most studied elements are still B, Ga, and Fe, but some reports also concern Zn, Sn, Ge and Ti.

Finally, Table 3 illustrates the isomorphous substitution of various elements into the remaining zeolitic structures.

In this review we shall focus on our works published on [B]-MFI, [Ga]-MFI, [Fe]-MFI, [Fe]-BEA, [Fe]-MCM-22, Zn-zeolite, and Cu-containing zeolites. Essentially, the various synthesis methods together with characterization techniques will be reviewed. The catalytic part will only be included, where it is considered essential.

2 Experimental

2.1 Synthesis Procedures

2.1.1 [B]-MFI

The gels were prepared by dissolving H_3BO_3 (Carlo Erba) in distilled water, adding the fluoride source (NH_4F, NaF, KF Carlo Erba; CsF, Aldrich) and tetrapropylammonium bromide, Fluka (TPABr) to the H_3BO_3 aqueous solution and then adding this solution to fumed silica (Serva) [181–183] (Testa F, Chiappetta R, Crea F, Aiello R, Fonseca A, Bertrand JC, Demortier G, Guth JL, Delmotte L, B.Nagy J, submitted for publication). The composition of the as-prepared gels was $9MF - xH_3BO_3 - 10SiO_2 - 1.25TPABr - 330H_2O$ with M = NH_4, Na, K and Cs and $x = 0.1$ and 10. Syntheses were carried out in Morey-type PTFE-lined 20 cm^3 autoclaves at 170 ± 2 °C, without stirring, under autogenous pressure. After being heated for various times required by the crystallization kinetics, the autoclaves were quenched in tap water, and the products were filtered, washed with distilled water until pH = 7 and dried overnight at 105 °C.

Table 1 Isomorphous substitution of MFI zeolites

Zeolite	Substituting element	Synthesis	Si/T or T content	Techniques of characterization	Precursors	Refs.
MFI	B	Theoretical Study	—	—	—	[85]
MFI membrane	B	Hydrothermal		Gas diffusion and permeance		[86]
MFI	B	Hydrothermal		^{11}B NMR, REDOR	$B(OH)_3$, $Na_2B_4O_7$	[87]
MFI	B	Hydrothermal	30–80	FTIR, TPD, C_2H_4 reaction	H_3BO_3	[88]
MFI	B	Hydrothermal	24	Catalysis		[89]
MFI	B	Hydrothermal	95	IR of OH groups, acidity IR	H_3BO_3	[90–92]
MFI	B	Hydrothermal		Catalysis		[93]
MFI	B	Theoretical study		B-siting		[94]
MFI	B	Hydrothermal	1–2 B/u.c.	XRD, ^{11}B NMR, SEM, Sorption	H_3BO_3	[95]
MFI	B	Hydrothermal	25	TA, XPS, Catalysis	H_3BO_3	[96]
MFI	B	Hydrothermal	0.6, 1.3, 2.1/u.c.	^{11}B NMR, IR	H_3BO_3	[97]
MFI	B	Hydrothermal	0.1–0.5 wt % B	^{11}B NMR, XRD, SEM, FTIR, XPS, TPD, SAM	H_3BO_3	[98]
MFI	B	Hydrothermal	2.5/u.c.	FTIR, XRD, ^{1}H-, ^{11}B-, ^{29}Si NMR	H_3BO_3	[99]
MFI	B	Hydrothermal	6	XRD, MAS NMR, TA	H_3BO_3	[100]
MFI	B	Hydrothermal		XRD		[101]
MFI	B	Hydrothermal	0.40 wt %	Acidity, n-butene, isomerization	H_3BO_3	[102]
MFI	B	Hydrothermal	37	TPD, C_3H_8 oxidation	H_3BO_3	[103]
MFI	Fe-Mo-B	Hydrothermal CVD	4.6 wt % Mo 1.41 wt% Fe 0.15 wt % B	$C_6H_6 + N_2O = C_6H_5OH$	BET, TPD	[104]
MFI	B,Al	Hydrothermal	0.40 wt %	Acidity, n-butene, isomerization	H_3BO_3	[102]
MFI	Al	Hydrothermal	14–42 14–23; 45 9.25	^{27}Al NMR, FTIR		[105]
MFI	Ga	Theoretical Study				[85]
MFI membrane	Ga	Hydrothermal		Gas diffusion and permeance		[86]

Table 1 (continued)

Zeolite	Substituting element	Synthesis	Si/T or T content	Techniques of characterization	Precursors	Refs.
MFI	Ga	Hydrothermal	25	TA, XPS, Catalysis	GaO(OH)	[96]
MFI	Ga	Hydrothermal	280	TPD, C_3H_8 oxidation	$Ga(NO_3)_3$	[103]
MFI	Ga	Hydrothermal	30	IR, ESR, C_3H_8 aromatization	$Ga_2(SO_4)_3$	[106]
MFI	Ga	Hydrothermal	30–180	Acidity, Catalysis	$Ga(NO_3)_3$	[107]
MFI	Ga	Hydrothermal	$18–21 \times 10^{19}$ atom. g^{-1}	Diffusion of benzene		[108]
MFI	Ga	Hydrothermal	39–212	XRD, SEM, ion exchange	$Ga_2(SO_4)_3$	[109]
MFI	Ga	Exchange	1.7 wt % Ga	XRD, TA, Catalysis	Ga_2O_3	[110]
MFI	Ga	Impregnation	2, 4, 6 wt % Ga	Catalysis	$Ga(NO_3)_3$	[111]
MFI	Ga	Exchange	(1.80 wt %) 3–18	Catalysis	$NaGaO_2$	[112]
MFI	Ga	Hydrothermal	38, 61, 127	FAAS: framework and extra-framework Ga	$Ga(NO_3)_3$	[113]
MFI	Ga	Hydrothermal	Si/Ga=50	NMR, TPD, SEM, Catalysis		[114]
MFI	Ga	Exchange		XPS, IR	Ga_2O_3 + NaOH	[115]
MFI	Ga_2O_3	Mechanical	5–25 wt %	Catalysis		[116]
MFI	Ga	Hydrothermal or Modification		ESR, ^{71}Ga NMR, ^{29}Si NMR, Catalysis	Ga_2O_3	[117]
MFI	V	Hydrothermal	270	TPD, C_3H_8 oxidation	VCl_3	[103]
MFI	V	Hydrothermal	Si/V=50	UV-Vis, Catalysis		[118]
MFI	V	Hydrothermal		UV-Vis, EPR, NMR, IR, Raman	V_2O_5, NH_4VO_3, $NaVO_3$, $VO(OPri)_3$, $VO(acac)_2$	[119]
MFI(VS1)	V	Hydrothermal	95.977/0.023	ESR, ESEM	$VOSO_4$	[120–122]
MFI	V	Hydrothermal or Modification		ESR, ^{51}V NMR, ^{29}Si NMR, Catalysis	VCl_3; V_2O_5	[117]

Table 1 (continued)

Zeolite	Substituting element	Synthesis	Si/T or T content	Techniques of characterization	Precursors	Refs.
MFI	V	Hydrothermal	29.41 mmol $(100\,g)^{-1}$	ESR, Catalysis	$VOCl_3$	[123]
MFI	V	Hydrothermal	42	ESR, IR, adsorption catalysis	$VO(COO)_2$	[124]
MFI	V	Hydrothermal	98, 120	ESR, ^{51}V-, ^{29}Si NMR, EPMA	$VOSO_4$	[125]
MFI	V			XRD, NMR, ESR, TA, IR, XRF, SSIMS	$VOSO_4$	[126]
MFI	V	Exchange		ESR, IR, UV-Vis, TPD	V_2O_3 + HZSM-5	[127]
MFI	Zn	Hydrothermal		XRD, SEM, XPS		[128]
MFI	Sn	Hydrothermal	0.8–3.3/u.c.	^{29}Si-, ^{119}Sn NMR, Catalysis	$SnCl_4$	[129]
MFI	Fe	Hydrothermal	14–42 14–23; 45 9.25	^{27}Al NMR, FTIR		[105]
MFI membrane	Fe	Hydrothermal		Gas diffusion and permeance		[86]
MFI	Fe	Hydrothermal	9	Catalysis		[89]
MFI	Fe	Hydrothermal	25	TA, XPS, Catalysis	$Fe(NO_3)_3$	[96]
MFI	Fe	Hydrothermal	50	FTIR, UV-Vis, NO probe		[130]
MFI	Fe	Hydrothermal	0.08–0.40 mol kg^{-1}	^{29}Si NMR, Acidity (TPD), ESR, BET	$Fe(NO_3)_3$	[131]
MFI	Fe	Hydrothermal	0.08 wt % Fe_2O_3	$C_6H_6 + N_2O = C_6H_5OH$		[132]
MFI	Fe	Hydrothermal	250	TPD, C_3H_8 oxidation	$Fe(NO_3)_3$	[103]
MFI	Fe	Hydrothermal	50–150	SEM, IR, Acidity, Catalytic activity	$FeCl_3$	[133]
MFI	Fe	Hydrothermal		UV res. Raman, XRD, ESR		[134]
MFI	Fe	Hydrothermal	Si/Fe = 54	NMR, TPD, SEM, Catalysis		[114]
MFI	Ge	Hydrothermal	10–25	SEM, Catalysis	GeO_2	[144]
MFI (TS1)	Ti	Hydrothermal	2.8% TiO_2	Oxidation of C_3H_8, C_4H_{10}		[135]

Table 1 (continued)

Zeolite	Substituting element	Synthesis	Si/T or T content	Techniques of characterization	Precursors	Refs.
MFI	Ti	Gas-solid reaction		FTIR, UV-Vis, XRD, ICP-AES, Catalysis	$TiCl_4$	[136]
MFI	Ti	Hydrothermal	Si/Ti ~ 50	UV-Vis, Catalysis		[118]
MFI	Fe-Mo-B	Hydrothermal CVD 0.15% B	4.6% Mo 1.41% Fe	$C_6H_6 + N_2O = C_6H_5OH$	BET, TPD	[104]
MFI	Mo	Impregnation	3.6 wt % MoO_3	XRD, FTIR, TPR, SEM	MoO_3 + HZSM-5	[137]
MFI	Be	Treatment of NH_4-ZSM-5	2280	^{9}Be NMR	$(NH_4)_2BeF_4$	[138]
MFI	In	Exchange	~ 3 wt % In_2O_3	SEM	In_2O_3	[139]
MFI	In	Hydrothermal or Modification		ESR, ^{115}In NMR, ^{29}Si NMR, Catalysis	In_2O_3	[117]
MFI	Cr	Hydrothermal	370	ESR	Cr_2O_3 (NaF medium)	[140]
MFI	Pt-Ga	Exchange		XPS, IR	NH_4-[Ga]-ZSM-5 + $Pt[NH_3)_4]Cl_2$	[115]
MFI	Mn	Exchange	0.3–0.6 mmol g^{-1}	IR, ESR, Mass Spectrometry	$MnCl_2$, $MnSO_4$, Mn_3O_4, $Mn(CH_3COO)_2$	[141]
MFI	Ru	Hydrothermal	10–33	XRD, IR, SEM, XPS	K_2RuO_4, Pr_4NRuO_4	[142]

Table 2 Isomorphous substitution in BEA zeolites

Zeolite	Substituting element	Synthesis	Si/T or T content	Techniques of characterization	Precursors	Refs.
BEA	B	Alkaline Exchange	36.6–37.6 14.0–15.7	XRD, FTIR, MAS NMR, BET	$NaBO_2$	[143]
BEA	B	Hydrothermal		^{11}B NMR, REDOR	$B(OH)_3$, $Na_2B_4O_7$	[87]
BEA	B	Hydrothermal	0.85 wt %	Acidity, *n*-butene, isomerization	H_3BO_3	[102]
BEA	Al	Hydrothermal	14–42 14–23; 45 9.25	^{27}Al NMR, FTIR		[105]
BEA	B, Al	Hydrothermal	0.85 wt %	Acidity, *n*-butene, isomerization	H_3BO_3	[102]
BEA	Ga, Al	Hydrothermal	13.4–24.0	XRD, TA, SEM, IR	$Ga(NO_3)_3$	[145]
BEA	Ga	Hydrothermal	10, 28, 57	FTIR, TPD, adsorption	$Ga_2(SO_4)_3$	[146]
BEA	Ga	Exchange (Galliation)	12.4, 16.1, 18.7	^{29}Si NMR, ^{27}Al NMR, XRD, FTIR	Ga_2O_3 + HZSM-5	[147]
BEA	Ga	Alcaline Exchange	14.0–15.7	XRD, FTIR, MAS NMR, BET	$NaGaO_2$	[143]
BEA	V	Adsorption in hydroxyl nests	0.05–0.5–1.5 wt %	UV-Vis, Photoluminescence	NH_4VO_3	[148]
BEA (CIT-6)	Zn	Hydrothermal	32	Multi NMR, BET	$Zn(CH_3COO)_2$	[149]
BEA	Sn	Hydrothermal	1.6 wt %	IR, Baeyer-Villiger oxidation	$SnCl_4$	[150]
BEA	Fe	Hydrothermal	20–23	IR, SEM, Sorption		[151]
BEA	Fe	Hydrothermal	14–42 14–23; 45 9.25	^{27}Al NMR, FTIR		[105]
BEA	Ti	Hydrothermal		XRD, Raman, IR, UV-Vis, XPS, adsorption	$Ti(SO_4)_2$	[152]
BEA	Ge	Hydrothermal		Adsorption Synchrotron Powder Diffraction		[153]

Table 3 Isomorphous substitution in various zeolites

Zeolite	Substituting element	Synthesis	Si/T or T content	Techniques of characterization	Precursors	Refs.
TsG-1	Ga	Hydrothermal	22/10	Sr^{2+} Ion Exchange		[154]
AST	Ge	Hydrothermal, F-medium		XRD, MAS NMR, SEM	GeO_2	[155]
AV-1	Y	Hydrothermal	8	Multi NMR, FTIR	$Y_2(SO_4)_3$	[156]
CIT-6 (BEA)	Zn	Hydrothermal	32	Multi NMR, BET	$Zn(CH_3COO)_2$	[149]
ETS-10	V	Hydrothermal	V/Ti = 0.87	MAS NMR, XRD, UV-Vis	$VOSO_4$	[157]
ETS-10	Cr	Hydrothermal	Si/(Cr + Ti) = 4.5	EDX, XRD, SEM, BET, UV-Vis, EPR, MAS NMR	Cr_2O_3	[158]
EUO	B	Hydrothermal	13–15	XRD, TG, MAS NMR	H_3BO_3	[159]
FER	Ga	Hydrothermal	2–7 Ga/u.c.	^{27}Al-, ^{29}Si-, ^{71}Ga NMR, XRD, SEM	$Ga(NO_3)_3$	[160]
FER	B	Hydrothermal	4–19	XRD, MAS NMR	H_3BO_3	[179]
H,Na-Y	Fe	Ion Exchange		ESR	$FeCl_3$	[161, 162]
LEV	B	Hydrothermal, F-medium	9–27	XRD, MAS NMR, SEM	B_2O_3	[163]
LTL	Fe	Hydrothermal	27.3–3.5	IR, TA, XPS, EM, Magnetic susceptibility	K_2FeO_4	[164]
MEL	B-B,Al		0.25 wt %			[102]
MEL	V	Hydrothermal	100–400	^{51}V NMR, ESR, UV-Vis, IR	$VOSO_4$	[165]
MEL	V	Hydrothermal	40, 80, 160	XRD, IR, ESR, NMR, Catalysis	$VOSO_4$	[166, 167]
MOR	B, Al, Ga	Theoretical Study	—	—	—	[168]
MOR	Al, Fe	Hydrothermal	9.25	^{27}Al NMR, FTIR	—	[105]
MTT	Al, Fe	Hydrothermal	14–42	^{27}Al NMR, FTIR	—	[105]
MTW	Ga	Hydrothermal	70	XRD, IR, ^{29}Si-, ^{74}Ga NMR, Acidity	$Ga(NO_3)_3$	[169]

Table 3 (continued)

Zeolite	Substituting element	Synthesis	Si/T or T content	Techniques of characterization	Precursors	Refs.
MTW	Fe	Hydrothermal	40–90	XRD, FTIR, SEM	$Fe(NO_3)_3$	[170]
NCL-1	V	Hydrothermal	80–250	IR, ESR, NMR, Catalysis		[171]
Sil-1	Mn	Hydrothermal	258	FTIR, XANES, EXAFS	MnO	[172]
SSZ-24	B	Hydrothermal		^{11}B NMR, REDOR	$B(OH)_3$, $Na_2B_4O_7$	[87]
TON	Al, Fe	Hydrothermal	14–23; 45	^{27}Al NMR, FTIR	—	[105]
TS-1	Ti	Hydrothermal	8–256	XRD, FTIR, MAS NMR, UV-Vis, ESR	$Ti(OBu)_4$	[173]
Y	Ce, Co, Cu, Cr, La	Exchange		Acylative cyclization	chlorides	[174]
Y	Fe	CVD	—	XRD, FTIR, TA, BET	$[(C_5H_5)Fe(CO)_2]_2$	[175]
Y	Si	Solid-Solid	—	IR, MAS NMR, XRD, TA	$(NH_4)_2SiF_6$	[176]
Y	Ga	Hydrothermal	1.85–2.18	MAS NMR, FTIR	Ga_2O_3	[177]
ZSM-12	V	Hydrothermal	88, 158, 322	ESR, NMR (^{29}Si, ^{51}V, ^{1}H)	$VOSO_4$	[178]

2.1.2
[Ga]-MFI

The initial gels have a general composition of $10SiO_2 - xMF - yGa(NO_3)_3 - 1.25TPABr - 330H_2O$ where M = NH_4, Na, K or Cs with $x = 9$, 18 and $y = 0.1$ and 0.3. The reactants were mixed in the following order: MF (M = NH_4, Na, K, Carlo Erba and M = Cs, Sigma), distilled water, $Ga(NO_3)_3$ (Aldrich), tetrapropylammonium bromide (TPABr, Fluka) and fumed silica (Sigma) [183, 184].

2.1.3
[Fe]-MFI

2.1.3.1
Fluoride Route

The reactants were ammonium fluoride (Carlo Erba, RPE), sodium fluoride (Carlo Erba, RPE), potassium fluoride (Carlo Erba, RPE), cesium fluoride (Aldrich, purum), tetrapropyl ammonium bromide (TPABr, Fluka, purum), iron nitrate ($Fe(NO_3)_3 \cdot 9H_2O$, Merck, purum), fumed silica (Sigma) and distilled water [185, 186]. The molar composition of the starting mixture was:

$$10SiO_2 - xFe(NO_3)_3 \cdot 9H_2O - yMF - 1.25TPABr - 300H_2O$$

where $0.1 \leq x \leq 0.3$, $y = 3 - 24$ and M = NH_4, Na, K or Cs.

The starting mixtures were prepared by adding the fluoride salt, iron nitrate, TPABr and the fumed silica to the distilled water in this order. After complete homogenization the resulting gels were placed in PTFE-lined 25 cm^3 stainless-steel autoclaves. The samples were obtained by hydrothermal synthesis at 170 °C for predetermined times. After quenching the autoclaves, the products were recovered, filtered, washed with distilled water and dried at 80 °C overnight. The crystallization time was ca. 7 days for the sample with the smallest crystallization rate.

2.1.3.2
Alkaline Route

Batch composition: $1.0SiO_2 - (xAl_2O_3 + yFe_2O_3) - 0.13Na_2O - 52.0H_2O - 0.125$ template for [Fe,Al]-ZSM-5: $x = 0.0025$; $y = 0.0025$, 0.005, 0.01 and 0.002, for [Al]-ZSM-5: $x = 0.01$; $y = 0$, for [Fe]-ZSM-5: $x = 0$; $y = 0.01$ and 0.01333 [187–192].

Source materials: distilled water; waterglass (Aldrich, vide supra); colloidal silica (Aldrich, Ludox 40); iron(III) oxide (Aldrich, 99.98%); iron(III) oxide enriched in ^{57}Fe to 80%; iron(III) nitrate nonahydrate (Aldrich > 98%); aluminum powder (Aldrich, 20 micron, > 99%); aluminum oxide trihydrate

(Hungarian origin); sodium hydroxide (Reanal, reagent grade); TPA bromide as template (Aldrich, 98%); perchloric acid, 70% (Aldrich, reagent grade); sodium perchlorate (Aldrich, 99%); hydrochloric acid 37% (Reanal, reagent grade); sulphuric acid 97% (Reanal, reagent grade).

Batch preparation recipe (for about 12–14 g volatile-free zeolite): Prepare stock-solutions of iron (dissolve known amount of iron (III) oxide, or iron tracer in concentrated HCl at 120 °C in a closed Teflon autoclave; or dissolve iron(III) nitrate in acidified water) and aluminum (dissolve aluminum powder in HCl diluted to 1 : 1). Save the Fe- and Al-contents, respectively, per gram of solution.

In order to simplify the description of recipes the exact amounts for the variable components will be summarized in Table 4 for various Si/Fe, etc., ratios.

Sixteen grams Aldrich-type waterglass (see Aldrich catalogue) + 140.9 g distilled water; stir and keep it in refrigerator overnight (solution 2). Blend appropriate amounts of stock-solution for iron (or aluminum, or iron and aluminum) with 5.8 g of 70% perchloric acid; keep in refrigerator overnight (solution 3). Under vigorous mixing add to (solution 3) so much of (solution 2) to reach pH = 4.5 (use pH paper). Admix to remaining (solution 2) 19.2 g cool Ludox 40 and continue the neutralization of (solution 3) while maintaining agitation. The meanwhile solidified gel should be broken up. Add 6.7 g TPA bromide, 2 g sodium chloride as solid and 0.2 g ZSM-5 seed (the best choice is silicalite-1 in the as-synthesized (AS) state, ground carefully in a porcelain mortar using a few drops of water).

The pH of the slurry should be between 10.5 and 11.0. Adjust the pH by either a few drops of perchloric acid or 50% sodium hydroxide. In a stainless-steel shaker, mill the slurry for at least 5 h using similar balls. Check (and adjust if necessary) the pH.

Crystallization vessel: Teflon-lined autoclave(s); temperature: 160 °C; time: depending on the degree of substitution 6–12 h, without agitation. Product recovery: after cooling, filter and wash with warm water (to pH = 10.5). Dry

Table 4 Amounts (g) of Al and/or Fe per batch at various slurry compositions

Sample	Fe [g]				Al [g]	
	Si/Fe = 200	Si/Fe = 100	Si/Fe = 50	Si/Fe = 37.5	Si/Al = 200	Si/Al = 50
[Fe,Al]-ZSM-5	0.057	—	—	—	0.027	—
	—	0.114	—	—	0.027	—
	—	—	0.228	—	0.027	—
[Fe]-ZSM-5	—	—	0.228	—	—	—
	—	—	—	0.342	—	—
[Al]-ZSM-5	—	—	—	—	—	0.108

in air thermostat at 110 °C. Save the zeolite product as such or apply "caring calcination" [193]. Besides the zeolitic crystals, the cool mother liquor contains 2–3 cm-long needle-like crystals of sparingly soluble TPA perchlorate. It is easy to separate them by a tea-strainer. Another method of preparation was used in the alkaline route [188].

Two different hydrogel systems were studies. The first system was prepared from the following initial reaction mixture:

$$\text{(A)}\quad x\mathrm{Na_2O}{-}y\mathrm{TPABr}{-}z\mathrm{Al_2O_3}{-}\mathrm{SiO_2}{-}q\mathrm{Fe_2O_3}{-}p\mathrm{HA}{-}20\mathrm{H_2O},$$

where $x = 0.1 - 0.32$; $y = 0.02 - 0.08$; $z = 0 - 0.05$; $q = 0.005 - 0.025$; the ratio $p/q = 3$ and HA stands for $H_2C_2O_4$ or H_3PO_4.

The second system was prepared from the following initial reaction mixture:

$$\text{(B)}\quad 0.16\mathrm{Na_2O}{-}x\mathrm{TPABr}{-}y\mathrm{EG}{-}z\mathrm{Al_2O_3}{-}\mathrm{SiO_2}{-}q\mathrm{Fe_2O_3}{-}p\mathrm{HA}{-}10\mathrm{H_2O},$$

where $x = 0 - 0.08$, $y = 0 - 6.0$; $z = 0 - 0.05$; $q = 0.005 - 0.1$; the ratio $p/q = 3$ and HA stays for $H_2C_2O_4$ or H_3PO_4.

2.1.4
[Fe]-BEA

The following reagents were used: 20% tetraethylammonium hydroxide (TEAOH), in H_2O (Aldrich), 25% TEAOH, in methanol (Aldrich), tetraethyl orthosilicate (TEOS, Aldrich) as silicon source, $Fe(NO_3)_3 \cdot 9H_2O$ (Merck), $Al(OH)_3$ (Pfaltz & Bauer), NaOH (Carlo Erba) and ultradistilled water [194]. The following general composition was used for [Fe]-BEA:

$40SiO_2 - xFe(NO_3)_3 \cdot 9H_2O - y\mathrm{TEAOH} - 4NaOH - 676H_2O$, with $x = 0.38$, 0.40, 0.44, 0.45, 0.49, 0.51, 0.60, 1.02, and $y = 10.88$, 13.6, 16.3 and 19.04.

The experimental procedure was carried out as follows: solution A was prepared by adding TEOS to aqueous solution of 20% TEAOH under magnetic stirring during 1 h. Solution B was prepared by adding NaOH to 20% aqueous TEAOH solution under magnetic stirring: Solution C was obtained by introducing $Fe(NO_3)_3 \cdot 9H_2O$ into ultradistilled water under magnetic stirring during 10 min.

Solution A is added dropwise to solution C. It is important to fully dissolve solution A in solution C. The obtained solution is solution D. Finally, solution D is added slowly to solution B under magnetic stirring leading to a light yellow gel: The system is stirred for 24 h. This allowed the ethanol formed during the hydrolysis of TEOS to evaporate.

2.1.5
[Fe]-MOR

The zeolite samples [Al]-BEA and [Fe,Al]-BEA were also prepared starting from the following molar composition [195]:

(B) $x\mathrm{Na_2O}{-}0.2\mathrm{TEAOH}{-}y\mathrm{Al_2O_3}{-}z\mathrm{Fe_2O_3}{-}w\mathrm{C_2H_2O_4}{-}\mathrm{SiO_2}{-}20\mathrm{H_2O}$,

where TEAOH stands for tetraethylammonium hydroxide, $C_2H_2O_4$ is oxalic acid; $0.1 \leq x \leq 0.15$; $0.015 \leq y \leq 0.02$; $0 \leq z \leq 0.015$, w/z was constantly equal to 3.

The zeolite samples of [Al]-MOR and [Fe,Al]-MOR were prepared starting from the following molar composition:

(A) $x\mathrm{Na_2O}{-}0.2\mathrm{TEABr}{-}y\mathrm{Al_2O_3}{-}z\mathrm{Fe_2O_3}{-}w\mathrm{C_2H_2O_4}{-}\mathrm{SiO_2}{-}20\mathrm{H_2O}$,

where TEABr stands for tetraethylammonium bromide, $C_2H_2OO_4$ is oxalic acid; $0.1 \leq x \leq 0.27$; $0.015 \leq y \leq 0.025$; $0 \leq z \leq 0.025$, w/z was constantly equal to 3.

The hydrogels were obtained by adding precipitated silica (BDH) to a previously prepared homogeneous solution consisting of the organic compound (Fluka), sodium hydroxide (Carlo Erba), sodium aluminate (Carlo Erba) and distilled water. In another beaker, a solution of an iron complex with oxalic acid was prepared, starting from iron nitrate (Carlo Erba) and oxalic acid (Carlo Erba). This solution was slowly added to the hydrogel, which, after 1 h of homogenization, was transferred into autoclaves.

2.1.6
[Co]-MFI

Two different gels were prepared [196, 197]. For the first series of gels (A), sodium silicate was used and the final batch compositions were: $35\mathrm{SiO_2} - x\mathrm{Na_2O} - y\mathrm{Co(CH_3COO)_2} \cdot 4\mathrm{H_2O} - 3.4\mathrm{TPABr} - 8.4\mathrm{H_2OSO_4} - 808\mathrm{H_2O}$ with $x = 11$, 12.9 and 15 and $y = 0.5$, 1, 1.5 and 2. The second type of gels (B) was prepared using colloidal silica with a global composition: $35\mathrm{SiO_2} - x\mathrm{Na_2O} - \mathrm{Co(CH_3COO)_2} - 3.4\mathrm{TPABr} - 808\mathrm{H_2O}$ with $x = 0$, 3, 6, 9 and 12. The gels, after complete homogenization, were placed in PTFE-lined 25 cm^3 stainless-steel autoclaves. The Co-containing samples were obtained by hydrothermal synthesis at 170 °C after 2 days. After quenching the autoclaves, the products were recovered, filtered, washed with distilled water and finally dried at 80 °C overnight. The samples were characterized by various physico-chemical techniques such as XRD, chemical and thermal analysis, XPS and diffuse reflectance UV-visible spectroscopy.

2.1.7
[Zn]-MFI

The zincosilicate of MFI structure has been synthesized according to conventional recipes [198–201]. The main differences consisted in using $Zn(NO_3)_2$ instead of aluminum compounds. Another modification comprised the use of phosphoric acid as crystallization promoter [202]. The initial gel was formed by mixing NaOH (Carlo Erba) solution with tetrapropylammonium bromide (TPABr) (98%, FLUKA) and precipitated silica gel (BDH, Laboratory reagent) followed by adding solutions of $Zn(NO_3)_2\,6H_2O$ (Carlo Erba) and H_3PO_4 (85%, Fluka). The pH of the initial gel was always ca. 11, and was adjusted using H_3PO_4. The Zn/Si ratio of the initial mixtures varied in the 0.01–0.1 range. The gel was crystallized at 170 °C in a Teflon-lined autoclave for various periods of time (23–140 h). The synthesis was quenched every day and small samples were taken for XRD analysis to follow the progress in crystallization. The hydrothermal syntheses were stopped when the intensity of the XRD reflections of the products attained the maximum and did not increase any longer. The intensities of the respective XRD reflections were compared with those of zinc-free silicalite. The resulting products were washed with water, dried and eventually calcined in air at 450 °C in order to remove the organic template.

2.1.8
Cu-TON

The zeolite syntheses were carried out starting from the following reaction mixtures:

$$x\mathrm{Na_2O} - y\mathrm{Al_2O_3} - 15\mathrm{CH_3OH} - \mathrm{SiO_2} - 10\mathrm{H_2O}$$

where $x = 0.04$ or 0.05 and y ranges from 0.00038 to 0.03 [203]. The syntheses were performed in a magnetically stirred stainless-steel reactor, while the reaction temperature range was 140 to 160 °C under autogenous pressure. The crystallization time varied mainly as a function of Al content in the hydrogel. The products were cooled to room temperature, washed and dried at 105 °C overnight. Three different procedures of copper introduction were applied:

1. direct ionic exchange of Na-TON with a 0.1 M copper (II) acetate solution (stirred for 2 h at 50 °C, followed by calcination at 600 °C),
2. calcination of Na-TON at 600 °C overnight, preliminary exchange with a 1 M NH_4Cl solution (twice at 60 °C) in order to obtain the NH_4^+ – TON form followed by calcination at 600 °C, and finally,
3. exchange with copper (II) acetate solution and calcination, as mentioned in point 1. Solid-state exchange reaction was performed by mixing $CuCl_2$ (5 wt %) with H-TON and treating the mechanical mixture for 8 h at 600 °C.

2.2 Characterization

2.2.1 General Characterization

Products were identified, and their crystallinities were determined by powder X-ray diffraction, using CuK_α radiation (Philips Model PW 1730/10 X-ray generator equipped with a PW 1050/70 vertical goniometer). For the calculation of the crystallinity of the intermediate samples, the intensity of the main peak at $d = 3.85$ Å (or 23.10°2Θ) for, e.g., MFI zeolite was compared with the intensity of a reference sample, which was obtained by ultrasonic treatment (Soniprep 150) of the most crystalline sample for each reaction investigated. Crystal morphology and size were determined by a scanning electron microscope (SEM), Jeol JSTM 330A. The amount of gallium and alkali cations in the crystals was determined by atomic absorption spectroscopy (Shimadzu AA-660). Each sample was first treated with ultrasound in order to eliminate the remaining amorphous phase, and then calcined in air at 550 °C for 3 h in order to eliminate the organic ions. Some 250 mg of each sample were then dissolved in 5 ml of HF (for analysis) and 2 ml of H_2OSO_4 (for analysis). The hydrofluoric acid was eliminated by heating the solution at 50 °C. The remaining solution was then diluted with distilled water for a precise volume. The F-content of the samples was analyzed using a specific F^--electrode, Orion model 94-09. The samples were prepared as follows. Some 0.1 g of a sample was mixed with ten times more Na_2CO_3 in a platinum crucible, and the mixture was heated with a Bunsen burner until complete fusion or the appearance of a transparent solution. The content of the crucible was then cooled to room temperature and diluted with distilled water. If a solid residue still remained in the crucible, it was dissolved by adding several drops of hydrochloric acid. The final solution was filtered and adjusted to a precise volume. Before analysis, this solution (50 vol %) was mixed with a solution TISAB (Orion application solution with CDTA-*trans*-1,2-diaminocyclohexane-*N*,*N*,*N*,*N*-tetraacetic acid) in order to decomplex the fluoride ions. The amount of TPA^+ ions trapped in the crystals was obtained by TG analysis. DSC curves were used to evaluate the decomposition path of the organic molecules. TG and DSC analyses were carried out with a Netzsch STA 409 instrument in N_2 atmosphere from 20 to 650 °C with a heating rate of 10 °C min^{-1} and a flow rate of 10 ml min^{-1}.

The amount of iron in the crystals was determined by atomic absorption spectrometry (Shimadzu AA-660). The amounts of the various elements (Fe, Si, F, Na, K and Cs) were determined by PIXE (Proton Induced X-ray Emission) and PIGE (Proton Induced γ-ray Emission) methods [204]. Mössbauer spectra were recorded at room temperature. The template was removed in a separate equipment by "caring" calcination prior to the measurements, i.e., heating to 450 °C in Ar at a ramp of 2 K s^{-1}, calcination in 5% O_2/He mixture at 500 °C, 1 h. Evacuation was performed down to a pressure of 4×10^{-1} Pa at 350 °C for 2.5 h (spectra were recorded afterwards under vacuum). Reducing treatments were carried out in a CO flow at 350 °C for 1.5 h (spectra were obtained in CO atmosphere). The spectra were fitted by assuming Lorentzian line-shape; no positional parameters were constrained. The estimated accuracy of data is ±0.03 mm s^{-1}.

2.2.2
The Cu-TON Obtained by Ion Exchange

The samples after calcination were modified with Cu^{2+} and NH_4^+ cations. The ion-exchange procedure was conducted at 50 °C. Aqueous solutions (0.5 M) of NH_4Cl and of $CuCl_2$, were added (15 ml per 1 g of the sample), and the mixture was magnetically stirred for 16 h. The procedure was repeated three times with fresh aliquots of solutions, and then the samples were washed and dried. The samples modified with copper were blue. The ammonium-exchanged samples were calcined at 450 °C to obtain the H-form. The samples were characterized by standard methods such as XRD (Philips 1730/70), FTIR (KBr, Bruker Vector 22), Raman spectroscopy (FTIR Raman Nicolet 760 with Nd YAG beam). Considering a possible reduction of incorporated zinc in hydrogen atmosphere or with another reducing agent, TPR data for selected samples were also measured. The TPR curves were recorded with an Asap Chemsorb 2705 apparatus. The flow rate of hydrogen (10%)/argon mixture was 108 cm^3 min^{-1}. The temperature ramp (20–900 °C) was 10 °C min^{-1}. A TPR measurement (Fig. 48) was performed for zincosilicate (Zn/Si = 0.05) after template removal as well for its copper cation-modified version. The Cu content was ~ 0.6 wt %. The samples were calcined at 400 °C in a helium stream before measurement. The catalytic activity of the samples was examined in propan-2-ol decomposition and in cumene cracking. The tests were conducted in a pulse microreactor combined with a gas chromatograph. The catalyst powder samples (15 mg) were activated in a he-

Table 5 NMR parameters

Nucleus	ν [MHz]	t_{pulse} [µs]	Θ [°]	Recycling time [s]	NMR Reference
^{11}B	128.3	1.0	$\pi/12$	0.2	$BF_3 \cdot OEt_2$
^{27}Al	104.3	1.0	$\pi/12$	0.1	$Al(H_2O)_6^{3+}$
^{71}Ga	121.99	1.0	$\pi/12$	0.1	$Ga(H_2O)_6^{3+}$
^{29}Si	39.7	6.0	$\pi/2$	6.0	$Si(CH_3)_4$
^{19}F	282.4	3.4	$\pi/2$	5.0	NaF
^{13}C	50.3	4.5	$\pi/2$	4.0	$Si(CH_3)_4$

lium stream at 400 °C for 30 min prior to the catalytic tests. Cumene cracking was carried out at 350 °C; decomposition of propan-2-ol at 200 °C with H-forms and at 250 °C with Cu-forms, respectively. The volume of injected substrate was 1 μl. For the multinuclear NMR analysis the NMR parameters are reported in Table 5.

3 Results and General Discussion

3.1 [B]-MFI

Borosilicalites of MFI or ZSM-5 structure were obtained with all the different cations used and over the whole range of concentrations. Table 6 shows the physico-chemical characterization data and Table 7 the ^{11}B NMR data for the different borosilicalite samples obtained with different amounts of H_3BO_3 and different alkali cations. Figure 1 shows typical ^{11}B NMR spectra of a precursor and the corresponding calcined samples. The framework tetrahedral boron (T) is characterized by a chemical shift of c.a. – 3.9 ppm with respect to BF_3-OEt_2 [205]. In the presence of Cs and 10 moles of H_3BO_3 in the initial gel, up to 9.4 tetrahedral B/u.c. can be incorporated in the structure.

However, during calcination a large amount of boron is eliminated from the structure (Table 7) and the relative amount of boron in the tetrahedral configuration decreases. The extra-framework boron is in a tetrahedral configuration in most cases, characterized by chemical shift of – 2.0 ppm.

Table 6 Physicochemical characterization of precursor samples of borosilicalites synthesized from MF – xH_3BO_3 – $10SiO_2$ – 1.25TPABr – 330 H_2O at 170 °C

x	M	B/u.c. [a]	M/u.c. [b]	TPA/u.c. [c]
0.1	NH_4	0.8	—	3.8
	Na	1.2	0.5	3.7
	K	1.0	0.7	3.4
	Cs	1.2	0.1	3.8
10	NH_4	4.4	—	3.5
	Na	6.3	2.0	3.7
	K	8.4	3.2	2.8
	Cs	9.4	3.7	2.4

[a] NMR values;
[b] Atomic absorption values;
[c] Thermal analysis values

Table 7 ^{11}B NMR data of borosilicalites synthesized from 9 MF – xH_3BO_3 – 10 SiO_2 – 1.25 TPABr – 330 H_2O at 170 °C

x	M	Sample [a]	δ [ppm]	I [%]	B_{tet}/u.c. [b]	δ [ppm]	I [%]	B trigonal	
								δ [ppm]	I [%]
0.1	NH_4	pr	– 4.2	100	0.8				
		c	– 4.0	27	0.2	– 2.0	18	5.6–17	55
	Na	pr	– 4.1	100	1.2				
		c	– 3.7	20	0.4	– 2.0	22	5.2–17	45
	K	pr	– 4.2	100	1.0				
		c	– 3.7	63	0.5	– 2.0	37		
	Cs	pr	– 3.3	100	1.2				
		c	– 4.0	12	0.24	– 2.0	47	5.2–17	34
10	NH_4	pr	– 4.1	80	3.5	– 2.0	20		
		c	—			—			
	Na	pr	– 4.0	73	4.6	– 2.0	22	5.5–17	5
		c	– 3.6	53	3.3	– 2.0	6	5.5–17	41
	K	pr	– 3.9	70	5.9	– 2.0	30		
		c	– 3.9	17	1.0	– 2.0	34	1.0–17	49
	Cs	pr	– 3.9	32	2.9	– 2.0	68	5.5–17	11
		c	– 3.8	54	5.1	– 2.0	26	5.5–17	20

[a] pr: precursor; c: calcined;
[b] Corresponding to the amount of boron at ca. – 4 ppm

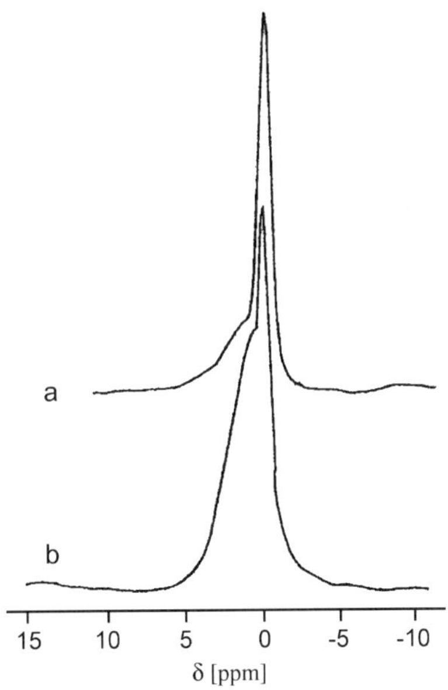

Fig. 1 ^{11}B MAS NMR spectra of **a** K-borosilicalite precursor and **b** calcined samples obtained with 10 moles of H_3BO_3

Sometimes, some of the extra-framework boron can also assume trigonal configuration as in the various borates. In this case, a broad NMR line is situated between 5.5 and 17 ppm. Quantitative determination of the trigonal boron was made by considering an average value of the quadrupole coupling constant of ~2.5 MHz [206] and using the corrections for the line intensities depending on the value of $\nu_Q^2/\nu_L \cdot \nu_{rot}$ with $\nu_Q = 1.25$ MHz, $\nu_L = 128.3$ MHz and $\nu_{rot} = 9$ kHz [207, 208]. The measured intensities were corrected by a factor of $1/0.83 = 1.2$. No correction was made for the intensities of the lines of the tetrahedral boron ($Q_{cc} = 0.2$ MHz [206]). The total amount of TPA/u.c. is equal to 3.4–3.8 for samples synthesized with 0.1 moles of H_3BO_3. For the K- and Cs-borosilicalite samples, the amount of TPA/u.c. decreases to 3.2 and 2.7 for samples synthesized with 4 moles of H_3BO_3 and to 2.8 and 2.4 for samples obtained with 10 moles of H_3BO_3. The decrease of TPA/u.c. is also indicative of boron incorporation into the MFI structure. Indeed, it was previously observed that the increase of Al in the zeolitic framework was accompanied by a decrease of TPA/u.c. [209]. The M/u.c. value remains quite low for low B-containing samples (Table 6). It varies from 0.1 to 0.7. As more boron is incorporated into the structure, this amount increases to 3.7 for the Cs-borosilicalite. If boron is incorporated in the zeolitic framework, its presence leads to a contraction of the unit cell because the atomic radius of the B atom (0.98 Å) is smaller than that of the Si atom (1.32 Å). The cell parameters and the unit cell volume decrease monotonously as a function of B/u.c. [210]. The decrease is largest for the K- and Cs-borosilicalites. From a correlation between the unit cell volume and B/u.c. reported in the literature [211], it can be predicted that some 5–6 boron atoms can be incorporated into the MFI structure using K^+ ions. As was already mentioned above, Cs^+ at high H_3BO_3 concentrations behaves quite peculiarly. For example, the B/u.c. in the framework increases during calcination with $x = 10$ moles (Table 7 and Fig. 2).

During calcination, the NMR line at -3.9 ppm increases, while the one at -2.0 ppm decreases. There is also some increase in the 5.5–17 ppm line. It seems as if the -2.0 ppm line, which was attributed to extra-framework tetrahedral boron is transformed predominantly to the -3.9 ppm line, i.e., the boron in the structure is in tetrahedral configuration, and partially it is in non-framework trigonal boron at 5.5–17 ppm. Hence we have to modify the attribution of the -2.0 ppm line. It is possible that the "extra-framework" tetrahedral boron in the precursor is really not an extra-framework boron, but characterizes a tetrahedral boron, which is still partially linked to the structure. If this is the case, a rather high amount of SiOH defect groups should be present in these samples. This is confirmed by the ^{29}Si NMR spectra, where a high concentration of SiOX groups (where X = Na, K, Cs, NH_4, H) at -103 ppm is detected (Fig. 3), indeed. Interestingly, the K-borosilicate samples do not show any anomaly (Table 7 and Fig. 4). No SiOX defect groups were detected in the precursor samples at a high boron concentration.

The attribution of the -2.0 ppm line in the ^{11}B NMR spectra to partially deformed framework tetrahedral boron such as $[(SiO)_3BOH]^-$ (i.e., having non-bridging oxygen in the structure) was also suggested for reedmergnerite [206]. From this study it can be concluded that more than four B/u.c. can be introduced into silicalite-1 using a fluoride-containing medium in the presence of either K^+ or Cs^+ ions. We can see now that the maximum of four B/u.c. observed in previous studies carried out in both alkaline [212] and fluoride [213] media is essentially linked to Na^+, which was the inorganic cation

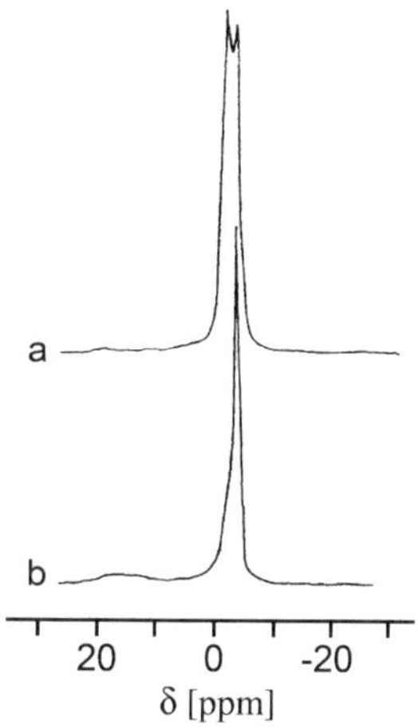

Fig. 2 ^{11}B MAS NMR spectra of **a** precursor and **b** calcined Cs-borosilicalites synthesized with 10 moles of H_3BO_3

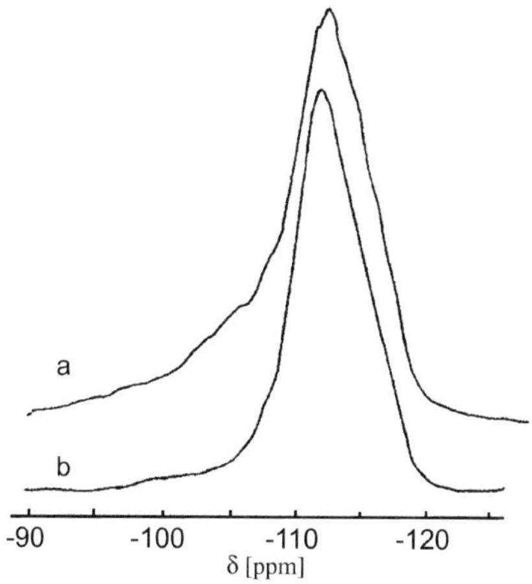

Fig. 3 ^{29}Si MAS NMR spectra of **a** precursor and **b** calcined Cs-borosilicalites synthesized with 10 moles of H_3BO_3

used. It was shown, indeed, in [212] that Na^+ preferentially accompanied Al in the structure, while B preferred TPA^+. It was also demonstrated that the boron species incorporated into the zeolite structure were in a trigonal form, i.e., $B(OH)_3$ in the alkaline medium. As only a maximum of 4 TPA^+/u.c. can be included in the channel, the maximum B/u.c. also equals four. The preferential interaction between $[SiOAl]^-$ and Na^+ on the one hand and between $[SiOB^-]$ and TPA^+ on the other can be understood on the basis of the hard and soft acid-base interaction. It is well known that hard acids accompany better hard bases and soft acids link preferentially to soft bases. As Na^+ is a harder acid than TPA^+, $[SiOAl]^-$ is also a harder base than $[SiOB]^-$. The preferential interactions lead then to the TPA^+-$[SiOB]^-$ pairs, as was demonstrated previously [212]. If, however, the Na^+ ions are replaced by either K^+ or Cs^+, which are softer acids than Na^+, the presence of these ions could also favor the introduction of boron into the zeolite structure, since preferential $[SiOB]^-K^+$ or $[SiOB]^-Cs^+$ pairs can be formed. The presence of either K^+ or Cs^+ in the channels lowers the possibility of introducing four TPA/u.c. Indeed, it is found for samples having more than four B/u.c. that the value of TPA/u.c. decreases. The thus-created available free space can then be occupied by the other "soft" counter-cations (K^+ or Cs^+), and no defect groups have to be created in the structure. Hence K^+ or Cs^+ are behaving towards $[SiOB]^-$ as Na^+ does towards $[SiOAl]^-$. In the latter case, it is possible to introduce up to 8–10 Al/u.c. [209, 212]. Hence, in the presence of K^+ or Cs^+ more than four B/u.c. can be obtained.

3.2 [Ga]-MFI

Highly crystalline [Ga]-ZSM-5 samples were obtained from gels with an initial composition of $10SiO_2 - x MF - y Ga(NO_3)_3 - 1.25TPABr - 330H_2O$ with $x = 9, 18$ and $y = 0.1$ and 0.3. Table 8 shows the results of the chemical analysis for various [Ga]-ZSM-5 samples where Ga/u.c. and M/u.c. were determined by atomic absorption, TPA/u.c. from thermogravimetric analysis, and F/u.c. from the use of the specific F-selective electrode. Gallium can be introduced quite efficiently in the presence of $NH_4{}^+$ ions as noted previously [28, 29, 32]. Indeed, the Si/Ga ratios of the initial gels and those of the final zeolitic samples are very close. Ga/u.c. values increase with increasing $Ga(NO_3)_3$ in the initial gels. K^+ ions are somewhat less effective to introduce gallium in tetrahedral positions of the lattice, the Na^+ ions are even less effective, and finally, the Cs^+ ions are the least effective. The amount of alkali cations used does not seem to have a large influence on the incorporation of gallium, meaning that saturation is already reached with nine MF moles in the reaction mixture.

This behavior is quite different from the influence of alkali cations on the incorporation of aluminum into the framework position of ZSM-5 in fluorine-containing media [30]. Indeed, it was found that K^+ ions were the

Table 8 Physicochemical characterization of [Ga]-ZSM-5 samples obtained from 10 SiO_2 – x MF – y $Ga(NO_3)_3$ – 1.25 TPABr – 330 H_2O at 170 °C

MF	$Ga(NO_3)_3$ initial	Ga/u.c.	M/u.c.	F/u.c.	TPA/u.c. tot	TPA/u.c. ~420 °C	TPA/u.c. ~470 °C
9 NH_4F	0.1	1.0	—	1.1	3.8	2.0	1.8
	0.3	2.9	—	0.5	3.3	1.3	2.0
18 NH_4F	0.1	1.1	—	1.5	3.9	2.2	1.7
	0.3	3.2	—	0.4	3.4	0.4	3.0
9 NaF	0.1	0.9	0.03	3.2	3.8	1.4	2.4
	0.3	1.9	0.07	1.2	3.8	1.8	2.0
18 NaF	0.1	0.5	0.08	2.7	3.8	2.0	1.8
	0.3	1.8	0.2	1.5	3.8	1.4	2.4
9 KF	0.1	0.9	0.05	1.5	3.8	2.3	1.5
	0.3	2.5	0.4	0.9	3.4	0.9	2.5
18 KF	0.1	0.7	0.5	1.3	3.8	2.0	1.8
	0.3	2.5	0.8	1.2	3.8	1.0 (0.3) [a]	2.5
9 CsF	0.1	1.0	0.9	4.3	3.8	2.0	1.8
	0.3	1.3	1.6	3.9	3.8	2.0	1.4 (0.4) [b]
18 Cs	0.1	0.9	1.5	4.6	3.8	2.0	1.2 (0.6) [b]
	0.3	1.7	1.6	4.8	3.8	1.9	1.4 (0.5) [b]

[a] Estimated amount stemming from a DTA peak at ca. 448 °C
[b] Stemming from a DTA peak at ca. 530 °C

most effective, followed by Na^+ and Cs^+ ions and the $NH_4{}^+$ ions were the least effective. In addition, while the influence of the alkali cation concentration was positive for $NH_4{}^+$, K^+ and Cs^+ ions, increasing concentration of $NH_4{}^+$ led to lower incorporation of aluminum into the MFI frame work [41].

The TPA/u.c. values are equal to ca. 3.8 (Table 8), and are close to the maximum possible 4 TPA/u.c., where the TPA^+ ions occupy the channel intersections, and the four propyl chains extend both the linear and zigzag channels. The amount of H_2O/u.c. is close to zero for all the samples studied. This value is in agreement with the results of the [Al]-ZSM-5 samples, where, with the 3.8 TPA/u.c. samples, no water was detected either [30].

The F/u.c. content generally decreases with increasing Ga/u.c. (Table 8). Its value is less than 2 for the samples synthesized in the presence of either NH_4F or KF. Higher values can be found in samples synthesized with NaF. For the CsF samples, values greater than 4 are found, suggesting the presence of CsF in the MFI channels, as shown previously for the borosilicalite samples synthesized in fluorine-containing media (Testa F, Chiappetta R, Crea F, Aiello R, Fonseca A, Bertrand JC, Demortier G, Guth JL, Delmotte L, B.Nagy J, submitted for publication). The total negative charges of the framework (tetrahedral Ga + F^-+ defect SiO^- groups) will be compared with the totally available positive charges (TPA^+ + M^+ + H^+) in the MFI channels. The ^{71}Ga NMR spectra are all

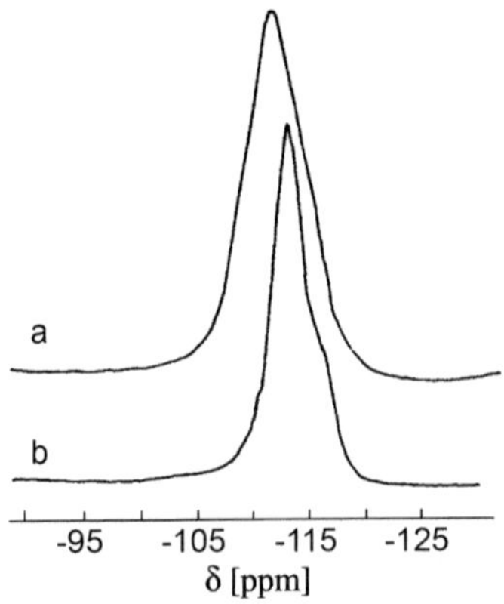

Fig. 4 ^{29}Si MAS NMR spectra of K-borosilicalite samples of **a** precursor and **b** calcined form obtained with 10 moles of H_3BO_3

characterized by a single NMR line at ca. 152 ppm vs. an aqueous 1 M $Ga(NO_3)_3$ solution, used as an external reference (Fig. 5a).

The apparent chemical shift shows that gallium is in the tetrahedral form in the MFI framework, and no octahedral gallium is detected at ca. 0 ppm [4, 5, 17, 18]. The NMR linewidths are all similar, and they only vary from 3850 to 4350 Hz. The average value is 4000 Hz. All the ^{29}Si NMR spectra of the as-made samples are characteristic of an MFI structure in an orthorhombic form [214] (Fig. 5b). The spectra were decomposed using 50% Gaussian and 50% Lorentzian lines in three contributions: one at – 103 ppm, and two lines at – 112 and – 116 ppm. The latter two stem from Si(0Ga) configurations, whereas the first one includes both the Si(1Ga) configuration and the SiOX defect groups (X = NH_4, Na, K or Cs, TPA and H): $L_{-103\ \mathrm{ppm}} = I_{\mathrm{SiOX}} + I_{\mathrm{Si(1Ga)}}$. As Loewenstein's rule is also obeyed for gallium incorporation, the $I_{\mathrm{Si(1Ga)}}$ was computed from chemical analysis data, Ga being only in tetrahedral framework position: $I_{\mathrm{Si(1Ga)}} = 100/0.25$ Si/Ga. The computed SiOX/u.c. values are listed in Table 9. If the framework negative charges (Ga + F)/u.c. are compared with the possible counterions (TPA + M)/u.c. (M = NH_4, Na, K and Cs), one can see that some of the data (almost one half) fit quite well. In most cases (more than half), the (TPA + M)/u.c. values are higher than the (Ga + F)/u.c. values, suggesting the presence of SiO^- defect groups to neutralize the excess positive charges.

The ^{29}Si NMR data show that the amount of defect groups is higher than that required by charge neutralization. This means that a rather high amount of SiOH/u.c. is also present in the structure. The latter was computed as the difference between the total amount of defect groups SiOX/u.c. and the contribution of SiOTPA (the amount of TPA decomposed at low [420 °C] temperatures) and of SiOM (M = NH_4, Na, K, Cs) (Table 9). Indeed, the SiOH/u.c. values are rather high, but their contribution decreases

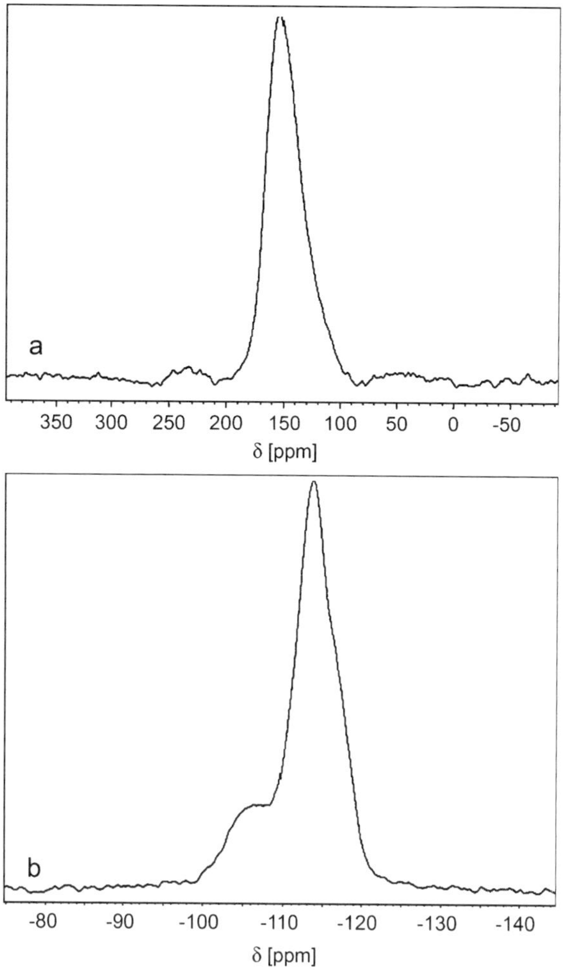

Fig. 5 **a** A typical ^{71}Ga MAS NMR spectrum of [Ga]-ZSM-5 sample synthesized from $10SiO_2 - 18CsF - 0.3Ga(NO_3)_3$ $-1.25TPABr-330H_2O$ at 170 °C (Ga/u.c. = 1.7). **b** A typical ^{29}Si NMR spectrum of a [Ga]-ZSM-5 sample synthesized from $10SiO_2 - 18CsF - 0.1Ga(NO_3)_3 - 1.25TPABr - 330H_2O$ at 170 °C (Ga/u.c. = 0.9)

with increasing Ga/u.c. (Fig. 6). The SiOH/u.c. values are higher for the K-containing samples than for the Na- and Cs-containing ones. The TPA^+ ions are incorporated intact into the zeolitic channels. The ^{13}C NMR spectra are characteristic of TPA^+ in a well-crystallized ZSM-5 zeolite synthesized either in F-containing media [215, 216] or in alkaline media [217]. The chemical shifts and linewidths are as follows: N–CH_2O–: $\delta = 62.7$ ppm ($\Delta H = 298$ Hz) with a shoulder at 65 ppm; –CH_2O–: $\delta = 16.4$ ppm ($\Delta H = 215$ Hz) and CH_3–: $\delta = 11.4$ ppm ($\Delta H = 100$ Hz) and $\delta = 10.2$ ppm ($\Delta H = 100$ Hz). The decomposition of the TPA^+ cations occurs either in one or two steps, similarly

Table 9 SiOX/u.c. (X = H, M and TPA), SiOTPA/u.c., SiOH/u.c., sum of positive charges (TPA+M)/u.c. (M = NH_4, Na, K and Cs), and sum of negative charges (Ga+F)/u.c. of [Ga]-ZSM-5 samples

MF	$Ga(NO_3)_3$	Si/Ga	SiOX [a] /u.c	SiOTPA [b] /u.c	SiOH [c] /u.c	(TPA + M) /u.c	(Ga + F) /u.c
$9NH_4F$	0.1	95.0	7.3	2.0	—	—	2.1
	0.3	32.1	2.4	1.3	—	—	3.3
$18NH_4F$	0.1	86.3	5.9	2.1	—	—	2.6
	0.3	29.0	2.0	0.4	—	—	3.7
9NaF	0.1	105.7	6.5	1.4	5.1	3.8	4.1
	0.3	49.5	3.6	1.8	1.8	3.9	3.1
18NaF	0.1			2.0		3.9	3.2
	0.3	52.3	1.5	1.4	0	4.0	3.3
9KF	0.1	105.7	14.8	2.3	12.5	3.8	2.4
	0.3	37.4	7.8	0.9	6.5	3.8	3.4
18KF	0.1	136.1	15.3	2.0	13.3	4.3	2.0
	0.3			1.2		4.5	3.7
9CsF	0.1	95.0	11.3	2.0	8.4	4.7	5.3
	0.3	72.8	7.4	2.0	3.8	5.4	5.2
18CsF	0.1	105.7	10.4	2.0	6.9	5.3	5.5
	0.3	55.5	7.7	1.9	4.2	5.4	6.5

[a] X = H, M and TPA, where M = NH_4, Na, K, Cs
[b] SiOTPA/u.c. = TPA/u.c. decomposed at low temperature (420 °C)
[c] SiOH/u.c. = SiOX/u.c. – (SiOTPA/u.c. + SiOM/u.c.)

to that of the [Al]-ZSM-5 samples (Fig. 7) [213, 216]. In one sample synthesized in the presence of 18 NH_4F and 0.3 $Ga(NO_3)_3$, essentially one high-temperature peak at ca. 480 °C is observed, characteristic of the decomposition of TPA^+ (SiOGa)-ion pairs. In this case, the value of Ga/u.c. (3.2) is close to that of TPA/u.c. (3.0) decomposed at 480 °C (Table 8). At lower Ga content, two decomposition peaks are observed at ca. 420 and 480 °C. The low-temperature peak can be attributed, as in the case of [Al]-ZSM-5 samples, to TPA^+ ions linked to SiO^- defect groups or to TPA^+ F^- (or OH^-) ion pairs [213, 216, 217]. This interpretation has nevertheless to be taken with some caution. Indeed, when the TPA/u.c. decomposed at higher temperature and is plotted as a function of Ga/u.c., the experimental points are rather scattered, and the difference between TPA/u.c. and Ga/u.c. becomes greater for low Ga/u.c. values. This suggests, that the high-temperature decomposition peak is related not only to the decomposition of TPA^+ ... $(SiOGa)^-$ species but also to the decomposition of the products that were formed during the decomposition of the TPA^+ F^- ion pairs. Several peculiarities have also been observed in the presence of 18 KF and 18 CsF (Fig. 7). An intermediate DTA peak was also observed at ~450 °C in the presence of 18 KF. With 18 CsF

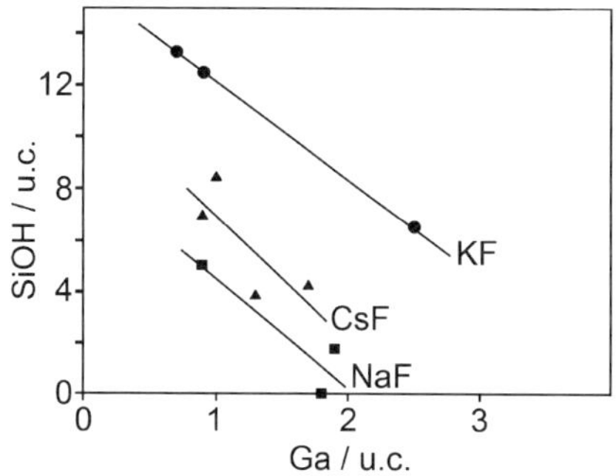

Fig. 6 Variation of SiOH/u.c. as a function of Ga/u.c. for the various [Ga]-ZSM-5 samples synthesized with NaF, KF and CsF

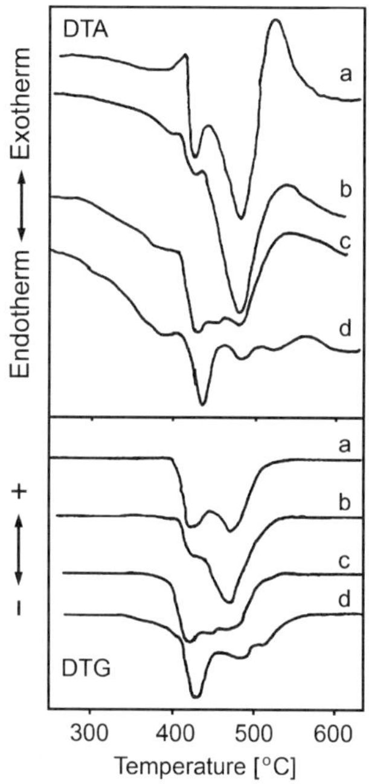

Fig. 7 DTG and DTA of [Ga]-ZSM-5 samples synthesized from $10SiO_2$-xMF-y $Ga(NO_3)_3$-1.25TPABr -330 H_2O at 170 °C: a $x = 9NH_4F$, $y = 0.3$; b $x = 18NH_4F$, $y = 0.3$; c $x = 18$ KF, $y = 0.1$; d $x = 18$ CsF, $y = 0.1$

(and also with 9 CsF) (Fig. 7), a new DTA peak also appeared at ~530 °C. This peak is characteristic of silicalite-1 and is due to the decomposition of the products occurring in the temperature range of the first DTA peak. The presence of this peak suggests that the crystallites are not homogeneous and contain highly siliceous regions, as it was shown already in [223].

3.3
Influence of Alkali Cations on the Incorporation of Al, B and Ga Into the MFI Framework

The T atoms are essentially incorporated in the zeolite framework as it is clearly shown by ^{11}B, ^{27}Al and ^{71}Ga NMR measurements [181, 184, 219]. Indeed, the position of the maxima are at – 4.0 ppm vs. $Et_2O{\cdot}BF_3$ for ^{11}B NMR, + 52 ppm vs. $Al(H_2O)_6^{5+}$ for ^{27}Al NMR spectra and + 152 ppm vs. $Ga(H_2O)_6^{3+}$ for ^{71}Ga NMR spectra. Table 10 shows the composition of the final [T]-MFI phases synthesized from $10SiO_2 - 9MF - yT - 1.25TPABr - 330H_2O$ at 170 °C, with M = NH_4, Na, K or Cs; T = 0.1 H_3BO_3, 0.16 $Al(OH)_3$ and 0.1 $Ga(NO_3)_3$. At such a low initial T atom content of the gels, the introduction of the T^{III} atoms is quite efficient. Indeed, the Si/T ratio of the initial gels is equal to 100 for B and Ga, and 62.5 for Al, while the Si/T ratios in the final samples vary between 119 and 73.

The fluorine contents are similar in all samples and vary between 1.1 and 4.8 F/u.c. The M/u.c. values are rather low showing that the tetrapropylammonium (TPA^+) ions neutralize the framework negative charges associated with the presence of T^{III} atoms in the framework. The TPA/u.c. values are equal to 3.8, which is close to the maximum theoretical value of 4. In this case all the channel intersections are occupied by TPA^+ ions. The amount of T/u.c. in the structure increases with increasing amount of T atoms in the initial gels. Up to 8 B/u.c. can be introduced into the framework starting from gel compositions of $10SiO_2 - 6H_3BO_3 - 9KF - 1.25TPABr - 330H_2O$, up to 6.5

Table 10 Comparison of the final zeolitic [T]-ZSM-5 phases (T = B, Al, Ga) synthesized from gels 10 SiO_2 – y T – 9 MF – 1.25 TPABr – 330 H_2O at 170 °C (X = H, M and/or TPA, and M = NH_4, Na, K, Cs)

9MF-0.1H_3BO_3	Si/B	B/u.c.	M/u.c.	F/u.c.	SiOX/u.c.
NH_4	119	0.8	—	3.0	5
Na	79	1.2	0.5	2.7	4
K	95	1.0	0.7	2.9	0
Cs	79	1.2	0.1	2.8	6
9MF-0.16$Al(OH)_3$	Si/Al	Al/u.c.	M/u.c.	F/u.c.	SiOX/u.c.
NH_4	73	1.3	—	2.5	5.2
Na	86	1.1	0.8	3.0	24.6
K	480	0.2	0.2	3.7	15.4
Cs	86	1.1	0.5	2.7	11.9
9MF-0.1$Ga(NO_3)_3$	Si/Ga	Ga/u.c.	M/u.c.	F/u.c.	SiOX/u.c.
NH_4	95	1.0	—	1.1	7.3
Na	106	0.9	0.03	3.2	6.5
K	106	0.9	0.05	1.5	14.8
Cs	95	1.0	0.9	4.8	11.3

Al/u.c. starting from $10SiO_2$ – 15KF – $6Al(OH)_3$ – 1.25TPABr – $330H_2O$, up to 3.2 Ga/u.c. starting from $10SiO_2$ – $18NH_4F$ – $0.3Ga(NO_3)_3$ – 1.25TPABr – $330H_2O$. As it can be seen from these results, K^+ ions are the preferred cations for the introduction of boron and aluminum, while NH_4^+ ions are more efficient in the case of Ga. (This sequence is not followed at low T^{III}/u.c. and these values are generally close to 1.) From the comparison of the initial and final Si/T ratios (Si/B_{gel} = 1.7 and Si/B of the zeolite is equal to 1.1; Si/Al_{gel} = 10 and $Si/Al_{zeolite}$ = 14; Si/Ga_{gel} = 33 and $Si/Ga_{zeolite}$ = 29) the following conclusion can be drawn. Ga is introduced with the highest yield, followed by Al and the introduction of B is the less effective one for these maximal conditions. The amount of the SiOX defect groups (X = H, M and/or TPA) was determined from ^{29}Si NMR measurements. It is much higher for [Al]- and [Ga]-ZSM-5 samples, while only a small amount of SiOX is found in [B]-ZSM-5 zeolites. Moreover, the NH_4-[T]-ZSM-5 zeolites show generally the smallest value of SiOX/u.c. The relative SiOX/u.c. values do not seem to be linked to the initial and final pH values being in the range of 8–9 for most cases. It is also interesting to compare both the relative amount and the temperature of decomposition of the two different TPA^+ species, whether the counter-cations balanced the negative charges of the $TOSi^-$ (high-temperature DTA peak) or the SiO^- defect groups (low-temperature DTA peak). While the relative amount depends on the T/u.c. values, the decomposition temperature is greatly influenced by the M/u.c. values of the counter-cations. When the temperatures of the extrema of the two peaks and of the single peak are plotted as a function of the Al/u.c., interesting variations can be noted (Fig. 8). At low Al content two peaks were observed merging into a single peak at high Al content (around 7–8 Al/u.c. curves a and b). A similar behavior was observed previously for samples synthesized in the presence of NH_4F. This relationship only holds, however, for samples prepared in the presence of NH_4F and for those prepared with a low amount of aluminum and hence containing only a low amount of M/u.c. If the samples contain more than 1.5 M/u.c. (there is only one exception), the two DTA peaks are merging much faster as a function of Al/u.c. (curve c in Fig. 8). Indeed, if the data for NaF, KF and CsF are plotted as a function of M/u.c., a clear-cut variation can be observed (Fig. 9). At low M/u.c. values two peaks are well detected (the only exception is that obtained with CsF, Al/u.c. = 2.5 and Cs/u.c. = 0.5). These two peaks merge into one for M/u.c. $\geq$ 1.5. This means that the alkali cations neutralize more preferentially the F^- ions and/or the defect SiO^- groups, while the remaining TPA^+ ions interact better with the negative charges $(SiOAl)^-$ of the framework. Hence, with a high M/u.c. content no TPA^+ ions are directly interacting with either F^- or SiO^- defect groups. The decomposition of TPA^+ cations in Ga-containing samples occurs either in one or two steps, similarly to that of the [Al]-ZSM-5 samples [213, 216] (Fig. 10).

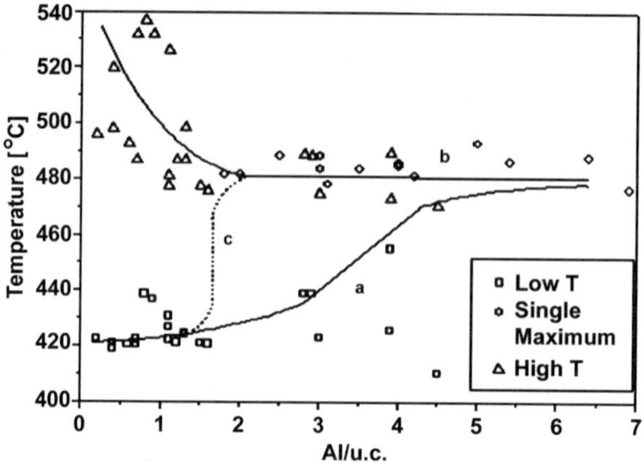

Fig. 8 Variation of the temperature of the DTA peaks as a function of Al/u.c. of the [Al]-ZSM-5 samples

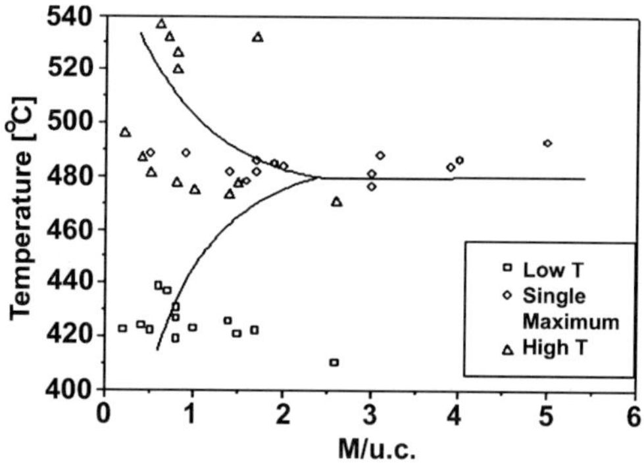

Fig. 9 Variation of temperature of the DTA peaks as a function of M/u.c. of the [Al]-ZSM-5 samples

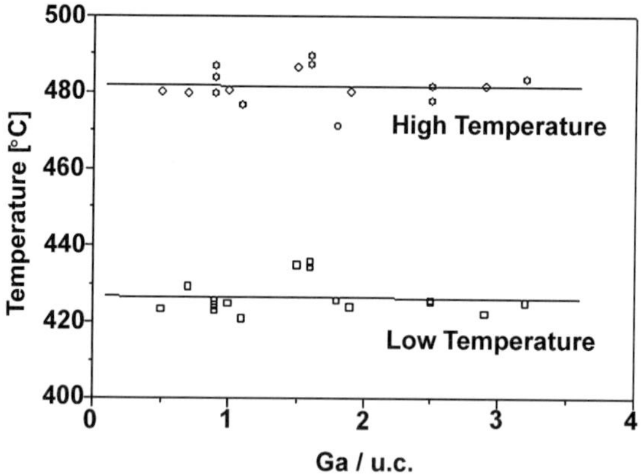

Fig. 10 Variation of the temperature of the DTA peaks as a function of Ga/u.c. of the [Ga]-ZSM-5 samples

When the temperature of the extrema of the two peaks are plotted as a function of Ga/u.c., no variation can be found, but the experimental points are rather scattered (Fig. 10). The variation of the same extrema as a function of M/u.c. is more informative (Fig. 11). Indeed, a slight increase is observed in the temperature of both decomposition peaks with increasing M/u.c. It is generally observed that the temperature of decomposition increases with increasing stability of the TPA^+ ions occluded in the MFI channels [217]. One could suggest that the increase of M/u.c. is accompanied by a decrease in defect groups SiOM/u.c. This is, however, not the case. It is also possible that both TPA^+ and M^+ ions compete for the same sites carrying negative charges, such as $(SiOGa)^-$, SiO^- and F^-. The SiOTPA/u.c. values do not seem to vary with M/u.c. either. Finally, the presence of MF in the MFI channels could also modify the decomposition temperature of the TPA^+ ions. Further work is necessary for a clear-cut picture of the various specific interactions between the available cations and anions. The synthesis in fluoride media leads to high crystallinity and regular morphology of the crystals. The crystal dimensions depend more on the nature of the T atoms than on the nature of the M ions. Large prismatic crystals of an average dimension of 150 μm × 50 μm × 40 μm are obtained for [B]-ZSM-5, while smaller prisms of average dimensions 50 μm × 25 μm × 15 (or 20) μm characterize the crystals of [Al]- and [Ga]-ZSM-5. The variation of both $\log 1/t_{ind}$ (t_{ind} being the induction time in hours) and $\log R$ (R being the crystallization rate in % per hour) as a function of the inverse of the radius of bare M nuclei ($1/r$) shows a complex behavior with a minimum for NH_4^+ ions and a maximum for K^+ ions (Fig. 12). Analyzing in more detail the crystallization rate only, the following observations can be made for the syntheses. The sequence Al > B > Ga is observed in the presence of Cs^+ and NH_4^+ ions, Ga > B > Al for K^+ ions and B > Al > Ga for

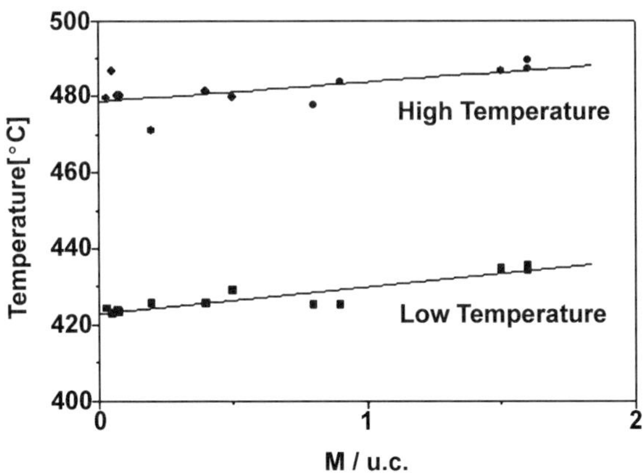

Fig. 11 Variation of the temperature of the DTA peaks as a function of M/u.c. of the [Ga]-ZSM-5 samples

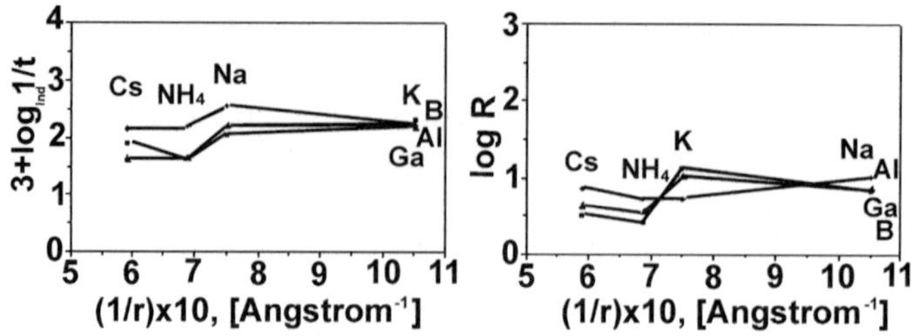

Fig. 12 Variation of the induction rate ($1/t_{\mathrm{ind}}$) and of the crystallization rate (R) as a function of $1/r$ (r: radius of the bare cation) for the [B]-ZSM-5, [Al]-ZSM-5 and [Ga]-ZSM-5 zeolites synthesized in fluoride-containing media

Na^+ ions. Note that the sequence found for K^+ ions follows the sequence of Sanderson electronegativities, i.e., Ga > B > Al [224]. On the other hand, the solubility of MF fluorides gives the sequence $NaF < KF < NH_4F < CsF$. Only the crystallization rate of [Al]-ZSM-5 follows this sequence. For the majority of the other cases (both induction and crystallization rates), KF influences the reaction rates the most. The effect of both the amount of T and the amount of MF was systematically explored on both the induction and crystallization rates. It can be concluded that the M^+ ions not only exert their coulombic stabilization effect (which is inversely proportional to r), but they also interact specifically with the various anionic species at the solid-liquid interface, where both nucleation and crystallization take place. Alkali cations deeply influence the incorporation of trivalent (T) atoms in metallosilicalite crystals. In particular K^+ ions are preferential cations for the introduction of boron and aluminum, while $NH_4{}^+$ ions are more efficient for Ga. The physico-chemical properties of metallosilicalites are also markedly influenced by alkali cations.

3.4
[Fe]-MFI

3.4.1
Fluoride Route

The gels with the composition $10SiO_2$–$x$$Fe(NO_3)_3 \cdot 9H_2O$–$y$MF–1.25TPABr–$300H_2O$, where $x = 0.1$, 0.2 and 0.3, $y = 3$, 6, 9, 12, 15, 18, 21 and 24 and M = NH_4, Na, K and Cs all led to crystalline MFI structures at 170 °C except for $x = 0.1$, $y = 3$ and M = Cs, where the gels remained amorphous. All the prepared gels were white before heating, because the iron species were not precipitated as hydroxides, but they formed soluble complexes with fluoride ions. Note that this is not the case for gels prepared in alkaline media, which were generally pale yellow or even "rusty" [225]. The as-synthesized products were all white powders, with a light greenish hue; only in some cases could brownish crystals be detected. In most cases, Fe/u.c. increases steadily as the amount of $Fe(NO_3)_3$ in the gel increases. The increase of Fe/u.c. in the final samples could be due to the ease of solubilization of $FeF_x^{(3-x)+}$ complexes and their transport to the nucleation and crystal growth sites. Also note that the Fe/u.c. values do not depend on the amount of CsF used in the gel. For NH_4F and KF, either independence or a slight increase is observed as a function of increasing F, while for NaF, a slight decrease was registered. NH_4F and CsF are the most efficient salts for the introduction of Fe into the tetrahedral positions of the structure. Indeed, the Si/Fe ratios of the final crystalline samples are always lower than or equal to the Si/Fe ratios of the corresponding gels. KF is less effective, as, in most cases, the Si/Fe ratios of the final crystalline samples are higher than those of the corresponding gels. Finally, NaF is the least effective salt, leading sometimes to three times higher Si/Fe values in the final samples than in the initial gels. The TPA/u.c. values determined by TG are close to 4, when the amount of Fe/u.c. is low (Fig. 13). As the amount of tetrahedrally incorporated Fe increases, TPA/u.c. decreases starting from approximately 2.5 Fe/u.c. for NH_4-[Fe]-silicalite-1 and 1.5 Fe/u.c. for (Cs, Fe)-silicalite-1. A similar decrease of TPA/u.c. values as a function of increasing tetrahedrally incorporated atoms has already been observed for [Al,TPA]-ZSM-5 [226, 227], [B,TPA]-silicalite-1 (Testa F, Chiappetta R, Crea F, Aiello R, Fonseca A, Bertrand JC, Demortier G, Guth JL, Delmotte L, B.Nagy J, submitted for publication) and [Ga,TPA]-silicalite-1 [184]. The DSC curves of Cs-[Fe]-silicalite-1 samples are similar to those obtained for the other M-[Fe]-silicalite-1 samples [228]. At low Fe/u.c. content, two endothermic peaks are observed at ~ 400 – 420 °C and at 440–460 °C (Fig. 14). Note that the high-temperature peak is located at lower temperature than it was reported for [Al]-ZSM-5 [226, 227, 229], [B]-silicalite-1 (Testa F, Chiappetta R, Crea F, Aiello R, Fonseca A, Bertrand JC,

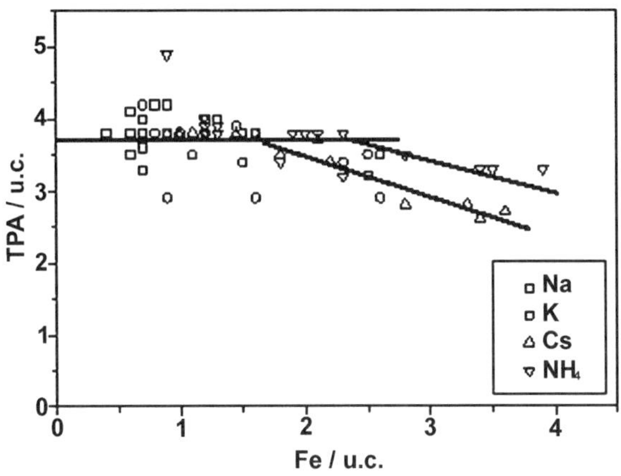

Fig. 13 Variation of TPA/u.c. as a function of Fe/u.c.

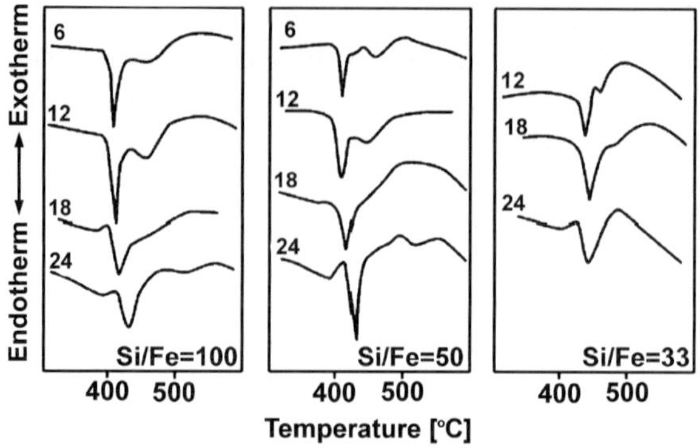

Fig. 14 DSC curves of Cs-[Fe]-silicalite-1 samples prepared with various amounts of CsF from the gel $10SiO_2 - xFe(NO_3)_3 \cdot 9H_2O - yCsF - 1.25TPABr - 300H_2O$ at 170 °C with $y = 6$, 12, 18 and 24

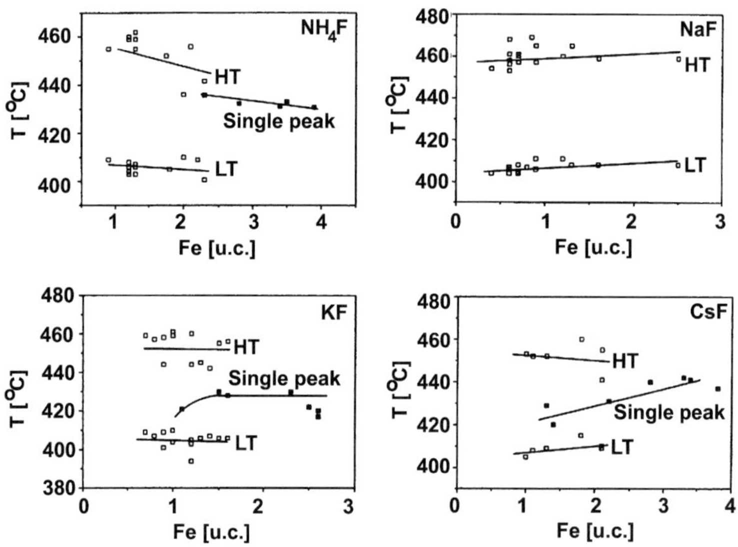

Fig. 15 Variation of the temperatures of the peaks in the DSC curves as a function of Fe/u.c.

Demortier G, Guth JL, Delmotte L, B.Nagy J, submitted for publication) or [Ga]-silicalite-1 [184]. The variation of the temperature of both the high- and low-temperature DSC peaks as a function of Fe/u.c. leads to interesting observations (Fig. 15). For the NH_4-[Fe]-silicalite-1 samples both the high-and low-temperature peaks decrease slightly with increasing Fe/u.c. On the other hand, for the (Na, Fe)-silicalite-1 samples both the low-temperature and the high- temperature peaks remain quasi-constant with increasing Fe/u.c. For NH_4^+ and Na^+ cations the interaction of TPA^+ with the framework is constant and is not much influenced by the presence of cations. This suggests that the M/u.c. is low for both NH_4^+ and Na^+ cations. Indeed, some 1.5 Na/u.c. were determined by the PIGE technique and by wet chemical analysis. The low-temperature peak was previously explained as being due to TPA^+ cations neutralizing the SiO^- defect groups in the structure [217, 226, 227, 231]. The behavior of the high-temperature peak is more controversial. It might be due to the decomposition of TPA^+ ions neutralizing the negative framework charges introduced by the presence of tetrahedral Fe(III) or other trivalent atoms. This is the case for the [Al]-ZSM-5 samples [226, 229, 231]. However, for silicalite-1 or for boro- and gallosilicalite-1, this peak also includes the products of decomposition stemming from the low-temperature peak (tripropylamine, dipropylamine, etc.) [184, 217, 226, 227, 231] (Testa F, Chiappetta R, Crea F, Aiello R, Fonseca A, Bertrand JC, Demortier G, Guth JL, Delmotte L, B.Nagy J, submitted for publication). The K-[Fe]-and Cs-[Fe]-silicalite-1 samples exhibit a different behavior (Fig. 15). At low Fe/u.c. values both the high-temperature peak and the low-temperature peak remain quasi-constant as Fe/u.c. increases. At higher Fe/u.c. values, however, only a single peak at an intermediate temperature is detected (Fig. 15). As there is some overlap between the Fe/u.c. values for the two DSC peaks and the single DSC peak due to decomposition of TPA, this suggests that not only the Fe/u.c. values are important, but the M/u.c. values as well, which greatly influence the decomposition temperature. The influence of the cations has already been shown for the [Al]-ZSM-5 [228, 229] and the borosilicalite-1 [226, 231] samples. In the case of the large K^+ and Cs^+ cations, due also to their higher amount in the samples, their influence can clearly be detected (Fig. 15). Their amount rises to ~ 3.0/u.c. The specific influence of the cations can also be observed, if the amounts of TPA/u.c., which have decomposed at high (HT) and low (LT) temperatures, are plotted as a function of Fe/u.c. (Fig. 16). For the NH_4-[Fe]-silicalite-1 samples, TPA/u.c. of the HT peak is always larger than that corresponding to the LT peak. They both decrease with increasing Fe/u.c. values, the TPA/u.c. values of the single peak are also decreasing with increasing Fe/u.c. as they represent the total amount of TPA/u.c. For the Na-[Fe]-silicalite-1 samples, the TPA/u.c. values of the LT peaks are always higher than those of the HT peaks and remain approximately constant as a function of Fe/u.c. On the other hand, for the K-[Fe]-silicalite-1 samples, at low Fe/u.c. values, the HT peak is larger than the LT peak and their TPA/u.c. values do not depend much on the Fe/u.c. values. Finally, for Cs-[Fe]-silicalite-1, the HT peak decreases, while the LT peak increases with increasing Fe/u.c. Of course, the TPA/u.c. values are decreasing with increasing Fe/u.c. for the single peak. Note that for [Al]-ZSM-5 [228], borosilicalite-1 [228] (Testa F, Chiappetta R, Crea F, Aiello R, Fonseca A, Bertrand JC, Demortier G, Guth JL, Delmotte L, B.Nagy J, submitted for publication) and gallosilicalite-1 [184], the specific influence of the cations M^+ could not be distinguished; the alkali cation only influenced the behavior of the TPA^+ ions

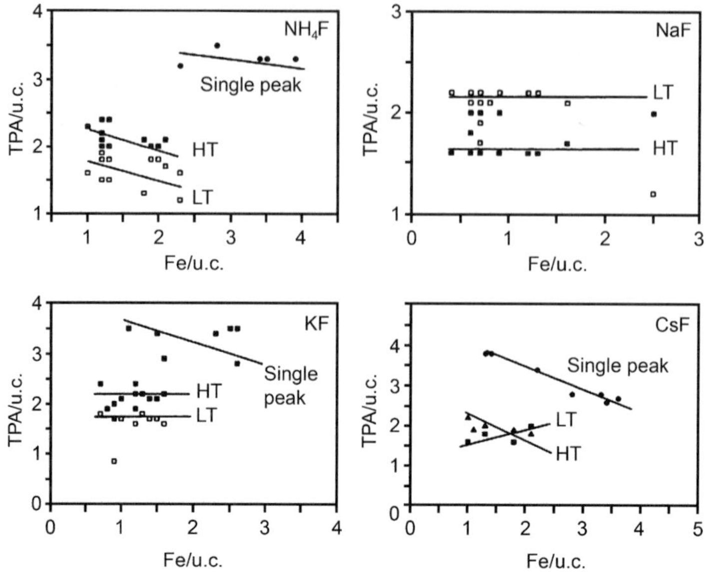

Fig. 16 Variation of TPA/u.c. decomposed at high (HT) and low (LT) temperatures as a function of Fe/u.c.

through their relative quantities. At present, no clear-cut explanation can be given for this specific behavior of the various cations. In order to better understand their role in the fluorine-route synthesis of zeolites a study is in progress to identify the various Fe-species, before and after calcination of the samples, together with the identification of the various intermediates produced during the partial decomposition of the TPA^+ ions.

All the ^{29}Si NMR spectra are characteristic of ZSM-5 in its orthorhombic form, the corresponding lines are, however, broader due to the presence of paramagnetic Fe(III)ions in the framework. All the experimental spectra were decomposed by using four NMR lines centered at approximately – 102, – 107, – 113 and – 117 ppm. (In one half of the samples a fifth line was added at approximately 110 ppm) The first two lines include the contribution of the SiOM (M = TPA and H) defect groups and that of $Si(OFe)(OSi)_3$ groups. The latter two lines stem directly from the MFI framework as $Si(OSi)_4$ groups. The amount of SiOM/u.c. was computed from both the – 102 and – 107 ppm lines. As these include both SiOM and Si(1Fe) groups: $I_{-102,-107\ \mathrm{ppm}} = I_{\mathrm{SiOM}} + I_{\mathrm{SiOM(1Fe)}}$. $I_{\mathrm{Si(1Fe)}}$ is computed from the Si/Fe ratio: $I_{\mathrm{Si(1Fe)}} = 100/0.25$ Si/Fe. The values computed this way are reported in Table 11. Contrary to what has been reported for [Al]-ZSM-5 [227] and [Ga]-ZSM-5 samples [184], no correlation could be found between the Fe/u.c. and the SiOM/u.c. values. The amount of defect groups is rather high, i.e., much higher than in the Al- and Ga-containing MFI structures. Only the Cs-[Fe]-ZSM-5 sample does not show any defect groups with 3.4 Fe/u.c. An extension of this work is necessary to

Table 11 Tetrahedral framework Fe and defect groups per unit cell in [Fe]-ZSM-5 samples obtained from the system $10SiO_2 - x\,Fe(NO_3)_3 \cdot 9H_2O - 15\,MF - 1.25TPABr - 300H_2O$ at 170 °C for 7 days

x	M	Fe/u.c.	SiOM/u.c. [a]
0.1	NH_4	1.2	10.8
	Na	0.7	21.2
	K	0.7	9.5
	Cs	1.3	13.1
0.2	NH_4	1.9	10.1
	Na	0.7	10.5
	K	1.6	24.1
	Cs	1.8	31.1
0.3	NH_4	3.9	11.9
	Na	1.6	12.1
	K	1.5	12.1
	Cs	3.4	0

[a] SiOM defect groups, where M = alkali cations, NH_4, H or TPA

Fig. 17 SEM pictures of some final [Fe]-silicalite-1 samples for Si/Fe ratio of 33 in the presence of various cations

get more details on both the as-made and the calcined [Fe]-ZSM-5 samples. Figure 17 shows SEM pictures of some of the final crystalline samples obtained at an Si/Fe ratio of 33 in the slurry in the presence of various cations. The morphology of the crystals depends highly on both the amount of Fe/u.c.

in the structure and its nature and amount in the gel. The Na-[Fe]-silicalite-1 samples contain essentially prismatic crystals, the length of which can reach 130 μm. The Fe/u.c. is below 1 for most of the samples. Even at high Fe/u.c. values, the morphology remains prismatic.

For the NH_4-[Fe]-silicalite-1 samples, the morphology is prismatic at low Fe/u.c. (up to 1.3), but it becomes ovoidal for higher Fe/u.c. values (equal to or higher than 2). Note again that the morphology remains ovoidal for more than 3 Fe/u.c., too. When the framework Fe/u.c. increases further, the morphology becomes spherical. These spheres are, in fact, agglomerates of small crystals. This habit can be observed for the K-[Fe]- and Cs-[Fe]-silicalite-1 samples as well. For these samples, the change in the morphology from prismatic to ovoidal, then to spherical occurs at lower Fe/u.c. values, showing that the nature and probably the amount of M^+ cations in the samples also influence the crystal morphology. The aspect (length to width) ratio decreases with increasing Fe-content in the initial gels. A rather complex behavior is detected if the aspect ratios are plotted as a function of $1/r$, where r is the radius of the bare exchanged cation M. The $1/r$ function is thus proportional to the electrostatic energy of gel stabilization by the various alkali counter-cations.

This will be discussed in more detail in connection with the study of the crystallization rate (see below). The specific effects of the alkali cations are also clearly visible, if the variation of lengths (L), widths (W) and aspect ratios (L/W) are plotted as a function of $\log R$, where R is the crystallization rate in % h^{-1} (Fig. 18). Both L and W are decreasing very rapidly and almost linearly as $\log R$ increases (Fig. 18a and b). NH_4-[Fe]- and Na-[Fe]-silicalite-1 samples show different variations, although the slopes of the lines are similar. On the other hand, K-[Fe]-and Cs-[Fe]-silicalite-1 samples show a large similarity. Interestingly, the aspect ratios, L/W, are increasing with increasing $\log R$ values. The pairs (NH_4, Fe)-(Na, Fe) and (K, Fe)-(Cs, Fe) silicalite-1 show similar behavior (Fig. 18c). All these results show convincingly that one of the most characteristic features in the physical appearance of zeolitic crystals, namely the aspect ratio, does not only depend on the rate of crystallization, but also on the specific interactions of the exchange cations with the various crystal faces. The crystallization curves representing the percentage of crystallinity as a function of time are shown in Fig. 19. The peak at $2\theta = 23.3°$ in the X-ray diffractogram was used as a measure of crystallinity, taking an ultrasonically treated final crystalline sample as standard for the 100% crystalline sample. When $\log(1/t_{ind})$ is plotted as a function of $\log R$ (t_{ind} being the induction time, i.e., the time necessary for the appearance of 4% crystallinity), a general tendency can be observed: the rate of induction increases with the increasing rate of crystallization. The points are rather scattered, suggesting that the possible influence of the various factors (pH, $Fe(NO_3)_3$-content and MF-content of the gels) is not exactly the same on the rates of induction and crystallization. In the following, general trends will only be indicated for the

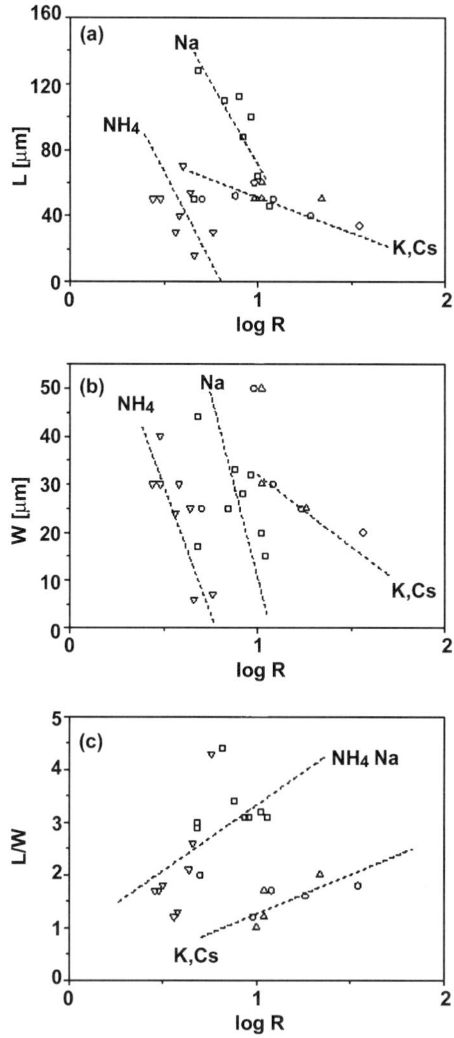

Fig. 18 Variation of the length L (**a**), the width W (**b**), and the L/W ratio (**c**) of the crystals as a function of $\log R$ (R is the crystallization rate in % h^{-1})

variation of the induction and crystallization rates. Due to the roughly parallel variation of $\log(1/t_{ind})$ and $\log R$, only the variation of $\log R$ will be analyzed in more detail. From the observation that the pH values of the mother liquor of the initial slurry (pH_i) are always higher than the final pH (pH_f) it follows that OH^- ions are consumed during the dissolution of the solid phase and the final condensation of the building units (BU_s) to form the crystalline [Fe]-silicalite-1. $\log R$ increases with increasing pH, although the experimental points are scattered, here too. A slope of 0.4 can be proposed for the general

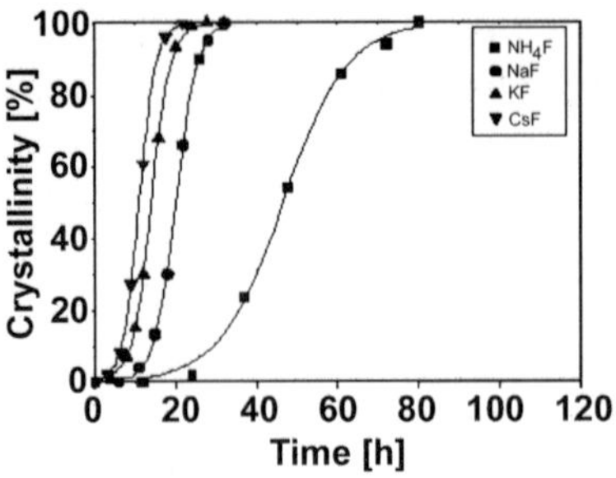

Fig. 19 Crystallization curves of [Fe]-silicalite-1 samples obtained from the mixture $10SiO_2 - 0.2Fe(NO_3)_3 \cdot 9H_2O - 15MF - 1.25TPABr - 300H_2O$ at 170 °C

trend showing the positive influence of the OH^- ions on the crystallization rate. A similar slope was determined (0.3) for the synthesis of silicalite-1 in fluoride-containing media (1 mol NaF in a given slurry) within the pH range of 2.5–6.5 [232] and for the synthesis of gallosilicalite-1 within the pH range of 7.5–10 [184]. However, increasing the NaF content (to 10 mol, under otherwise identical conditions) a steeper slope was obtained for silicalite-1 within the pH range of 6–7.5 [233] and for borosilicalite-1 within the pH range of 7–9.5 (Testa F, Chiappetta R, Crea F, Aiello R, Fonseca A, Bertrand JC, Demortier G, Guth JL, Delmotte L, B.Nagy J, submitted for publication). The positive role played by the OH^- ions emphasizes the fact that even in fluoride-containing media the OH^- ions play an important mineralizing role. $\log R$ slightly decreases with increasing $Fe(NO_3)_3$-content in the gel. Sometimes, a minimum appears as in the case of NH_4-[Fe]-silicalite-1. A similar decrease in the crystallization rates with trivalent substituting elements (like Al, B, Ga) has previously been observed showing the difficulty of the replacement of silicon by foreign substituents in tetrahedral framework positions [184, 227, 234] (Testa F, Chiappetta R, Crea F, Aiello R, Fonseca A, Bertrand JC, Demortier G, Guth JL, Delmotte L, B.Nagy J, submitted for publication).

The variation of the crystallization rate as a function of the amount of MF salts in the slurry shows an interesting behavior. $\log R$ decreases with increasing MF content for NH_4-[Fe]-silicalite-1 samples. As the F^- ions are mineralizing agents in these syntheses, the decrease might be attributed to a particular behavior of the NH_4^+ ions. Indeed, for the other two cations, Na^+ and K^+, $\log R$ increases with increasing MF content. For the $KF{:}0.1Fe(NO_3)_3$ system even a minimum is found for 15 KF in the gel. The specific influence of the various alkali cations on the crystallization rate can also be recognized, if $\log R$ is plotted as a function of $1/r$ for 15 moles MF in the slurry. A similar variation is found for the system $24MF{:}0.3Fe(NO_3)_3$. Cs^+ is the most effective cation for accelerating crystallization, followed by K^+ and Na^+. NH_4^+ is the least effective cation. As mentioned above, the function $1/r$ is proportional to the electrostatic stabilization energy of a negatively charged surface by the positive cations. As no variation in parallel is found, this means that the electrostatic stabilization energy is not the main factor influencing the crystallization rate for various exchange cations. Note that as a rough increase is observed for $\log R$ as a function of pH_i, the pH_i values show a similar variation as in the function of $1/r$ (not shown). This shows once again that the cations act more through their chemical characteristics than by solely physical (electrostatic) interaction in the fluoride-containing media. It is worth mentioning that the pH_i values are the lowest in the presence of NH_4F. As both NH_4^+ and F^- ions stem from a weak base and a weak acid, respectively, in dilute aqueous solutions the approximate pH is given by $pH = (pK_{NH4} + pK_{HF})/2 = 5.6$. This value is equal to 5.6 because $pK_{NH4} = 9.25$ and $pK_{HF} = 2.0$. The observed values are between 6.5 and 7.0. If, instead of $Fe(NO_3)_3$, $Ga(NO_3)_3$ is used in similar synthesis slurries, the initial pH_i is markedly higher, reaching 6.7–8.5. Thus, specific characteristics of the substituting ions (like their propensity to hydrolysis of the respective F^--containing ions, or their specific adsorption/reaction on the silica component) and not least the resulting ionic strength in the fairly concentrated slurries, all influence greatly, and in a fairly complex way, the initial pH_i of the slurry and with it the induction and crystallization rates as well. These results emphasize the need for further investigation of the "solution chemistry" of the initial crystallization slurry. Whereas in the fluoride route, using slightly acidic or neutral media, the complex stability of various soluble Fe(III) fluoro-complexes determine the (near-)equilibrium concentrations of various iron species in the slurry, their chemical identification is currently hampered, because the temperature of synthesis and the resulting ionic strength are too high. Nevertheless, one point seems to be important and worth noting. Irrespective of the chemical identity of the Fe(III)species (subsequently referred to as BU), it is certain that this BU is a mononuclear iron species, because the products are absolutely free of extra-framework (EFW) iron oxide/hydroxide which, when present even in trace amounts, imparts rusty coloration of various grades (from buff to a bright orange-red rusty color). The mononuclearity of the BU is strongly supported by Mössbauer spectroscopy studies carried out on zeolite samples synthesized within the frame of this research work (vide infra). This behavior is in contrast with synthesis routes carried out in alkaline media. In these methods, gelation of iron silicate is always preceded by the preparation of a strongly acidic Fe(III) salt solution (using sulphuric or rather perchloric acid). After cooling, the iron silicate zeolite-precursor gel is formed by the addition of cool waterglass solution, rich in monomeric silica species. However, earlier magnetic [235], spectroscopic [236] and potentiometric [237] measurements showed that a number of complex ions (like $[Fe(OH)(H_2O)_5]^{2+}$, $[(H_2O]_5Fe\text{-}O\text{-}Fe(H_2O)_5]^{4+}$, $[Fe_2(OH)_2(H_2O)_8]^{4+}$ where the two Fe(III)-ions of octahedral, O_h, coordination share common edge, etc.) are formed when the pH is changed. Whilst at lower pH (< 0.5) the whole iron appears to be monomeric, above pH = 2 at least 90% of the Fe(III)ions form dimeric species. It must not be forgotten that, in order to attain gelation, the pH has to be raised at least to 4.2–4.4. In previous publications [41, 225], using XP and Mössbauer spectroscopies, we showed that these binuclear iron species cannot be separated into monomeric ones; they are present in the iron silicate gel too, and because only one of them can be built into the zeolitic framework, a relatively high percentage (using sulphuric acid at the onset, about 30%) of iron is present in EFW positions from the very beginning. In other words, these samples, depending on the total iron content, are more or less rust-colored. It is surprising, indeed, that we find hints in the literature [238] that in strongly alkaline media, at or above 180 °C, these colored gels (in the presence of TMA cations) crystallize into white sodalite (SOD) specimens. This seems to be possible only if the binuclear iron species get separated from each other (probably via forming ferrate anions) during the arduous conditions of the (sodalite) synthesis. We believe that this mechanism cannot be operative in the synthesis of ZSM-5 (MFI) zeolites. From the point of view of catalytic activity both the fluoride-and alkaline routes might result in excellent catalysts; nonetheless, for scientific (e.g., spectroscopic) studies the fluoride-route should be preferred, because it leads to products with framework (FW) iron in unique tetrahedral, T_h coordination.

Table 12 The synthesized MFI products through the variation of the starting hydrogel composition, in the type-A and type-B systems

Sample	Si/Fe_{gel}	Si/Al_{gel}	TPABr	EG	Time [days]	Si/Al_{cryst}	Si/Fe_{cryst}
1A	100	∞	0.08	0.0	1	—	85.5
2A	10	∞	0.08	0.0	3	—	15.4
3A	100	100	0.08	0.0	1	97.7	73.8
4A	10	∞	0.02	0.0	3	—	—
6A	10	10	0.02	0.0	4	25.0	307
7A	∞	25	0.02	0.0	2	16.5	—
8B	20	∞	0.08	3	1	—	258
9B	6.67	∞	0.04	3	2	—	252
10B	6.67	∞	0.04	3	5	—	—
11B	20	10	0.04	3	2	23.5	418
12B	10	10	0.04	3	2	23.0	459
13B	6.67	10	0.04	3	6	24.1	442
14B	20	∞	0.02	3	2	—	504
17B	20	20	0.02	3	2	—	—
18B	20	10	0.02	3	5	—	—
19B	10	10	0.02	3	5	29.1	403
21B	∞	20	0.0	3	3	13.7	—
22B	20	20	0.0	3	2	35.0	423
23B	20	10	0.0	3	5	18.1	369

EG: ethylene glycol; A: without EG; B: with EG

3.4.2
Alkaline Route

In Table 12 the products obtained for both systems A and B in the presence of oxalic acid or phosphoric acid are reported.

3.4.2.1
[Fe]-MFI Synthesized by the Alkaline Route

[Fe]-MFI products were obtained in alkaline media using either oxalic or phosphoric acid as complexing agents [35]. In addition, the initial gels were prepared either without ethylene glycol (system A) or in the presence of ethylene glycol (system B) (Table 12).

The general compositions of the gels were: $x\mathrm{Na_2O} - y\mathrm{TPABr} - z\mathrm{Al_2O_3} - \mathrm{SiO_2} - q\mathrm{Fe_2O_3} - p\mathrm{HA} - 20\mathrm{H_2O}$ where $x = 0.1 - 0.32$; $y = 0.02 - 0.08$; $z = 0 - 0.05$; $q = 0.005 - 0.025$; the ratio $p/q = 3$ and HA stands for $H_2C_2O_4$ or H_3PO_4. The second system was prepared from the following initial reaction mixture: $0.16\mathrm{Na_2O} - x\mathrm{TPABr} - y\mathrm{EG} - z\mathrm{Al_2O_3} - \mathrm{SiO_2} - q\mathrm{Fe_2O_3} - p\mathrm{HA} - 10\mathrm{H_2O}$ where $x = 0.08$; $y = 0 - 6.0$; $z = 0 - 0.05$; $q = 0.005 - 0.1$; the ratio $p/q = 3$ and HA stands for $H_2C_2O_4$ or H_3PO_4 and EG for ethylene glycol.

First, it can be observed that, if only TPABr is present in the initial gel (system A), [Fe]-MFI zeolite can be obtained in a large Si/Fe range and reaction time of 1–4 days (samples 1A–7A in Table 12). The obtained [Fe]-MFI products are thermally stable. Indeed, if the reaction time is increased, the transformation of the MFI structure into more stable phases is not observed. Even after a thermal treatment up to 850 °C, the MFI is the sole phase detected. The crystallization time increases with increasing iron and/or aluminum content of the starting hydrogel (compare samples 1A and 2A, and samples 4A and 6A, respectively).

In system B, the formation of [Al]-MFI is very easy in the presence of ethylene glycol (EG) only (sample 21B). This result confirms the data reported in [239]. In the presence of EG, only Al-rich hydrogels were studied, because no pure MFI zeolite could be synthesized in a silica-rich system [213]. EG is also able to direct the formation of the MFI structure in the presence of both Al and Fe (samples 22B and 23B). If TPABr is also present in the initial gel, the reaction time decreases (see samples 11B and 18B, and 12B and 19B). In addition, the stability of the final MFI products also increases with increasing TPABr content of the gel. Indeed, a pure [Fe]-MFI sample is obtained with 0.08 TPABr (sample 8B) even after 6 days of crystallization, while sample 14B synthesized in the presence of 0.02 TPABr shows the co-crystallization of MFI with quartz and cristobalite after 4 days of crystallization. The increasing TPABr content in the initial hydrogel results in decreasing reaction time (see samples 14B and 8B). On the other hand, the increase in the initial Al or Fe content in the hydrogels leads to the increase in reaction time: compare samples 9B and 13B for the Al-content and samples 11B and 13B for the influence of Fe-content.

For both systems, the pH value plays an important role in the crystallization process. As a matter of fact, pure MFI zeolite is formed, when the pH value of the initial hydrogel is between 9 and 12.5.

The chemical analyses data are also reported in Table 12 for both systems A and B. It can be noticed that for type-A synthesis system the amount of iron incorporated into the zeolitic structure is related to its content in the TPABr-rich hydrogels (see samples 1A and 2A). For these TPABr-rich hydrogels, when both iron and aluminum are present (sample 3A), iron is incorporated preferentially compared to aluminum. However, if the TPABr content decreases to 0.02, iron cannot be incorporated so efficiently (sample 6A) and aluminum is preferentially incorporated.

This behavior can be explained by the different interactions between soft and hard acids with iron and aluminum atoms. It is known that hard acids accompany better hard bases and soft acids prefer soft bases. Since the Na^+ is a harder acid than the TPA^+, they interact well with $[Si-O-Al]^-$ and $[Si-O-Fe]^-$ groups, respectively, since $[Si-O-Al]^-$ is a harder base than the $[Si-O-Fe]^-$. These kinds of interactions have been demonstrated by different authors for B-MFI-type zeolite (see above). For system type-B (with EG) we always detect low amounts of iron in the MFI samples, even in the presence of high TPABr contents (see sample 8B, Table 12). In contrast, the amount of aluminum in the zeolitic framework is related to its content in the starting hydrogel, and it is almost always a high amount. This confirms that the hydrogel systems that contain both sodium and alcohol, such as EG,

Fig. 20 Scanning electron micrograph of [Fe]-MFI sample 1A (see Table 12)

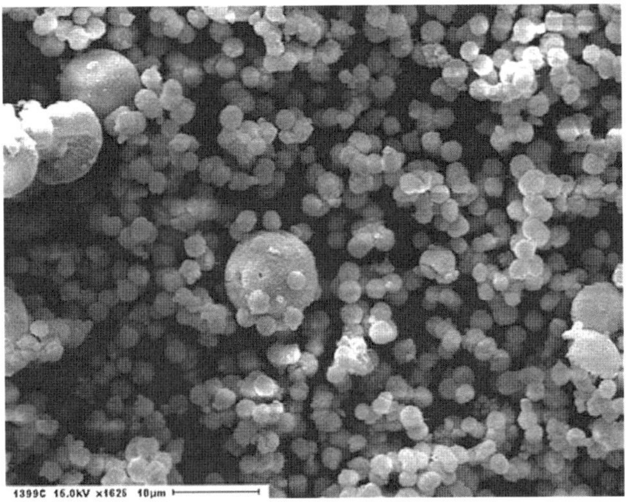

Fig. 21 Scanning electron micrograph of [Fe]-MFI sample 2A (see Table 12)

show a high amount of sodium in the crystals [84, 212] and, consequently, they prefer the interaction with the $[Si-O-Al]^-$ groups and favor Al incorporation. Probably the EG molecules compete with the TPA^+ cations and they fill the zeolitic channels as detected by thermal analysis; in fact, we found EG molecules intact in the MFI channels.

The size of the crystals obtained from system A decreases when the amount of iron incorporated in the zeolite increases, and it changes from 10 μm (samples 1A, Table 12) to 2.3 μm (sample 2A, Table 12). The morphology for the samples also changes from brick-like to spheres (see Figs. 20 and 21).

These observations give indirect confirmation of the iron incorporation into the MFI framework. The presence of aluminum in the zeolite allows reduction in the crystal size compared to that of the Al-free samples. In fact, sample 3A shows a spherical habitus with a diameter of 5 μm, whereas sample 2A evidences the same morphology with smaller size (2.7 μm) due to the high amount of aluminum and iron incorporated into the zeolitic framework. The morphology of the samples obtained from system B shows a brick-like shape for both iron and iron-aluminum systems. The crystal size also decreases when the amount of Al incorporated into the zeolite increases and changes from 12 to 1 μm.

3.4.2.2
Fe-Content in ZSM-5 Zeolites

The synthesis methods of Fe-containing ZSM-5 zeolites in alkaline media are characterized by complexing the Fe(III) with monomeric, short-chain silicic acids in strongly acidic solutions (like sulphuric or perchloric acid, at the value – 4 to – 3 of the Hammett constant) in order to avoid formation of iron hydroxides/oxides. By adding sodium aluminate and/or sodium hydroxide, the gelation of the seemingly homogeneous, colorless colloid solution begins at pH 4.0–4.5. In the literature there are hints (see, e.g., [241]) that once the ferrisilicates are formed the pH of the synthesis slurry can be safely increased to values of the gelation limit or even further without formation of rust-red iron hydroxides. In a previous paper, it was argued that by increasing the pH part of the Fe(III) might be released from the silicate bonding leading to yellow or even rust-red coloration of the gel [242]. The color is due to (magnetically interacting) bi-, ter- or polynuclear hydroxo- or oxo-complexes of iron exhibiting mainly octahedral (O_h) co-ordination. In the alkaline solution of the synthesis slurry (pH 10.5–11.0) at the temperature of the synthesis (about 160 °C), the silicic acid component goes into solution and, because this solution is always supersaturated with respect to the crystalline phase, enters the crystals. The Fe(III) ions thus released temporarily from the silicate bonding will be or will not be incorporated into the lattice, depending on reasons not clearly understood yet. The incorporation of the substituting (transition metal or other) ions seems to follow interesting preferences: at very low concentrations in the gel (using the nomenclature of Wichterlova et al. [243]), e.g., the Co^{2+} ions are incorporated in γ positions first (mid-positions in the five-membered rings), followed by the occupation of the β (in the six-rings) and finally the α sites (in the main channels). (It would not be surprising to find that the γ and β sites in orthorhombic ZSM-5 zeolites are identical with the T4 and T12 sites according to the site notation of van Koningsveld et al. [244].) Concerning Fe(III) ions as substituents, a similar preference can be imagined even when the necessary experimental details are currently lacking. Those Fe(III) ions which could not be incorporated form (at low concentrations) mainly mononuclear, non-interacting, therefore colorless extra-framework Fe(III) components, present already in the as-synthesized material. Calcination and/or heat-treatment, especially in the presence of water vapor, result in agglomeration of the so far well-separated EFW Fe(III) components and development of the characteristic rusty color, indicative of little, interacting Fe – O – Fe-clusters. This "mechanism" of isomorphous substitution seems to be valid primarily for Fe(III) as substituting ions, but it is believed, its validity is wider. The increased probability of Fe(III)–Fe(III) interactions leading to hydroxide/oxide formation during the gel dissolution can be made responsible for the decreasing percentage of iron incorporation as the Fe(III)-content in the gel increases. Till even today it was deemed that the "fluoride-route" of the synthesis produces a higher degree of Fe(III) incorporation, with less EFW iron dispersed in the voids and pores of the zeolitic crystals. This belief was based on the fact that, in the synthesis slurry-containing F^- ions, the iron is present mainly as a mononuclear Fe(III)—fluoride complex, whereby the immediate Fe(III)—Fe(III) interactions leading eventually to rust formation will be hindered. As will be seen later, this view turned out to be a fully erroneous assumption. In a previous publication [185] we dealt with the synthesis of Fe-containing ZSM-5 zeolites in fluoride-containing media. The samples crystallized from the slurries

(A) $10SiO_2 - 0.2Fe(NO_3)_3 \cdot 9H_2O - 15NH_4F - 1.25TPABr - 300H_2O$,

and

(B) $10SiO_2 - 0.2Fe(NO_3)_3 \cdot 9H_2O - 18CsF - 1.25TPABr - 300H_2O$

exhibited white color with a weak green hue. The Mössbauer spectra revealed participation for the T_h coordinated Fe(III) species in excess of 94% in both samples (IS $< 0.3\,mm\,s^{-1}$, and after calcination QS $< 1.2\,mm\,s^{-1}$). The Fe-content of the sample prepared with NH_4F addition was 1.81 wt % (using AAS) and that synthesized with CsF had a 1.66 wt % Fe-content. The conclusion is unequivocal: the EFW iron component is either missing or present only in trace amounts. We could not believe that in spite of white color and Mössbauer parameters characteristic of tetrahedral Fe(III), these samples contained surprisingly high amounts (sometimes exceeding even 30%) of EFW iron. Fe(III) has five unpaired electrons in weak ligand-field, high-spin state ($^6S_{5/2}$). In perfect T_h or O_h, i.e., "cubic" fields only one signal appears close to the $g = 2.0023$ value ($H = 0.3308$ T) of the free electron in the X-band EPR spectrum. If the symmetry of the field is lower, say axial, the powder spectrum becomes anisotropic and may have signals in the $2 < g \leq 6$ ($0.1104 \leq H < 0.3312$ T) interval. If the symmetry is reduced further, to fully rhombic, this results in a seemingly "isotropic" signal at $g = 4.29$. In very fine powders, the occurrence of axial- or lower-symmetry sites are inevitably due to the increased influence of solid surface tension (as concerns consequences, vide infra). In Fig. 22 two EPR X-band spectra are shown. In Fig. 23 the EPR spectra of the ZSM-5 sample F11 (ZSM-5 synthesized in NH_4F medium; University of Calabria; spectrum (a)) and Z4 (synthesis in alkaline medium as

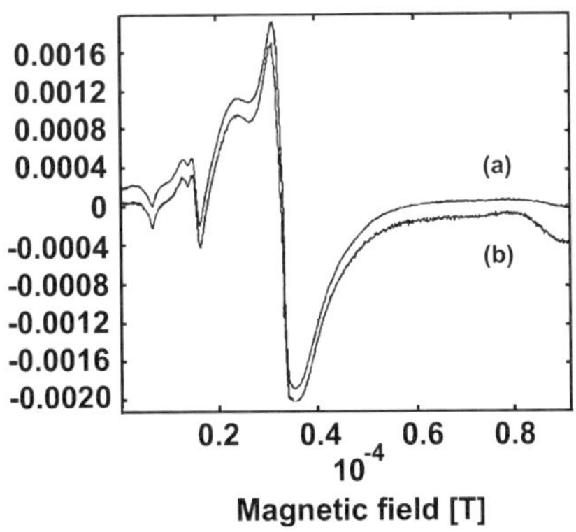

Fig. 22 X-band EPR spectra of [Fe]-MFI samples; (a) Fe content is 1.25 wt %, (b) Fe content is 1.42 wt %; both synthesized in slurries containing F^- ions

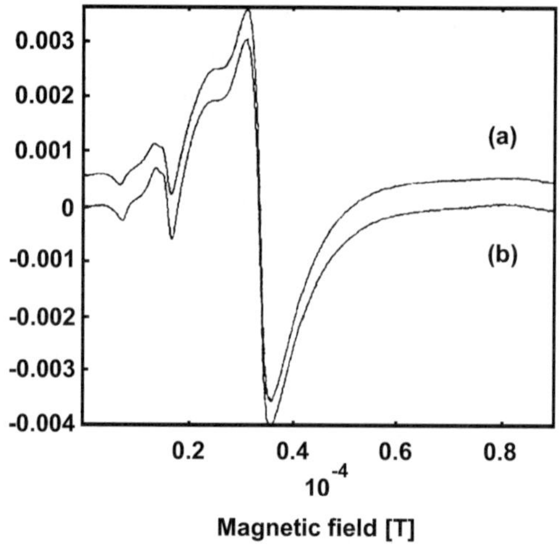

Fig. 23 X-band EPR spectra of ZSM-5 samples (a) synthesized in NH_4F-content slurry (sample F11, Fe: 3.58 wt %) and (b) prepared in alkaline slurry (sample Z4, Fe: 3.81 wt %)

described previously; Szeged University; spectrum (b)) are compared. Fe-contents: F11: 3.58 wt %, Z4: 3.81 wt %. These spectra, too, show perfect overlapping in the whole $0.1 \leq H/T \leq 0.8$ interval, moreover, as in the case of the previous pair, even the intensities of the EPR signals are comparable.

Maybe it is not surprising that two samples, synthesized in fluoride media and having comparable solid phase compositions, produce identical EPR spectra. However, for samples F11 and Z4 only the solid phase compositions are nearly identical, the pH-values of the slurries deviate considerably from each other (slightly acidic for F11 and fairly alkaline for Z4).

Identical EPR spectra mean nearly equal FW and EFW iron contents. Surprisingly, after "caring calcination" the deferrization of sample Z4 by hydrochloric acid + hydroxylamine hydrochloride resulted in about 33% EFW iron. Also, instead of the original u.c. composition ($Na_{\sim 4.08}Fe_{4.08}Si_{91.92}O_{196}$, Si/Fe = 22.5) corresponding to 3.81 wt % (total) iron content, the fitting u.c. formula is $Na_{\sim 2.70}Fe_{2.70}Si_{93.30}O_{196}(0.665\ Fe_2O_3)$, (Si/Fe = 36.4, i.e., Fe: 2.51 wt % in FW and 1.30 wt % in EFW position expressed as oxide (given in parenthesis). The amount of Na is a computed value. Due to imperfect washing after synthesis this composition is always liable to distortions. In other words, the [Fe]-ZSM-5 samples synthesized in F^- ion-containing slurries contain comparable amounts of EFW iron as those prepared in alkaline media and this statement can be corroborated by direct deferrization experiments carried out on the respective sample. All of the Fe EPR spectra (even in the AS state) have a common feature: they exhibit a more or less expressed "hump" at about $g = 2.5–2.7$ (corresponding to $H = 0.265–0.245$ T magnetic fields) which, as generally believed, is characteristic of EFW iron hydroxide/oxide in the zeolitic voids. It would be important to reconcile the findings of EPR spectrometry with the Mössbauer data. First of all, how is it possible that, while the "hump" can clearly be seen in the EPR spectra, at the same time this highly dispersed EFW iron is fairly Mössbauer-silent in the AS sample? In Table 13 the abridged Mössbauer parameters of the ZSM-5 sample crystallized from the previous F^- ion-containing slurry (A) are compiled, in the AS state, after calcination at 500 °C and reduction by CO at 350 °C (i.e., "350/CO") [185].

The isomer shift, IS, of 0.26 mm s^{-1} is well below the IS = 0.30 mm s^{-1} value, the upper limit of T_h coordinated Fe(III) at ambient temperature. Yet, it can be surmised that the IS = 0.26 mm s^{-1} value hides mixed T_h and O_h co-ordinations, since in all those cases where the separation of a resultant spectrum into sub-spectra is not possible (and this is valid for Mössbauer studies applying AS zeolite samples), the computed parameters are necessarily averages [245]. Thus, there are surroundings for Fe(III) ions in T_h co-ordination characterized by IS < 0.30, as are others exhibiting O_h co-ordination with IS > 0.30. Unfortunately, the real FW/EFW ratio cannot be clearly seen, because various ligands (like water, TPA^+ cations, etc.) located in the immediate surroundings of tetrahedrally coordinated Fe(III) may let this appear octahedral

Table 13 Mössbauer parameters of an NH_4-[Fe]-ZSM-5 sample (synthesized from slurry (A), Fe: 1.81 wt %) derived from 25 °C in-situ spectra

Treatment	Component	NH_4-[Fe]-ZSM-5 IS	QS	FWHM	RI
As synthesized	${Fe_{T_h}}^{3+}$	0.26	—	1.80	94
	Fe^{2+}	1.02	1.93	0.47	6
500/calcined [c]	$(Fe_{O_h/T_h})^{3+}$	0.33	0.81	0.90 [a]	50
				1.07 [a]	50
350/evacuated [d]	${Fe_{T_h}}^{3+}$	0.23	1.66	0.80	89
	Fe^{2+}	1.09	2.31	0.57	11
350/CO [e]	${Fe_{T_h}}^{3+}$	0.24	1.89	0.38	38
	Fe^{2+}	1.02	1.76	0.82	30
	$(Fe\,C_x)$ [b]	- 0.59	0.67	2.38	32

IS: isomer shift (mm s^{-1}), related to α-iron; QS: quadrupole splitting (mm s^{-1}); FWHM: full line-width at half maximum (mm s^{-1}); RI: relative intensity (spectral area, %)

[a] During the decomposition different line-widths were allowed, while the same intensity was constrained

[b] Probably fraction of non-resolved iron carbide

[c] Calcined at 500 °C

[d] Evacuated at 350 °C

[e] Treated with CO at 350 °C

(from the point of view of co-ordination). The RI values obtained after evacuation at 350 °C (i.e., 350/evacuated) treatment (see Table 13.) reveal only 11% EFW iron, provided only this iron component has a tendency to autoreduction and the probabilities of Mössbauer effect for the various Fe species are nearly equal (see later). The iron carbide formation after "350/CO" treatment is a disturbing effect, nevertheless, the RI = 38 value for $Fe^{3+}_{T_h}$ indicates beyond doubt: these samples should have contained fairly large amounts of EFW iron before the calcination. The burning off of the organics and the concomitant steaming contributed to the agglomeration of the "single ion" Fe(III) component, and thereby the oxidic clusters produced became "visible" for Mössbauer spectroscopy, too. The immense literature on EPR spectroscopy of Fe(III) ions produced huge collection of both senseless and acceptable suggestions for explaining the extremely complex set of overlapping EPR signals. Let us see, what assignments are proposed by Ratnasamy [264]:

$g \approx 2.0$	T_h and O_h coordinated Fe(III) ions ("single ions") in perfectly cubic environments; (overlapping, seemingly) isotropic signals;
$2.0 \le g \le 6.0$	T_h and O_h coordinated Fe(III) ions in slightly distorted cubic (i.e., "axial") environments, provided powder spectra are registered; in this case g is a tensor;
$g \le 4.30$	T_h and O_h coordinated Fe(III) ions and/or their clusters in (strongly distorted) "fully rhombic" environments; the signal is (seemingly) isotropic.

From this compilation, the EPR signal of Fe-oxide/hydroxide clusters built up from a few ions (pairs, trimers, tetramers and agglomerations of higher degree) are obviously missing. Such little clusters, even though their sizes are far from being "infinite", can be well seen in EPR spectroscopy. According to Korteweg and van Reijen [241] even trimers and tetramers produce increasingly broad resonance signals due to Heisenberg-type exchange within clusters, but no interaction should be assumed between clusters far away from each other. On the basis of experimental observations (intensity change of the respective EPR line caused by hydrogenation, hydration, treatment by acids, etc.), it is generally believed that such EFW iron generates signals at about $g = 2.4$–2.6 (this is the "hump" mentioned previously) and at $g = 4.30$ mentioned by Ratnasamy as well. The trouble is that the assignment of the $g = 2.4$–2.6 EPR signal to EFW iron oxide clusters is very doubtful. If there are no axial Fe(III) environments (and by analyzing the spectra one gets the impression that they appear at higher (> 3 wt %) Fe(III) concentrations), the clusters produce a more or less smeared out signal at $g \approx 2.0$, but the superposition of (at least) two EPR lines of slightly different g-values, intensities and widths might result in lower magnetic fields in the "hump" at $g = 2.4$–2.6 strengthening the belief that the "center of gravity" of the EFW iron EPR signal is located there. In order to avoid such mistakes we utilized and implemented an idea of spectrum decomposition proposed by Lin et al. [246] to see the individual component parts of Fe(III) EPR spectra. For spectrum decomposition, a sophisticated computer program for non-linear least squares curve fitting and parameter-estimation was used. The EPR spectra were regarded as composites of individual (derivative) Gaussian lines of unknown number. In order to start with the program and avoid local minima of the object function the initial estimates for the number of sub-spectra and the respective parameters (location, intensity and width) have been determined by applying a random drawing method combined with a few steepest descent steps. The double integral of each sub-spectrum EPR line thus obtained has been computed by the exact formula

$$I = 1/8\sqrt{2\pi e}\,\mathrm{Amp}(\mathrm{Width})^2 = 0.5166\,\mathrm{Amp}(\mathrm{Width})^2 \quad (1)$$

where Amp is the peak-to-peak amplitude (in arbitrary units) of the EPR line, and Width is the peak-to-peak width, measured in Gauss. The result of such decomposition is shown in Fig. 24 for the [Fe]-ZSM-5 sample synthesized in alkaline medium as described in the respective section. An optimum in the fit could be reached assuming three sub-spectra: dF1c, dF2c and dF3c (see Fig. 24a); their sum, dFc, and the measured spectrum, dF, are compared in Fig. 24b. Its total Fe-content was 3.27 wt %, of which less than 60.9% was in FW and more than 39.1% in EFW state. These data were measured by XRF on the AS sample and on another, which underwent "caring calcination" followed by "exhaustive" (yet not for 100% effective) deferrization using hydrochloric acid and hydroxylamine. Thus, the u.c. composition is $Na_{\sim 2.09}Fe_{2.09}Si_{93.91}O_{192}$ (Si/Fe = 44.9). (The amount of Na is a computed value, and the $3.27 \times 0.391 = 1.28$ wt % Fe in the EFW state equals to 0.671 mol Fe_2O_3 per u.c.) The FW/EFW ratio of the sample is comparable with that of samples prepared earlier in our laboratory (cf. [225]). For dF1c, which is supposed to be an FW component (see later), one found $g_1 = 1.94$ and the width $W_1 = 422.79$ G. Similarly, for dF2c, being supposedly the oxidic, EFW

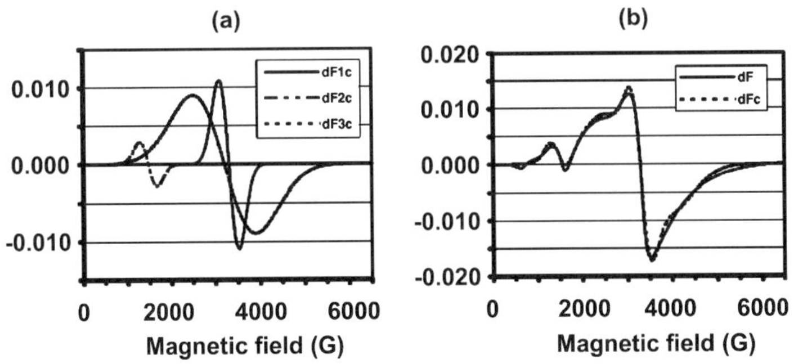

Fig. 24 Decomposition of the X-band spectrum of an [Fe]-ZSM-5 sample **a** dF1c: FW iron component ($g_1 = 1.94$); dF2c: main EFW iron component ($g_2 = 2.01$); dF3c: fully rhombic EFW iron component ($g = 4.29$). **b** Comparison of the measured (dF) and computed spectrum (dFc)

component, $g_2 = 2.01$ and $W_2 = 1392.51$ G were observed. The intensities computed by the ad-hoc formula are: $I_{FW} = 2036.3$ in a.u. and $I_{EFW} = 17976.4$ also in a.u. The ratio "R" of the intensities was $R = I_{EFW}/I_{FW} = 8.828$, thus considering the measured FW and EFW concentrations as well, the ratio of the "transition probabilities" EFW to FW, "r", was $r = R \times (1.99/1.28) = 13.72$. In other words, EPR spectra (under ambient conditions) are 13.72-times more sensitive for the EFW transitions than for the FW transitions. Unfortunately, this value is dependent on the temperature, the iron content and even on the dispersity of the oxide. Nevertheless, if the requirements are not too severe, this value can be used to make estimates for the respective amounts on the basis of EPR spectra alone, provided the total amount of iron is known. (We can immediately make a check: the corrected intensities for the FW and EFW components are: 2036.3/1 (60.9%), 17 976.4/13.72 = 1309.3 (39.1%), i.e., the principle works, indeed.) Inspecting Fig. 24 one gets the impression that the contribution of the dF3c component to the resultant EPR spectrum with its intensity $I_3 = 447.1$ is not negligible. However, this Fe(III) is an EFW species in fully rhombic environment, thus increasing I_{EFW} by 447.1/13.73, the corresponding percentage of Fe(III) producing this dF3c line is 1.0%, which equals to 0.03 wt % iron, i.e., it is well within the cumulated experimental error and can be neglected.

3.4.2.3 The Nature of the Iron Species

Mössbauer spectra of the [Fe]-MFI sample (at 77 K) and the related parameters after decomposition are reported in Fig. 25 and Table 14, respectively. The spectrum of the as-synthesized sample exhibits a broad Fe^{3+} singlet, with a comparatively low isomer shift (IS) value. The singlet (lack of quadrupole splitting, QS) indicates a symmetric environment, whereas the low isomer shift ($IS_{77\,K} < 0.4$ mm s^{-1}) is indicative of tetrahedral coordination. Thus, this spectrum is a typical one characteristic for the isomorphously substituted ferric ions in the as-synthesized sample, prior to the removal of template molecules used in the synthesis. The activation of the sample results in partial removal of iron from the framework. Combined $(Fe, Al)_{framework}$–O–$(Fe, Al)_{extra\text{-}framework}$ pairs may be formed as reflected in the appearance of quadrupole splitting. (The symmetry is extended to a distorted octahedral one, as shown by the increase of the IS and QS values as well.)

During long-term catalytic tests in benzene hydroxylation and repeated activation-reaction cycles, change in the iron species is noted which can be interpreted as a migration of iron from the initial positions in the zeolite framework to more stable locations (framework to extra-framework migration). This change is indicated by a decrease of line width (FWHM). Presence of an Fe^{2+} component detected in minor amount (RI ~ 4%) in the sample after exposure to the reaction mixture attests that a reversible $Fe^{2+} \leftrightarrow Fe^{3+}$

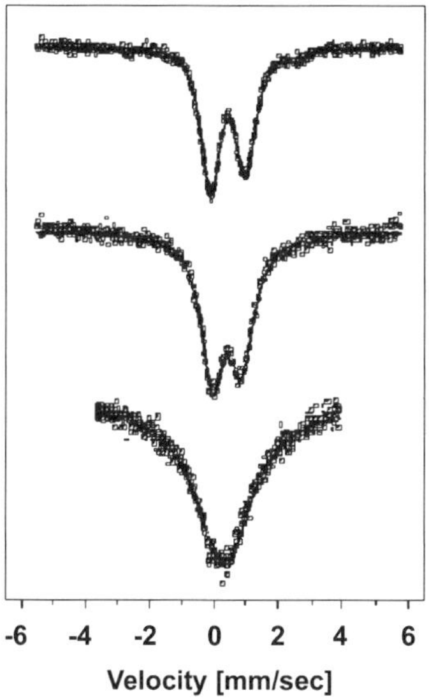

Fig. 25 77 K Mössbauer spectra of as-synthesized [Fe]-MFI sample (*bottom*); after three cycles of activation/reaction (*middle*); and exposed for 3 h of reaction (*top*)

Table 14 Mössbauer parameters extracted from – 196 °C spectra

Sample	Component	IS	QS	FWHM	RI
7 as-made	$Fe^{3+}{}_{tetr}$	0.33	—	2.10	100
7 after 3 h	$Fe^{3+}{}_{oct}$	0.45	1.08	0.68	96
of reaction	Fe^{2+}	1.10	3.04	0.34	4
7 after three cycles of reaction/regeneration	$Fe^{3+}{}_{oct}$	0.42	0.92	0.87	100

IS = isomer shift, related to α-iron, mm s^{-1};
QS = quadrupole splitting, mm s^{-1};
FWHM = full line width at half maximum, mm s^{-1};
RI = relative spectral contribution, %

cycle is involved in the reaction. It may be noted that the samples were exposed to air after being removed from the catalytic reactor and, therefore, partial reoxidation of Fe^{2+} probably occurred. More complete information is expected from in-situ Mössbauer measurements, but the detection of Fe^{2+} suggests that iron ions are reduced during the catalytic reaction.

A further aspect worth to be noted is the absence of magnetically split components in the spectra. This indicates that there is no superparamagnetic relaxation at 77 K, i.e., neither presence of extended oxidic (antiferromagnetic) Fe – O – Fe clusters, nor carbidic (ferromagnetic) Fe – C – Fe chains are detected. It may be noted that the threshold size of particles sufficient to exhibit a magnetic component is a few nanometers. Thus, high dispersion of iron ions, most probably close to ionic ones, is additionally proven.

3.4.3 Catalysis

Details concerning the method of preparation of the [Fe,Al]-MFI samples have been previously reported [247, 248]. Fe was introduced both during hydrothermal synthesis of the zeolite, or by different post-synthesis methods. The characteristics of the samples and the code used for indicating them throughout the text are summarized in Table 15. In the code, the subscript after the Fe symbol indicates the wt% iron content of the zeolite and the subscript after the formula indicates the method used to introduce the iron into the zeolite structure: I.E. stands for "ion exchange", CVD for "Chemical Vapor Deposition" (often indicated as sublimation of $FeCl_3$), and S.S.R.

Table 15 Characteristics of the [Fe,Al]-MFI catalysts used for the tests of benzene hydroxylation

Sample code#	Method of addition of Fe	Si/Al	Al/Fe	Fe (wt%)	Note
H-ZSM-5-C	—	13	≈ 200	< 0.05	Commercial samples from Alsi Penta (SN27)
H-MFI-S	—	> 1000	—	—	Pure Fe-free H-silicalite-1
$Fe_{1.1}MFI_{H.T.}$-S	hydrothermal synthesis	> 1000	—	1.1	Pure Al-free [Fe]-silicalite-1
$Fe_{1.1}MFI_{H.T.}$-A55	hydrothermal synthesis	55	1.48	1.1	[Fe,Al]-ZSM-5 sample
$Fe_{2.2}MFI_{H.T.}$-A54	hydrothermal synthesis	54	0.75	2.2	[Fe,Al]-ZSM-5 sample
$Fe_{2.3}MFI_{H.T.}$-A90	hydrothermal synthesis	90	0.91	2.2	[Fe,Al]-ZSM-5 sample
$Fe_{1.8}MFI_{I.E.}$-A13	ion exchange	13	3.71	1.8	Parent zeolite is
$Fe_{3.6}MFI_{I.E.}$-A13	ion exchange	13	1.85	3.6	commercial Alsi
$Fe_{0.4}MFI_{CVD}$-A13	CVD*	13	16.7	0.4	Penta (SN27)
$Fe_{0.8}MFI_{S.S.R.}$-A62	solid-state reaction	62	1.97	0.8	$FeCl_3$ is mixed with the zeolite and heated in air

* $FeCl_3$ sublimation
# C indicates commercial sample; S silicalite and A indicates the Si/Al ratio

for "solid-state reaction", while H.T. indicates that iron was directly introduced during hydrothermal synthesis. Post-synthesis introduction of iron was made, using as parent zeolite a commercial NH_4^+-MFI zeolite (from Alsi-Penta) with a SiO_2/Al_2O_3 ratio of 25. H-ZSM-5 was obtained from the same parent zeolite by calcination. The iron content in these samples is around 250 ppm. Usual practice was to calcine the dried catalysts at 550 °C in air.

Prior to the catalytic tests, the [Fe,Al]-MFI samples were activated in-situ at temperatures ranging from 600 to 700 °C in the presence or absence of steam. The catalytic tests were made in a fixed-bed reactor at a temperature typically of 400 °C feeding a mixture containing (i) 3 mol % benzene and 6 mol % N_2O in helium or (ii) 20 mol % benzene and 3 mol % N_2O in helium. Sometimes steam (up to about 3 mol %) was also added to the feed. The total flow rate was 3 L h^{-1} and the amount of catalyst 0.5 g (contact time of 0.6 s g ml^{-1}). The feed was prepared using a calibrated mixture of N_2O in helium and adding benzene using an infusion pump and a vaporizer chamber. The feed could be sent either to the reactor or to a bypass for its analysis. The feed coming out of the reactor or from the bypass could be sent to vent or to one of two parallel absorbers containing pure toluene as the solvent (plus calibrated amounts of tetrahydrofuran as the internal standard) cooled at about – 15 °C in order to condense all organic products. The line to the absorbers was heated at about 200 °C in order to prevent condensation of the products. The vent, after condensation of the organic products, was sent to a sampling valve for analysis of the residual gas composition. The reactor outlet stream was sent alternatively to the two parallel absorbers for a given time (typically 3 or 5 min), in order to monitor the change in the catalytic activity averaged over this time. N_2O, O_2, N_2 and total oxidation products (CO and CO_2) were analyzed using TCD-gas chromatography and a 60/80 Carboxen-1000 column, whereas benzene and phenol (as well as other minor aromatic by-products) were determined by FID-gas chromatography using a ECONO-CAP SE-30 "wide bore" column or a Mass-GC equipped with a capillary Chrompack CP-Sil 5CB-MS fused silica column.

3.4.4 Role of the Catalyst Composition

Reported in Fig. 26 is the catalytic behavior (productivity and selectivity to phenol in benzene hydroxylation with N_2O) as a function of time on stream over a series of [Fe,Al]-MFI-type catalysts prepared by hydrothermal synthesis. The characteristics and composition of these samples are reported in Table 15. The samples were selected in order to summarize the influence of catalyst composition on the activity and rate of deactivation. A commercial H-ZSM-5 sample (H-ZSM-5-C) showed good activity in agreement with literature data. However, it deactivated very quickly, and in about 1 h the productivity decreased by a factor of about 50. The selectivity to phenol progressively decreased from an initial value of around 65–70% to less than 10%. It should be noted that similar to most commercial samples, H-ZSM-5-C contains traces of iron (see Table 15) due to contamination during the industrial preparation. When a pure Fe-free silicalite-1 catalyst (H-MFI-S) was tested, no formation of phenol could be observed at all (Fig. 26). Silicalite-1 has the same MFI structure as ZSM-5 but is Al-free. The synthesis of silicalite leads to the presence of structural defects (hydroxyl nests), as confirmed by the presence of silanol groups evidenced by FTIR spectroscopy. Pretreatment of the sample at 700 °C in helium before the catalytic tests results in dehydroxylation of these hydroxyl nests forming oxygen vacancies, which may

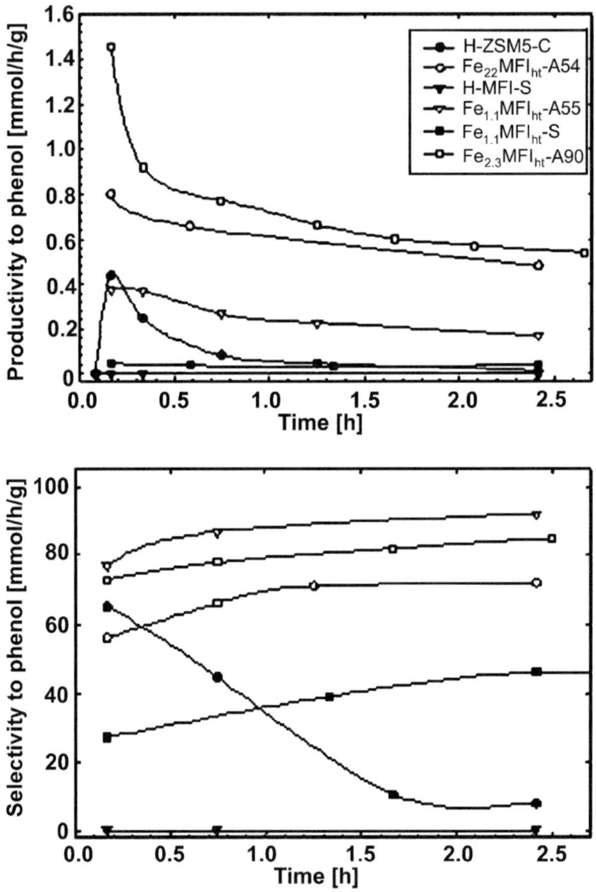

Fig. 26 Behavior of the catalysts as a function of the time on stream (TOS) with respect to the productivity and selectivity to phenol by benzene hydroxylation with N_2O. Catalysts pretreated at 700 °C in helium for 2 h. Reaction conditions: $T = 400$ °C, 3 vol% benzene, 6 vol% N_2O, balance He

activate N_2O, because they behave as F-centers. In agreement, the catalyst shows an initial activity in N_2O decomposition (around 22% N_2O conversion after 5–10 min of time on stream). This suggests that defects in the zeolite can activate N_2O, but are unable to either form active oxygen species in benzene hydroxylation and/or to activate the organic substrate (benzene) for this reaction.

When an Al-free [Fe]-silicalite ($Fe_{1.1}MFI_{H.T.}$-S) is used, the productivity to phenol is low but quite stable. No change in the productivity was observed up to about 20 h of time on stream. The selectivity to phenol is initially low and progressively increases up to final constant values of about 50%. The conversion of N_2O is low (about 5%) and nearly constant with increasing time on stream. Therefore, in the absence of Al the activity of the catalyst towards phenol synthesis is low, but remarkably stable, suggesting that Al sites may play a role in the synthesis of phenol, but also in the side-reactions leading to catalyst deactivation. When both Fe and Al are present in the zeolite ($Fe_{1.1}MFI_{H.T.}$-A55 and $Fe_{2.2}MFI_{H.T.}$-A54) the productivity to phenol increases by a factor of about 10 or 20, and the selectivity to phenol also markedly increases reaching values higher than 90% for $Fe_{1.1}MFI_{H.T.}$-A55. The productivity to phenol is approximately proportional to the amount of iron (compare phenol productivities for $Fe_{1.1}MFI_{H.T.}$-A55 and $Fe_{2.2}MFI_{H.T.}$-A54; the latter has twice the iron content and initially twice the productivity than that of the former, while the Si/Al content is the same). The selectivity to phenol, on the contrary, is higher in the case of the sample having lower iron content. The trend with time on stream is quite similar for the two catalysts. The productivity to phenol decreases, although much less dramatically than in the case of H-ZSM-5-C, while the selectivity to phenol increases. The conversion of N_2O is initially around 40% and then decreases to about 10–15% for $Fe_{1.1}MFI_{H.T.}$-A55. When this result is compared with that of the sample having a comparable iron content, but no aluminum ions ($Fe_{1.1}MFI_{H.T.}$-S), it is evident that the presence of Al ions (or probably better sites related to Al, such as Al-OH Brønsted acid) together with iron ions is a condition necessary for efficient activation of N_2O. However, sites for the decomposition of N_2O are also present which progressively become inactivated by the carbonaceous-type species formed on the catalyst, causing its deactivation. In fact, the rate of N_2O conversion in $Fe_{1.1}MFI_{H.T.}$-A55 decreased by a factor of about 6–8 during the 2.5 h of the experiments (Fig. 26), while the productivity to phenol decreased by a factor of about 2. The side decomposition of N_2O produces O_2 (which is also detected at the reactor outlet), and probably O_2 determines the total combustion of benzene, phenol or reaction intermediates and thus the formation of CO_2 (CO is usually observed only in traces). This explains why in parallel to the decrease in N_2O conversion, an increase in the selectivity to phenol is observed. With increasing iron content ($Fe_{2.2}MFI_{H.T.}$-A54) the initial conversion of N_2O is about 65% and then decreases to about 20–25% after 2.5 h of time on stream. This indicates that the increase in iron content increases the number of active sites for phenol synthesis (the productivity is almost doubled), but probably also increases the formation of a second type of iron species. These species are probably small iron-oxide particles within the zeolite cavities formed in the process of partial migration of iron from framework to non-framework positions during the initial catalyst pretreatment. The increased amount of iron favors the aggregation of these iron species and thus, a higher amount of aggregated iron oxide is thought to be responsible for the decomposition of N_2O to $N_2 + O_2$ instead of $N_2 + \alpha$-O. As a consequence, the productivity of phenol in $Fe_{2.2}MFI_{H.T.}$-A54 is higher than in $Fe_{1.1}MFI_{H.T.}$-A55, but the selectivity to phenol is lower. Decreasing the amount of Al, while maintaining the amount of iron (compare $Fe_{2.3}MFI_{H.T.}$-A90 with $Fe_{2.2}MFI_{H.T.}$-A54) constant leads to an increase in both the productivity and selectivity, although the general trend remains again similar. This indicates that probably an Al/Fe ratio close to 1 is the optimal compromise to maximize both productivity and selectivity to phenol, suggesting that the active sites for phenol synthesis comprise both Al and Fe.

3.4.5 Role of Methodology in Iron Introduction in [Fe]-MFI Catalysts

In order to analyze in a concise way the effect of different methods of addition of iron into [Al]-MFI samples on the catalytic performances (activity and deactivation), catalytic data have been summarized in Table 16. The following parameters are reported: (i) the maximum observed phenol productivity, which is generally obtained at the beginning of the reaction or after about 20 min of time on stream (see Fig. 26), (ii) phenol productivity after 2 h, indicating catalyst stability, and (iii) the selectivity in correspondence with the maximum phenol productivity and after 2 h time on stream. Table 16 compares the behavior of some selected samples in which Fe was introduced post-synthetically by (i) an ion-exchange method ($Fe_{1.8}MFI_{I.E.}$-A13 and $Fe_{3.6}MFI_{I.E.}$-A13) using an aqueous solution of iron-ammonium-sulfate, (ii) CVD (contacting the dehydrated zeolite at 300 °C with a flow of $FeCl_3$ in N_2) and (iii) solid-state reaction ($FeCl_3$ and the zeolite are mixed homogeneously and heated to 400 °C). For better comparison, data for two samples prepared by the hydrothermal method and containing both Fe and Al ($Fe_{1.1}MFI_{H.T.}$-A55 and $Fe_{2.2}MFI_{H.T.}$-A54) as well as data for the parent zeolite used for ion exchange and CVD (H-ZSM-5-C) are also reported.

The post-synthetic introduction of iron in the Fe-[Al]-MFI catalysts generally leads to lower selectivities and productivities to phenol, although in some cases (e.g., $Fe_{3.6}MFI_{I.E.}$-A13) good initial behavior is obtained. With respect to the behavior of the parent zeolite (H-ZSM-5-C), the productivity to phenol increases more than twice, and an increase in the selectivity to phenol could also be noted. However, similarly to H-ZSM-5-C, after 1 h the productivity to phenol becomes negligible. When the iron content is lower,

Table 16 Phenol productivity and selectivity for [Fe,Al]-MFI samples in which iron was introduced using different methods. Before benzene oxidation the catalysts were pretreated at 700 °C in helium for two hours. Reaction conditions as described in the text

Sample	Phenol productivity Max [1]	Phenol productivity 2 h [2]	Selectivity [3] Max [3]	Selectivity [3] 2 h [3]
$Fe_{1.1}MFI_{H.T.}$-A54	0.38	0.18	81	92
$Fe_{2.2}MFI_{H.T.}$-A55	0.81	0.62	56	69
H-ZSM-5-C	0.44	0.02	61	11
$Fe_{1.8}MFI_{I.E.}$-A13	0.21	0.01	28	12
$Fe_{3.6}MFI_{I.E.}$-A13	1.03	0 [a]	78	17 [a]
$Fe_{0.4}MFI_{CVD}$-A13	0.29	0.13 [b]	38	12
$Fe_{0.8}MFI_{S.S.R.}$-A62	0.04	0 [a]	7	1

[1] Maximum phenol productivity (mmol h^{-1} g^{-1}) at 400 °C
[2] Phenol productivity (mmol h^{-1} g^{-1}) at 400 °C after 2 h of reaction
[3] Phenol selectivity (%) at 400 °C expressed on the total product basis in correspondence to the maximum productivity and after 2 h of reaction.
[a] The catalyst is completely deactivated just after 1 h of reaction. The selectivity corresponds therefore to a reaction time of 50 min
[b] The productivity to phenol does not further decrease in tests up to 20 h of time on stream

much poorer performances were observed, because of the higher rate of side decomposition of N_2O. Therefore, the addition of iron by ion exchange could lead to catalysts with an initial activity better than that of the parent zeolite, but the effect of fast deactivation discussed before did not change. The addition of iron by CVD leads to not very selective catalysts due to a high rate of N_2O decomposition, but stable activity was noted after an initial decrease in phenol productivity. According to the previous characterization, the CVD (sublimation) method leads to the formation of isolated or binuclear iron species [249], probably with characteristics similar to those of the active species responsible for the generation of the selective α-O species during the interaction with N_2O. However, especially when the feed or the zeolite is not fully dehydrated, nanoparticles of iron oxides also form [250] being responsible for the decomposition of N_2O and lowering selectivity. In agreement, the sample prepared by solid-state reaction—in which the formation of the latter species is enhanced—leads to very selective and active catalysts in phenol formation.

3.5 [Fe]-BEA

Using the reagents shown in the experimental part the synthesis of [Fe]-BEA proposed in [158] could not be reproduced. Indeed, starting from a gel composition of $40SiO_2 - 1.02Fe(NO_3)_3 \cdot 9H_2O - 19.04TEAOH - 4NaOH - 676H_2O$ using 24 h ageing time only an amorphous phase was obtained at 120 °C after 12 days. Similarly unsuccessful was the attempt when the ageing was done at 273 °C for 3 h. The reaction temperature of 150 °C did not lead to any [Fe]-BEA zeolite. Finally, even if a solution of 25% TEAOH in methanol was used, no [Fe]-BEA zeolite could be obtained following the method proposed in [159]. In order to find a well-reproducible method using our reagents, we have systematically varied the amount of iron source, the amount of TEAOH, the ageing time of the gel, the reaction time and the temperature of the reaction [194]. The general composition of the gel was the following: $40SiO_2 - xFe(NO_3)_3 \cdot 9H_2O - yTEAOH - 4NaOH - 676H_2O$ with $x = 0.38$, 0.49, 0.51, $y = 19.04$, ageing time = 2, 18, or 24 h, reaction time = 7, 9, 10, 20, 21, 22, 26, 28, or 30 days. Only one synthesis led to [Fe]-Beta, with $x = 0.49$, $y = 19.04$, 24 h ageing and 20 days of synthesis time at 120 °C. Even this synthesis was not reproducible. Note that using fumed silica instead of TEOS as silica source, some unidentified layered compounds were obtained. For $x = 0.45$, 0.49 and 0.60 and $y = 10.88$, 13.6 and 16.3, ageing time = 2 or 24 h, reaction time = 16, 18, 20, 21, 22, 25, 29 or 40 days, $T = 120$ °C, [Fe]-BEA cocrystallized in most cases with an unknown phase having a diffraction peak at $2\theta = 5.6°$. Note that a similar peak was obtained during the synthesis of Beta zeolite with low Al-content [251]. This peak disappears during calcination at 450 °C. As the reaction temperature of 120 °C was not adequate to obtain pure [Fe]-BEA in a reproducible way, the reaction temperature was raised to 150 °C. At this temperature the amounts of $Fe(NO_3)_3 \cdot 9H_2O$ and TEAOH were also varied in order to optimize the synthesis conditions. The data are reported in Table 17.

It is clearly seen from Table 17 that under certain conditions pure [Fe]-BEA can be obtained in a reproducible manner at 150 °C. The conditions are 0.45 or 0.60 $Fe(NO_3)_3 \cdot 9H_2O$, 13.6 or 16.3 TEAOH, 24-h ageing time and 4–8 days reaction time. The narrow crystallization fields are reported in Fig. 27, where the crystallization field of [Fe]-BEA is surrounded at higher Fe content by a phase where [Fe]-BEA coexists with an unknown phase U. Note that at low

Table 17 Synthesis conditions for the [Fe]-BEA obtained from gels of composition $40\,SiO_2 - xFe(NO_3)_3 \cdot 9H_2O - yTEAOH - 4NaOH - 676H_2O$ at 150 °C[a]

Sample	x	y	Reaction time [days]	Product
7	0.60	16.3	4, 5, 6, 7, 8	[Fe]-BEA
8	0.45	16.3	8	[Fe]-BEA
10	0.45	13.6	8	[Fe]-BEA

[a] Ageing time: 24 h;
[b] Unidentified phase

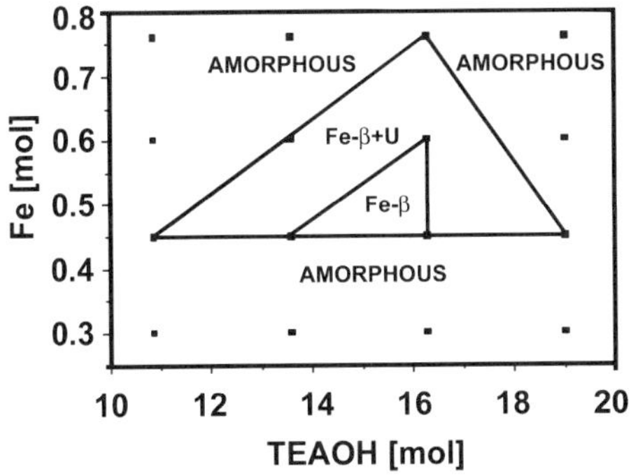

Fig. 27 Crystallization fields of [Fe]-BEA (β = BEA) zeolite from gels of composition $40SiO_2 - xFe(NO_3)_3 \cdot 9H_2O - y$TEAOH – 4NaOH – $676H_2O$ at 150 °C, aging time 24 h

ageing time of the gels (2 h), only an amorphous phase was obtained in all cases. The white color of all the final crystalline [Fe]-BEA zeolite samples suggests that Fe(III) occupies tetrahedral framework sites in the structure. In the zone of crystallization of [Fe]-BEA—i.e., 0.60 $Fe(NO_3)_3 \cdot 9H_2O$ and 16.3 TEAOH—[Fe,Al]-BEA and [Al]-BEA were also synthesized maintaining the sum of moles of iron and aluminum source equal to 0.6. The data are reported in Table 18. It is seen that only the crystalline phases [Fe,Al]-BEA or [Al]-BEA were obtained in all synthetic runs already at 3 or 4 days of crystallization time. These results reinforce the existence of the zone where pure [Fe]-BEA zeolite samples could be obtained.

The [Fe]-BEA zeolite was also obtained in the presence of oxalic acid (see experimental part). In Table 19 the results on the synthesis of BEA zeolite-type are reported. First, when the reaction temperature increases from 140 to 170 °C, the crystallization time is shortened, but the thermal stability of the product is decreased. As a matter of fact, from the reaction system B2 the BEA-type zeolite was obtained after 7 days at 140 °C and after 4 days at 170 °C, but if the reaction time was prolonged to 8 days at 170 °C, the products were BEA and quartz. On the contrary, at 140 °C only the BEA phase was

Table 18 Synthesis conditions for [Fe,Al]- and [Al]-BEA obtained from gels of composition $40SiO_2 - xFe(NO_3)_3 \cdot 9H_2O - zAl(OH)_3$ – 16.3TEAOH – 4NaOH – $676H_2O$ at 150 °C[a]

Sample	x	y	Reaction time [days]	Product
13	0.30	0.30	4	[Fe,Al]-BEA
14	0.18	0.42	4, 5, 6	[Fe,Al]-BEA
15	0.06	0.54	3, 4, 5	[Fe,Al]-BEA
16	0	0.60	4, 5, 6	[Al]-BEA

[a] Aging time: 24 h

Table 19 Products obtained from the starting hydrogel B (see experimental)*

Sample	x	y	z	Time [days]	Products	Si/Al_{zeo}	Si/Fe_{zeo}
B2	0.10	0.02	—	7	BEA	23.7	—
B3	0.10	0.02	0.005	10	BEA	28.1	59.4
B4	0.10	0.02	0.010	18	BEA	28.5	38.9

* Moles of reactants per mol of SiO_2

formed even after 18 days. The BEA-type zeolite is also obtained in the presence of a mixture of TEAOH and TEABr, but the preliminary experiments carried out with this mixture showed that for a complete crystallization it is necessary to double the crystallization time.

As in the case of the MOR-type zeolite (see below), it was not possible to obtain the Al-free (or [Fe]-BEA) starting from the studied hydrogel system. Consequently, the experiments for the iron introduction were carried out at the lowest Al content in the hydrogel (equal to 0.02). First, it must be noted that the pH values play an important role in the formation of BEA-type zeolite. As in the case of the MOR-type zeolite, the pH value of the starting hydrogel must be higher than 13. The crystallization field of Fe, [Al]-BEA is more limited than that of the corresponding form of the MOR zeolite. In fact, the iron-containing [Al]-BEA zeolite crystallizes from the hydrogels having a Si/Fe ratio ranging from 100 to 50. Higher amounts of iron in the initial reaction mixture (Si/Fe = 33.3) did not lead to crystallization of BEA, in fact, after long reaction times the only detected crystalline phase was the mordenite. The crystallization time is a function of the iron content in the starting hydrogel. As a matter of fact, it increases with the iron content, e.g., the reaction time was 7 days in absence of iron, (sample B2) and 18 days with (Si/Fe = 50, sample B4).

3.6
[Fe]-MOR

At the beginning it is useful to underline that an unsuccessful attempt to obtain a pure iron form of BEA- and MOR-type zeolites was undertaken, but evidently the absence of aluminum in the initial hydrogel does not allow the formation of crystalline phases even after long reaction times.

In Table 20 the results of the synthesis of the [Al]-MOR and [Al,Fe]-MOR zeolites are shown [195]. As expected, the amount of Al in the starting hydrogel is the crucial parameter during the crystallization process. Indeed, the zeolite MOR was obtained from the mixtures with a Si/Al ratio lower than 33.3. The alkali content influences the nature of the products. Starting from low Al content, the dense phase (quartz) co-crystallizes with a zeolitic phase

Table 20 Products obtained starting from the hydrogel A (see experimental)*

Sample	Na_2O	Al_2O_3	Fe_2O_3	Time [days]	Products	Si/Al_{zeo}	Si/Fe_{zeo}
M4	0.15	0.02	—	2.5	MOR	24.1	—
M5	0.27	0.02	—	2.5	MOR	—	—
M6	0.15	0.025	—	4	MOR	—	—
M7	0.15	0.02	0.005	5	MOR	24.7	59.1
M8	0.15	0.02	0.010	8	MOR	24.9	37.5
M9	0.17	0.02	0.015	11	MOR	23.9	26.6

* Moles of reactants per mol of SiO_2

changing into zeolite BEA when the amount of sodium hydroxide in the starting hydrogel decreased, and a layered phase appeared when the lowest alkali content was tested. This confirms that a large amount of sodium hydroxide is necessary for the synthesis of MOR-type zeolite [252]. In any case, for the formation of [Al]- or [Fe,Al]-MOR zeolite the pH value of the starting hydrogel must be higher than 13. In the presence of iron only, this reaction system leads to the formation of quartz and in the presence of aluminum (Al_2O_3) ranging from 0.005 to 0.015 no crystalline phases were detected even after long reaction times. Since with this hydrogel system it was impossible to obtain mordenite in the absence of aluminum, the iron incorporation was tested in the system containing the lowest amount of Al_2O_3 (equal to 0.02). The maximum amount of iron that could be introduced into the zeolite was equal to 0.015 Fe_2O_3 (Si/Fe = 26.6). The crystallization time became longer when the iron content in the starting hydrogel increased. In the system without iron (sample M4), MOR crystallized in 2.5 days, on the contrary for a higher Si/Fe ratio (equal to 26.6, sample M9) 11 days were required. Only for the highest iron loading it was necessary to increase the amount of sodium hydroxide in order to maintain the pH of the hydrogel higher than 13, the value which was previously found to be essential to obtain pure MOR-type zeolite.

3.7 [Fe]-TON, [Fe]-MTW

[Fe]-MTW and [Fe]-TON zeolites were also synthesized [189]. For [Fe]-MTW the initial hydrogel composition was: $x\mathrm{Na_2O} - 0.2\mathrm{MTEABr} - \mathrm{SiO_2} - q\mathrm{Fe_2O_3} - p\mathrm{C_2H_2O_4} - 20\mathrm{H_2O}$ where $x = 0.1–0.16$; $q = 0.005–0.02$ and the ratio $p/q = 3$. MTEABr stands for methyltriethylammonium bromide. [Fe]-TON was obtained from gels: $x\mathrm{Na_2O} - 15\mathrm{CH_3OH} - \mathrm{SiO_2} - q\mathrm{Fe_2O_3} - p\mathrm{C_2H_2O_4}$.

The effects of Si/Fe ratio, pH values and crystallization time on the nature of the products are reported in Table 21. The results show that it is possible to obtain Fe-zeolites in a wide range of OH^-/SiO_2 and Si/Fe ratios using the iron-oxalate complex. The various types of zeolite show different behavior regarding the crystallization process.

Due to the observation that the nature of iron complex does not affect the nature of the crystalline products and the reaction time, for the syntheses of the [Fe]-MTW and [Fe]-TON zeolites only oxalic acid is used. In the case of the [Fe]-MTW zeolite it can be observed that it crystallizes in a narrower range of Si/Fe ratios, compared to the [Fe]-MFI zeolite. The lower limit of the Si/Fe ratio is about 30, as a matter of fact for a lower ratio (25) no crystalline products are found after a very long reaction time (32 days). Another observation is that in this kind of synthesis the pH value plays an important role. The [Fe]-MTW crystallizes in this system in a very narrow range of the pH value. For pH value greater than 12 the MTW-type zeolite co-crystallizes with other crystalline phases such as magadiite. For a pH value smaller than 11.5 (sample C4 in Table 21) no crystalline product is found after a long reaction time (27 days). The [Fe]-TON-type zeolite crystallizes in a range of the Si/Fe ratio larger than the range of [Fe]-MTW zeolite but more restricted than for the [Fe]-MFI zeolite. In fact, it is possible to obtain [Fe]-TON for a ratio of Si/Fe = 25, for lower ratio the formation of the [Fe]-TON zeolite is not completed even after a long synthesis time (20 days). Similar to [Fe]-MTW zeolite crystallization, the pH values play an important role in the formation of [Fe]-TON. As a matter of fact, higher pH values—higher than 12—lead to the co-crystallization of TON with a dense phase such as cristobalite. Lower pH values, smaller than 11, do not lead to the formation of any crystalline phases even after long reaction times (22 days). All the samples of iron zeolites, MFI-, MTW- and TON-type, show white color even after overnight calcination at 600 °C. This is an indication that no extra-framework iron is present in the

Table 21 Influence of Si/Fe ratio, sodium hydroxide content and pH value of the initial reaction mixture on the nature of the final products

Sample	Si/Fe_{gel}	Na_2O	pH_{gel}	Time [days]	Product
C1	100	0.10	11.8	6	MTW
C2	50	0.15	12.2	6	MTW+magadiite
C3	50	0.11	11.7	9	MTW
C4	33.3	0.13	11.3	27	amorphous
C5	33.3	0.15	11.6	21	MTW
C6	25	0.16	11.7	32	amorphous
D1	100	0.07	12.2	8	TON+cristobalite
D2	100	0.06	11.5	8	TON
D3	66.6	0.08	11.8	8	TON
D4	33.3	0.14	11.5	11	TON
D5	25	0.18	11.6	11	TON
D6	20	0.21	11.5	20	TON+amorphous

Table 22 Chemical analysis, morphology and crystal dimensions of MFI, MTW and TON zeolites

Sample	Zeolite	Si/Fe_{gel}	$Si/Fe_{zeolite}$
C7	MTW	200	210.7
C1	MTW	100	102.5
C3	MTW	50	47.4
C5	MTW	33.3	31.5
D2	TON	100	132.0
D3	TON	66.6	67.6
D4	TON	33.3	35.4
D5	TON	25	27.2

zeolite samples. The results of the sample characterizations are reported in Table 22. It can be observed that for the two zeolite types the incorporation of iron is related to its content in the starting hydrogel. Hence, with this preparation procedure it is possible to incorporate different amounts of iron related to the initial reaction mixture. The only limitation in the incorporation is the range of the Si/Fe ratio that each type of zeolite is able to accommodate in its framework as described in the comments of Table 21. The differences between the three zeolite types are also in the morphology and in the crystal size. The [Fe]-MFI crystals show spherical morphology with inhomogeneous crystal size distribution varying from 2 to 20 μm. The [Fe]-MTW samples present rice-like shape morphology with a length of about 1.6 μm and a width of about 1 μm, and finally, the [Fe]-TON-type crystals show needle agglomerates with a linear dimension smaller than 1 μm. In all cases the iron zeolites present morphologies similar to those of the corresponding Al forms.

3.8
[Fe,Al]-MCM-22

The X-ray powder diffraction shows clearly that the as-synthesized solid has the structure of MCM-22 zeolite (Fig. 28). Tables 23 and 24 show the comparison between the line positions and the relative intensities for the sample [Fe,Al]-MCM-22 and the literature data [253]. Before calcination most lines are broad and overlap. The calcined [Fe,Al]-MCM-22 sample also shows the characteristics of calcined MCM-22 [256] (Fig. 28 and Tables 23 and 24).

SEM photographs of the as-synthesized [Fe,Al]-MCM-22 samples 2, 3 and 5 are shown in Fig. 29. They are very similar to those reported for the MCM-22 samples [254, 255]. The thin platelets of ca. 1–2 μm diameter frequently form spherical aggregates of ca. 10–15 μm, and the morphology does not depend on the Fe content. Calcination of the samples did not affect the morphology of the crystals.

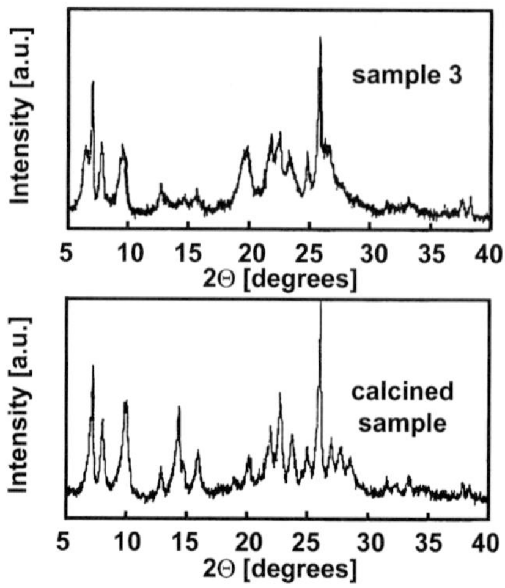

Fig. 28 XRD patterns of the as-made [Fe,Al]-MCM-22 sample 3 and of the calcined [Fe]-MCM-22 sample 5

Table 23 Comparison between X-ray powder diffraction patterns of as-made sample 3 and precursor MCM-22

As-made d(hkl)	I_{rel}	MCM-22 [a] d(hkl)	I_{rel}
13.63	57.6	13.53	36
12.51	70.9	12.38	100
11.30	51.9	11.13	34
9.17	48.4	9.14	20
6.94	34.7	6.89	6
6.76	29.7	6.68	4
6.08	28.3	6.02	5
4.44	53.5	4.47	22
4.08	57.4	4.05	19
3.95	58.8	3.95	21
3.79	48.0	3.77	13
3.58	48.2	3.57	20
3.44	100	3.43	55
3.39	53.8	3.36	23
3.07	31.2	3.06	4
2.69	29.8	2.68	5
2.38	29.7	2.38	5
2.34	27.3	—	—

[a] From [160]

Table 24 Comparison between X-ray powder diffraction patterns of calcined sample 5 and calcined MCM-22

Calcined d(hkl)	I_{rel}	MCM-22 calcined [a] d(hkl)	I_{rel}
12.34	71.0	12.34	100
11.07	50.3	11.02	51
8.85	58.5	8.90	46
6.16	55.8	6.17	35
5.54	37.7	5.55	13
4.38	36.5	4.38	10
4.10	39.9	4.11	10
4.05	47.6	4.06	20
3.91	61.3	3.82	29
3.75	45.9	3.75	15
3.56	40.4	3.56	14
3.42	100	3.42	61
3.30	43.6	3.30	13
3.21	40.5	3.21	9
3.06	28.2	3.12	6
2.68	28.4	2.68	6
2.37	26.4	2.37	4
2.34	25.2	—	—

[a] From [160]

(a) (b) (c)

Fig. 29 Scanning electron micrographs of [Fe, Al]-MCM-22 samples 2 (**a**) and 3 (**b**) and of [Fe]-MCM-22 sample 5 (**c**)

The iron content, aluminum and sodium determined by atomic absorption and the content of the organic molecule trapped in the zeolite channels determined by thermogravimetric analysis are reported in Table 25.

The values are expressed in moles per 100 g of activated samples, i.e., the sample from which the water and the organic molecules were removed. The sum of iron and aluminum incorporated into the MCM-22 structure is quite constant and the Fe/Al ratio is always larger than in the reaction mixture (4.6, 1.3 and 0.5 for samples 2, 3 and 4, respectively). The sodium content is

Table 25 Chemical analysis of the as-made [Fe,Al]-MCM-22 and [Fe]-MCM-22 samples[a,b] and ion exchange capacities of the corresponding calcined samples

Sample	Fe[a] [$\times 10^2$ mol%]	Al[a] [$\times 10^2$ mol%]	Si/ Al+Fe	Na[a] [$\times 10^2$ mol%]	K/ Al+Fe	Fe_{tetr} [%]	HMI[b] [$\times 10^2$ mol%]	DSC peaks (°C)	HMI[b] [$\times 10^2$ mol%]
1	7.8	—	19.9	0.28	0.40	40	19	377.6	8
								466.3	11
2	7.4	1.6	17.1	0.56	0.48	36	20	384.1	10
								479.4	10
3	5.3	4.1	16.4	0.91	0.71	49	20	385.7	9
								490.8	11
4	3.5	6.3	15.8	0.30	0.83	80	20	384.4	11
								502.8	9
5	9.3	—	16.5	0.69	0	0	20	376.7	8
								464.6	12

[a] Atomic absorption spectroscopy
[b] Thermal gravimetry

very low compared to the trivalent atom content. The amount of HMI is equal to 0.20 mol % and is not dependent on the trivalent atoms in the structure. As the amount of (Fe + Al) is equal to ca. 0.09 mol % and the Na amount is much smaller than this value (it varies from 0.003 to 0.009 mol %) (Table 25), some 0.081 mol % of protonated HMI has to neutralize the framework negative charges linked to both Al and Fe (see below).

The two different HMI species cannot be shown by the ^{13}C NMR spectrum, because it is not well resolved due to the presence of the paramagnetic Fe(III) in the framework. Note that only the ^{13}C NMR spectrum of sample 4 containing a low amount of Fe could be measured. The Al atoms are well introduced into the structure in a tetrahedral form. Indeed, the chemical shift of $\sim$ 54 ppm characterizes tetrahedral species of $Al(OSi)_4$ configuration (Table 26). It is not possible to detect different Al species in the structure, as was the case for MCM-22 zeolite [256, 257], because of the line broadening due to the Fe(III) species, although some asymmetry can be recognized in the spectra.

The ^{27}Al NMR linewidth increases with increasing Fe(III) content in the samples (Fig. 30). The linewidths do not differ very much for the calcined samples (Table 26). However, for samples 2 and 3 with higher Fe contents a new broad NMR line appears at ca. 30 ppm, which could be either due to extra-framework Al species (very much broadened due to extra-framework Fe species), or due to deformed Al species in the structure. Work is in progress to identify the various Al species in the [Fe]-MCM-22 samples. While the Fe/u.c. value increases from 1.7 to 4.5, that of Al/u.c. decreases from 3.0 to 0.8 and their sum remains constant for the different samples. This means that the Fe species and the Al species are in competition and the introduction of a total of 4.3 = (Fe + Al)/u.c. is favored under our reaction conditions.

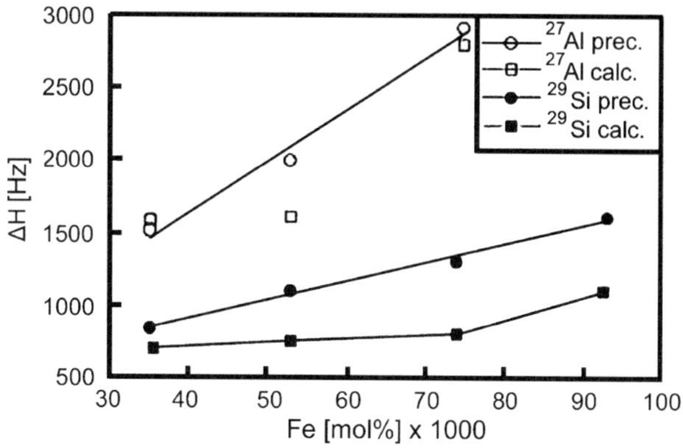

Fig. 30 Variation of the ^{27}Al and ^{29}Si NMR linewidths as a function of Fe content of the [Fe,Al]-MCM-22 samples

Table 26 ^{27}Al NMR data of precursor [Fe]-MCM-22 samples (a) and calcined [Fe]-MCM-22 samples (b)

Sample	(a) δ [ppm]	ΔH [Hz]	(b) δ [ppm]	ΔH [Hz]
1	—	—	—	—
2	54	2930	55.6 ($\sim$ 30)	2800 (7000)
3	53.4	2000	55.6 ($\sim$ 30)	1600 (7000)
4	55.0	1520	55.0	1550
5	—	—	—	—

EPR results were used to characterize the Fe species in the zeolite samples. One typical spectrum (sample 3) is recorded in Fig. 31, and Table 27 shows the g values of the various Fe-containing species. Despite the great importance of the EPR technique in characterizing the various paramagnetic species, no clear-cut interpretation has been given until now. For example, species characterized by a g factor of 2.0 are considered octahedral [258, 259], while such species are described as tetrahedral framework species in [260]. The signal at $g = 2.3$ is considered as the $g_{||}$ contribution of the octahedral species [258, 259]. Finally, at $g = 4.3$ can be found the tetrahedral framework species, the intensity of which is higher at low temperatures [259]. This

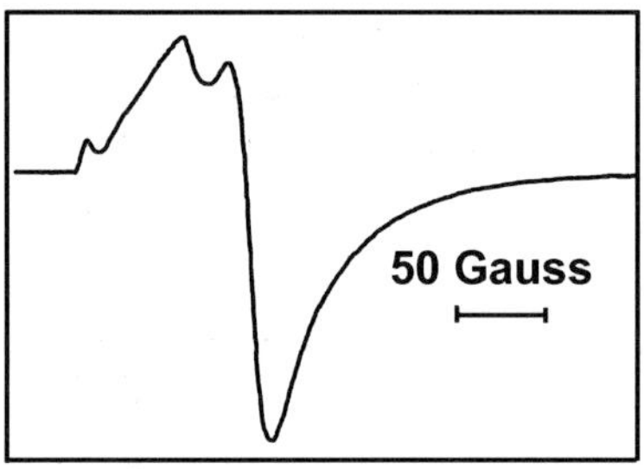

Fig. 31 Room temperature EPR spectrum of as-made [Fe,Al]-MCM-22 sample 3

Table 27 EPR results of as-made [Fe,Al]-MCM-22 zeolite samples taken at room temperature

Sample	g values		
1	4.263	2.307	1.985
2	4.559	2.440	2.017
3	4.497	2.449	2.011
4	4.381	2.437	2.012
5	4.446	2.451	2.169

species can be included in the EPR signal observed at low magnetic field, where only the middle of the spectrum was computed to be between 4.23 and 4.56. The latter species are considered as deformed tetrahedral species. However, the correct interpretation is the one given above for the [Fe]-MFI samples. That way, the iron species with $g \approx 2.0$ is attributed to tetrahedral framework iron, and the $g \approx 4.3$ value is characteristic of iron in rhombic configuration. The Mössbauer spectra of the as-made [Fe]-MCM-22 samples confirm the presence of tetrahedral Fe(III) ions. The spectra of samples 1 and 5 were decomposed by attempting to use two doublets, one for the tetrahedral configuration of Fe and one for the octahedral configuration. The isomer shift of tetrahedrally coordinated Fe(III) is equal to 0.18 mm s^{-1} for both samples. Octahedrally coordinated Fe(III) is detected in sample 5 and is characterized by higher isomer shift (0.32 mm s^{-1}) and quadrupolar splitting (0.93 mm s^{-1}). These values are in good agreement with those reported in the first genuine publication by Meagher et al. [261] and later confirmed by others [262, 263]. It is interesting to note that despite the brownish color of sample 5, the amount of tetrahedral Fe(III) is still high (some 87%).

3.9
[Co]-MFI

Figure 32 illustrates the XRD patterns of three samples (see Table 28) [196, 197]. It can be seen that sample 1 shows a well-crystallized MFI structure

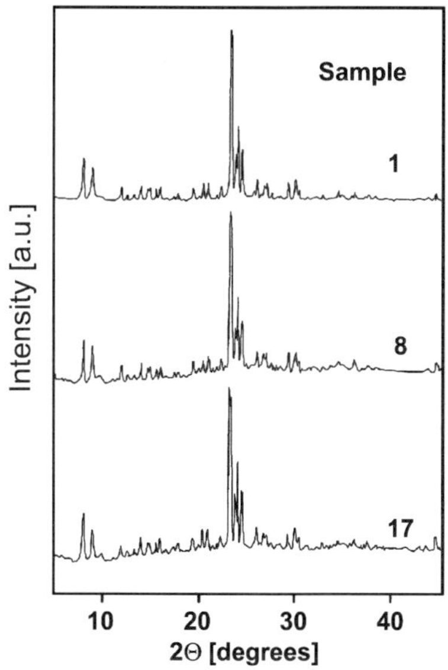

Fig. 32 XRD patterns of samples Nos. 1, 8 and 17 (see Table 28)

Table 28 Chemical composition of the as-made Co-containing MFI samples. a) Samples obtained from gels A of composition $35SiO_2 - xNa_2O - yCo(CH_3COO)_2 - 3.4TPABr - 8.4H_2OSO_4 - 808H_2O$ at 170 °C after two days of synthesis. b) Samples obtained from gels B of composition: $35SiO_2 - Co(CH_3COO)_2 - xNa_2O - 3.4TPABr - 808H_2O$ at 170 °C after two days of synthesis

No	x	y	Co/u.c.	Na/u.c.	H_2O/u.c.	TPA/u.c.	TPA/u.c. – T [°C] LT[a]	HT[b]
a)								
1	11	0.5	2.7	8.3	0	3.6	1.8–402	1.8–456
3	11	1.5	5.0	18.7	3.5	3.7	2.0–398	1.7–468
5	12.9	0.5	6.3	12.9	3.5	4.0	2.1–407	1.9–456
8	12.9	2	12.7	16.5	7.0	3.4	1.8–398	1.6–456
9	15	0.5	7.0	18.3	3.0	3.7	3.7–406 (450[c])	
12	15	2	10	15.6	8.3	3.1	3.1–406 (450[c])	
b)								
14	3		2.7	4.3	5	4.2	1.9–404	2.3–438
15	6		5.6	10	10.8	3.8	1.8–406	2.0–459
16	9		7.2	8.4	11.2	4.0	1.7–410	2.3–459
17	12		6.7	8.4	11.4	3.6	3.6–451	

[a] LT: low temperature peak;
[b] HT: high temperature peak;
[c] shoulder

of orthorhombic symmetry, while sample 8 shows the presence of some amorphous phase as well. In the first series A, where the [Co]-MFI samples were obtained from a sodium silicate source, the crystallinity of the samples decreases both with the increase of alkalinity (x) and the Co-content (y) in the gel. The decrease in crystallinity with increasing alkalinity is also observed in series B, where colloidal silica was used as a silica source.

Note that high alkalinity is required for the [Co]-ZSM-5 synthesis. Indeed, when no NaOH was added to the reaction mixture, no crystalline material was obtained after 2 days of synthesis. In addition, the initial gel was pink showing the presence of octahedrally coordinated Co, while for the gels prepared with high Na_2O content, the gels were all blue having the characteristic color of $Co(OH)_4^-$ ions. The chemical analysis by atomic absorption spectroscopy (AAS) yielded the values of Co/u.c. and Na/u.c., while thermal analysis led to the TPA/u.c. and H_2O/u.c. values (Table 28). The Co/u.c. values are rather high and suggest that the larger part of the cobalt is in extra-framework positions. These values increase with increasing Co-content of the gel, and they also increase with increasing alkalinity (Table 28a and b). The Na/u.c. values are also rather high and suggest that most of the Co(II) ions are in extra-framework species. The H_2O/u.c. varies in a random manner and does not seem to be linked to any Co/u.c. or Na/u.c. variation. The DSC

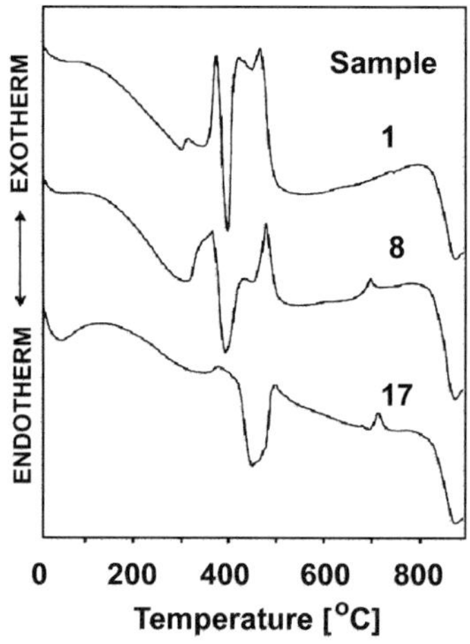

Fig. 33 DSC curves of samples Nos. 1, 8, and 17 (see Table 28)

results are very revealing with respect to the possible incorporation of Co into the tetrahedral framework. Figure 33 illustrates the DSC curves of three samples. When the framework Co content is low (sample 1), two peaks characterize the decomposition of the occluded TPA^+ ions: the low temperature peak (LT) at ca. 400 °C and the high temperature peak (HT) at ca. 455 °C. The temperature of decomposition is not influenced by the formal Co/u.c. values. The TPA/u.c. does not vary much as a function of Co/u.c. and is close to an average value of 3.7/u.c. Approximately half of it is decomposed at low temperature and the other half at high temperature. Only at high Co content (samples No. 8 and 12), there is some decrease in the TPA/u.c. values suggesting the incorporation of some of Co into the samples. At high alkalinities of the samples of series B (No. 17), only the high temperature peak is detected, suggesting again that some of the Co could be incorporated into the structure. Indeed, this behavior of the TPA^+ ions was observed previously in the [Al]-ZSM-5 [226, 265], B-ZSM-5 (Testa F, Chiappetta R, Crea F, Aiello R, Fonseca A, Bertrand JC, Demortier G, Guth JL, Delmotte L, B.Nagy J, submitted for publication) and [Fe]-ZSM-5 [266] samples. Diffuse reflectance (DR) spectroscopy of Co^{2+} permits the observation of d–d transitions in the near infrared and visible region and charge transfer (CT) transitions in the ultraviolet region. Co^{2+} is the only common d^7 ion, and because of its stereochemistry the respective spectra have been extensively studied [267]. The two representative cases: Co^{2+} ion in octahedral (O_h) and tetrahedral (T_h) [267] crystal field can be interpreted in the same way as the octahedral d^2 ion (e.g., V^{3+}) and octahedral d^3 ion (e.g., Cr^{3+}), respectively. (The corresponding Tanabe-Sugano diagrams can be seen in [267]). As all d–d transitions for octahedral complexes possessing a symmetry center are symmetry-forbidden, these bands are exceptionally weak in the spectra. Thus, the DR spectra of zeolites, in particular

ZSM-5 zeolites, with framework (FW) and extra-framework (EFW) Co^{2+} content are dominated by intense $O^{2-} \rightarrow Co^{2+}$ charge transfer transitions, characteristic essentially of Co^{2+} in T_h coordination around 200–230 nm; the most intense T_h transition, $\nu 3$ [corresponding to $^4A2(e^4t_2^3) \rightarrow {}^4T1(P)$], appears around 590–630 nm and a weaker one, $\nu 2$, already in the near infrared (1400–1500 nm).

(The third d–d transition, $\nu 1$, is also in the infrared region [2000–3300 nm], but its examination is hindered due to overlapping vibrational bands). Even though the T_h d–d transitions are symmetry- and spin-allowed (no center of symmetry and $\Delta S = 0$), the most intense $\nu 3$ band in the visible region always appears as a broad band split into three components due to symmetry perturbations, spin-orbit coupling and other reasons. Figure 34 exhibits in the 460–700 nm visible region (from blue to red) at least four poorly resolved bands (at 486, 530, 584 and 643 nm), of which the last three are T_h d–d transitions. This triplet is indicative of high-spin (d^7) Co^{2+} in a tetrahedral crystal field. The weak 486 nm band is caused by the most intense $\nu 3$ transition of high-spin Co^{2+} in octahedral coordination. The nearly invisible 357 nm band is supposed to be due to high-spin Co^{3+} in unknown coordination [268].

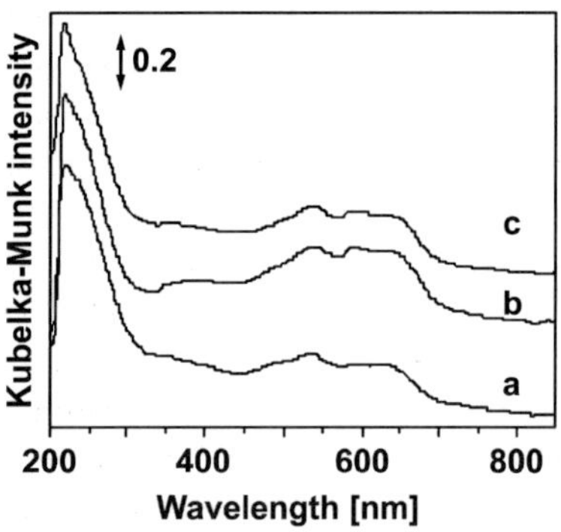

Fig. 34 Diffuse reflectance UV-Vis spectra of samples Nos. 1 (a), 5 (b) and 9 (c) recorded at a resolution of 2 nm (see Table 28)

Table 29 Visible diffuse reflectance spectra of selected as-made [Co]-ZSM-5 samples

Sample	Band intensities [a.u.] 480 nm Co octahedral	530 + 590 + 640 nm Co tetrahedral	340 nm Co(III)
1	11	21	25
5	12	35	16
9	13	22	17

Table 29 compares the relative intensities of the 480 nm band (octahedral Co(III)), the sum of 530, 590 and 640 nm bands (tetrahedral Co(II)) and the 340 nm band (Co(III)). The values are in arbitrary units, and only their variations are meaningful for the state of Co ions in the ZSM-5 samples. The octahedral Co(II) ions should be extra-framework ions, the tetrahedral Co(II) ions could be due to both framework and extra-framework species and the Co(III) ions are considered to be essentially extra-framework species. It can be seen that tetrahedral Co(II) ions are present in all studied samples in roughly similar amounts. Only sample 5 shows a higher amount of tetrahedral Co(II) ions. The amount of extra-framework Co(II) ions are similar in samples 1, 5, and 9. The amount of extra-framework Co(III) ions are similar in samples 5 and 9. No systematic variations could be detected as a function of the formal Co/u.c. content (Table 28). XPS or electron spectroscopy for chemical analysis (ESCA) is used for characterizing surface species. For Co^{2+}, the 2p electron transitions are observed which, after ionization, become separated into two levels: $2p_{3/2}$ and $2p_{1/2}$. The $2p_{1/2}$ peak is shifted upwards in energy by about 15.7 eV. Each peak has one, more or less intense, shoulder, a so-called satellite. Thus, we do not see two, but rather four peaks (three peaks and one strong shoulder) in the XP spectrum.

The appearance of these "excess" peaks is due to multiplet splitting of the respective electron orbitals. It would have been very informative to take the XP spectra of each sample, however, the well-known general trend of such syntheses in alkaline media (resulting only in partial substitution), and the registered UV-Vis spectra (see above) revealed unequivocally that it is impossible to attain complete incorporation of Co^{2+}, thus, only three samples (1, 8 and 12) have been chosen for closer examination. In a previous study we succeeded in detecting a slight, but unmistakable 1.9 eV difference in the locations of $2p_{3/2}$ XP signals of FW and EFW Fe^{3+} ions in heat-treated [Fe]-ZSM-5 samples [225]. Even though the topic was not treated in the literature so far, it was deemed to be interesting to try finding a small shift for FW and EFW Co^{2+}, too. This would have been indicative of mixed coordination, i.e., the presence of both tetrahedral and octahedral Co ions. The XP spectra of the Co^{2+} photoelectron region for the [Co]-ZSM-5 samples 1, 8, and 12 are given in Fig. 35. Table 30 provides information on the binding energies of the $Co(2p_{3/2})$ and $Co(2p_{1/2})$ core levels (with respect to Si(2p)) and on the location of the satellites. If we add a fourth [Co]-ZSM-5 sample to the selected ones (synthesized from magadiite in the presence of Co-pyrocatecholate by P. Fejes) (in Table 30 marked as reference), it is perhaps not an overstatement that the increase of the Co^{2+} content (cf. Table 29) brings about a little upward shift in the location of the $2p_{3/2}$ peaks, which do not attain 781.5 eV, typical of free Co^{2+}, even for sample 8, having the largest Co^{2+} content (12 Co^{2+}/u.c., corresponding to Si/Co = 6.56). It is believed that the binding energies of core level electrons for Co^{2+} in tetrahedral coordination (similar to the respective Fe^{3+} levels) are smaller than those in octahedral symmetry and

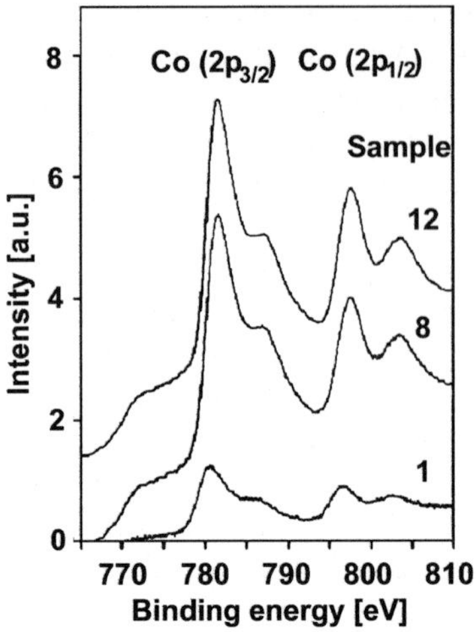

Fig. 35 XPS spectra of $Co(2p_{3/2})$ and $Co(2p_{1/2})$ of samples Nos. 1, 8 and 12 (see Table 30)

Table 30 XPS data of selected [Co]-ZSM-5 samples

Sample	Binding energies [eV] Co $(2p_{3/2})$	Co $(2p_{1/2})$	Satellites [eV]	
Reference [Co]-ZSM-5 *	780.90	796.80	785.90	802.85
1	780.95	796.95	785.95	803.05
8	781.40	797.25	786.75	803.35
12	781.05	797.10	786.60	803.35
21	781.05	797.10	786.00	803.20

* provided by P. Fejes, see text

in mixed coordination. The spectrometer—not able to resolve the two peaks (separated by a few tenths of eV)—registers only one peak at the weighted average of the two binding energies. This manifests itself as an upward energy shift when the percentage of octahedral Co^{2+} increases with the cobalt content.

3.10
Calcination Using Ozone: Preservation of Framework Elements

The [Co]-ZSM-5, [Co,Al]-ZSM-5 and [Ga]-MCM-22 samples were synthesized in alkaline media, while the B- and [Ga]-ZSM-5 samples were obtained in fluorine-containing media [269]. The ozone treatment was carried out as follows. The sample dried at room temperature was placed in a glass boat in thin layer and placed into a glass tube and heated up to 373 K for dehydration. After 1 h of treatment in oxygen, the oxygen flow was changed for an ozone-containing stream. Ozone in an oxygen stream was produced using an Ozmatic Modular 2 ozone generator. The flow rate of the ozone/oxygen mixture was 2 $dm^3\ h^{-1}$. The temperature was adjusted to 483 K. Burning off the template was completed after 3 h. The oxygen treatment was performed in the usual way. The samples were placed into ceramic jars in layers not exceeding a height of 4 mm and heated up in an O_2-rinsed electric furnace to 373 K for 1 h for dehydration. After that, the temperature was raised to 820 K at a rate of 3 $K\ min^{-1}$, while maintaining a steady 2 $d\,m^3\ h^{-1}$ flow of O_2. After 3 h at 820 K burning off the template was completed, and the samples were let to cool in air. Finally, IR spectroscopy was used to check the samples for complete removal of the template. The as-prepared, the oxygen-treated and the ozone-treated [Co]-ZSM-5 samples were characterized by XRD measurements, multinuclear NMR and UV-visible diffuse reflectance spectroscopies. The XRD spectra show that no significant alteration occurred either during the oxygen or the ozone treatment. The diffuse reflectance spectra of the as-made [Co]-ZSM-5 (sample 1), the oxygen and the ozone treated samples are reported in Fig. 36. Table 31 shows all the results of the diffuse reflectance spectra.

The more intense band at ca. 200–230 nm is due to $O^{2-} \rightarrow Co^{2+}$ charge transfer transitions. The poorly resolved triplet at 530, 590 and 640 nm stems from d–d transitions of tetrahedral Co(II). The weak 480 nm band resulted from the d–d transition of high-spin Co(II) in octahedral coordination.

Finally, the 340 nm band is due to high-spin Co(III) in unknown coordination. Table 31 shows the relative intensities of the various bands in arbitrary units. It can be seen that the smallest changes occur during the O_3 treatment. Indeed, the extra-framework octahedral Co(II) species are transferred into extra-framework Co(III) species (Table 31a and c, and Fig. 36a and c). The relative intensities of the tetrahedral Co(II) species, although somewhat de-

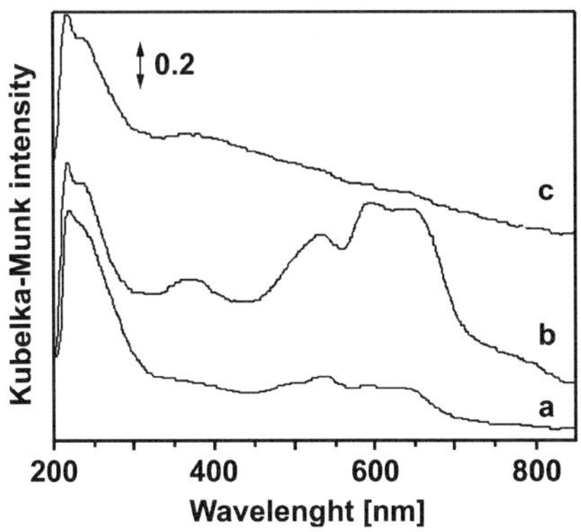

Fig. 36 Diffuse reflectance UV-Vis spectra of samples No. 1 (see Table 31) as-made (a), calcined in O_2 (b) and calcined in O_3 (c), [Co]-ZSM-5 and [Co,Al]-ZSM-5 samples

Table 31 Visible diffuse reflectance spectra of as-made (a), oxygen- (b) and ozone-treated (c) [Co]-ZSM-5 and [Co,Al]-ZSM-5 samples

No.	Co/u.c.	Al/u.c.	Band intensities [a.u.] 530 + 590 + 640 nm Co(II) octahedral	480 nm Co(II) tetrahedral	340 nm Co(III)
a) as-made samples					
1	2.7	—	11	21	25
5	6.3	—	12	35	16
9	7.0	—	13	22	17
18	2.0	2.4	8	17	13
20	1.8	3.2	0	19	46
21	1.6	4.2	17	20	26
b) O_2-treated samples					
1	2.7	—	1	41	46
5	6.3	—	0	54	24
9	7.0	—	0	51	12
18	2.0	2.4	18	8	41
20	1.8	3.2	7	26	36
21	1.6	4.2	13	31	32
c) O_3-treated samples					
1	2.7	—	2	15	65
5	6.3	—	0	18	60
9	7.0	—	0	17	64
18	2.0	2.4	0	10	58
20	1.8	3.2	0	10	70
21	1.6	4.2	0	18	63

creasing, do not change much, suggesting that possible framework species remain essentially in the framework. We would like to emphasize that these measurements are not quantitative and only some trends of band intensities can be shown. Note also that a definite proof for tetrahedral Co(II) species in the zeolite framework is still missing, although XPS, XRD and diffuse reflectance spectra suggest that part of the Co(II) ions occupy tetrahedral positions [197]. The diffuse reflectance spectra of the O_2-treated samples are more controversial. Indeed, the intensities of the Co(II) tetrahedral lines are increasing in almost all samples but sample 18. Note also that the increase is much higher when aluminum is not present in the structure (Fig. 36a and b).

More quantitative data are necessary to understand these phenomena. The ^{27}Al NMR spectra show the presence of both tetrahedral framework (in large amount) and octahedral extra-framework (in small amount) aluminum species (Table 32). Oppositely to the changes observed, in the Co visible diffuse reflectance spectra, the relative amounts of tetrahedral and octahedral species do not change during either O_2 or O_3 treatment. Only the highest Al-content sample shows some decrease in the framework tetrahedral aluminum species. The ^{29}Si NMR data are reported in Table 33, and Fig. 37

Table 32 ^{27}Al NMR data of as-made, oxygen- and ozone-treated [Co,Al]-ZSM-5 samples

No.	Co/u.c.	Al/u.c.	As-made Al_T	As-made Al_O	O_2-treated Al_T	O_2-treated Al_O	O_3-treated Al_T	O_3-treated Al_O
18	2.0	2.4	78	22	79	21	—	—
20	1.8	3.2	87	13	86	14	85	15
21	1.6	4.2	91	9	84	16	86	14

Al_T: Al in tetrahedral coordination
Al_O: Al in octahedral coordination

Table 33 ^{29}Si NMR data of as-made, O_2-treated and O_3-treated [Co]-ZSM-5 samples

No.	Line intensities [%] SiOM + SiOCo + SiOAl (– 102 ppm) + (– 106 ppm)	Si(OSi)$_4$ (– 112 ppm) + (– 115 ppm)
a) as-made samples		
1	21	79
5	25	75
9	24	76
18	20	80
20	32	68
21	27	73
b) O_2-treated samples		
1	0	100
5	0	100
9	0	100
18	13	87
20	21	79
21	15	85
c) O_3-treated samples		
1	10	90
5	16	84
9	10	90
18	18	82
20	24	76
21	16	84

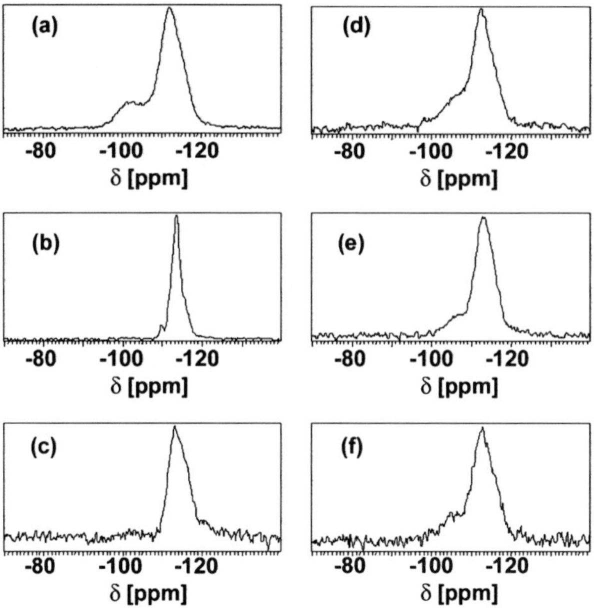

Fig. 37 ^{29}Si NMR spectra of [Co]-ZSM-5 sample No. 1 (see Table 31) as-made (**a**), O_2-treated (**b**), O_3-treated (**c**) and of [Co,Al]-ZSM-5 sample No. 21 as-made (**d**), O_2-treated (**e**), O_3-treated (**f**)

illustrates the spectra of samples 1 ([Co]-ZSM-5) and 21 ([Co,Al]-ZSM-5) in the as-made (a), O_2-treated (b), O_3-treated (c) forms. The O_2-treated samples show completely different spectra with respect to the as-made samples when no aluminum is present in the samples (samples 1, 5 and 9). Indeed, in these cases, the – 102 ppm line is absent in the O_2-treated samples, and the spectrum shows high resolution (Fig. 37b). The latter is characteristic of completely siliceous materials, that means no Co is found in the calcined samples. In addition, due to the presence of extra-framework cobalt oxide species, no cross-polarized spectra could be taken. The O_3-treated samples always show the presence of the – 102 ppm line even in the absence of aluminum. In the Al-containing samples (18, 20 and 21) the presence of tetrahedral aluminum is always detected confirming the ^{27}Al NMR results (Fig. 37c).

The ^{71}Ga NMR of the [Ga]-MCM-22 samples shows clearly the advantage of the ozone treatment over the calcinations in oxygen. Indeed, in the O_3-treated samples, only the tetrahedral framework gallium is detected at ca. 150 ppm as in the as-made sample (Fig. 38).

Oppositely, during the O_2 treatment some 22% of the gallium is found as octahedral extra-framework species at ca – 10 ppm. The ^{29}Si NMR spectra are illustrated in Fig. 39. The spectrum of the as-made [Ga]-MCM-22 (Si/Ga = 16) is quite similar to the previously published spectrum of aluminosilicate

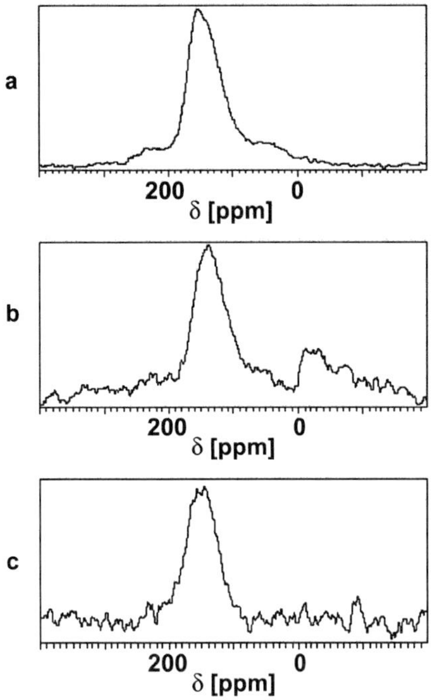

Fig. 38 ^{71}Ga NMR spectra of as-made (**a**), O_2-treated (**b**), O_3-treated (**c**) [Ga]-MCM-22 sample

MCM-22 [270]. The lines at – 94 ppm and at – 100 ppm are due to $=Si(OM)_2$ and =Si(OM) defect groups, respectively (M = H, Na and hexamethylendiamine). The number of the other NMR lines is eight if the decomposition is made following a P6/mmm model of the structure, or thirteen if it is made following the C/mmm model [270]. It is clearly seen that the as-made and the O_3-treated samples are similar and the O_3 treatment leaves the structure quasi intact (Fig. 39a and b). In contrast, the sample calcined in air shows a very different ^{29}Si NMR spectrum (Fig. 39c), where the $= Si(OM)_2$ and $\equiv$ SiOM groups are drastically reduced. Work is in progress to interpret more deeply the decomposed ^{29}Si NMR spectra.

The ^{11}B NMR results are also very clear concerning the state of boron in the various samples. In the as-made samples, boron is essentially at tetrahedral framework positions characterized by an NMR line at – 3.7 ppm. The O_3-treated samples basically preserve the tetrahedral framework boron. Sometimes a supplementary NMR line is also found at – 2 ppm what we interpret as stemming from deformed tetrahedral boron species, where one B–O–Si bond is broken: $(SiO)_3BOH^-$. When the samples are calcined in air, some 50% of the boron is found as trigonal extra-framework species (δ = ca. 6 ppm, quite a broad line). Fig. 40 illustrates the ^{11}B NMR spectra of (NH_4,

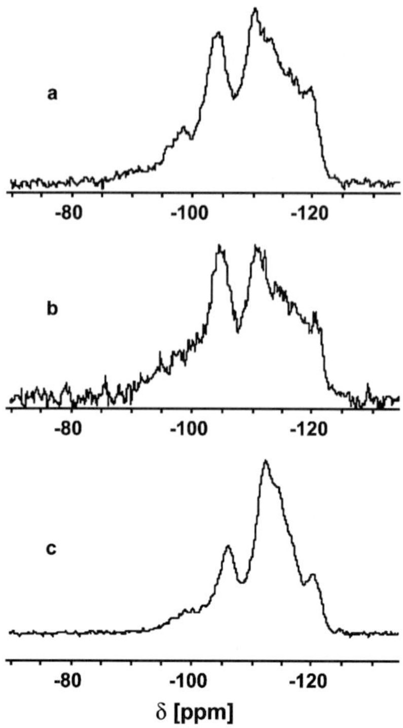

Fig. 39 ^{29}Si NMR spectra of [Ga]-MCM-22 (Si/Ga = 16) of as-made (**a**), O_3-treated (**b**), and O_2-treated (**c**) samples

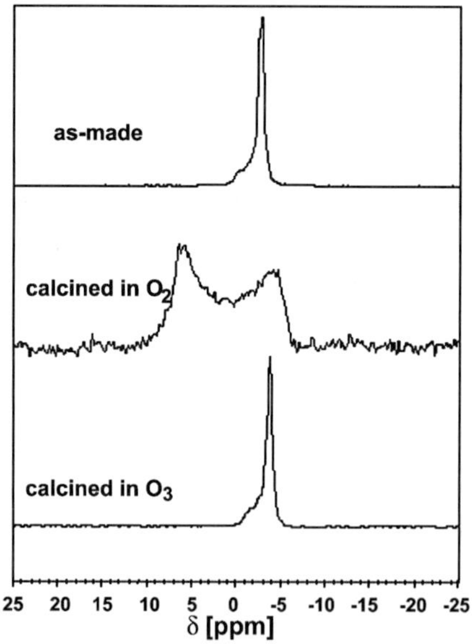

Fig. 40 ^{11}B NMR spectra of $(NH_4 - TPA)$-[B]-ZSM-5 samples (Si/B = 20.8)

TPA)-[B]-ZSM-5 samples (Si/B = 20.8). The as-made and O_3-treated samples only show the tetrahedral framework boron at −3.7 ppm. Oppositely, the sample calcined in air shows the presence of a large amount of trigonal extra-framework boron at ca. 6 ppm.

3.11 Cu-TON

The main characteristics of the Cu-TON samples, chemical analysis and EPR results are reported in Table 34. It can be observed that the amount of copper introduced into the zeolite depends on the exchange procedure. A small content of Cu is introduced during double ionic exchange (Na-TON → NH_4^+ − TON → H^+ − TON → Cu-TON). Following Wichterlova et al. [271] and due to the low Al-content of the samples, the Cu^{2+} ions occupy essentially cation exchange sites close to individual $(Al-O-Si)^-$ negative charges. These Cu species were readily reduced as it was shown elsewhere [272]. On the contrary, higher amount of copper is detected when the exchange process is carried out by the solid-state reaction method. In both cases, i.e., in the direct exchange and the solid-state reaction, we observed "over-exchange". Moreover, by applying the direct exchange procedure not all sodium ions were removed, thus, they block the openings of the TON pore system. This is confirmed by nitrogen adsorption measurements of the micropore volume: Na-TON = 0.06 ml g^{-1}, H^+ − TON = 0.91 ml g^{-1}. The presence of sodium is also proved by ^{23}Na NMR spectroscopy. Data of Table 34 show that the solid-state reaction is one of the best techniques for introducing the transition metals. The EPR results indicate that the Cu^{2+} ions have octahedral symmetry in the Cu-TON zeolite ($g_{||} = 2.34$, $g_{\perp} = 2.09$ and $A_{||} = 12.7$ mT) according to the interpretation of the data found by Kucherov et al. [273].

The spectrum of the as-made sample is shown in Fig. 41. It can be seen that it is composed of three lines: the −111.5 ppm is attributed to T_1 sites (eight sites per u.c.), the −113.2 ppm line to T_2 sites (eight sites per u.c.) and

Table 34 Chemical analysis of the Cu-TON samples (Si/Al = 50) prepared in three different ways (Cu = direct exchange, NH_4^+ − Cu = double exchange and $CuCl_2$ = solid-state reaction) and results of EPR ($A||$ is given in mT)

Sample	Al [wt%]	Na [wt%]	Cu [wt%]	$g\perp$	$g\|\|$	$A\|\|$
As-made	0.83	0.85	—	—	—	—
Calcined	0.75	0.80	—	—	—	—
Cu	0.76	0.35	4.82	2.082	2.345	12.7
NH_4^+-Cu	0.71	not detec.	0.52	2.098	2.345	12.9
$CuCl_2$	0.74	not detec.	6.13	2.091	2.348	12.7

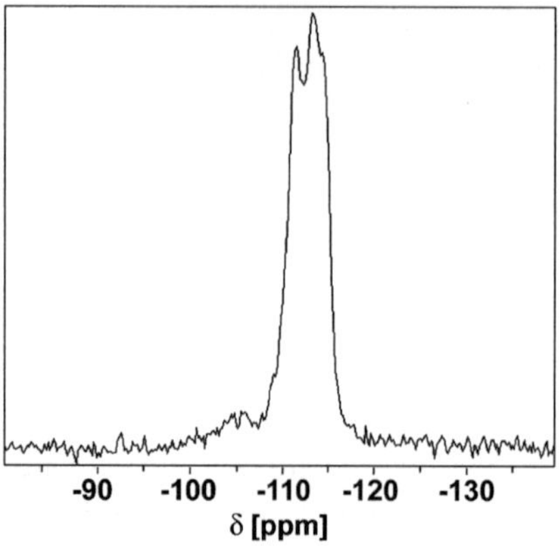

Fig. 41 ^{29}Si NMR spectrum of the as-made Cu-TON sample

the -114.3 ppm to T_3 and T_4 sites (four sites per u.c. for each type). The amount of defect SiOH groups at -105 ppm is quite small. The spectra do not change for the differently treated samples. Only the solid-state exchanged sample, containing a high amount of copper, shows quite broad NMR lines. As after the thermal treatment all the aluminum is in tetrahedral coordination, this is an indirect evidence of the high thermal stability of TON-type zeolite

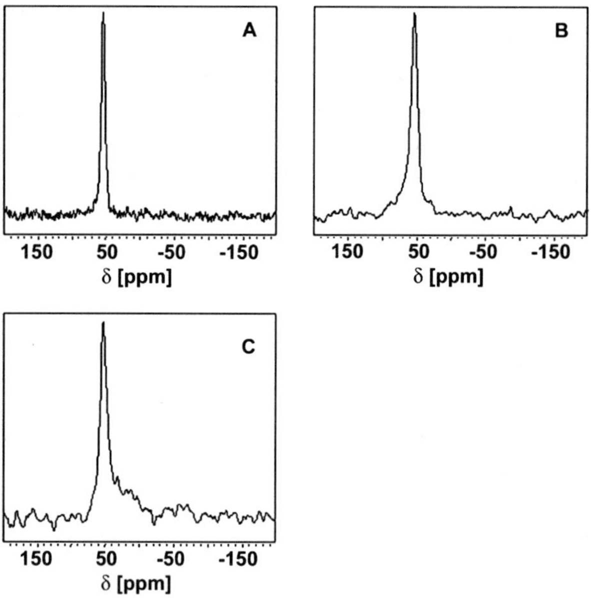

Fig. 42 ^{27}Al MAS NMR spectra of three different TON zeolite samples. **a** as-made form; **b** calcined form and **c** Cu-TON obtained by the solid-state reaction

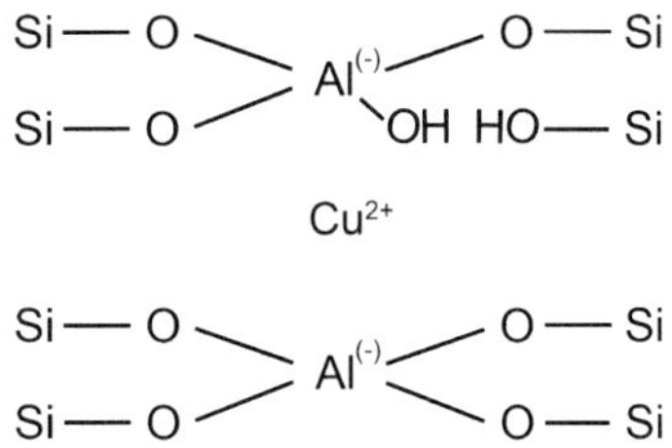

Scheme 1 Schematic representation of the deformed aluminum site

synthesized by the methanol procedure. Figure 42 shows the ^{27}Al MAS NMR spectra of the three different samples of TON zeolite confirming the absence of aluminum in octahedral coordination. A very low amount of octahedral aluminum is detected for the Cu-TON prepared by the solid-state reaction. We suppose that this small quantity of aluminum is due to the procedure of copper introduction.

There is a correlation between the amount of the copper introduced into the zeolite framework and the deformation of Al tetrahedral sites. This is also confirmed by chemical analysis. When Cu is over-exchanged by the solid-state reaction procedure, the low intensity NMR line at 0 ppm assigned to extra-framework aluminum is observed. In the case of the two solution-mediated ionic-exchange procedures, only slight deformation of the Al_T sites is observed.

The nature of the deformed tetrahedral aluminum site is still a question of controversy. In many dealumination processes, this line (at ca. 30 ppm) has been attributed to the tetrahedral framework atoms where at least one Al–O–Si bond is broken. The 2D 3QMAS NMR data unambiguously show a chemical shift at ca. 70 ppm confirming its attribution. In the present case, we can hypothesize that one Al–O–Si bond is broken, which leads to the deformed tetrahedral aluminum sites (see Scheme 1).

3.12
[Zn]-MFI

As indicated in Fig. 43, zincosilicate samples with zinc contents show high crystallinity up to a Zn/Si ratio of 0.07. There are no reflections that could be attributed to a ZnO admixture. The higher zinc loading results in a noticeable contribution of quartz and amorphous phase in the products. The scanning electron microscopic photographs (Fig. 44) show regular crystallites of the samples. The particle size depends on zinc loading decreasing with growing Zn content (3–10 μm). The IR spectra show two bands in the range 1000–1100 cm^{-1} for the samples with low zinc loading, whereas the zinc-rich sample exhibits only one band at 1050 cm^{-1} (Fig. 45a). The reported spectra of other metallosilicates [19] indicate bands attributed to the Si – O – Me

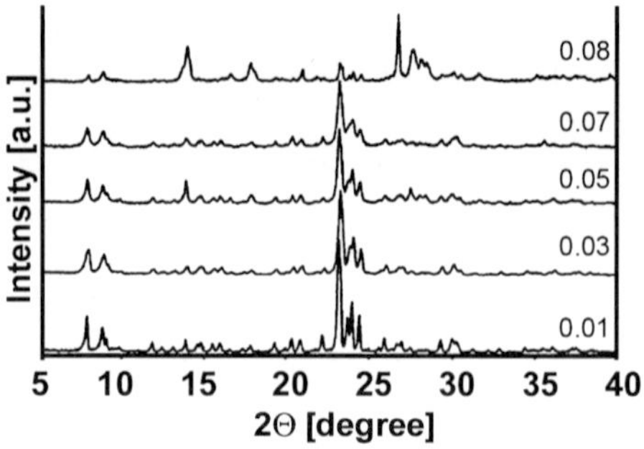

Fig. 43 XRD patterns of selected zincosilicate samples (Zn/Si indicated)

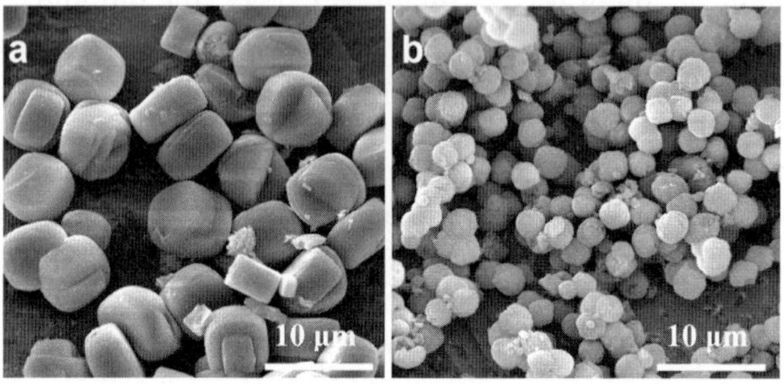

Fig. 44 Scanning electron micrograph of the samples with: **a** Zn/Si = 0.01 and **b** Zn/Si = 0.07

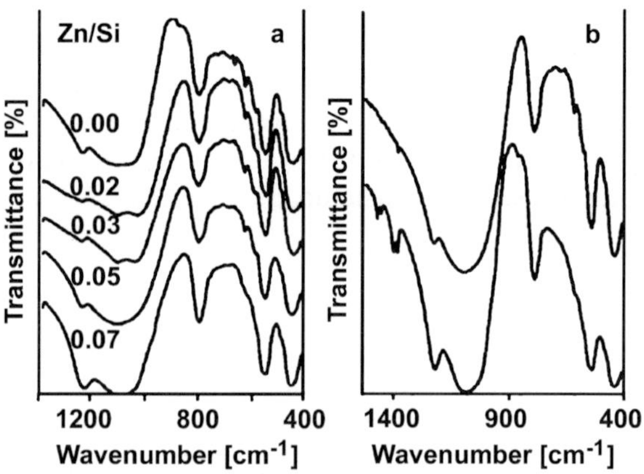

Fig. 45 IR spectra (KBr pellets) of the indicated samples (**a**), and of the sample with Zn/Si = 0.05 (**b**) (*upper spectrum*) and its ammonium modification (*lower spectrum*)

bonds. The bands are always observed at $\sim 900\ cm^{-1}$. If the stretching bands of lower wavenumbers ($\sim 1020\ cm^{-1}$) were assigned to Zn – O it would be difficult to explain their absence in the Zn-rich samples (Fig. 45a). The spectrum of the sample modified with NH_4 cations (Fig. 45b) clearly indicates the presence of N – H band (1400 cm^{-1}), which proves the ion-exchange properties of the prepared zincosilicates.

The ion-exchange ability results from the negative lattice charge caused by the introduced tetrahedral zinc. Our attempts to employ Raman spectroscopy [274] indicated the presence of the band at ca. 300 cm^{-1} assigned to tetrahedral Zn [275]. Its intensity decreased on heating (Fig. 46), which indicates removal of some zinc from the tetrahedral lattice position. The band at ca. 380 cm^{-1} attributed to octahedral Zn became predominant after calcination. The solid-state ^{29}Si MAS NMR data [199] indicate the predominant signal at ca. - 114 ppm, corresponding to a silicon tetrahedron surrounded by four adjacent Si tetrahedra.

Another smaller signal at ca. - 103 ppm results from silicon connected with three Si tetrahedra. The fourth adjacent neighbor is most likely the ZnO_4 unit. The intensity of the latter signal decreases markedly after calcination of the sample indicating that part of the zinc is released from the lattice position. However, the ability of the calcined samples to undergo cation-exchange modification (Fig. 45) clearly indicates the presence of some zinc in the lattice.

The correlation between the zinc content and the temperature of DTG peaks (Fig. 47) shows that the temperature of template removal is remarkably affected by the presence of zinc introduced into the MFI structure.

An increase in the DTG temperature is particularly significant for the samples with the lowest zinc content. The increased temperature of template removal for the zincosilicates can result from stronger bonding of the TPA cations to the negatively charged lattice. Deeper interpretation of thermal analysis results requires further study. So far we can observe a distinctive difference in thermal behavior of the samples of low (up to Zn/Si = 0.03) and high zinc contents. This corroborates the observed differences in properties of the samples with low and high zinc content (catalytic activity, ion-

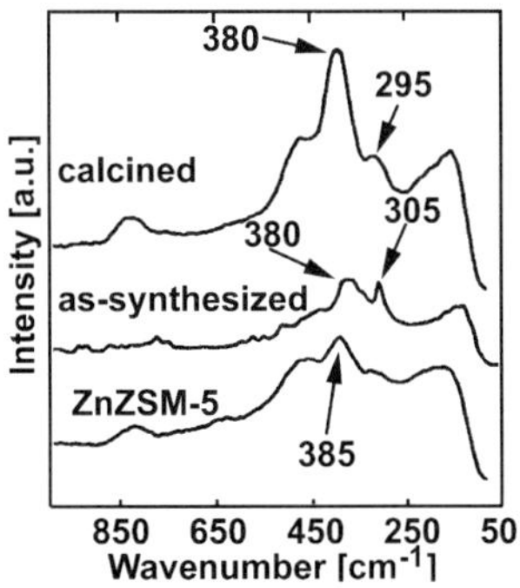

Fig. 46 Raman spectra of the indicated samples

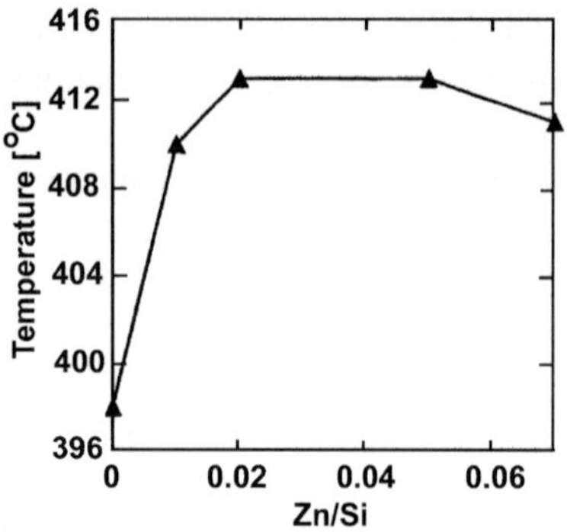

Fig. 47 Correlation between the Zn/Si ratio of the samples and temperature of the DTG peaks

exchange capacity). The latter are probably less ordered and contain more structure defects. The measurement of temperature-programmed reduction (TPR) in hydrogen indicates (Fig. 48) relatively high resistance of the zincosilicate in the MFI structure to reduction, regardless of the relatively high zinc loading (Zn/Si = 0.05). Some consumption of hydrogen starts after exceeding the temperature of 500 °C. A small peak is noticeable at ca. 580 °C and two distinct large reduction peaks are observed at temperatures of 730 and 850 °C. The TPR profiles of the samples with lower Zn content do not differ very much from that indicated in Fig. 48. The peak at 580 °C can be assigned to Zn cations [276], and we noticed that it is more pronounced for the sample modified with Zn cations. The other larger peaks result from the reduction of the lattice zinc. The modification of zincosilicate MFI with copper cations (0.6 wt %) significantly affects the reducibility of the sample. The Cu^{2+} cations undergo reduction already at ca. 200 °C. The distinctive and large peak attains the maximum at ca. 320 °C.

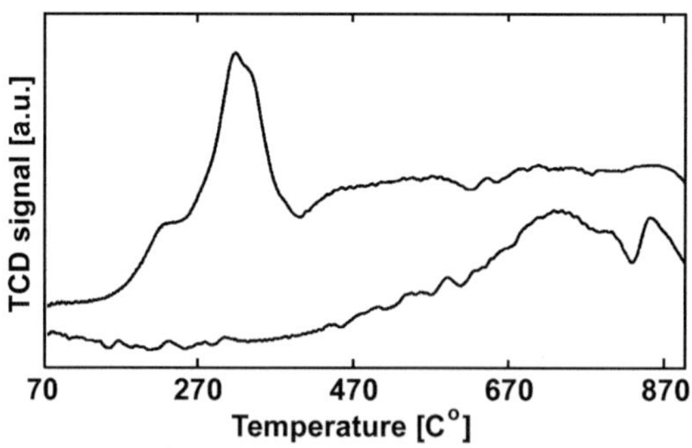

Fig. 48 TPR profiles for the sample with Zn/Si = 0.05 after template removal (*lower curve*) and for its Cu modification (*upper curve*)

Further temperature rise does not result in very distinct peaks, although the consumption of hydrogen is considerable. This indicates that lattice Zn is involved in the reduction. The peak attributed to the Zn cations is still noticeable, although shifted above 600 °C. The TPR curve declines above 900 °C, which can suggest that the zinc reduction is complete at this temperature. The shape of TPR profiles differs very much from those for silicalite impregnated with ZnO described by Wan [277], where the reduction of zinc oxide already started at 200 °C, as well as from zeolite MFI modified with Zn cations (introduced either by ion-exchange or impregnation), where the main TPR peak appeared at 550 °C [276].

Zinc can be introduced into the MFI structure by conventional hydrothermal synthesis. The presence of phosphate anions in the initial mixtures is advantageous. The resulting MFI zincosilicates show high crystallinity for Zn/Si ratios below 0.07. Further increase in the Zn content results in formation of amorphous products or non-porous crystalline phases (quartz). The crystallization time depends on the zinc content. Higher Zn loading requires longer crystallization. The properties of the resulting samples are distinctively affected by the introduced "heteroatoms". There are considerable differences in properties of the samples with low zinc contents (up to Zn/Si = 0.03) and those exceeding this ratio. Thermal analyses of the as-synthesized products show similar curves for the samples with low Zn contents (up to Zn/Si = 0.03). The introduction of Zn into the lattice results in enhanced temperature of template removal, compared with zinc-free silicalite. The negative lattice charge of zincosilicates results in stronger interaction with the TPA cations and increases the thermal stability of the latter. Thermal analyses do not indicate structure degradation up to 600 °C, which is consistent with XRD data.

The ^{29}Si MAS NMR spectra of zincosilicates under study exhibit a signal at ca. – 102 ppm, which results from silicon surrounded by three Si and one Zn tetrahedra. The intensity of the above signal declines on heating because of the removal of some Zn from the lattice. The Raman spectra also exhibit a band at ca. 300 cm^{-1}, which could be assigned to tetrahedral lattice Zn. The intensity of this band decreases after thermal treatment, because some of the Zn atoms are released from the lattice. Regardless of that, the calcined samples can be modified with various cations. TPR measurements indicate quite high resistance of zincosilicate to reduction with hydrogen (up to ca. 600 °C). The possibility of modifying the obtained zincosilicates by ion exchange and their noticeable catalytic activity (contrary to the zinc-free silicalite-1) in cumene cracking and in propan-2-ol decomposition strongly supports the localization of the introduced zinc in the lattice. The acid-site strength in H-modifications of zincosilicates is much lower than that in H-ZSM-5, but the modification of these materials with transition metal cations (e.g., Cu) could be promising in the red-ox reactions.

3.13 Dealumination of Levyne – Characterization of Framework and Extra-Framework Species

For the synthesis of levyne, the global composition of the initial gel was $4Na_2O - 2K_2O - 6MeQI - Al_2O_3 - 30SiO_2 - 500H_2O$ where MeQ is the methylquinuclidinium ion. The reaction gel was prepared by mixing of 30 wt % aqueous solution of NaOH (pellets EPR, Carlo Erba), MeQI, $Al(OH)_3$ (dry gel, Pfaltz and Bauer), distilled water and SiO_2 (fumed silica, Serva). MeQI was prepared by mixing quinuclidine (1-azabicyclo[2,2,2]octane, Aldrich) and iodomethane. The reaction mixture was heated at 150 °C under autogenous pressure in static conditions for programmed times, using modified Morey-type autoclaves (8 cm^3). The calcination of the sample was carried out in air, heating the sample from 30 to 700 °C at a rate of 10 °C min^{-1} under a 15 ml min^{-1} air flow. The 1D NMR spectra were recorded either on a Bruker MSL-400 or a Bruker CXP-200 spectrometer. For ^{29}Si (39.7 MHz), a 6.0 μs ($\theta = \pi/2$) pulse was used with a repetition time of 6.0 s. For ^{27}Al (104.3 MHz), a 1.0 μs ($\theta = \pi/12$) was used with a repetition time of 0.2 s. The 3QMAS experiments at 9.4 T were performed on a Bruker ASX-400 using a recently developed 4 mm MQMAS probehead (Bruker) spinning at 15 kHz, with a radio-frequency field, ν_{RF}, estimated at ca. 280 kHz. The pulse sequence was composed of three pulses corresponding to the Z-filter MQMAS method [278], which yields pure absorption spectra. The pulse lengths were experimentally adjusted to 1.75 μs, 0.6 μs ($\nu_{RF} = 280$ kHz) and 8 μs ($\nu_{RF} = 10$ kHz). The recycle delay was the same as for 1D ^{27}Al MAS experiments. The delay, t_1, between the first and second pulse was regularly incremented by 67 μs, according to the method of rotor synchronization [279]. This allows the removal of the spinning sidebands, generally appearing along the isotropic axis, and to reduce significantly the acquisition time. A total of 576 and 2304 scans per increment were used for the as-made and calcined levyne samples, respectively.

The MQMAS method has been previously described [280–283]. We may recall that in addition to the total elimination of the second-order quadrupolar interactions, this technique yields a separation of the different species by both their isotropic chemical shift, δ_{CS}, and their second-order quadrupolar interaction $PQ = CQ(1 + \eta^{2/3})^{1/2}$. Nevertheless, it remains difficult to deduce quantitative information from the MQMAS spectra, since the efficiency for the excitation of multiple quantum transitions strongly depends on the quadrupolar coupling constant of each species [283]. Indeed, it is shown that the intensity detected in MQMAS is likely to be underestimated for sites experiencing very weak and very strong quadrupolar interactions, whereas for sites with similar quadrupolar parameters, the direct comparison of the isotropic projection may give a good approximation of their relative population. However, using PULSAR [284], a home made simulation software, which calculates the response of a nucleus to a series of pulses, taking into account both the quadrupolar parameter and the experimental radio-frequency field, one can predict the actual spectral intensity for each site. Thus, a method to recover the correct population of aluminum on each site is to compare the experimental MQMAS spectra with the theoretically calculated data. This is easy to perform when each site is well-characterized by a pure quadrupolar lineshape and, thus, a unique set of (PQ, δ_{CS}) parameters [282]. But in the case of a distribution of parameters, such a comparison becomes extremely difficult and inaccurate, and it is advantageous to use the method recently developed by Zwanziger [285] for the analysis of DAS spectra. A detailed description of the inverse theory and regularization method used for this analysis and their effective application to MQMAS is outside the scope of this paper. This method has already been successfully applied to provide precise knowledge of the relative population of aluminum on five sites in a well-crystallized AlPO-11 aluminophosphate [286].

The Si/Al ratio of the as-synthesized sample is close to 15, the ratio of the initial gel [180]. This equality means that the incorporation of aluminum into the levyne structure is quite effective [180, 220]. Indeed, the ^{27}Al MAS NMR spectrum of the as-made sample clearly shows that only one NMR line at 53.8 ppm is detected, which is characteristic of tetrahedral coordination (Fig. 49a). During the calcination at 700 °C, some of the tetrahedral framework aluminum leaves the structure, and this extra-framework aluminum becomes octahedral, giving rise to a signal at ca. 0 ppm (Fig. 49b).

Previous ^{13}C NMR measurements of the occluded MeQ^+ ions have shown that they are incorporated intact in the levyne channels [220]. However, the thermal analysis of the precursor samples still containing the MeQ^+ ions has demonstrated that two different MeQ^+ ions were present in the channels [220]. The ones that are decomposed at lower temperature (460 °C) are neutralizing some of the SiO^- defect groups (2.7/u.c.), while the ones decomposed at higher temperature (590 °C) are neutralizing the negative charges linked to the presence of the tetrahedral aluminum in the structure, the $(SiOAl)^-$ groups (3.4/u.c.). The high-resolution solid-state ^{29}Si NMR spectra of the as-made and calcined levyne samples are shown in Fig. 50. As the levyne structure contains two crystallographically different tetrahedral sites, 36 T1 and 18 T2 sites [221], care has to be taken in interpreting the NMR spectra. The relative intensities of the various lines of both the as-made and the calcined samples are reported in Table 35. The NMR line at – 115 ppm is assigned to a Si(0Al) configuration of the T2 sites [218, 220]. The other NMR lines are tentatively assigned as follows. The – 108 ppm line could stem from a Si(1Al) configuration on T2 sites and Si(0Al) configuration of T1 sites. The – 103 ppm line could correspond to the sum of Si(2Al)T2 and Si(1Al)T1 configurations. Finally, the – 97 ppm line could stem from the sum of Si(3Al)T2 and Si(2Al)T1 configurations [220]. However, this hypothesis cannot lead to a quantitative interpretation of the NMR spectra.

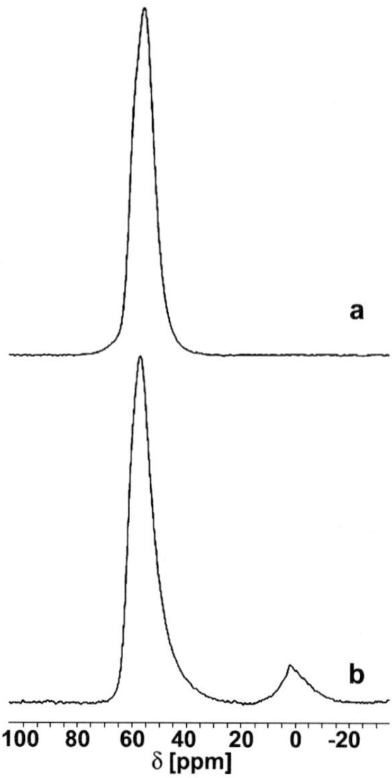

Fig. 49 ^{27}Al MAS NMR spectra of the as-made levyne sample (**a**) and of the sample calcined at 700 °C (**b**)

Table 35 ^{29}Si MAS NMR data of the as-made and levyne samples calcined at 700 °C

Sample	δ [ppm]	*I* [%]	δ [ppm]	*I* [%]	δ [ppm]	*I* [%]	δ [ppm]	*I* [%]
as-made	– 98.4	14	– 103.1	26	– 108.3	40	– 114.9	20
calcined	– 96.1	8	– 102.8	21	– 109.0	46	– 115.6	25

The discrepancy very probably stems from the presence of a high amount of defect groups SiOM (M = Na, K, H and/or MeQ). It can be seen from Fig. 50 that the intensity of the – 103 ppm line decreases during calcination and that the intensity of the – 97 ppm line is drastically reduced. It is rather well known that the former could include the -SiOM defect groups, while the latter includes either = $Si(OM)_2$ or = Si(OAl)(OM) defect groups [214].

The presence of two crystallographically different sites raises interesting questions concerning the siting of aluminum on the T1 and T2 sites. Is the distribution of aluminum random or specific [230, 240]? Is the dealumination random or specific? The high-resolution ^{27}Al NMR spectra give valuable in-

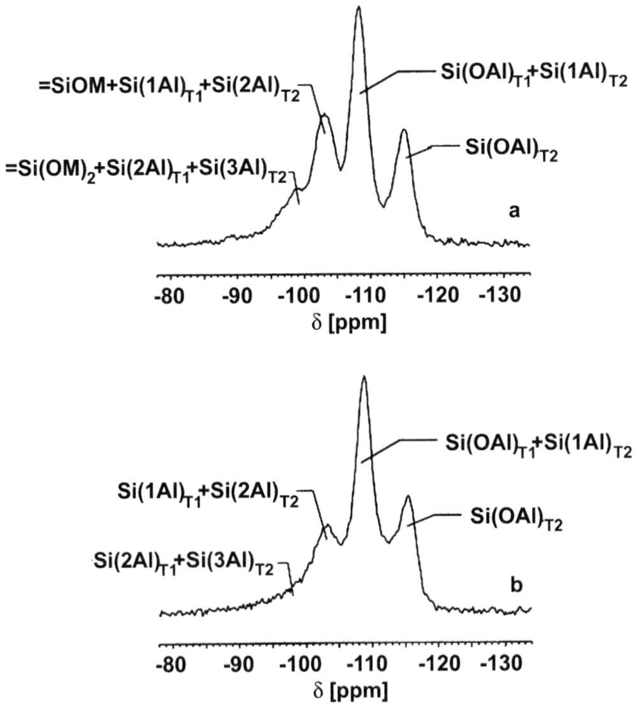

Fig. 50 ^{29}Si MAS NMR spectra of the as-made levyne sample (**a**) and of the sample calcined at 700 °C (**b**)

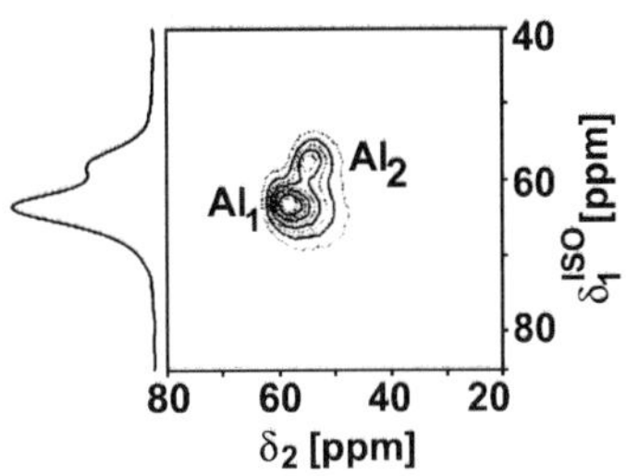

Fig. 51 ^{27}Al 3QMAS NMR spectrum of the as-made levyne sample (tetrahedral region)

formation and helps in answering the above-mentioned questions. The 1D ^{27}Al MAS NMR spectra (Fig. 49) already suggest the presence of two different tetrahedral aluminum atoms in the structure. In order to better characterize the levyne samples, 2D multiple-quantum ^{27}Al NMR experiments have been carried out. Indeed, it has already been shown that the use of three-quantum transitions greatly increases the resolution. The 3QMAS ^{27}Al NMR spectra of the as-made and calcined samples are shown in Figs. 51 and 52, respectively.

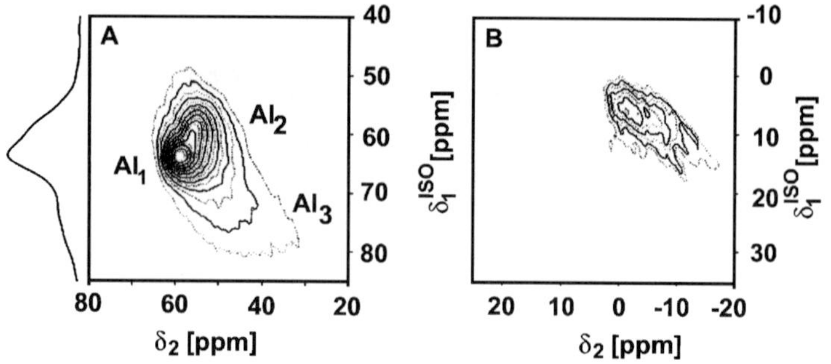

Fig. 52 ^{27}Al 3QMAS NMR spectrum of the levyne sample calcined at 700 °C: **a** tetrahedral region; **b** octahedral region

Table 36 3QMAS ^{27}Al NMR data of the as-made and levyne samples calcined at 700 °C

Sample	Tetrahedral region									Octahedral region		
	Al_1			Al_2			Al_3			O		
	δ [ppm]	PQ [MHz]	I [%]	δ [ppm]	PQ [MHz]	I [%]	δ [ppm]	PQ [MHz]	I [%]	δ [ppm]	PQ [MHz]	I [%]
as-made	62	2.7	69	56	1.9	31	—	—	—	—	—	—
calcined	62	2.6	53	57	1.8	31	64	4.8	12	4	3.5	4

The 3QMAS NMR spectrum of the as-made sample clearly shows two different species at $\delta = 62$ ppm (Al_1) and $\delta = 56$ ppm (Al_2) (Fig. 51, and Table 36). Moreover, the relative intensities of the two lines are, respectively 2 : 1, suggesting that the aluminum distribution is random in the structure. Indeed, this ratio corresponds to the ratio of the crystallographically different tetrahedral sites in the levyne structure, T1/T2= 2. The attribution of the two different tetrahedral aluminum species is hence obvious. The 62 ppm line (Al_1) characterizes the tetrahedral atoms on sites T1, while the 56 ppm line (Al_2) stems from the tetrahedral A1 atoms on sites T2. The determined quadrupolar coupling constants PQ being similar (1.9 MHz and 2.7 MHz, respectively), the relative intensities on the isotropic projection are regarded as correct. This assumption is confirmed by the computation of the 3QMAS spectrum (Fig. 53), which gives an accurate result for the relative populations (Table 36).

During calcination, a small amount of octahedral extra-framework aluminum species (O site) appears around 4 ppm (Fig. 52). In addition, a new tetrahedral species (Al_3) is also detected at $\delta = 64$ ppm, which could be due to deformed tetrahedral aluminum atoms. The PQ of the Al species is much larger than for the other two tetrahedral sites and is equal to 4.8 MHz.

The qualitative analysis of the 3QMAS spectrum (Fig. 52) yields a better understanding of the dealumination process. Table 36 reports all the data for

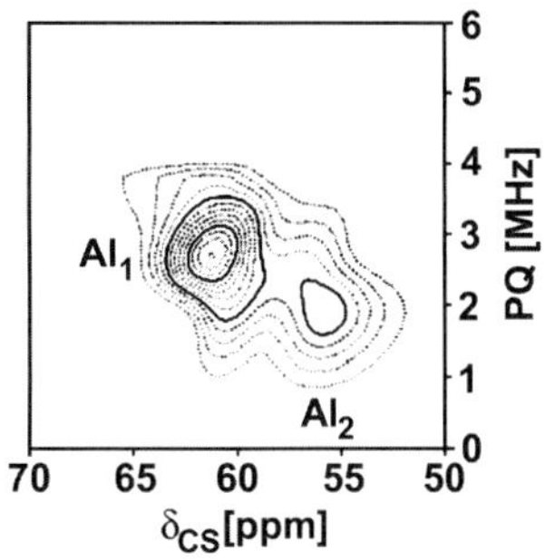

Fig. 53 Simulated spectrum of the as-made levyne sample using the inversion method

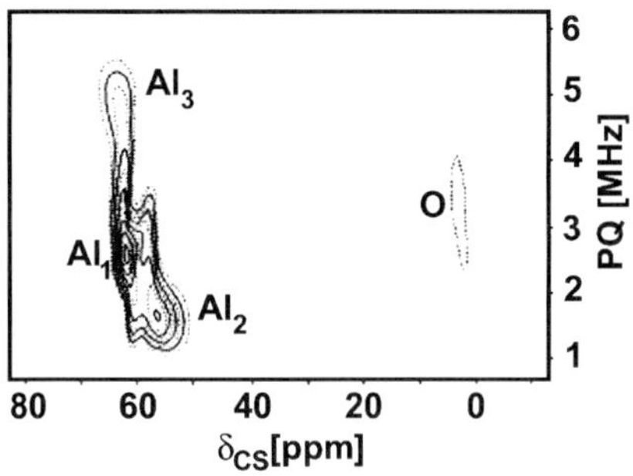

Fig. 54 Simulated spectrum of the calcined levyne sample using the inversion method

the three tetrahedral species at 57, 62 and 64 ppm, and the octahedral species at around 4 ppm (Fig. 54).

As the relative population of Al_2 remains constant throughout calcination, it seems clear that the third tetrahedral species Al_3 and the octahedral species appearing during calcination are generated from the Al_1 species. Moreover, Al_1 and Al_3 species have close isotropic chemical shifts and differ only in their PQ. This is an indication that the Al_1 atoms, which occupy the T1 site in the structure, undergo distortion to yield the Al_3 species. Hence, one can conclude that the tetrahedral Al atoms on sites T1 with signals at 62 ppm (Al_1) are partially transformed into deformed tetrahedral species with signals at 64 ppm (Al_3).

On the other hand, the 4% octahedral species formed during calcination seem to stem also from Al_1 species. These results strongly suggest that the dealumination occurring during calcination is highly specific, involving only the crystallographic tetrahedral T1 sites. A more systematic work on dealumination should be carried out in order to check the specific dealumination from the T1 sites. This represents a very interesting result and its importance could be checked in catalytic reactions. Specific dealumination could also be suggested from the framework of offretite (OFF) [70] using the changes in the ^{27}Al MAS NMR and the ^{29}Si MAS NMR spectra during dealumination [222].

It is shown that the ^{27}Al NMR studies of levyne-type zeolites can be improved using the 3QMAS NMR experiment. This technique appears to be complementary to ^{29}Si MAS NMR experiments as it totally removes the ambiguity on the distribution of Al on the crystallographically different sites. Quantitative analysis of the MQMAS spectra has been performed using an appropriate method of spectral inversion and has allowed the determination of the relative populations of the different species. It is concluded that the distribution of Al is random in the as-made sample. Dealumination yielding octahedral aluminum species seems to affect only the Al1 species, which occupy the T1 sites. In addition, some of the original tetrahedral Al atoms on site T1 are transformed into deformed tetrahedral species and to extra-framework octahedral species.

4 Conclusions

The isomorphous substitution of silicon in tetrahedral coordination in zeolite framework for other elements results in metal silicates of peculiar behavior dissimilar to the parent material. Insertion of ions with three positive oxidation states, such as Fe^{3+}, B^{3+}, Ga^{3+}, etc., leads to metal silicates having acidic character of lower or higher strengths than the aluminum-containing structures. In this way, new compositions with zeolite structures can be obtained. These modified zeolites find their applications in chemical industry mainly for the production of fine chemicals. Those derivatives which have tetravalent elements, such as Ti^{4+}, Ge^{4+} exhibit interesting catalytic properties, too. Isomorphous substitution plays an indispensable role in the preparation of catalytic materials with zeolite structure, which may thus preserve the shape selectivity as the most intrinsic phenomenon observed for molecular sieve catalysts.

Several synthesis procedures have been elaborated in various laboratories working in the field of zeolite chemistry. Both the alkaline and the fluorine route are used in the synthesis of isomorphously substituted materials. In some cases, both result in well-organized crystal structures, and the selection between the options depends on the decision of the experimenter. The synthesis options have been discussed and offered for application to the readers.

Characterization of the structures is always the most intriguing part of the post-synthetic work. Almost all instrumental analytical techniques are applied for this purpose for newly prepared matters. Comparing the measured XRD profile with those simulated by various computer programs gives the most useful data of structures. This technique provides data also on the positions of substituting elements in the framework. For those nuclei, where transition of the nuclei spins is expected, the NMR technique gives basic information on the coordination of framework constituents. IR and Raman spectroscopy and ESR measurements also supply information on the position of substituted elements. Nevertheless, even with the availability of these techniques, the coordination of the substituting elements is a matter of deep discussion even nowadays. We showed an example to this point discussing the position and the stability of Fe^{3+} ions in different zeolite structures.

The thermal and hydrothermal behavior of substituting elements in framework positions is a crucial problem as it concerns the industrial application of isomorphously substituted materials. The stability problems start already in the template removal step, which is necessary to open the pore structure of microporous matter. The general method is to burn the organic template compounds off above 800 K, using air or oxygen stream, for several hours. The solvent extraction is regarded as a less severe treatment, preserving the constituents in tetrahedral coordination. The efficiency of this treatment is not complete in some cases. Research work led to the use of ozone as oxidant. For this method much lower temperature is required to burn the template off, while preserving the tetrahedral coordination of the substituting elements. When a framework element is removed from the original tetrahedral position it may deposited in the channel system resulting in an additional adsorption and catalytic center. The dealumination leads to this sort of modification of the acidity of zeolites. The stability of the framework of isomorphously substituted zeolites has been reported in the review on the basis of our own experimental data.

Acknowledgements The authors are indebted to Prof. Karge for the very careful reading of the manuscript and for his helpful suggestions. The authors thank Maria Hodos, Monika Urban and Eva Molnar for the technical assistance. ZK acknowledges the support of the Hungarian Janos Bolyai Fellowship.

References

1. Coudurier G, Auroux A, Vedrine JC, Farlee RD, Abrams L, Shannon RD (1987) J Catal 108:1
2. Millini R, Perego G, Bellussi G (1999) Top Catal 9:13
3. Kokotailo GT, Lawton SL, Olson DH, Meier WM (1978) Nature (London) 272:437
4. Taramasso M, Perego G, Notari B (1983) U.S. Patent 4,410,501
5. Bordiga S, Coluccia S, Lamberti C, Marchese L, Zecchina A, Boscherini F, Buffa F, Genoni F, Leofanti G, Petrini G, Vlaic G (1994) J Phys Chem 98:4125
6. Tozzola G, Mantegazza MA, Ranghino G, Petrini G, Bordiga S, Ricchiardi G, Lamberti C, Zulian R, Zecchina A (1998) J Catal 179:64
7. Millini R, Massara EP, Perego G, Bellussi G (1992) J Catal 137:497
8. Ricchiardi G, Damin A, Bordiga S, Lamberti C, Spano G, Rivetti F, Zecchina A (2001) J Am Chem Soc 123:11409
9. Bordiga S, Damin A, Bonino F, Ricchiardi G, Lamberti C, Zecchina A (2002) Angew Chem Int Ed 41:4734

10. Bayense CR, Kentgens APM, de Haan JW, van de Ven LJM, van Hooff JHC (1992) J Phys Chem 96:775
11. Liu X, Klinowski J (1992) J Phys Chem 96:3403
12. Arean CO, Palomino GT, Geobaldo F, Zecchina A (1996) J Phys Chem 100:6678
13. Lamberti C, Palomino GT, Bordiga S, Zecchina A, Spano G, Arean CO (1999) Catal Lett 63:213
14. Fricke R, Kosslick H, Lischke G, Richter M (2000) Chem Rev 100:2303
15. Szostak R, Thomas TL (1986) J Catal 100:555
16. Bordiga S, Buzzoni R, Geobaldo F, Lamberti C, Giamello E, Zecchina A, Leofanti G, Petrini G, Tozzola G, Vlaic G (1996) J Catal 158:486
17. Zecchina A, Geobaldo F, Lamberti C, Bordiga S, Palomino GT, Arean CO (1996) Catal Lett 42:25
18. Berlier G, Spoto G, Bordiga S, Ricchiardi G, Fisicaro P, Zecchina A, Rossetti I, Selli E, Forni L, Giamello E, Lamberti C (2002) J Catal 208:64
19. Szostak R (1989) Mol Sieves. Van Nostrand Reinhold, New York, p 205
20. Chu CTW, Chang C (1985) J Phys Chem 89:1569
21. Strodel P, Neyman KM, Knözinger H, Rösch N (1995) Chem Phys Lett 240:547
22. Stave MS, Nicholas JB (1995) J Phys Chem 99:15046
23. Parrillo DJ, Lee C, Gorte RJ, White D, Farneth WE (1995) J Phys Chem 99:8745
24. Vezzalini G, Quartieri S, Galli E, Alberti A, Cruciani G, Kvick A (1997) Zeolites 19:323
25. Inui T, Makino Y, Okazumi F, Nagano S, Miyamoto A (1987) Ind Eng Chem Res 26:647
26. Petit L, Bournonville JP, Raatz F (1989) In: Jacobs PA, van Santen RA (eds), Zeolites: Facts, Figures, Future; Proc 8th Int Zeolite Conf, Amsterdam, The Netherlands, July 10–14, 1989; Stud Surf Sci Catal vol 49, Elsevier, Amsterdam, 1989, p 1163
27. Gnep NS, Doyemet JY, Guisnet M (1989) In: Karge HG, Weitkamp J (eds), Zeolites As Catalysts, Sorbents And Detergent Builders; Proc Int Symp, Würzburg, Germany, September 4–8, 1988; Stud Surf Sci Catal vol 46, Elsevier, Amsterdam, 1989, p 153
28. Gabelica Z, Mayenez C, Monque R, Galiasso R, Giannetto R (1992) In: Occelli ML, Robson HE (eds) Mol Sieves, vol I. Van Nostrand, New York, p 190
29. Gabelica Z, Giannetto R, Dos Santos F, Monque R, Galiasso R (1993) In: von Ballmoos R, Higgins JB, Treacy MMJ (eds) (1993) Proc 9th Int Zeolite Conf, Montreal, Canada, July 5–10, 1992. Butterworth-Heinemann, Boston, p 231
30. Aiello R, Crea F, Nigro E, Testa F, Mostowicz R, Fonseca A, B.Nagy J (1999) Micropor Mesopor Mater 28:241
31. Kessler H, Patarin J, Schott-Darie C (1994) In: Jansen JC, Stöcker M, Karge HG, Weitkamp J (eds), Advanced Zeolite Science And Applications; Stud Surf Sci Catal vol 85, Elsevier, Amsterdam, 1994, p 75
32. Guth JL, Kessler H, Higel JM, Lamblin JM, Patarin J, Seive A, Chezeau JM, Wey R Occelli ML, Robson HE (eds) (1989) Zeolite synthesis, ACS Symposium Series No. 398. Am Chem Soc, Washington, DC, p 176
33. Nigro E, Mostowicz R, Crea F, Testa F, Aiello R, B.Nagy J (1997) In: Chon H, Ihm S-K, Uh YS (eds), Progress In Zeolite And Microporous Materials; Proc 11th Int Zeolite Conf, Seoul, Korea, August 12–17, 1996; Stud Surf Sci Catal vol 105, Elsevier, Amsterdam, 1997, p 309
34. Mostowicz R, Crea F, B.Nagy J (1993) Zeolites 13:678
35. Crea F, Mostowicz R, Testa F, Aiello R, Nastro A, B.Nagy J (1993) In: von Ballmoos R, Higgins JB, Treacy MMJ (eds) (1993) Proc 9th Int Zeolite Conf, Montreal, Canada, July 5–10, 1992. Butterworth-Heinemann, Boston, p 147

36. Mostowicz R, Nastro A, Crea F, B.Nagy J (1991) Zeolites 11:732
37. Testa F, Chiapetta R, Crea F, Aiello R, Fonseca A, B.Nagy J (1995) In: Beyer HK, Karge HG, Kiricsi I, B.Nagy J (eds), Catalysis By Microporous Materials; Proc ZEOCAT '95, Szombathely, Hungary, July 9–13, 1995; Stud Surf Sci Catal vol 94, Elsevier, Amsterdam, 1995, p 349
38. Testa F, Chiappetta R, Crea F, Aiello R, Fonseca A, B.Nagy J (1996) J Coll Surf A 115:223
39. Testa F, Crea F, Aiello R, Teti S (1994) In: Orsini PG (eds) (1994) AIMAT'94, Proc 2° Convegno Nazionale sui Materiali per l'Ingegneria, Trento, Italy, September 19–21, 1994. Servizio Stamperia e Fotoriproduzione dell'Università degli Studi di Trento, Italy, p 165
40. Mostowicz R, Testa F, Crea F, Nastro A, Aiello R, Fonseca A, B.Nagy J (1994) In: Weitkamp J, Karge HG, Pfeifer H, Hölderich W (eds), Zeolites And Related Microporous Materials: State Of The Art 1994; Proc 10th Int Zeolite Conf, Garmisch-Partenkirchen, July 17–22, 1994; Stud Surf Sci Catal vol 84, Elsevier, Amsterdam, 1994, p 171
41. Mostowicz R, Testa F, Crea F, Aiello R, Fonseca A, B.Nagy J (1997) Zeolites 18:308
42. Joyner RV, Stockenhuber M (1998) In: Treacy MMJ, Marcus BK, Bisher ME, Higgins JB (eds) (1998) Proc 12th Int Zeolite Conf, Baltimore, Maryland, USA, July 5–10. Mater Res Soc, Warrendale, Pennsylvania, USA, p 1367
43. Smirnov AV, Di Renzo F, Lebedeva OE, Brunel D, Chiche B, Tavolaro A, Romanovsky BV, Giordano G, Fajula F, Ivanova I (1997) In: Chon H, Ihm S-K, Uh YS (eds), Progress In Zeolite And Microporous Materials; Proc 11th Int Zeolite Conf, Seoul, Korea, August 12–17, 1996; Stud Surf Sci Catal vol 105, Elsevier, Amsterdam, 1997, p 1325
44. Sobolev VI, Pavon GI, Kharitonov AS, Romannikov VN, Volodin AM, Ione KG (1993) J Catal 139:435
45. Szostak R, Thomas TL (1986) J Chem Soc Chem Commun 113
46. Fejes P, B.Nagy J, Halasz J (2000) Appl Catal A 190:117
47. Chandwadkar AJ, Bhat RN, Ratnasamy P (1991) Zeolites 11:42
48. Kumar R, Thangaraj A, Bhat RN, Ratnasamy P (1990) Zeolites 10:85
49. Kumar R, Ratnasamy P (1989) J Catal 116:440
50. Borade RB, Adnot A, Kaliaguine S (1991) Zeolites 11:710
51. Testa F, Crea F, Diodati GD, Pasqua L, Aiello R, Terwagne G, Lentz P, B.Nagy J (1999) Micropor Mesopor Mater 30:187
52. B.Nagy J, Aiello R, Crea F, Testa F (1998) In: Treacy MMJ, Marcus BK, Bisher ME, Higgins JB (eds) (1998) Proc 12th Int Zeolite Conf, Baltimore, Maryland, USA, July 5–10. Mater Res Soc, Warrendale, Pennsylvania, USA, p 1619
53. Patarin J, Guth JL, Kessler H, Coudurier H, Raatz F (1986) Fr. Patent 8617711
54. Inui T, Kim JB, Takeguchi T (1997) Zeolites 17:354
55. Kagawa K, Ichikawa Y, Iwamoto S, Inui T (1998) Micropor Mesopor Mater 25:15
56. Verberckmoes A, Weckhuysen BM, Schoonheydt RA (1998) Micropor Mesopor Mater 22:165
57. Gier TE, Stucky GD (1991) Nature 349:508
58. Harrison WTA, Gier TE, Moran KL, Nicol JM, Eckert H, Stucky GD (1991) Chem Mater 3:27
59. Harrison WTA, Nenoff TM, Gier TE, Stucky GD (1994) J Solid State Chem 113:168
60. Nenoff TM, Harrison WTA, Gier TE, Stucky GD (1991) J Am Chem Soc 113:378
61. Novak-Tusar N, Kaucic V (1995) Zeolites 15:708

62. Kumar N, Lindfors LE (1995) In: Beyer HK, Karge HG, Kiricsi I, B.Nagy J (eds), Catalysis By Microporous Materials; Proc ZEOCAT '95, Szombathely, Hungary, July 9–13, 1995; Stud Surf Sci Catal vol 94, Elsevier, Amsterdam, 1995, p 325
63. Kumar N, Lindfors LE, Byggningsbacka R (1996) In: Andreev A, Petrov L, Bonev C, Kadinov G, Mitov I (eds) (1996) Heterogeneous catalysis. Proc 8th Int Symp Heterogeneous Catalysis, Varna, Bulgaria, October 5–9, 1996. Pensoft Publishers, Bulgaria, p 249
64. Annen MJ, Davis ME, Higgins JB, Schlenker JL (1991) J Chem Soc Chem Commun 1175
65. Camblor MA, Davis ME (1994) J Phys Chem 98:13151
66. Freyhardt CC, Lobo RF, Khodabandesh S, Lewis JE, Tsapatsis M, Yoshikawa M, Camblor MA, Pan M, Helmkamp MM, Zones SI, Davis ME (1996) J Am Chem Soc 118:7299
67. Röhrig C, Gies H (1995) Angew Chem Int Ed Engl 34:63
68. Valange S, Gabelica Z, Onida B, Garrone E (1998) In: Treacy MMJ, Marcus BK, Bisher ME, Higgins JB (eds) (1998) Proc 12th Int Zeolite Conf, Baltimore, Maryland, USA, July 5–10. Mater Res Soc, Warrendale, Pennsylvania, USA, p 2711
69. Kowalak S, Jankowska A, Baran Z (2001) Chem Commun 575
70. Baerlocher C, Meier WM, Olson DH (2000) Atlas of zeolite framework types, 5th edn. Elsevier, Amsterdam
71. Rössner F, Hagen A, Mroczek U, Karge HG, Steinberg KH (1993) In: Guczi L, Solymosi F, Tétényi P (eds), New Frontiers In Catalysis; Proc 10th Int Congress on Catalysis, Budapest, Hungary, July 19–24, 1992; Stud Surf Sci Catal vol 75, Elsevier, Amsterdam, 1993, p 1707
72. Skrzypek J, Sloczńki J, Ledakowicz S (1994) Methanol synthesis. Polish Scientific Publishers, Warszawa
73. Wan BZ, Chu HM (1992) J Chem Soc Faraday Trans 88:2943
74. Berndt H, Lietz G, Lücke B, Völter J (1996) Appl Catal A 146:351
75. Berndt H, Lietz G, Völter J (1996) Appl Catal A 146:365
76. Inui T, Makino Y, Okazumi F, Miamoto A (1988) In: Grobet PJ, Mortier WJ, Vansant EF, Schulz-Ekloff G (eds), Innovation In Zeolite Materials Science; Proc Int Symp, Nieuwpoort, Belgium, September 13–17, 1987; Stud Surf Sci Catal vol 37, Elsevier, Amsterdam, 1988, p 487
77. Shelef M (1995) Chem Rev 95:209
78. Tabata T, Kokitsu M, Ohtsuka H, Okada O, Sabatino LMF, Bellussi G (1996) Catal Today 27:91
79. Ciambelli P, Corbo P, Gambino M, Iacoponi S, Minelli G, Moretti G, Porta P (1995) In: Bonneviot L, Kaliaguine S (eds), A Refined Tool For Designing Catalytic Sites; Proc Int Symp, Quėbec, Canada, October 15–20, 1995; Stud Surf Sci Catal vol 97, Elsevier, Amsterdam, 1995, p 295
80. Ciambelli P, Corbo P, Gambino M, Migliardini F, Minelli G, Moretti G, Porta P (1995) In: Aiello R (ed) (1995) Proc III Convegno Nazionale di Scienza e Tecnologia delle Zeoliti, Cetraro, Italy, September 28–29, 1995. De Rose Montalto, Cosenza, Italy, p 217
81. Karge HG, Beyer HK (1991) In: Jacobs PA, Jaeger NI, Kubelková L, Wichterlová B (eds), Zeolite Chemistry And Catalysis; Proc Int Symp, Prague, Czechoslovakia, September 8–13, 1991; Stud Surf Sci Catal vol 69, Elsevier, Amsterdam, 1991, p 43
82. Varga J, Fudula A, Halasz J, Schöbel Gy, Kiricsi I (1995) In: Beyer HK, Karge HG, Kiricsi I, B.Nagy J (eds), Catalysis By Microporous Materials; Proc ZEOCAT'95, Szombathely, Hungary, July 9–13, 1995; Stud Surf Sci Catal vol 94, Elsevier, Amsterdam, p 665
83. Kucherov AV, Slinkin AA (1986) Zeolites 6:175

84. Giordano G, Di Renzo F, Remoué F, Fajula F, Plee D, Schultz P (1993) In: Rouquerol J, Rodriguez-Reinoso F, Sing KSW, Unger KK (eds), Characterization Of Porous Solids III; Proc IUPAC Symp (COPS III), Marseille, France, May 9–12, 1993; Stud Surf Sci Catal vol 87, Elsevier, Amsterdam, 1993, p 141
85. Deka RC, Tajima N, Hirao K (2001) J Mol Struct–Theochem 535:31
86. Tuan VA, Falconer JL, Noble RD (2000) Micropor Mesopor Mater 41:269
87. Fild C, Shantz DF, Lobo RF, Koller H (2000) Phys Chem Chem Phys 2:3091
88. Xu WQ, Suib SL, O'Young CL (1993) J Catal 144:285
89. Hölderich W (1986) In: Murakami Y, Iijima A, Ward JW (eds), New Developments In Zeolite Science And Technology; Proc 7th Int Zeolite Conf, Tokyo, Japan, August 17–22, 1986; Stud Surf Sci Catal vol 28, Elsevier, Amsterdam, Kodansha, Tokyo, 1986, p 827
90. Datka J, Piwowarska Z (1989) J Chem Soc Faraday Trans I 85:837
91. Datka J, Kowalek M (1993) J Chem Soc Faraday Trans I 89:1829
92. Datka J, Piwowarska Z (1989) J Chem Soc Faraday Trans I 85:47
93. Hölderich W, Eichhorn N, Lehnert R, Marosi L, Mross W, Reinke R, Ruppel W, Schlimper H (1983) In: Olson D, Bisio A (eds) (1984) Proc 6th Int Zeolite Conf, Reno, USA, July 10–15, 1983, Butterworths, Guildford, Surrey, UK, p 545
94. Vetrivel R (1992) Zeolites 12:424
95. Schwieger W, Bergk KH, Unger B, Wendlandt KP, Toufar H, Brunner E (1992) React Kinet Catal Lett 46:271
96. Kan QB, Wu ZY, Xu RR, Riu XS, Peng SY (1992) J Mol Catal 74:223
97. de Ruiter R, Jansen JC, van Bekkum H (1992) Zeolites 12:56
98. Simon MW, Nam SS, Xu WQ, Suib SL, Edwards JC, O' Young CL (1992) J Phys Chem 96:6381
99. de Ruiter R, Kentgens APN, Grootendorst J, Jansen JC, van Bekkum H (1993) Zeolites 13:128
100. Perego G, Bellussi G, Millini R, Alberti A, Zanardi S (2003) Micropor Mesopor Mater 58:213
101. Dong WY, Ren Y, Zhou WZ, Long YC (2003) Acta Chim Sinica 61:251
102. Bianchi D, Simon MW, Nam SS, Xu WQ, Suib SL, O'Young CL (1994) J Catal 145:551
103. Zatorski LW, Centi G, Lopez Nieto J, Trifiro F, Bellussi G, Fattore V (1989) In: Jacobs PA, van Santen RA (eds), Zeolites: Facts, Figures, Future; Proc 8th Int Zeolite Conf, Amsterdam, The Netherlands, July 10–14, 1989; Stud Surf Sci Catal vol 49, Elsevier, Amsterdam, 1989, p 1243
104. Yoo JS, Sohail AR, Grimmer SS, Choifeng C (1994) Catal Lett 29:299
105. Bortnovsky O, Melichar Z, Sobalik Z, Wichterlova B (2001) Micropor Mesopor Mater 42:97
106. Khodakov AY, Kustov LM, Bondarenko TN, Dergachev AA, Kazansky VB, Minachev KM, Borbély G, Beyer HK (1990) Zeolites 10:603
107. Giannetto G, Montes A, Gnep NS, Florentino A, Cartraud P, Guisnet M (1994) J Catal 145:86
108. Bülow M, Micke A (1994) J Chem Soc Faraday Trans 90:2585
109. Handreck GP, Smith TD (1989) J Chem Soc Faraday Trans I 85:3215
110. Dooley KM, Chang CC, Price GL (1992) Appl Catal A 84:17
111. Gnep NS, Doyemet JY, Seco AM, Ribeiro FR, Guisnet M (1988) Appl Catal A 43:155
112. Shevchenko DP, Dergachev AA, Zhurablev GI, Kustov LM, Tkachenko OP, Shpiro ES, Kazanskii VB, Minachev KM (1991) Izv Akademi Nauk 12:2726
113. Monque R, Parisi A, Gonzalez S, Giannetto G (1992) Zeolites 12:806
114. Parikh PA, Subrahmanyam N, Bhat YS, Halgeri AB (1992) Appl Catal A 90:1

115. Shpiro ES, Shevchenko DP, Kharson MS, Dergachev AA, Minachev KM (1992) Zeolites 12:670
116. Barre M, Gnep NS, Magnoux P, Sansare S, Choudhary VR, Guisnet M (1993) Catal Lett 21:275
117. Fejes P, Halász J, Kiricsi I, Kele Z, Tasy Gy, Hannus I, Fernandez C, B.Nagy J, Rockenbauer A, Schöbel Gy (1992) In: Guczi L, Solymosi F, Tétényi P (eds), New Frontiers In Catalysis; Proc 10th Int Congress on Catalysis, Budapest, Hungary, July 19–24, 1992; Stud Surf Sci Catal vol 75, Elsevier, Amsterdam, 1992, p 421
118. Marchal C, Tuel A, Ben Taarit YS (1993) In: Guisnet M, Barbier J, Barrault J, Bouchoule C, Duprez D, Pérot G, Montassier C (eds), Heterogeneous Catalysis And Fine Chemicals III; Proc 3rd Int Symp, Poitiers, France, April 5–8, 1993; Stud Surf Sci Catal vol 78, Elsevier, Amsterdam, 1993, p 447
119. Kornatowski J, Wichterlova B, Jirkovsky J, Löffler E, Pilz W (1996) J Chem Soc Faraday Trans 92:1067
120. Prakash AM, Kevan L (2000) J Phys Chem 104:6860
121. Abe T, Taguchi A, Iwamoto M (1995) Chem Mater 7:1429
122. Gao X, Wachs IE (2000) J Phys Chem B 104:1261
123. Whittington BI, Anderson JR (1993) J Phys Chem 97:1032
124. Fejes P, Marsi I, Kiricsi I, Halász J, Hannus I, Rockenbauer A, Tasy Gy, Koracz L, Schöbel Gy (1991) In: Jacobs PA, Jaeger NI, Kubelková L, Wichterlová B (eds), Zeolite Chemistry And Catalysis; Proc Int Symp, Prague, Czechoslovakia, September 8–13, 1991; Stud Surf Sci Catal vol 69, Elsevier, Amsterdam, 1991, p 173
125. Rigutto MS, van Bekkum H (1991) Appl Catal 68:L1
126. Kornatowski J, Sychev M, Goncharuk V, Baur WH (1990) In: Öhlmann G, Pfeifer H, Fricke R (eds), Catalysis And Adsorption On Zeolites; Proc ZEOCAT'90, Leipzig, Germany, August 20–23, 1990; Stud Surf Sci Catal vol 65, Elsevier, Amsterdam, 1991, p 581
127. Petras M, Wichterlova B (1992) J Phys Chem 96:1805
128. Al-Dacus MA, Al-Arfaj AA, Klein UKA, Barri SAI, Shirokoff J (1999) Arab J Sci Eng 24:57
129. Mal NK, Ramaswamy V, Ganapathy S, Ramaswamy AV (1994) J Chem Soc Chem Commun 1933
130. Spoto G, Zecchina A, Berlier G, Bordiga S, Clerici MG, Basini L (2000) J Mol Catal A 158:107
131. Katada N, Miyamoto T, Begum HA, Naito N, Niwa M, Matsumoto A, Tsutsumi K (2000) J Phys Chem B 104:5511
132. Panov GI, Sheveleva GA, Kharitonov AS, Romanikov VN, Vostrikova LA (1992) Appl Catal A 82:31
133. Kim GJ, Ahn WS (1991) Appl Catal 71:55
134. Yu Y, Xiong G, Li C, Xiao FS (2000) J Catal 194:487
135. Clerici MG (1991) Appl Catal 68:249
136. Zhang SD, Xu CH, Feng LR, Qiu FL (2004) Acta Chim Sinica 62:381
137. Agudo AL, Benitez A, Fierra JLG, Palacios JM, Neira J, Cid R (1992) J Chem Soc Faraday Trans 88:385
138. Han S, Schmitt KD, Shihabi DS, Chang CD (1993) J Chem Soc Chem Commun 1287
139. Kanazirev V, Valtchev V, Tarasov MP (1994) J Chem Soc Chem Commun 1043
140. Nakamura O, Mambrin JST, Pastore HO, Vichi EJS, Gandra FG, da Silva EC, Vargas H, Pelzl J (1992) J Chem Soc Faraday Trans 88:2071
141. Beran S, Wichterlova B, Karge HG (1990) J Chem Soc Faraday Trans 86:3033
142. Latham K, Thompsett D, Williams CD, Round CI (2000) J Mater Chem 10:1235

143. Yang C, Xu QH, Hu C (2000) Mat Chem Phys 63:55
144. van der Water LGA, van der Waal JC, Jansen JC, Maschmeyer T (2004) J Catal 223:170
145. Camblor MA, Perez-Pariente J, Fornes V (1992) Zeolites 12:280
146. Hedge SG, Abdullah RA, Bhat RN, Ratnasamy P (1992) Zeolites 12:951
147. Kanazirev V, Price GL, Tyuliev G (1992) Zeolites 12:846
148. Dzwigaj S, Matsuoka M, Anpo M, Che M (2000) J Phys Chem B 104:6012
149. Takewaki T, Beck LW, Davis ME (1999) J Phys Chem B 103:2674
150. Corma A, Nemeth LT, Renz M, Valencia S (2001) Nature 412:423
151. Joshi N, Joseph EM, Shiralkar VP (1994) J Chem Soc Faraday Trans I 90:387
152. Zhou W, Chun Y, Xu QH, Wang C, Dong JL (2004) Chem J Chinese Univ 25:16
153. Corma A, Rey F, Valencia S, Jorda JR, Rius J (2003) Nature Mater 2:493
154. Lee Y, Kim SJ, Schoonen MAA, Parise JB (2000) Chem Mater 12:1597
155. Wang YX, Song JQ, Gies H (2003) Solid State Sci 5:1421
156. Rocha J, Ferreira P, Lin Z, Brandao P, Ferreira A, de Jesus JDP (1998) J Phys Chem B 102:4739
157. Brandao P, Valente A, Rocha J, Anderson MW (2002) In: Aiello R, Testa F, Giordano G (eds), Impact Of Zeolites And Other Porous Materials On The New Technologies At The Beginning Of The New Millennium; Proc 2nd INT FEZA Conf, Taormina, Italy, September 1–5, 2002; Stud Surf Sci Catal vol 142, Elsevier, Amsterdam, 2002, p 327
158. Brandao P, Philippou A, Valente A, Rocha J, Anderson M (2001) Phys Chem Chem Phys 3:1773
159. Millini R, Carluccio LC, Carati A, Parker WO (2001) Micropor Mesopor Mater 46:191
160. Camblor MA, Martens JA, Grobet PJ, Jacobs PA (1990) In: Öhlmann G, Pfeifer H, Fricke R (eds), Catalysis And Adsorption On Zeolites; Proc ZEOCAT'90, Leipzig, Germany, August 20–23, 1990; Stud Surf Sci Catal vol 65, Elsevier, Amsterdam, 1991, p 613
161. Wichterlova B, Kubelkova L, Jiru P, Kolikova D (1980) Coll Chech Chem Commun 45:2143
162. Wichterlova B, Jiru P (1980) React Kinet Catal Lett 13:197
163. De Luca P, Violante D, Vuono D, Catanzaro L, B.Nagy J, Nastro A (2004) Micropor Mesopor Mater 71:39
164. Latham K, Round CI, Williams CD (2000) Micropor Mesopor Mater 38:333
165. Sen T, Ramaswamy V, Ganapathy S, Rajamohanan PR, Sivasanker S (1996) J Phys Chem 100:3809
166. Hari Prasad Rao PR, Ramaswamy AV (1992) J Chem Soc Chem Commun 1245
167. Hari Prasad Rao PR, Ramaswamy AV, Ratnasamy P (1992) J Catal 137:225
168. Yuan SP, Wang JG, Li YW, Peng SY (2001) Acta Physico-Chim Sinica 17:811
169. Zhi YX, Tuel A, Ben Taarit Y, Naccache C (1992) Zeolites 12:138
170. Kosova G, Cejka J (2002) Coll Czech Chem Commun 67:1760
171. Hari Prasad Rao PR, Ramesh Reddy A, Ramaswamy AV, Ratnasamy P (1993) In: Guisnet M, Barbier J, Barrault J, Bouchoule C, Duprez D, Pérot G, Montassier C (eds), Heterogeneous Catalysis And Fine Chemicals III; Proc 3rd Int Symp, Poitiers, France, April 5–8, 1993; Stud Surf Sci Catal vol 78, Elsevier, Amsterdam, 1993, p 385
172. Tusar NN, Logar NZ, Arcon I, Thibault-Starzyk F, Ristic A, Rajic N, Kaucic V (2003) Chem Mater 15:4745
173. Li G, Wang XS, Guo XW, Liu S, Zhao Q, Bao XH, Lin LW (2001) Mat Chem Phys 71:195
174. Rao YVS, Kulkarni SV, Sabrahmanyam M, Rao AVR (1993) J Chem Soc Chem Commun 1456

175. Mohamed MM, Eissa NA (2003) Mater Res Bull 38:1993
176. Pál-Borbély G, Beyer HK (2003) Phys Chem Chem Phys 5:2145
177. Occelli ML, Schwering G, Fild C, Eckert H, Auroux A, Iyer PS (2000) Micropor Mesopor Mater 34:15
178. Moudrakovski IL, Sayari E, Ratcliffe CI, Ripmeester JA, Preston KF (1994) J Phys Chem 98:10895
179. Perego G, Bellussi G, Millini R, Alberti A, Zanardi S (2003) Micropor Mesopor Mater 56:193
180. Tuoto CV, B.Nagy J, Nastro A (1995) In: Bonneviot L, Kaliaguine S (eds), A Refined Tool For Designing Catalytic Sites; Proc Int Symp, Québec, Canada, October 15–20, 1995; Stud Surf Sci Catal vol 97, Elsevier, Amsterdam, 1995, p 551
181. Testa F, Chiappetta R, Crea F, Aiello R, Fonseca A, B.Nagy J (1995) In: Beyer HK, Karge HG, Kiricsi I, B.Nagy J (eds), Catalysis By Microporous Materials; Proc ZEO-CAT '95, Szombathely, Hungary, July 9–13, 1995; Stud Surf Sci Catal vol 94, Elsevier, Amsterdam, 1995, p 349
182. Testa F, Chiappetta R, Crea F, Aiello R, Fonseca A, B.Nagy J (1996) Colloid Surf A 115:223
183. B.Nagy J, Aiello R, Crea F, Testa F (1998) In: Treacy MMJ, Marcus BK, Bisher ME, Higgins JB (eds) (1998) Proc 12th Int Zeolite Conf, Baltimore, Maryland, USA, July 5–10. Mater Res Soc, Warrendale, Pennsylvania, USA, p 1619
184. Nigro E, Crea F, Testa F, Aiello R, Lentz P, B.Nagy J (1999) Micropor Mesopor Mater 30:199
185. Testa F, Pasqua L, Crea F, Aiello R, Lázár K, Fejes P, Lentz P, B.Nagy J (2003) Micropor Mesopor Mater 57:57
186. Testa F, Crea F, Aiello R, Lázár K, Fejes P, Lentz P, B.Nagy J (2001) In: Galarneau A, di Renzo F, Fajula F, Vedrine J (eds), Zeolites And Mesoporous Materials At The Dawn Of The 21st Century; Proc 13th Int Zeolite Conf, Montpellier, France, July 8–13, 2001; Stud Surf Sci Catal vol 135, Elsevier, Amsterdam, 2001, p 04-P-17
187. Fejes P, Kiricsi I, Lázár K, Marsi I, Rockenbauer A, Korecz L, B.Nagy J, Aiello R, Testa F (2003) Appl Catal A 242:247
188. Giordano G, Katovic A (2001) In: Gamba A, Colella C, Coluccia S (eds), Oxide-Based Systems At The Crossroads Of Chemistry; Proc 2nd Int Workshop, October 8–11, 2000, Como, Italy; Stud Surf Sci Catal vol 140, Elsevier, Amsterdam, 2001, p 297
189. Giordano G, Katovic A, Fonseca A, B.Nagy J (2001) In: Galarneau A, di Renzo F, Fajula F, Vedrine J (eds), Zeolites And Mesoporous Materials At The Dawn Of The 21st Century; Proc 13th Int Zeolite Conf, Montpellier, France, July 8–13, 2001; Stud Surf Sci Catal vol 135, Elsevier, Amsterdam, 2001, p 04-P-17
190. Giordano G, Katovic A, Perathoner S, Pino F, Centi G, B.Nagy J, Lázár K, Fejes P (2002) In: Aiello R, Testa F, Giordano G (eds), Impact Of Zeolites And Other Porous Materials On The New Technologies At The Beginning Of The New Millennium; Proc 2nd INT FEZA Conf, Taormina, Italy, September 1–5, 2002; Stud Surf Sci Catal vol 142, Elsevier, Amsterdam, 2002, p 477
191. Perathoner S, Pino F, Centi G, Giordano G, Katovic A, B.Nagy J, Lázár K, Fejes P (2002) In: Aiello R, Testa F, Giordano G (eds), Impact Of Zeolites And Other Porous Materials On The New Technologies At The Beginning Of The New Millennium; Proc 2nd INT FEZA Conf, Taormina, Italy, September 1–5, 2002; Stud Surf Sci Catal vol 142, Elsevier, Amsterdam, 2002, p 503
192. Perathoner S, Pino F, Centi G, Giordano G, Katovic A, B.Nagy J (2003) Top Catal 23:125
193. Fejes P, B.Nagy J, Kovacs K, Vanko G (1996) Appl Catal A 145:155

194. Aloi D, Testa F, Pasqua L, Aiello R, B.Nagy J (2002) In: Aiello R, Testa F, Giordano G (eds), Impact Of Zeolites And Other Porous Materials On The New Technologies At The Beginning Of The New Millennium; Proc 2nd INT FEZA Conf, Taormina, Italy, September 1–5, 2002; Stud Surf Sci Catal vol 142, Elsevier, Amsterdam, 2002, p 469
195. Giordano G, Katovic A, Caputo D (2001) In: Gamba A, Colella C, Coluccia S (eds), Oxide-Based Systems At The Crossroads Of Chemistry; Proc 2nd Int Workshop, October 8–11, 2000, Como, Italy; Stud Surf Sci Catal vol 140, Elsevier, Amsterdam, 2001, p 307
196. Nigro E, Testa F, Aiello R, Lentz P, Fonseca A, Oszkó A, B.Nagy J (2001) In: Galarneau A, di Renzo F, Fajula F, Vedrine J (eds), Zeolites And Mesoporous Materials At The Dawn Of The 21st Century; Proc 13th Int Zeolite Conf, Montpellier, France, July 8–13, 2001; Stud Surf Sci Catal vol 135, Elsevier, Amsterdam, 2001, p 04-P-18
197. Nigro E, Testa F, Aiello R, Lentz P, Fonseca A, Oszkó A, Fejes P, Kukovecz A, Kiricsi I, B.Nagy J (2001) In: Gamba A, Colella C, Coluccia S (eds), Oxide-Based Systems At The Crossroads Of Chemistry; Proc 2nd Int Workshop, October 8–11, 2000, Como, Italy; Stud Surf Sci Catal vol 140, Elsevier, Amsterdam, 2001, p 353
198. Giordano G, Katovic A, Szymkowiak-Janiszewka E, Kowalak S (2003) Collect Czech Chem Commun 68:1149
199. Katovic A, Szymkowiak-Janiszewka E, Giordano G, Kowalak S, Fonseca A, B.Nagy J (2001) In: Galarneau A, di Renzo F, Fajula F, Vedrine J (eds), Zeolites And Mesoporous Materials At The Dawn Of The 21st Century; Proc 13th Int Zeolite Conf, Montpellier, France, July 8–13, 2001; Stud Surf Sci Catal vol 135, Elsevier, Amsterdam, 2001, p 04-P-14
200. Katovic A, Giordano G, Kowalak S (2002) In: Aiello R, Testa F, Giordano G (eds), Impact Of Zeolites And Other Porous Materials On The New Technologies At The Beginning Of The New Millennium; Proc 2nd INT FEZA Conf, Taormina, Italy, September 1–5, 2002; Stud Surf Sci Catal vol 142, Elsevier, Amsterdam, 2002, p 39
201. Robson H (ed) Verified syntheses of zeolitic materials. Elsevier, Amsterdam, 2001
202. Kumar R, Bhaumik A, Ahedi RK, Ganapathy S (1996) Nature 381:298
203. Cosentino F, Katovic A, Giordano G, Lentz P, B.Nagy J (1999) In: Kiricsi I, Pál-Borbély G, B.Nagy J, Karge HG (eds), Porous Materials In Environmentally Friendly Processes; Proc 1st Int FEZA Conf, Eger, Hungary, September 1–4, 1999; Stud Surf Sci Catal vol 125, Elsevier, Amsterdam, 1999, p 109
204. Debras G, Derouane EG, Gilson JP, Gabelica Z, Demortier G (1983) Zeolites 3:37
205. Gabelica Z, B.Nagy J, Bodart P, Debras G (1984) Chem Lett 1059
206. Turner GL, Smith KA, Kirkpatrick RJ, Oldfield E (1986) J Magn Reson 67:544
207. Massiot D, Bessada C, Coutures JP, Taulelle F (1990) J Magn Reson 90:231
208. Hunger M, Engelhardt G, Koller H, Weitkamp J (1993) Solid State Nucl Magn Reson 2:111
209. Debras G, Gourgue A, B.Nagy J, de Clippeleir G (1985) Zeolites 5:377
210. Testa F, Chiappetta R, Crea F, Aiello R, Fonseca A, B.Nagy J (1995) In: Beyer HK, Karge HG, Kiricsi I, B.Nagy J (eds), Catalysis By Microporous Materials; Proc ZEOCAT '95, Szombathely, Hungary, July 9–13, 1995; Stud Surf Sci Catal vol 94, Elsevier, Amsterdam, 1995, p 349
211. Cichocki A, Lasocha W, Michalik M, Sawlowicz Z, Bus M (1990) Zeolites 10:583
212. de Ruiter R, Jansen JC, van Bekkum H (1992) Zeolites 12:56
213. Soulard M, Bilger S, Kessler H, Guth JL (1987) Zeolites 7:463
214. Engelhardt G, Michel D (1987) High-resolution solid-state NMR of silicates and zeolites. Wiley, Chichester, UK
215. Chezeau JM, Delmotte L, Guth JL, Soulard M (1989) Zeolites 9:78

216. Fonseca A, B.Nagy J, El Hage-Al Asswad J, Mostowicz R, Crea F, Testa F (1995) Zeolites 15:259
217. El Hage-Al Asswad J, Dewaele N, B.Nagy J, Hubert RA, Gabelica Z, Derouane EG, Crea F, Aiello R, Nastro A (1988) Zeolites 8:221
218. Caullet P, Patarin J, Faust AC (1995) In: Aiello R (ed) (1995) Proc III Convegno Nazionale di Scienza e Tecnologia delle Zeoliti, Cetraro, Italy, September 28–29, 1995. De Rose Montalto, Cosenza, Italy, p 55
219. Nigro E, Mostowicz R, Crea F, Testa F, Aiello R, B.Nagy J (1997) In: Chon H, Ihm S-K, Uh YS (eds), Progress In Zeolite And Microporous Materials; Proc 11th Int Zeolite Conf, Seoul, Korea, August 12–17, 1996; Stud Surf Sci Catal vol 105, Elsevier, Amsterdam, 1997, p 309
220. Tuoto CV, B.Nagy J, Nastro A (1997) In: Chon H, Ihm S-K, Uh YS (eds), Progress In Zeolite And Microporous Materials; Proc 11th Int Zeolite Conf, Seoul, Korea, August 12–17, 1996; Stud Surf Sci Catal vol 105, Elsevier, Amsterdam, 1997, p 213
221. Meier WM, Olson DH (1992) Atlas of zeolites structure types. Heinemann, Butterworth, p 114
222. Carvalho AP, de Carvalho MB, Ribeiro FR, Fernandez C, B.Nagy J, Derouane EG, Guisnet M (1993) Zeolites 13:462
223. Guth JL, Kessler H, Higel JM, Lamblin JM, Patarin J, Seive A, Chezeau JM, Wey R (1989) In: Occelli ML, Robson HE (eds) Zeolite synthesis. ACS Symposium Series 398. Oxford University Press, Washington, DC, USA, p 176
224. Tielen M, Geelen M, Jacobs PA (1985) Acta Phys Chem 31:1
225. Fejes P, B.Nagy J, Halász J, Oszkó A (1998) Appl Catal A 175:89
226. Debras G, Gourgue A, B.Nagy J, de Clippeleir G (1985) Zeolites 5:377
227. Aiello R, Crea F, Nigro E, Testa F, Mostowicz R, Fonseca A, B.Nagy J (1999) Micropor Mesopor Mater 28:241
228. Testa F, Crea F, Aiello R, B.Nagy J (1999) In: Kiricsi I, Pál-Borbély G, B.Nagy J, Karge HG (eds), Porous Materials In Environmentally Friendly Processes; Proc 1st Int FEZA Conf, Eger, Hungary, September 1–4, 1999; Stud Surf Sci Catal vol 125, Elsevier, Amsterdam, 1999, p 165
229. B.Nagy J, Aiello R, Crea F, Testa F (1998) In: Treacy MMJ, Marcus BK, Bisher ME, Higgins JB (eds) (1998) Proc 12th Int Zeolite Conf, Baltimore, Maryland, USA, July 5–10. Mater Res Soc, Warrendale, Pennsylvania, USA, p 1619
230. Raatz F, Roussel JC, Cantini R, Ferre G, B.Nagy J (1987) In: Grobet PJ, Mortier WJ, Vansant EF, Schulz-Ekloff G (eds), Innovation In Zeolite Materials Science; Proc Int Symp, Nieuwpoort, Belgium, September 13–17, 1987; Stud Surf Sci Catal vol 37, Elsevier, Amsterdam, 1988, p 301
231. Fonseca A, B.Nagy J, El-Hage-Al Asswad J, Mostowicz R, Crea F, Testa F (1995) Zeolites 15:259
232. Tavolaro A, Mostowicz R, Crea F, Nastro A, Aiello R, B.Nagy J (1992) Zeolites 12:756
233. Mostovicz R, Crea F, B.Nagy J (1993) Zeolites 13:678
234. Golemme G, Nastro A, B.Nagy J, Subotic B, Crea F, Aiello R (1991) Zeolites 11:776
235. Bose A (1985) Proc Ind Acad Soc A 1:754
236. Mulay LN, Selwood PW (1955) J Am Chem Soc 77:2693
237. Hedström BOA (1938) Arkiv Khemi 6
238. Goldfarb D, Bernardo M, Strohmaier KG, Vaughan DEW, Thomann H (1994) J Am Chem Soc 116:6344
239. Giordano G, Di Renzo F, Fajula F (1993) In: Hattori T, Yashima T (eds), Zeolites And Microporous Crystals; Proc Int Symp on Zeolites and Microporous Crystals,

Nagoya, Japan, August 22–25, 1993; Stud Surf Sci Catal vol 83, Elsevier, Amsterdam, 1994, p 41
240. Bodart P, B.Nagy J, Debras G, Gabelica Z, Jacobs PA (1986) J Phys Chem 90:5183
241. Korteweg GA, van Reijen LL (1981) J Magn Res 44:159
242. Fejes P, Kiricsi I, Kovács K, Lázár K, Marsi I, Oszkó A, Rockenbauer A, Schay Z (2002) Appl Catal A 223:147
243. Wichterlova B, Dedecek J, Sobalik Z (1998) In: Treacy MMJ, Marcus BK, Bisher ME, Higgins JB (eds) (1998) Proc 12th Int Zeolite Conf, Baltimore, Maryland, USA, July 5–10. Mater Res Soc, Warrendale, Pennsylvania, USA, p 941
244. van Koningsveld H, Jansen JC, van Bekkum H (1990) Zeolites 10:235
245. Calis G, Frenken P, de Boer E, Swolfs A, Hefni MA (1987) Zeolites 7:319
246. Lin DH, Coudurier G, Vedrine JC (1989) In: Jacobs PA, van Santen RA (eds), Zeolites: Facts, Figures, Future; Proc 8th Int Zeolite Conf, Amsterdam, The Netherlands, July 10–14, 1989; Stud Surf Sci Catal vol 49, Elsevier, Amsterdam, 1989, p 1431
247. Centi G, Perathoner S, Romeo G (2001) In: Galarneau A, di Renzo F, Fajula F, Vedrine J (eds), Zeolites And Mesoporous Materials At The Dawn Of The 21st Century; Proc 13th Int Zeolite Conf, Montpellier, France, July 8–13, 2001; Stud Surf Sci Catal vol 135, Elsevier, Amsterdam, 2001, p 31-O-03
248. Giordano G, Katovic A, Perathoner S, Pino F, Centi G, B.Nagy J, Lázár K, Fejes P (2002) In: Aiello R, Testa F, Giordano G (eds), Impact Of Zeolites And Other Porous Materials On The New Technologies At The Beginning Of The New Millennium; Proc 2nd INT FEZA Conf, Taormina, Italy, September 1–5, 2002; Stud Surf Sci Catal vol 142, Elsevier, Amsterdam, 2002, p 477
249. Wichterlowa B, Dedecek J, Sobalik Z (2001) In: Centi G, Wichterlowa B, Bell A (eds) (2001) Catalysis by unique metal ion structures in solid matrices, NATO Sci Ser II, vol 13. Kluwer, Dordrecht, The Netherlands, p 31
250. Centi G, Vazzana F (1999) Catal Today 53:683
251. Hari Prasad Rao PR, Leony Leon CA, Ueyama K, Matsukata M (1998) Micropor Mesopor Mater 21:305
252. Jacobs PA, Martens JA (eds) (1984) Synthesis of high-silica aluminosilicate zeolites. Elsevier, Amsterdam
253. Rubin MK, Chu P (1990) US Patent 4,954,325
254. Corma A, Corell C, Pérez-Pariente J (1995) Zeolites 15:2
255. Ravishankar R, Bhattacharya R, Jacob NE, Sivasanker S (1995) Micropor Mater 4:83
256. Lawton SL, Fung AS, Kennedy GJ, Alemany LB, Chang CD, Hatzikos GH, Lissy DN, Rubin MK, Timken HKC, Steuernagel S, Woessner DE (1996) J Phys Chem 100:3788
257. Hunger M, Ernst S, Weitkamp J (1995) Zeolites 15:188
258. Wichterlova B, Kubelkova L, Jiru P, Kolihova D (1980) Collect Czech Chem Commun 45:2143
259. Varga J, Halasz J, Horvath D, Méhn D, B.Nagy J, Schöbel G, Kiricsi I (1997) In: Kruse N, Frennet A, Bastin JM (eds) (1997) Heterogeneous catalysis. Proc 4th Int Congress on Catalysis and automotive Pollution, Brussels, Belgium, April 9–11, 1997. Universite Libre de Bruxelles, Belgium, p 191
260. Catana G, Pelgrims J, Schoonheydt RA (1995) Zeolites 15:475
261. Meagher A, Nair V, Szostak R (1988) Zeolites 8:3
262. Lázár K, Borbély G, Beyer H (1991) Zeolites 11:214
263. Hagen A, Roessner F, Weingart I, Spliethov B (1995) Zeolites 15:270
264. Ratnasamy P, Kumar R (1991) Catal Today 9:329
265. Aiello R, Crea F, Nigro E, Testa F, Mostowicz R, Fonseca A, B.Nagy J (1999) Micropor Mesopor Mater 28:241

266. Testa F, Crea F, Aiello R, B.Nagy J (1999) In: Kiricsi I, Pál-Borbély G, B.Nagy J, Karge HG (eds), Porous Materials In Environmentally Friendly Processes; Proc 1st Int FEZA Conf, Eger, Hungary, September 1–4, 1999; Stud Surf Sci Catal vol 125, Elsevier, Amsterdam, 1999, p 165
267. Verberckmoes A, Weckhuysen BM, Schoonheydt RA (1998) Micropor Mesopor Mater 22:165
268. Kagawa K, Ichikawa Y, Iwamoto S, Inui T (1998) Micropor Mesopor Mater 25:15
269. Mehn D, Kukovecz A, Kiricsi I, Testa F, Nigro E, Aiello R, Daelen G, Lentz P, Fonseca A, B.Nagy J (2001) In: Galarneau A, di Renzo F, Fajula F, Vedrine J (eds), Zeolites And Mesoporous Materials At The Dawn Of The 21st Century; Proc 13th Int Zeolite Conf, Montpellier, France, July 8–13, 2001; Stud Surf Sci Catal vol 135, Elsevier, Amsterdam, 2001, p 11-P-27
270. Aiello R, Crea F, Testa F, Demortier G, Lentz P, Wiame M, B.Nagy J (2000) Micropor Mesopor Mater 35–36:585
271. Wichterlova B, Dedecek J, Sobalik Z (1995) In: Beyer HK, Karge HG, Kiricsi I, B.Nagy J (eds), Catalysis By Microporous Materials; Proc ZEOCAT '95, Szombathely, Hungary, July 9–13, 1995; Stud Surf Sci Catal vol 94, Elsevier, Amsterdam, 1995, p 641
272. Ciambelli C, Sonnino D, Corbo P, Cosentino F, Katovic A, Giordano G (1997) In: Haber J, Dyrek K, Lagan JM (eds) (1997) EUROPACAT III. Proc 3rd European Congress on Catalysis, Krakow, Poland, 31 August–6 September 1997. Polish Academy of Sciences, Poland, p 93
273. Kucherov AV, Slinkin AA, Kondratev DA, Bondarenko TN, Rubinstein AM, Minachev KM (1985) Zeolites 5:320
274. Kowalak S, Szymkowjak E, Jankowska A, Giordano G (2000) In: Petrov L, Bonev C, Kadinov G (eds) (2000) Heterogeneous catalysis. Proc 9th Int Symp Heterogeneous Catalysis, Varna, Bulgaria, September 23–27, 2000. Pensoft Publishers, Bulgaria, p 229
275. Rudolph WW, Pye CC (1999) Phys Chem Chem Phys 1:4583
276. Berndt H, Lietz G, Voelter J (1996) Appl Catal A 146:365
277. Wan BZ, Chu HM (1992) J Chem Soc Faraday Trans 88:2943
278. Amoureux JP, Fernandez C, Steuernagel S (1996) J Magn Reson A 123:116
279. Massiot D (1996) J Magn Reson A 122:240
280. Frydman L, Harwood JS (1995) J Am Chem Soc 117:5367
281. Fernandez C, Amoureux JP (1995) Chem Phys Lett 242:449
282. Fernandez C, Amoureux JP, Chezeau JM, Delmotte L, Kessler H (1996) Micropor Mater 6:331
283. Amoureux JP, Fernandez C, Frydman L (1996) Chem Phys Lett 259:347
284. Amoureux JP, Fernandez C, Dumazy Y (1995) J Chim Phys Phys Chim Biol 92:1939
285. Zwanziger JW (1994) Solid State NMR 3:219
286. Fernandez C, Delevoye L, Amoureux JP, Lang DP, Pruski M (1997) J Am Chem Soc 119:6858

Author Index Volumes 1–5

Subject Index

Printing: Krips bv, Meppel
Binding: Stürtz, Würzburg